199/III Current Topics in Microbiology and Immunology

Springer
Berlin
Heidelberg
New York
Barcelona
Budapest
Hong Kong
London
Milan
Paris
Tokyo

The Molecular Repertoire of Adenoviruses III

Biology and Pathogenesis

Edited by W. Doerfler and P. Böhm

With 51 Figures

Springer

WALTER DOERFLER
PETRA BÖHM
Institut für Genetik
der Universität zu Köln
Weyertal 121
50931 Köln
Germany

Cover illustration: In contrast to the usual CGR representation of DNA sequences, a regular grid with 64^2 cells was constructed. All points whose coordinates fall into a cell, i.e., presenting all subsequences of a length of six nucleotides, were counted. In this graphic representation the regular grids of Ad12 and Ad2 DNA sequences were subtracted after being normalized. The color of each sphere reflects the origin of a certain hexanucleotide, i.e., blue to red, predominant in Ad12 or Ad2, respectively. It can be seen that the sequence suffixes tt and aa were found more often in Ad12, whereas all dinucleotides ending in c and g seem to be more equally distributed in Ad2. Green-colored spheres represent equally distributed hexanucleotides in both adenovirus genomes. (This computer analysis was performed by Jörg Sprengel, Cologne, Germany.)

Cover design: Künkel+Lopka, Ilvesheim

ISBN-13: 978-3-642-79588-6 e-ISBN-13: 978-3-642-79586-2
DOI: 10.1007/978-3-642-79586-2

Softcover reprint of the hardcover 1st edition 1995
Library of Congress Catalog Card Number 15-12910

Typesetting: Thomson Press (India) Ltd, Madras
SPIN: 10484002 27/3020/SPS – 5 4 3 2 1 0 – Printed on acid-free paper.

Preface

When asked whether they wanted to contribute a chapter to these three volumes on the *Molecular Repertoire of Adenoviruses* almost all of my colleagues working in the field responded favorably and supplied fine chapters on their areas of expertise.

In 1995, adenoviruses continue to provide a major and highly developed experimental system for biochemical, cell biological, genetic, virological, and epidemiological investigations in mammalian molecular biology. As a considerable body of information has become available thanks to the continued efforts of many, this virus system has become particularly useful to those who want to address details of biological mechanisms and their relation to structure and function.

We have tried to cover as wide a field as possible in current adenovirus research and to encompass the entire gamut of adenovirology and adenovirus molecular biology. In spite of the tendencies that come from many corners to seduce researchers into directing their efforts towards applied molecular biology, which of course has its place and merits, there are still sturdy groups who pursue their interests in basic molecular biology and in particular in adenovirus research, an "eternal archetype," as one of the experts chose to name the virus.

In the first of the three volumes, we present an overview of adenovirus research and go on to cover the topics of the structure and assembly of adenoviruses, viral infection, and viral gene products. In the following two volumes, we turn our attention to topics such as DNA replication, recombination and integration, post-transcriptional control, transformation and E1A, adenovirus genetics, pathogenesis, and gene therapy.

The three volumes appearing now have also been initiated to mark with some, perhaps unavoidable, delay the occasion of the 40th anniversary of the discovery of adenoviruses by W.P.

Rowe and R.J. Huebner and, independently, by M.R. Hilleman and J.H. Werner in 1953/1954.

I was encouraged to undertake the task of editing the current three volumes on adenovirus molecular biology by visits to many libraries in different countries on several continents during the last decade and by the observation that the three volumes on *The Molecular Biology of Adenoviruses*, which I had edited for Springer-Verlag's series *Current Topics in Microbiology and Immunology* in 1983 and 1984, had apparently been diligently worked through by many readers. The craftsmanship of the book covers had barely resisted the intellectual assault and showed that the books had been consulted with curiosity.

There are few viral genomes that have not been developed into useful vector systems. An increasing number of research reports have been devoted to adenoviruses as potential vectors for human somatic gene therapy. Should that virus system indeed prove its worth for this important application, researchers will have to appreciate the subtleties of the biology and the molecular biology of this virus system. As someone who has not been directly involved in research on gene therapy, it appears to me that considerable efforts will have to be expended before all the problems concerning the therapeutic application of the adenovirus system or, more likely, of even better systems will be solved. The importance of this goal justifies and will require a great deal of very active research.

We wish to thank all the contributors, the editors of the *Current Topics* series, and in particular Springer-Verlag for their help in getting these three volumes to press.

Cologne

WALTER DOERFLER
PETRA BÖHM

List of Contents

List of Contents of Companion Volume 199/I

List of Contents of Companion Volume 199/II

List of Contributors

(Their addresses can be found at the beginning of their respective chapters.)

Modulation of Gene Expression by Adenovirus Transformation

A. Zantema and A.J. van der Eb

1 Introduction

Human adenoviruses (Ad) cause a productive infection in human cells, while in rodent cells in general they bring about only an abortive infection, which occasionally can lead to morphological transformation. In the transformed cells, viral DNA is integrated into the genome and at least the genes of the viral E1 region are expressed. This is consistent with the observation that the transforming activity of the virus is localized in region E1. Rodent cells, e.g., from rat, mouse, or hamster, can be transformed efficiently by Ad DNA, but human cells, including fibroblasts and epithelial cells, are extremely refractory to transformation. Only a few transformed cell lines have been isolated from human embryonic kidney cultures, in spite of considerable efforts (Graham et al. 1977; Byrd et al. 1982). In contrast, human embryonic retina cells can be transformed reproducibly with Ad DNA (Whittaker et al. 1984), although less efficiently than rodent embryo or kidney cultures. The E1 region consists of two transcriptional

---

Department of Molecular Carcinogenesis, Wassenaarseweg 72, 2333 AL Leiden, The Netherlands

units, E1A and E1B. For complete morphological transformation both regions are needed, but the E1A region by itself can immortalize rodent cells (HOUWELING et al. 1980).

The various Ad serotypes differ in their ability to induce tumors upon inoculation into newborn hamsters; for example, Ad type 5 (Ad5) is non-oncogenic, whereas Ad12 is highly oncogenic. However, all Ad serotypes or their DNA can transform rodent cells. The transformed cells exhibit a similar oncogenic behavior in syngeneic animals as the intact Ads.

The transcriptional regions E1A and E1B each code for two or three proteins. In transformed cells there are two prominent E1A mRNAs, which differ in size due to a differential use of splice donor sites. These mRNAs, 12S and 13S, code for proteins of 243 and 289 amino acids, respectively (for Ad5). The larger product contains an internal stretch of amino acids which is absent in the smaller E1A protein. In addition, in lytic infection mRNA of 9S, 10S, and 11S are found. In the case of Ad12 E1A a 9.5S mRNA has been described, which is not found for Ad5 E1A (BROCKMANN et al. 1990). This 9.5S mRNA codes for a 52-amino acid (aa) protein, of which the first 29aa are also found in the other E1A proteins while the rest is unique. Whether this exclusive Ad12 protein is important for the differences between Ad5 and Ad12 remains to be established. In transient assays the 9.5S product has been shown to inhibit the basal expression level of the collagenase 12-*O*-tetra-decanolylphorbol-13-acetate (TPA)-responsive element (TRE) but not the TPA-induced expression (BROCKMANN et al. 1994). The E1B region codes for one 22S mRNA in transformed cells, which is translated into two proteins, initiating at the first or second AUG (BOS et al. 1983). The open reading frame (ORF) of the first AUG codes for a protein of about 20 kDa and the ORF of the second AUG for a 55-kDa protein. In lytically infected cells more E1B mRNAs are found, including a late 12S mRNA coding only for the 20-kDa protein and other species coding for an intact 20-kDa protein and 55-kDa derivatives with internal deletions (ANDERSON et al. 1984; LEWIS and ANDERSON 1987).

2 Role of the E1A Proteins

The role of the E1A proteins in transformation is based to a large extent on sequestration of a number of cellular protein from their normal cellular association partners. The domains in E1A required for transformation coincide with those required for binding to cellular proteins. These domains include the E1A N terminus and two regions conserved between the various Ad serotypes: CR1 and CR2. A third conserved region, CR3, is present only in the E1A 13S product and is involved in transcription activation but not (directly) in transformation (reviewed in FLINT and SHENK 1989; BERK 1986). Since the E1A 12S product can replace the wild-type E1A in transformation assays, it is clear that activation of gene expression due to the CR3 region is not essential for transformation of cells. The association to cellular proteins of the various domains of E1A and the properties associated with these domains are reviewed elsewhere (PEEPER and ZANTEMA 1993) and summarized in Fig. 1. Although most attention, also in this review, is usually focused on the E1A conserved regions and the N terminus, it should be noted that intactness of the second exon may also be required for transformation. Recent studies indeed showed contributions of the second exon to gene transcription (BONDESSON et al. 1992; LINDER et al. 1992), while binding of a 48-kDa protein to the C terminus has also been demonstrated (BOYD et al. 1993).

The domains of the E1A proteins that are required for transformation (N terminus, CR1, and CR2) also seem to influence transcription of cellular genes by associating with cellular proteins. The 300-kDa protein (p300) which associates with the N-terminal region of E1A was shown to be a component of TATA-binding protein complexes (ABRAHAM et al. 1993). Therefore, p300 probably is involved in transcription, and interaction with the E1A proteins could thus alter this process. The proteins associating with the CR1 and CR2 regions of E1A can normally bind to transcription factor E2F. The binding of Rb and p107 to E2F has been shown to inhibit the E2F transcriptional activity (HELIN et al. 1993a; CRESS et al. 1993). When E1A proteins are present, they bind to Rb, p107, and the p107-related p130 and thereby release E2F from these inhibitors, which leads to an increased transcription activation by E2F (BAGCHI et al. 1990; RAYCHAUDHURI et al. 1991; HELIN et al. 1992; KAELIN et al. 1992). During a normal cell cycle, E2F is released from the inhibiting Rb protein by phosphorylation of Rb during the G_1 phase of the cell cycle. The active E2F transcription factor then stimulates expression of genes required for the S phase of the cell cycle, such as dihydrofolate reductase (DHFR) and thymidine kinase. The E2F polypeptide alone is not active as a transcription

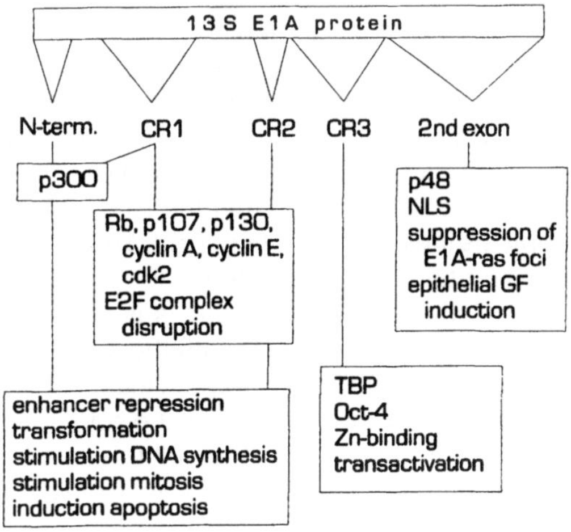

Fig. 1. E1A 13S protein. The cellular proteins which bind to certain domains and the functions in which those domains are involved are indicated. The 12S E1A protein essentially is different only because it lacks CR3 due to differential splicing. *TBP*, TATA-binding protein; *NLS*, nuclear localization signal; *GF*, growth factor

factor: the active E2F transcription factor consists of a heterodimer of E2F and DP-1 (HELIN et al. 1993b; BANDARA et al. 1993). The E2F component is one of a family of transcription factors (LEES et al. 1993; IVEY-HOYLE et al. 1993) to which Rb and p107 bind with different efficiencies (DYSON et al. 1993). The E2F factors are probably just an example of many transcription factors regulated by proteins such as Rb, as is illustrated by the inhibition of the transcription factor Elf-1 by the underphosphorylated form of Rb (WANG et al. 1993). In addition to Rb, p107, and p130, immunoprecipitates of E1A also contain cyclin A, cyclin E, and cdk2. These latter components most likely do not bind to E1A directly, but via p107 and p130, resulting in the quaternary complexes E1A–p107–cyclin A–cdk2, E1A–p107–cyclin E–cdk2, E1A–p130–cyclin A–cdk2, and E1A–p130–cyclin E–cdk2 (FAHA et al. 1993).

The binding to cellular proteins by the N terminus, CR1, and CR2 of E1A coincides with the ability of E1A to stimulate DNA synthesis. This suggests that transcription modulation via association with these domains of E1A is needed for progression through the cell cycle, as is indeed the case for the activation of the E2F transcription factor.

The E1A proteins can induce apoptosis. The domains involved in apoptosis (WHITE et al. 1991) are also the regions that cause the strongest induction of DNA synthesis. For the induction of apoptosis wild-type p53 is required, which is accompanied, at least in some cells, by stabilization of p53 (LOWE and RULEY 1993; DEBBAS and WHITE 1993). This apoptotic effect leads to massive cell death, unless apoptosis-inhibiting proteins such as the Ad E1B gene products or bcl-2 are present (RAO et al. 1992).

3 Role of the E1B Proteins

Complete morphological transformation of primary cells requires, in addition to E1A, the Ad E1B region. Both proteins encoded by the E1B region seem to be involved in controlling some of the effects of E1A. The importance of the E1B 21-kDa protein as an inhibitor of the E1A-induced apoptosis is illustrated by the fact that the capacity of 21-kDa mutants to inhibit apoptosis parallels their contribution to transformation (SUBRAMANIAN et al. 1993) and that the E1B 21-kDa protein can be replaced by another apoptosis inhibitor, bcl-2 (RAO et al. 1992). The inhibition of apoptosis by the E1B 21-kDa protein is not restricted to E1A-induced apoptosis, as also tumor necrosis factor (TNF)-α-induced apoptosis is inhibited (WHITE et al. 1992). In addition, the E1B 55-kDa protein also inhibits the apoptotic effect of E1A (RAO et al. 1992; DEBBAS and WHITE 1993). Therefore, at least part of the transforming effect of E1B can be attributed to inhibition of the apoptotic effect caused by E1A. Thus, the expression of E1B renders the cell tolerant for levels of the transforming E1A protein that are required for cell transformation (LOWE and RULEY 1993).

The presence of the E1B 55-kDa protein also leads to a stabilization of p53. In the case of the Ad5 protein, this is accompanied by a direct interaction between the two proteins; for the Ad12 protein no interaction has been demonstrated

(Sarnow et al. 1982; Zantema et al. 1985). As wild-type p53 is a transcription factor, it is possible that this E1B protein inhibits the transcription activation potential of p53, as indeed has been shown (Yew and Berk 1992). This inhibition is found for both the Ad5 55-kDa and the Ad12 54-kDa proteins, suggesting either that the Ad12 54-kDa protein has a (much weaker) interaction with p53 or that an unknown mechanism exerted by the E1B proteins is responsible for this inhibition of the p53-mediated transactivation. The inhibition of transcription by the E1B 55-kDa protein was shown not to be limited to p53-induced transcription, suggesting that the E1B 55-kDa protein is a general repressor of transcription (Yew et al. 1994). Therefore, the role of the E1B 55-kDa protein might be to inhibit the expression of genes activated by p53, so that cell cycle arrest or apoptosis are prevented.

4 Differences in Gene Expression Related to Differentiation

In this review we will focus on differences in gene expression found in Ad-transformed cells compared to the nontransformed parental cells and discuss experiments performed to understand these differences in gene expression. As most regulations of gene expression are mediated by E1A rather than E1B proteins, we will focus on effects of the E1A proteins. The differences in expression can be either a direct effect of the Ad gene products or an indirect effect, as transformation may lead to a change in the differentiation state of the cell or to alterations in the regulation of the cell cycle progression.

Effects of E1A on the differentiation state of a cell have been described for embryonic carcinoma cells or their differentiated derivatives after transformation by E1A (Montano and Lane 1987; Velcich and Ziff 1989; Weigel et al. 1990; Weigel and Nevins 1990; de Groot et al. 1991a, b). This probably means that E1A induces or inhibits genes which are essential for a certain state of differentiation or dedifferentiation of the cell. It is not known how E1A alters the differentiation state of cells such as F9 or P19 embryonic carcinoma cells. A possible explanation for the E1A-induced differentiation of P19 cells might be the fact that E1A strongly activates c-*jun* transcription (van Dam et al. 1990; de Groot et al. 1991a, b) and that ectopic expression of c-*jun* can lead to differentiation of P19 cells (de Groot et al. 1990). Studies with mutant E1A could possibly clarify whether c-*jun* activation and P19 differentiation are casually related.

On the other hand, E1A has also been reported to inhibit the differentiation of certain cell types. For example, constitutive expression of E1A in PC12 cells results in a block of nerve growth factor (NGF)-induced neuronal differentiation (Maruyama et al. 1987). This effect of E1A was shown to be dependent on the CR1 region (Boulukos and Ziff 1993). In parallel with the inhibition of the NGF-induced neuronal differentiation, a CR1-dependent downregulation of the NGF receptors $p140^{trk}$ and $p75^{NGFR}$ was noted, as well as of the epidermal growth factor receptor (Boulukos and Ziff 1993), which explains why NGF can no longer induce differentiation of PC12 cells.

Moreover, the E1A proteins inhibit the in vitro differentiation (fusion) of myoblasts into myotubes, normally occurring at high cell density and low serum concentration (Webster et al. 1988). This inhibition of differentiation was shown to be dependent on the regions in E1A which are needed for association with the 300-kDa protein (Mymryk et al. 1992). Various muscle-specific genes are not expressed in the presence of E1A, and muscle-specific transcription factors are either downregulated in their expression, such as myoD and myogenin (Enkemann et al. 1990; Braun et al. 1992), or inactivated, such as myf5 (Braun et al. 1992). By using a limited set of E1A mutants, the latter study showed that these effects are dependent on an intact CR1 region. By use of additional mutants, it was shown that the inhibition of myoD expression was dependent on the regions of E1A responsible for the binding to p300. In addition, there is an inhibition of the transcriptional activity of myoD by E1A which is mediated by the CR1 and CR2 regions (Caruso et al. 1993). A study on the inactivation by E1A of the muscle-specific creatine kinase promoter in a rhabdomyosarcoma cell line showed a strict dependence on the N-terminal region of 12S E1A (*dl* 2–36 and *dl* 15–35 do not inhibit, while *dl* 38–87 does; therefore there is no correlation with p300 binding). In vitro binding studies showed that E1A can bind to the basic helix–loop–helix (bHLH) domains in the transcription factors myogenin and E12, with a specificity correlating with the downregulation of this creatine kinase enhancer (Taylor et al. 1993).

5 Effects on Expression of Serum-Inducible Genes

As stated before, the regulation of gene expression by E1A is considered to be mediated via its binding to cellular proteins. As a result of these associations, transcription factors E2F and Elf-1 are activated. Studies on the E2F transcription factor (for a review see Farnham et al. 1993) have shown that promoters of delayed-early- and late-response genes frequently have recognition sites for E2F, e.g., c-*myc*, N-*myc*, DHFR, thymidine kinase, DNA polymerase-α, *cdc*2, and thymidylate synthase. Some of these genes have been studied in transient assays and were shown to be induced by E1A via their E2F binding site: c-*myc*, N-*myc* and DHFR (Hiebert et al. 1989, 1991; Lipp et al. 1989; Thalmeier et al. 1989). Consistently, infection of baby rat kidney cells with Ad also leads to an E1A-dependent increase in expression of *cdc2* (Draetta et al. 1988). In contrast, c-*myc* is not induced after infection of quiescent 3T3 cells with Ad2 (Liu et al. 1985) and is even downregulated after Ad2 infection of KB cells (which normally express c-*myc*) (Rosahl and Doerfler 1992). Expression of c-*myc* also has been compared in Ad-transformed cells versus the nontransformed parental cells and was shown to be generally downregulated in the Ad-transformed cells (Timmers et al. 1988). This illustrates that gene regulation in transformed cells is not always mimicked in transient assays. These different effects in transient assays versus transformed or lytically infected cells might indicate that there are various ways in which the E1A gene product influences c-*myc* expression and that the relative contributions may vary in the various experiments. Indeed, it has been reported that the c-*myc* expression is positively regulated by YY1 (RIGGS et al. 1993), a transcription factor known to be affected by E1A (see below).

The downregulation of c-*myc* in Ad-transformed cells also illustrates that transformation is not caused by general activation of genes normally induced by growth factors. This would have been a simple explanation for serum-independent growth. Some other early- or late-response genes, such as JE and collagenase I (see Table 1), were also decreased in their expression (TIMMERS et al. 1988, 1989; VAN DAM et al. 1989), whereas others were increased, e.g., c-*jun* (VAN DAM et al. 1990), and c-*fos* was unaffected. Also the cyclins, the regulatory subunits of the cyclin-dependent kinases (cdk), which are normally increased during certain stages of the cell cycle, are differentially regulated by Ad E1A: cyclin A and cyclin E are increased and cyclin D decreased (BUCHOU et al. 1993; SPITKOVSKY et al. 1994). Stimulation of 3Y1 cells carrying a dexamethasone-inducible E1A 12S gene with either dexamethasone or serum showed that cyclin A induction was similar in both cases, cyclin E induction was delayed in the cell cycle when induced by E1A expression, while cyclin D1 expression was not induced by E1A expression (ISHII et al. 1993). These results suggest that E1A can bypass certain cell cycle events which normally occur in the early to mid G_1 phase. A similar situation is found for the early-response gene c-*myc*, which is induced by serum in quiescent 3T3 cells, but is not activated when S phase is induced by infection with Ad2 (LIU et al. 1985).

The mechanism by which E1A proteins alter expression of serum-inducible genes is still poorly understood, although some information is available for c-*myc*. As mentioned above, stimulation of c-*myc* expression in transient assays can be explained by the known association of E1A proteins with cellular proteins, but this cannot account for the inhibition of c-*myc* transcription in transformed cells or after viral infection. This illustrates that the association of E1A proteins with the known cellular proteins cannot yet explain the altered expression of cellular genes in Ad-transformed cells. Further studies on additional E1A-associating proteins and on the mechanism by which E1A alters the expression of genes will be necessary to clarify the mechanism of E1A-induced transformation. Data concerning the mechanism of gene regulation by the E1A proteins will be reviewed below. Alterations in gene expression by E1A might also influence the phosphorylation of transcription factors in Ad-transformed cells via modulation of expression of kinases or phosphatases. However, it should be realized that the E1A proteins also release the cyclin A/cdk2 and cyclin E/cdk2 kinases from their normal cellular substrates, although so far there have been no reports on altered phosphorylation activity of these kinases caused by the association with E1A proteins.

Table 1. Genes activated or repressed by adenovirus (Ad) E1A

Cellular genes	Expression	E1A region involved	Trans-formation	Trans-fection	Infection
c-*jun* [a]	Activated	CR1 or 13S	x	x	–
PCNA[b]	Activated	12S	x	–	–
Cyclin A[c]	Activated	CR1	x	x[q]	–
Cyclin D[d]	Repressed	CR1	x	x[q]	x
Cyclin E[e]	Activated	–	x[q]	x[q]	x
JE[f]	Repressed	CR1	x	–	–
Collagenase I[g]	Repressed	N-(4–25), CR1	x	x	–
Stromelysin[h]	Repressed	CR1	x	–	–
Fibronectin[i]	Repressed	CR1, CR2	x	x	–
Human hsp70[j]	Activated	N-(2–51) or CR3	x	x	x
c-*myc* [k]	Repressed	CR1	x	–	x
c-*myc* [l]	Activated	CR1, CR2	–	x	–
N-*myc* [m]	Activated	CR1, CR2	–	x	–
DHFR[n]	Activated	CR1, CR2	–	x	–
Muscle creatine kinase[o]	Repressed	N-(2–35)	–	x	–
β-Tubulin[p]	Activated	13S	–	–	x

The various domains of E1A involved in the gene regulation are indicated. The studies in which the alterations in gene expression were detected involved E1A-transformed cells, transient assays studying the effect of E1A on promoter/reporter constructs, or cells studying the endogenous gene expression after viral infection. DHFR, dihydrofolate reductase; PCNA, proliferating cell nuclear antigen.
[a]VAN DAM et al. 1990; KITABAYASHI et al. 1991; DE GROOT et al. 1991a.
[b]MORRIS and MATHEWS 1990; KANNABIRAN et al. 1993.
[c]BUCHOU et al. 1993; SPITKOVSKY et al. 1994.
[d]BUCHOU et al. 1993; SPITKOVSKY et al. 1994.
[e]SPITKOVSKY et al. 1994.
[f]TIMMERS et al. 1989; VAN DAM et al. 1989.
[g]OFFRINGA et al. 1990.
[h]OFFRINGA et al. 1988; VAN DAM et al. 1989.
[i]JOCHEMSEN et al. 1986; ROBERTS et al. 1985; NAKAMURA et al. 1992; NAKAJIMA et al. 1992.
[j]KRAUS et al. 1992; NEVINS 1982; KAO and NEVINS 1983; WILLIAMS et al. 1989; LUM et al. 1992.
[k]TIMMERS et al. 1988; VAN DAM et al. 1989; ROSAHL and DOERFLER 1992.
[l]HIEBERT et al. 1989.
[m]HIEBERT et al. 1991.
[n]HIEBERT et al. 1991.
[o]TAYLOR et al. 1993.
[p]STEIN and ZIFF. 1984.
[q]Effect of E1A-ER fusion protein after addition of estrogen.

6 Effects on Expression of Other Genes

The E1A proteins can alter transcription of cellular genes by various mechanisms. First, a general activation of many genes by only the 13S E1A protein can take place, which does not involve activation of specific transcription factors but rather concerns an increase of the overall transcription of genes (reviewed in BERK 1986). This effect is largely restricted to genes newly introduced into cells by transfection or virus infection. In general, transcription of endogenous genes is not activated, with a few exceptions (see for example the cellular genes in

Table 1 that are activated after virus infection). In this respect the interaction of E1A with the TATA-binding protein may be important (LEE et al. 1991). Also, it should be noted that in many transient assays the apparent activation of promoters was found to be caused by the stabilization of the reporter plasmids, due to the presence of the E1B 21-kDa protein (HERRMANN and MATHEWS 1989).

Analysis of the transcription activation by the CR3 region unique to the 13S E1A protein has revealed that this 50-aa domain has two functions. The N-terminal part of CR3, when linked to GAL4-DNA-binding domain, is a strong transcription activator, whereas the C-terminal part of CR3 does not have transcriptional activation potential but is required for the activation of E1A-inducible promoters (LILLIE and GREEN 1989). Therefore, the C-terminal part of the CR3 region is expected to target E1A to promoters by interaction with transcription factors. Indeed the C-terminal part of CR3 has been shown to be required in vitro for efficient association of the E1A protein with the DNA-binding domain of activating transcription factor (ATF)-2, c-Jun, Sp1 and USF (LIU and GREEN 1994). However, transcription mediated by these transcription factors is not in all cases activated by E1A. This is explained by the requirement of a transactivation domain in the transcription factors which can synergistically function with the transactivation domain of the E1A protein (LIU and GREEN 1994). The capacity of CR3 to interact by its C terminus to transcription factors and by its N terminus to the TATA-binding protein makes it suitable to activate transcription. This adaptor function is also found for E1A-induced activation of Oct-4-mediated transcription. (SCHOLER et al. 1991) and for the E1A requirement for the cooperative transactivation of the retinoic acid receptor and the TATA-binding protein (BERKENSTAM et al. 1992).

Another general effect of E1A, but now associated with both 12S and 13S E1A proteins, is the inhibition of several viral and cellular enhancers, such as those of SV40, polyoma (MICHALIDES et al. 1985), and the immunoglobulin (Ig)H genes (HEN et al. 1985). As for the general transcriptional activation, the mechanism does not seem to be mediated by a specific factor or interaction, since many different transcription factor binding sites can be targets for this repressing activity (ROCHETTE EGLY et al 1990). Furthermore, the effect is dependent on the cell type used, indicating that the identity of the transcription factors present plays an essential role. For example, it has been shown that the IgH enhancer, which is repressed by E1A in lymphoid cells, is activated by E1A in the mouse fibroblast L cells (BORRELLI et al. 1986) and the enhancer of a human BK virus is activated by E1A in a monkey kidney cell line, while it is repressed in HeLa cells, probably via the induction of a 98-kDa repressor (REIFEL MILLER et al. 1991).

Transformation is in general accompanied by alterations in cell adhesion, caused in part by altered expression of cell surface proteins (BOBER et al. 1988). One of the genes decreased in expression is the fibronectin gene, for example in cells expressing only Ad5 E1A 12S or 13S (ROBERTS et al. 1985) or expressing either Ad5 or Ad12 E1A together with the *ras* oncogene (JOCHEMSEN et al. 1986). Also in 3Y1 cells fibronectin is downregulated by the E1A gene products after stable transformation or in transient assays (NAKAMURA et al. 1992;

NAKAJIMA et al. 1992). Studies with E1A deletion mutants showed that this downregulation is mediated via the CR1 and CR2 regions, correlating with Rb binding. Analysis of the fibronectin promoter showed that perfect or almost perfect G_{10} stretches in the promoter are the target for inhibition by E1A. The binding of cellular factors to these elements is increased by E1A, suggesting that a repressor is induced (NAKAMURA et al. 1992; NAKAJIMA et al. 1992).

In Ad-transformed cells synthesis of several metalloproteases, e.g., stromelysin and collagenases, is also downregulated. Below, we will discuss the AP-1-dependent downregulation of the collagenase type I promoter. Not all metalloproteases are downregulated by the same mechanism, as is illustrated by the observation that the collagenase type IV gene is downregulated by repression of an enhancer at −1.6 kb via an AP-2-like site (FRISCH and MORISAKI 1990). This reduced expression of proteases by E1A has probably little effect on the transformed phenotype, but reduces the metastatic behavior of human tumor cells (FRISCH et al. 1990).

Introduction of the E1A gene into human tumor cells can sometimes reduce the tumorigenicity of the cells (FRISCH 1991). Apart from the downregulation of proteases, the 12S or 13S E1A proteins influence malignancy by the downregulation of the proto-oncogene *neu*. This repression requires an intact CR2 region of the E1A proteins and the element TGGAATG in the *neu* promoter (YU et al. 1990). Expression of E1A leads to a reduced tumorigenicity and metastatic potential in nude mice (YU et al. 1991, 1992). Also, human ovarian cancer cells which are highly malignant due to enhanced *neu* expression show reduced oncogenic behavior by expression of the E1A proteins (YU et al. 1993b). The E1A gene was shown to reduce the malignancy of the cells by different pathways. Overexpression of *neu* in cells in which endogenous *neu* expression had been reduced by E1A led to an increased tumorigenicity but hardly induced the metastatic phenotype. This indicates that E1A is capable of reducing the metastatic ability of the cells in other ways than by inhibiting *neu* expression (YU et al. 1993a).

The E1A proteins can also inhibit the effects of certain external signals received by cells. For example, interferon (IFN)-induced gene transcription is inhibited by E1A, which is dependent on the N terminus and the CR1 region. This inhibition is mediated by a reduction of one or more of the subunits of factor E (or ISGF3), a transcription factor binding to the IFN-stimulated response element (ISRE) (REICH et al. 1988; GUTCH and REICH 1991; KALVAKOLANU et al. 1991; ACKRILL et al. 1991). Another example of this category is the diminishing by E1A proteins of the growth-inhibitory effect of transforming growth factor (TGF)-β on keratinocytes. This effect is reduced with E1A mutants that fail to bind to the cellular proteins p300, p107, Rb, and cyclin A (MISSERO et al. 1991). The inhibition of transcription of the c-*myc* promoter by TGF-β is abrogated by E1A in transient assays, an effect that depends on the ability of E1A to bind to Rb (PIETENPOL et al. 1990). However, the abrogation by E1A of the reduced expression of *cdc2* by TGF-β is not dependent on the binding of E1A protein to Rb, in agreement with the observation that this mutant reduced the TGF-β inhibition as wild-type E1A in these experiments (ABRAHAM et al. 1992).

The induction of hsp70 expression is found in transformed human cells and in transient assays, although controversial data have been published as to the contribution of 12S and 13S E1A (Wu et al. 1986; Simon et al. 1987). Studies focusing on the 13S E1A effect showed the involvement of various transcription factor-binding sites, including the CCAAT element. In the context of a heterologous promoter, the CCAAT element can mediate the transactivation by E1A (Williams et al. 1989). The CCAAT transcription factor was shown to be able to bind in vitro to the 13S E1A protein (and weakly to the 12S E1A protein) (Lum et al. 1992). The studies focusing on the effect of the 12S E1A protein showed that an Ad5 12S E1A virus infection can activate the hsp70 promoter via a specific TATAA sequence (Simon et al. 1987, 1988). Similarly, c-*fos* expression is also induced via its TATAA sequence (Simon et al. 1990). Activation of the hsp70 promoter requires the N-terminal sequence and part of the CR1 region (aa 2–51) of 12S E1A (Kraus et al. 1992). It has been suggested that this activation is caused by the disruption by E1A of the binding of the TATA-binding protein with Dr1, an inhibitor of transcription (Inostroza et al. 1992).

Another factor influenced by E1A is YY1, which can function as a transcription initiator, directing RNA polymerase II to the promoter. On the other hand, upstream YY1 sites can repress transcription, which is counteracted by the E1A proteins (Shi et al. 1991). YY1 binds to both 12S and 13S E1A proteins (Usheva et al. 1994).

7 Activation of Transcription by 13S-E1A via ATF

The E1A gene products stimulate at least five different viral transcription units (region E1B, E2A, E3, E4, and the major late promoter, MLP) and, consequently, are essential for viral replication (for reviews see Shenk and Flint 1991; Jones 1992). In general, this transactivation requires the CR3 element uniquely present in the 13S E1A protein. Although various transcription factor-binding sites are involved, in particular the ATF-binding sites are required (E2A, E3, and E4 promoters). This ATF-binding site (core sequence CGTCA) is very similar, and in some cases identical, to the cyclic adenosine monophosphate (cAMP)-responsive (CRE)-binding site (TGACGTCA). A closely related site is the activating protein-1 (AP-1)-binding site (TGACTCA), to which members of the Jun and Fos family preferentially bind (for a review see Angel and Karin 1991). In all cases, the transcription factors bind as dimers to the DNA. This binding requires a leucine zipper domain for the dimerization and a basic region for the interaction with DNA characteristic for the basic leucine zipper, or bZip, family of transcription factors. Heterodimerization is not restricted to members within the mentioned families, as, for example, CREB can dimerize with ATF-1, c-Jun with ATF-2, and c-Jun with CREB. The various homo- and heterodimers differ in their affinity for various ATF/AP-1 DNA-binding sites and are differentially activated by stimuli such as TPA and cAMP.

ATF transcription factors that are inducible by E1A are ATF-2 (also called CRE-BP1) and the homologous ATFa. Both ATF-2 and ATFa are distinct from most ATF/AP-1 factors since they contain a putative Zn finger domain near the N terminus. Fusion proteins such as Gal4-ATF-2 or myb-ATF-2 are activated by E1A when bound to heterologous binding sites (LIU and GREEN 1990; MAEKAWA et al. 1991). ATF-2 has been shown to bind directly to E1A in vitro (ABDELHAFIZ et al. 1993; LIU and GREEN 1994). The ATFa cDNA was cloned by screening a λgt11 library with the ATF-binding site from the E2A promoter (GAIRE et al. 1990). Three different ATFa products were found which differ from each other due to differential splicing. None of the ATFa products on its own is capable of increasing the E2A promoter activity, but their presence increases the transactivation of this promoter by E1A. Furthermore, this factor is activated as Gal4-ATFa by the 13S E1A from a Gal4-binding site and was shown to bind in vitro to E1A (CHATTON et al. 1993). In both cases, the interaction of the 13S E1A protein with ATF requires both the putative Zn finger domain and the C-terminal part containing the bZip domain (CHATTON et al. 1993; ZU et al. 1991). The presence of the putative Zn finger, although important, does not automatically imply inducibility by E1A, as CRE-BPa, a recently cloned Zn finger-containing gene product homologous to ATF-2 and ATFa, is not inducible by E1A (NOMURA et al. 1993). The activation of ATF-2 and ATFa by E1A was studied with full-length cDNA or cDNA with small deletions due to differential splicing. However, also cDNA that lack a large internal domain with a potential transactivation element have been described (IVASHKIV et al. 1990; GEORGOPOULOS et al. 1992). In the case of the ATFa factors, it has been shown that this ATFa product is a dominant inhibitor of ATFa (PESCINI et al. 1994). Whether the effect of E1A on ATF is influenced by the relative amounts of the various splice forms is not yet known.

8 Downregulation of Cellular Gene Products via AP-1 Elements

Ad-transformed cells show a decreased expression of secreted proteases, including stromelysin and type I collagenase (FRISCH et al. 1990; OFFRINGA et al. 1988, 1990). The inhibition of the TPA-induced expression of the collagenase I gene was shown to be dependent on the CR1 region (OFFRINGA et al. 1990) and the N terminus (B.M. Hagmeyer, personal communication) of E1A, suggesting that the association with p300 is essential for this downregulation. The inhibition of the collagenase I promoter was shown to be dependent on the TRE (OFFRINGA et al. 1990), which normally binds c-Jun/c-Fos or c-Jun/c-Jun (for a review see ANGEL and KARIN 1991). Transcription induction by either c-Jun, c-Fos, or their viral counterparts is inhibited by E1A. Analysis of the E1A effect on hybrid Jun-GHF1 transcription factors showed that the inhibition is directed to the DNA-binding domain of c-Jun (HAGMEYER et al. 1993). Thus, E1A inhibits c-Jun via the domain that is most conserved among the AP-1 proteins and this clarifies the observation that transactivation by all Jun factors is inhibited by E1A (unpublished observations). The affinity of c-Jun dimers for a TRE sequence is dependent on the phosphorylation of a region close to the DNA-binding domain: stronger phosphorylation leads to a weaker binding of c-Jun to DNA (Boyle et al. 1991). However, expression of E1A does not effect the phosphorylation status of this region near the DNA-binding domain, suggesting that inhibition via the DNA-binding part of c-Jun by E1A is not caused by phosphorylation. Consistently, transactivation by c-Jun proteins in which the relevant Ser and Thr residues were mutated to Ala could still be fully inhibited by E1A (Hagmeyer et al. 1993). Therefore, it is still not clear how E1A inhibits collagenase expression. The contradictory observation that in Ad-transformed cells in vivo no footprint on the TRE is observed (Hagmeyer et al. 1993), while extracts from Ad-transformed cells do show in vitro binding activity to the collagenase TRE (Offringa et al. 1990), suggest that binding of AP-1 is altered in Ad-transformed cells due to a weak interaction with a cellular factor. In cell extracts, however, this putative factor is diluted so far that it can no longer inhibit the binding of AP-1 to the collagenase TRE. Two inhibitory proteins for AP-1 have been described, an activity designated IP-1 (Auwerx and Sassone-Corsi 1991, 1992) and a protein, called Jif-1, which binds to and thereby prevents Jun from binding to DNA (Monteclaro and Vogt 1993). In addition, a bHLH protein, called FIP, which binds to and activates c-Fos has been cloned (Blanar and Rutter 1992).

A difference between nontransformed and Ad-transformed cells is the association of a 21-kDa protein with c-Jun in Ad-transformed cells. This protein turned out to be identical to liver regeneration factor 1 (LRF-1) (B.M. Hagmeyer et al., in preparation). LRF-1 is a bZip transcription factor capable of dimerizing with c-Jun and JunB but not with c-Fos (Hsu et al. 1991). The various complexes containing LRF-1 differentially influence transcription activation (Hsu et al. 1992, 1993), and therefore induction of LRF-1 in Ad-transformed cells may influence the activity of AP-1-dependent promoters. However, LRF-1-containing complexes still bind to TRE sequences, while in vivo no footprint is found on the collagenase TRE. Therefore, it still has to be established whether the induction of LRF-1 in Ad-transformed cells is responsible for the downregulation of the collagenase expression. It is not known whether other c-Jun-interacting proteins (Chevray and Nathans 1992) or other members of the bZip transcription factor family, such as the maf factors (reviewed in Engel 1994) are altered in Ad-transformed cells.

9 Activation of Cellular Genes via AP-1/ATF Elements

Although the transcription activation of AP-1 complexes containing c-Jun is inhibited by E1A, the phosphorylation of residues 63 and 73 in the N terminus is strongly increased in Ad-transformed cells (Hagmeyer et al. 1993). Phosphorylation of

Ser-63 and Ser-73 in c-Jun is also found after TPA treatment of cells, which leads to an increase of the transactivation potential of c-Jun (PULVERER et al. 1991). This increase can also be caused by oncogenes such as *sis, src, ras,* and *raf* (BINETRUY et al. 1991; SMEAL et al. 1991, 1992) and is found at low serum concentrations in 3T3 cells transformed by *raf, src, ras,* or *erb*-B (KAMADA et al. 1994). In these cells, TRE reporter constructs are indeed expressed at high levels, in contrast to gene expression from TRE-containing promoters in E1A-transformed cells. These data indicate that increased phosphorylation of the transactivation domain of c-Jun is a property shared by E1A and other oncoproteins, whereas E1A has an additional effect on the DNA-binding domain of Jun, leading to transcription inhibition of factors such as c-Jun/c-Jun and c-Jun/c-Fos.

The inhibition of the transcriptional activity of c-Jun by E1A seems contradictory to the enhanced expression of the c-*jun* gene in E1A-transformed cells and in cells transiently transfected with a c-*jun* promoter/reporter together with E1A (VAN DAM et al.1990; KITABAYASHI et al.1991; DE GROOT et al.1991a). In transformed cells the E1A 12S product is capable of inducing transcription of c-*jun* mRNA (KITABAYASHI et al.1991), which requires an intact CR1 domain (VAN DAM et al. 1990). In transient assays contradictory results have been reported: either the induction was dependent on an intact CR1 region (VAN DAM et al.1990) or the effect was unique for the 13S E1A protein (DE GROOT et al.1991a). Analysis of the c-*jun* promoter showed that activation by E1A was actually mediated by two TRE sequences. In addition, a CCAAT transcription factor-binding sequence and the 5'untranslated region (5'-UTR) of the transcribed region contributed to a full inducibility by E1A (VAN DAM et al.1993). About the same regions in the c-*jun* promoter were found to be involved in c-*jun* induction by E1A in F9 cells (KITABAYASHI et al.1992). The involvement of multiple elements in the inducibility by E1A of the c-*jun* promoter can possibly be explained by interactions of the various transcription factors which bind to this promoter region. In addition, it was studied whether the TRE as a single element could mediate transactivation by E1A. The *jun*(2)-TRE could be activated weakly and a multimerized *jun*(2)-TRE was clearly induced by E1A. Therefore, the *jun*(2)-TRE appears to be the decisive element in the c-*jun* promoter for activation by E1A. The protein complex binding to this element was shown to be c-Jun/ATF-2 (VAN DAM et al.1993). This indicates that the E1A protein does not inhibit all c-Jun complexes and clarifies the apparent contradiction that the E1A proteins increase the c-Jun expression but at the same time inhibit its activity. The mechanism by which E1A activates c-Jun/ATF-2 is not completely clear, except that E1A is known to cause an increased phosphorylation of the transactivation domain of c-Jun, thereby activating c-Jun. Additional alterations may occur as well, and preliminary observations indicate that the c-*Jun*/ATF-2 complex is stabilized in Ad E1A-transformed cells (M.C.A. Duyndam, unpublished observations). This effect would be analogous to the dimer stabilization of the bZIP transcription factors by the HTLV-1 tax protein, resulting in an increased DNA binding (FRANKLIN et al. 1993; WAGNER and GREEN 1993).

Another gene which is activated via an ATF-like site by the 12S E1A product is the proliferating cell nuclear antigen (PCNA), the auxiliary factor of DNA polymerase δ. The domains of E1A that are involved in the activation of the PCNA promoter suggest that p300 and p107–cyclin A but not Rb contribute (Morris and Mathews 1990; Kannabiran et al. 1993). The activation actually occurs via an ATF-like element (Morris and Mathews 1991) in conjunction with upstream sequences, together called the PCNA E1A-responsive element (PERE) (Labrie et al. 1993). This element (-59/-45) is important for both basal and E1A-induced activity of the PCNA promoter. The multimerized PERE element in a heterologous promoter can also be activated by E1A (Morris et al. 1994). However, other elements in the PCNA promoter also contribute to the activation by E1A. When the PERE element in the minimal PCNA promoter was replaced by Gal4-binding sites, transcription was still activated when a Gal4 fusion protein with a weak transcriptional activator was present. These Gal4 fusion proteins are not activated by 12S E1A when the Gal4-binding sites are linked to the E1B basal promoter (Morris et al. 1994). This indicates that activation of a promoter by E1A depends on various transcription factors and their mutual interactions. Although elements can be found in the promoter which by itself can mediate activation by E1A, it seems impossible to identify unique targets for the activation by E1A in the entire promoter. Apart from the PCNA promoter, contributions from various elements was also found for the activation by 12S E1A of the c-jun promoter (van Dam et al. 1993) and the activation by 13S E1A of the hsp70 promoter (Williams et al. 1989).

10 Alterations in the AP-1/ATF Repertoire in Adenovirus-Transformed Cells

In summary, the data mentioned above indicate that transformation by E1A leads to an alteration in the repertoire of active AP-1/ATF transcription factors: at least part of the AP-1 factors are inhibited, whereas some heterodimers of AP-1 factors with members of the ATF family and also ATF homo- or heterodimers may be activated. These alterations are due in part to the inactivation of AP-1 by E1A, the mechanism of which has not yet been clarified, and by the activation of transcription of *c-jun* and LRF-1. The various dimeric bZip transcription factors have slight variations in their preference for recognition sequences, which may result in expression of different sets of genes. This is illustrated by the down-regulation of collagenase I expression and the increased expression of *c-jun*. Other examples of genes which probably are regulated by E1A via the AP-1/ATF transcription factors are the activation of the mouse DNA polymerase-β gene (Yamaguchi et al. 1992) and the repression of the phosphoenolpyruvate carboxykinase gene (Kalvakolanu et al. 1992). It is reasonable to assume that the decreased or increased expression of some of these genes will contribute to the transformed phenotype. The mechanism by which E1A alters the repertoire of active AP-1/ATF transcription factors is not clear and may require the cloning of additional genes coding for proteins which bind to the N terminus and CR1 of E1A.

In the discussion on gene regulation by E1A, no distinction has been made between the E1A proteins from different adenovirus serotypes. However, the cells obtained after transformation by Ad5 or Ad12 strongly differ in their oncogenic behavior. Although experimental evidence suggests that the E1A proteins from both serotypes bind to the same cellular proteins, there may still be differences that have not been detected so far. It has been well established that certain differences in gene expression exist between the Ad5- and Ad12-transformed cells, such as the downregulation of the major histocompatibility complex(MHC) class I antigens in Ad12-transformed cells. Another difference is the decreased expression of JunB in Ad12-transformed cells (van Dam et al. 1990), which implies that there is a difference in the repertoire of active AP-1/ATF transcription factors between Ad5- and Ad12-transformed cells.

11 Concluding Remarks

The E1A proteins can cause very diverse alterations in gene transcription. Apart from the general transcription activation by the 13S E1A protein, alterations in transcription via the domains required for transformation are also found. The observation that E1A associates with cellular proteins has led to the concept that these viral gene products function by sequestering cellular proteins (Harlow et al. 1986). Transcriptional effects are therefore generally explained by a direct interaction of E1A with transcription factors. Remarkably, interactions of the E1A proteins with many factors has been demonstrated in vitro or after overexpression, e.g., the TATA-binding protein, Oct-4, CTF, myogenin, E12, YY1, ATF-2, and ATFa. Whether all these interactions are indeed functional in vivo still needs to be established, but for the time being they could provide an explanation for the observed alterations in gene transcription.

Acknowledgments. The authors thank Dr. B.M. Hagmeyer and Dr. H. van Ormondt for their critically reading of the manuscript.

References

Abdelhafiz HAM, Chen CY, Marcell T, Kroll DJ, Hoeffler JP (1993) Structural determinants outside of the leucine zipper influence the interactions of CREB and ATF-2—interaction of CREB with ATF-2 blocks E1a-ATF-2 complex formation. Oncogene 8: 1161–1174

Abraham SE, Carter MC, Moran E (1992) Transforming growth factor-beta1 (TGFbeta1) reduces cellular levels of p34cdc2, and this effect is abrogated by adenovirus independently of the E1A-associated pRB binding activity. Mol Biol Cell 3: 655–665

Abraham SE, Lobo S, Yaciuk P, Wang HGH, Moran E (1993) p300, and p300-associated proteins, are components of TATA-binding (TBP) complexes. Oncogene 8: 1639–1647

Ackrill AM, Foster GR, Laxton CD, Flavell DM, Stark GR, Kerr IM (1991) Inhibition of the cellular response to interferons by products of the adnovirus type 5 E1A oncogene. Nucleic Acids Res 19: 4387–4393

Anderson CW, Schmitt RC, Smart JE, Lewis JB (1984) Early region 1B of adenovirus 2 encodes two coterminal proteins of 495 and 155 amino acid residues. J Virol 50: 387–396

Angel P, Karin M (1991) The role of Jun, Fos and the AP-1 complex in cell-proliferation and transformation. (review) Biochim Biophys Acta 1072: 129–157

Auwerx J, Sassone-Corsi P (1991) IP-1: a dominant inhibitor of Fos/Jun whose acitivity is modulated by phosphorylation. Cell 64: 983–993

Auwerx J, Sassone-Corsi P (1992) AP-1 (Fos-Jun) regulation by IP-1: effect of signal transduction pathways and cell growth. Oncogene 7: 2271–2280

Bagchi S, Raychaudhuri P, Nevins JR (1990) Adenovirus E1A proteins can dissociate heteromeric complexes involving the E2F transcription factor: a novel mechanism for E1A trans-activation. Cell 62: 659–669

Bandara LR, Buck VM, Zamanian M, Johnston LH, Lathangue NB (1993) Functional synergy between DP-1 and E2F-1 in the cell cycle-regulating transcription factor DRTF1/E2F. EMBO J 12: 4317–4324

Berk AJ (1986) Adenovirus promoters and E1A transactivation. Annu Rev Genet 20: 45–79

Berkenstam A, Ruiz MDV, Barettino D, Horikoshi M, Stunnenberg HG (1992) Cooperativity in transactivation between retinoic acid receptor and TFIID requires an activity analogous to E1A. Cell 69: 401–412

Binetruy B, Smeal T, Karin M (1991) Ha-Ras augments c-Jun activity and stimulates phosphorylation of its activation domain. Nature 351: 122–127

Blanar MA, Rutter WJ (1992) Interaction cloning: identification of a helix-loop-helix zipper protein that interacts with c-Fos. Science 256: 1014–1018

Bober FJ, Birk DE, Shenk T, Raska KJr (1988) Tumorigenicity of adenovirus-transformed cells: collagen interaction and cell surface laminin are controlled by the serotype origin of the E1A and E1B genes. J Virol 62: 580–585

Bondesson M, Svensson C, Linder S, Akusjärvi G (1992) The carboxy-terminal exon of the adenovirus E1a protein is required for E4F-dependent transcription activation. EMBO J 11: 3347–3354

Borrelli E, Hen R, Wasylyk C, Wasylyk B, Chambon P (1986) The immunoglobulin heavy chain enhancer is stimulated by the adenovirus type 2 E1A products in mouse fibroblasts. Proc Natl Acad Sci USA 83: 2846–2849

Bos JL, Jochemsen AG, Bernards R, Schrier PI, Van Ormondt H, van der Eb AJ (1983) Deletion mutants of region E1a of Ad12 E1 plasmids: effect on oncogenic transformation. Virology 129: 393–400

Boulukos KE, Ziff EB (1993) Adenovirus 5 E1A proteins disrupt the neuronal phenotype and growth factor responsiveness of PC12 cells by a conserved region 1-dependent mechanism. Oncogene 8: 237–248

Boyd JM, Subramanian T, Schaeper U, Laregina M, Bayley S, Chinnadurai G (1993) A region in the C-terminus of adenovirus-2/5 E1a protein is required for association with a cellular phosphoprotein and important for the negative modulation of T24-ras mediated transformation, tumorigenesis and metastasis. EMBO J 12: 469–478

Boyle WJ, Smeal T, Defize LH, Angel P, Woodgett JR, Karin M, Hunter T (1991) Activation of protein kinase C decreases phosphorylation of c-Jun at sites that negatively regulate its DNA-binding activity. Cell 64: 573–584

Braun T, Bober E, Arnold HH (1992) Inhibition of muscle differentiation by the adenovirus E1a-protein—repression of the transcriptional activating function of the HLH protein Myf-5. Gene Dev 6: 888–902

Brockmann D, Tries B, Esche H (1990) Isolation and characterization of novel adenovirus type 12 E1A mRNAs by cDNA PCR technique. Virology 179: 585–590

Brockmann D, Feng L, Kroener G, Tries B, Esche H (1994) Adenovirus type 12 early region 1A expresses a 52R protein repressing the trans-activating activity of transcription factor c-Jun/AP-1. Virology 198: 717–723

Buchou T, Kranenburg O, van Dam H, Roelen D, Zantema A, Hall FL, van der Eb A (1993) Increased cyclin-A and decreased cyclin-D levels in adenovirus-5 E1A-transformed rodent cell lines. Oncogene 8: 1765–1773

Byrd P, Brown KW, Gallimore PH (1982) Malignant transformation of human embryo retinoblasts by cloned adenovirus 12 DNA. Nature 298: 69–71

Caruso M, Martelli F, Giordano A, Felsani A (1993) Regulation of MyoD gene transcription and protein function by the transforming domains of the adenovirus E1A oncoprotein. Oncogene 8: 267–278

Chatton B, Bocco JL, Gaire M, Hauss C, Reimund B, Goetz J, Kedinger C (1993) Transcriptional activation by the adenovirus larger E1a product is mediated by members of the cellular transcription factor ATF family which can directly associate with E1a. Mol Cell Biol 13: 561–570

Chevray PM, Nathans D (1992) Protein interaction cloning in yeast: identification of mammalian proteins that react with the leucine zipper of Jun. Proc Natl Acad Sci USA 89: 5789–5793

Cress WD, Johnson DG, Nevins JR (1993) A Genetic Analysis of the E2F1-gene distinguishes regulation by Rb, p107, and adenovirus-E4. Mol Cell Biol 13: 6314–6325

Debbas M, White E (1993) Wild-type p53 mediates apoptosis by E1A, which is inhibited by E1B. Gene Dev. 7: 546–554

de Groot RP, Kruyt FA, van der Saag PT, Kruijer W (1990) Ectopic expression of c-jun leads to differentiation of P19 embryonal carcinoma cells. EMBO J 9: 1831–1837

de Groot RP, Foulkes N, Mulder M, Kruijer W, Sassone Corsi P (1991a) Positive regulation of jun/AP-1 by E1A. Mol Cell Biol 11: 192–201

de Groot RP, Meijer I, Vandenbrink S, Mummery C, Kruijer W (1991b) Differential regulation of JunB and JunD by adenovirus type-5 and type-12 E1A-proteins. Oncogene 6: 2357–2361

Draetta G, Beach D, Moran E (1988) Synthesis of p34, the mammalian homolog of the yeast cdc2+/CDC28 protein kinase, is stimulated during adenovirus-induced proliferation of primary baby rat kidney cells. Oncogene 2: 553–557

Dyson N, Dembski M, Fattaey A, Ngwu C, Ewen M, Helin K (1993) Analysis of p107-associated proteins—p107 associates with a form of E2F that differs from pRB-associated E2F-1. J Virol 67: 7641–7647

Engel JD (1994) Meticulous AP-1 factors. Nature 367: 516–517

Enkemann SA, Konieczny SF, Taparowsky EJ (1990) Adenovirus 5 E1A represses muscle-specific enhancers and inhibits expression of the myogenic regulatory factor genes, MyoD1 and myogenin. Cell Growth Differ 1: 375–382

Faha B, Harlow E, Lees E (1993) The adenovirus E1A-associated kinase consists of cyclin E-p33cdk2 and cyclin A-p33cdk2. J Virol 67: 2456–2465

Farnham PJ, Slansky JE, Kollmar R (1993) The role of E2F in the mammalian cell cycle (review). Biochim Biophys Acta 1155: 125–131

Flint J, Shenk T (1989) Adenovirus E1A protein paradigm viral transactivator. Annu Rev Genet 23: 141–161

Franklin AA, Kubik MF, Uittenbogaard MN, Brauweiler A, Utaisincharoen P, Matthews MA, Dynan WS, Hoeffler JP, Nyborg JK (1993) Transactivation by the human T-cell leukemia virus Tax protein is mediated through enhanced binding of activating transcription factor-2 (ATF-2) ATF-2 response and cAMP element-binding protein (CREB). J Biol Chem 268: 21225–21231

Frisch SM (1991) Antioncogenic effect of adenovirus E1A in human tumor cells. Proc Natl Acad Sci USA 88: 9077–9081

Frisch SM, Morisaki JH (1990) Positive and negative transcriptional elements of the human type IV collagenase gene. Mol Cell Biol 10: 6524–6532

Frisch SM, Reich R, Collier IE, Genrich LT, Martin G, Goldberg GI (1990) Adenovirus E1A represses protease gene expression and inhibits metastasis of human tumor cells. Oncogene 5: 75–83

Gaire M, Chatton B, Kedinger C (1990) Isolation and characterization of two novel, closely related ATF cDNA clones from HeLa cells. Nucleic Acids Res 18: 3467–3473

Georgopoulos K, Morgan BA, Moore DD (1992) Functionally distinct isoforms of the CRE-BP DNA-binding protein mediate activity of a T-cell-specific enhancer. Mol Cell Biol 12: 747–757

Graham FL, Smiley J, Russell WC, Nairn R (1977) Characteristics of a human cell line transformed by DNA from human adenovirus type 5. J Gen Virol 36: 59–74

Gutch MJ, Reich NC (1991) Repression of the interferon signal transduction pathway by the adenovirus E1A oncogene. Proc Natl Acad Sci USA 88: 7913–7917

Hagmeyer BM, Konig H, Herr I, Offringa R, Zantema A, Vandereb AJ, Herrlich P, Angel P (1993) Adeno virus E1A negatively and positively modulates transcription of AP-1 dependent genes by dimer-specific regulation of the DNA binding and transactivation activities of Jun. EMBO J 12: 3559–3572

Harlow E, Whyte P, Franza BR Jr, Schley C (1986) Association of adenovirus early-region 1A proteins with cellular polypeptides. Mol Cell Biol 6: 1579–1589

Helin K, Less JA, Vidal M, Dyson N, Harlow E, Fattaey A (1992) A cDNA encoding a pRB-binding protein with properties of the transcription factor E2F. Cell 70: 337–350
Helin K, Harlow E, Fattaey A (1993a) Inhibition of E2F-1 transactivation by direct binding of the retinoblastoma protein. Mol Cell Biol 13: 6501–6508
Helin K, Wu CL, Fattaey AR, Lees JA, Dynlacht BD, Ngwu C, Harlow E (1993b) Heterodimerization of the transcription factors E2F-1 and DP-1 leads to cooperative trans-activation. Gene Dev7: 1850–1861
Hen R, Borrelli E, Chambon P (1985) Repression of the immunoglobulin heavy chain enhancer by the adenovirus-2 E1A products. Science 230: 1391–1394
Herrmann CH, Mathews MB (1989) The adenovirus E1B 19-kilodalton protein stimulates gene expression by increasing DNA levels. Mol Cell Biol 9: 5412–5423
Hiebert SW, Lipp M, Nevins JR (1989) E1A-dependent trans-activation of the human MYC promoter is mediated by the E2F factor. Proc Natl Acad Sci USA 86: 3594–3598
Hiebert SW, Blake M, Azizkhan J, Nevins JR (1991) Role of E2F transcription factor in E1A-mediated transactivation of cellular genes. J Virol 65: 3547–3552
Houweling A, van den Elsen PJ, van der Eb AJ (1980) Partial transformation of primary rat cells by the leftmost 4.5% fragment of adenovirus 5 DNA. Virology 105: 537–550
Hsu JC, Laz T, Mohn KL, Taub R (1991) Identification of LRF-1, a leucine-zipper protein that is rapidly and highly induced in regenerating liver. Proc Natl Acad Sci USA 88: 3511–3515
Hsu JC, Bravo R, Taub R (1992) Interactions among LRF-1, JunB, c-Jun, and c-Fos define a regulatory program in the G1 phase of liver regeneration. Mol Cell Biol 12: 4654–4665
Hsu JC, Cressman DE, Taub R (1993) Promoter-specific trans-activation and inhibition mediated by JunB. Cancer Res 53: 3789–3794
Inostroza JA, Mermelstein FH, Ha I, Lane WS, Reinberg D (1992) Drl, a TATA-binding protein-associated phosphoprotein and inhibitor of class II gene transcription. Cell 70: 477–489
Ishii T, Shimizu M, Kanayama Y, Nakada S, Nojima H, Oda K (1993) Differential activation of cyclin and cyclin-dependent kinase genes by adenovirus E1A12S cDNA product. Exp Cell Res 208: 407–414
Ivashkiv LB, Liou HC, Kara CJ, Lamph WW, Verma IM, Glimcher LH (1990) mXBP/CRE-BP2 and c-Jun form a complex which binds to the cyclic AMP, but not to the 12-O-tetradecanoylphorbol-13-acetate, response element. Mol Cell Biol 10: 1609–1621
Ivey-Hoyle M, Conroy R, Huber HE, Goodhart PJ, Oliff A, Heimbrook DC (1993) Cloning and characterization of E2F-2, a novel protein with the biochemical properties of transcription factor E2F. Mol Cell Biol 13: 7802–7812
Jochemsen AG, Bernards R, van Kranen HJ, Houweling A, Bos JL, van der Eb AJ (1986) Different activities of the adenovirus types 5 and 12 E1A regions in transformation with the EJ Ha-ras oncogene. J Virol 59: 684–691
Jones NC (1992) The multifunctional products of the adenovirus E1A gene. In: Doerfler W, Bohm P (eds) Malignant transformation by DNA viruses. VCH, Weinheim, pp 87–113
Kaelin WG Jr, Krek W, Sellers WR, DeCaprio JA, Ajchenbaum F, Fuchs CS, Chittenden T, Li Y, Farnham PJ, Blanar MA et al (1992) Expression cloning of a cDNA encoding a retinoblastoma-binding protein with E2F-like properties. Cell 70: 351–364
Kalvakolanu DV, Bandyopadhyay SK, Harter ML, Sen GC (1991) Inhibition of interferon-inducible gene expression by adenovirus E1A proteins: block in transcriptional complex formation. Proc Natl Acad Sci USA 88: 7459–7463
Kalvakolanu DVR, Liu JS, Hanson RW, Harter ML, Sen GC (1992) Adenovirus-E1A represses the cyclic AMP-induced transcription of the gene for phosphoenolpyruvate carboxykinase (GTP) in hepatoma cells. J Biol Chem 267: 2530–2536
Kamada S, Toyoshima K, Akiyama T (1994) Serum-independent phosphorylation of c-Jun and alterations in AP-1 components by transformation with various oncogenes. J Biol Chem 269: 4565–4570
Kannabiran C, Morris GF, Labrie C, Mathews MB (1993) The adenovirus-E1A 12S product displays functional redundancy in activating the human proliferating cell nuclear antigen promoter. J Virol 67: 507–515
Kao HT, Nevins JR (1983) Transcriptional activation and subsequent control of the human heat shock gene during adenovirus infection. Mol Cell Biol 3: 2058–2065
Kitabayashi I, Chiu R, Gachelin G, Yokoyama K (1991) E1A dependent up-regulation of c-jun/AP-1 activity. Nucleic Acids Res 19: 649–655
Kitabayashi I, Kawakami Z, Chiu R, Ozawa K, Matsuoka T, Toyoshima S, Umesono K, Evans RM, Gachelin G, Yokoyama K (1992) Transcriptional regulation of the c-jun gene by retinoic acid and E1A during differentiation of F9-cells. EMBO J 11: 167–175
Kraus VB, Moran E, Nevins JR (1992) Promoter-specific trans-activation by the adenovirus-E1a12S product involves separate E1a domains. Mol Cell Biol 12: 4391–4399

Labrie C, Morris GF, Mathews MB (1993) A complex promoter element mediates transactivation of the human proliferating cell nucelar antigen promoter by the 243-residue adenovirus E1A oncoprotein. Mol Cell Biol 13: 1697–1707

Lee WS, Kao CC, Bryant GO, Liu X, Berk AJ (1991) Adenovirus E1A activation domain binds the basic repeat in the TATA box transcription factor. Cell 67: 365–376

Lees JA, Saito M, Vidal M, Valentine M, Look T, Harlow E, Dyson N, Helin K (1993) The retinoblastoma protein binds to a family of E2F transcription factors. Mol Cell Biol 13: 7813–7825

Lewis JB, Anderson CW (1987) Identification of adenovirus type 2 early region 1B proteins that share the same amino terminus as do the 495R and 155R proteins. J Virol 61: 3879–3888

Lillie JW, Green MR (1989) Transcription activation by the adenovirus E1a protein. Nature 338: 39–44

Linder S, Popowicz P, Svensson C, Marshall H, Bondesson M, Akusjarvi G (1992) Enhanced invasive properties of rat embryo fibroblasts transformed by adenovirus E1A mutants with deletions in the carboxy-terminal exon. Oncogene 7: 439–443

Lipp M, Schilling R, Bernhardt G (1989) Trans-activation of human MYC: the second promoter is target for the stimulation by adenovirus E1a proteins. Oncogene 4: 535–541

Liu F, Green MR (1990) A specific member of the ATF transcription factor family can mediate transcription activation by the adenovirus E1a protein. Cell 61: 1217–1224

Liu F, Green MR (1994) Promoter targeting by adenovirus E1a through interaction with different cellular DNA-binding domains. Nature 368: 520–525

Liu HT, Baserga R, Mercer WE (1985) Adenovirus type 2 activates cell cycle-dependent genes that are a subset of those activated by serum. Mol Cell Biol 5: 2936–2942

Lowe SW, Ruley HE (1993) Stabilization of the p53 tumor suppressor is induced by adenovirus-5 E1A and accompanies apoptosis. Gene Dev 7: 535–545

Lum LSY, Hsu S, Vaewhongs M, Wu B (1992) The hsp70 gene CCAAT-binding factor mediates transcriptional activation by the adenovirus E1a protein. Mol Cell Biol 12: 2599–2605

Maekawa T, Matsuda S, Fujisawa J, Yoshida M, Ishii S (1991) Cyclic AMP response element-binding protein, CRE-BP1, mediates the E1A-induced but not the Tax-induced trans-activation. Oncogene 6: 627–632

Maruyama K, Schiavi SC, Huse W, Johnson GL, Ruley HE (1987) myc and E1A oncogenes alter the responses of PC12 cells to nerve growth factor and block differentiation. Oncogene 1: 361–367

Michalides R, Wagenaar E, Weijers P (1985) Rearrangements in the long terminal repeat of extra mouse mammary tumor proviruses in T-cell leukemias of mouse strain GR result in a novel enhancer-like structure. Mol Cell Biol 5: 823–830

Missero C, Filvaroff E, Dotto GP (1991) Induction of transforming growth factor beta 1 resistance by the E1A oncogene requires binding to a specific set of cellular proteins. Proc Natl Acad Sci USA 88: 3489–3493

Montano X, Lane DP (1987) The adenovirus Ela gene induces differentiation of F9 teratocarcinoma cells. Mol Cell Biol 7: 1782–1790

Monteclaro FS, Vogt PK (1993) A Jun-binding protein related to a putative tumor suppressor. Proc Natl Acad Sci USA 90: 6726–6730

Morris GF, Mathews MB (1990) Analysis of the proliferating cell nuclear antigen promoter and its response to adenovirus early region 1. J Biol Chem 265: 16116–16125

Morris GF, Mathews MB (1991) The adenovirus E1A transforming protein activates the proliferating cell nuclear antigen promoter via an activating transcription factor site. J Virol 65: 6397–6406

Morris GF, Labrie C, Mathews MB (1994) Modulation of transcriptional activation of the proliferating cell nuclear antigen promoter by the adenovirus-E1A 243-residue oncoprotein depends on proximal activators. Mol Cell Biol 14: 543–553

Mymryk JS, Lee RWH, Bayley ST (1992) Ability of adenovirus 5 Ela proteins to suppress differentiation of BC3h1 myoblasts correlates with their binding to a 300-kda cellular protein. Mol Biol Cell 3: 1107–1115

Nakajima T, Nakamura T, Tsunoda S, Nakada S, Oda K (1992) E1A-responsive elements for repression of rat fibronectin gene transcription. Mol Cell Biol 12: 2837–2846

Nakamura T, Nakajima T, Tsunoda S, Nakada S, Oda K, Tsurui H, Wada A (1992) Induction of E1a-responsive negative factors for transcription of the fibronectin gene in adenovirus E1-transformed rat cells. J Virol 66: 6436–6450

Nevins JR (1982) Induction of the synthesis of a 70, 000 dalton mammalian heat shock protein by the adenovirus E1A gene product. Cell 29: 913–919

Nomura N, Zu YL, Maekawa T, Tabata S, Akiyama T, Ishii S (1993) Isolation and characterization of a novel member of the gene family encoding the cAMP response element-binding protein CRE-BP1. J Biol Chem 268: 4259–4266

Offringa R, Smits AM, Houweling A, Bos JL, van der Eb AJ (1988) Similar effects of adenovirus E1A and glucocorticoid hormones on the expression of the metalloprotease stromelysin. Nucleic Acids Res 16: 10973–10984

Offringa R, Gebel S, van Dam H, Timmers M, Smits A, Zwart R, Stein B, Bos JL, van der Eb A, Herrlich P (1990) A novel function of the transforming domain of E1a: repression of AP-1 activity. Cell 62: 527–538

Peeper DS, Zantema A (1993) Adenovirus-E1A proteins transform cells by sequestering regulatory proteins. Mol Biol Rep 17: 197–207

Pescini R, Kaszubska W, Whelan J, DeLamarter JF, Hooft van Huijsduijnen R (1994) ATF-a0, a novel variant of the ATF/CREB transcription factor family, forms a dominant transcription inhibitor in ATF-a heterodimers. J Biol Chem 269: 1159–1165

Pietenpol JA, Stein RW, Moran E, Yaciuk P, Schlegel R, Lyons RM, Pittelkow MR, Munger K, Howley PM, Moses HL (1990) TGF-beta 1 inhibition of c-myc transcription and growth in keratinocytes is abrogated by viral transforming proteins with pRB binding domains. Cell 61: 777–785

Pulverer BJ, Kyriakis JM, Avruch J, Nikolakaki E, Woodgett JR (1991) Phosphorylation of c-jun mediated by MAP kinases. Nature 353: 670–674

Rao L, Debbas M, Sabbatini P, Hockenbery D, Korsmeyer S, White E (1992) The adenovirus E1a proteins induce apoptosis, which is inhibited by the E1b 19-kda and bcl-2 proteins. Proc Natl Acad Sci USA 89: 7742–7746

Raychaudhuri P, Bagchi S, Devoto SH, Kraus VB, Moran E, Nevins JR (1991) Domains of the adenovirus E1A protein required for oncogenic activity are also required for dissociation of E2F transcription factor complexes. Genes Dev 5: 1200–1211

Reich N, Pine R, Levy D, Darnell JE Jr (1988) Transcription of interferon-stimulated genes is induced by adenovirus particles but is suppressed by E1A gene products. J Virol 62: 114–119

Reifel Miller AE, Berg DT, Grinnell BW (1991) Repression of enhancer activity by the adenovirus E1A tumor protein(s) is mediated through a novel HeLa cell nuclear factor. J Biol Chem 266: 13873–13882

Riggs KJ, Saleque S, Wong KK, Merrell KT, Lee JS, Shi Y, Calame K (1993) Yin-yang 1 activates the c-myc promoter. Mol Cell Biol 13: 7487–7495

Roberts BE, Miller JS, Kimelman D, Cepko CL, Lemischka IR, Mulligan RC (1985) Individual adenovirus type 5 early region 1A gene products elicit distinct alterations of cellular morphology and gene expression. J Virol 56: 404–413

Rochette Egly C, Fromental C, Chambon P (1990) General repression of enhanson activity by the adenovirus-2 E1A proteins. Genes Dev 4: 137–150

Rosahl T, Doerfler W (1992) Alterations in the levels of expression of specific cellular genes in adenovirus-infected and adenovirus- transformed cells. Virus Res 26: 71–90

Sarnow P, Ho YS, Williams J, Levine AJ (1982) Adenovirus E1b-58kd tumor antigen and SV40 large tumor antigen are physically associated with the same 54 kd cellular protein in transformed cells. Cell 28: 387–394

Scholer HR, Ciesiolka T, Gruss P (1991) A nexus between Oct-4 and E1A: implications for gene regulation in embryonic stem cells. Cell 66: 291–304

Shenk T, Flint J (1991) Transcriptional and transforming activities of the adenovirus E1A proteins. Adv Cancer Res 57: 47–85

Shi Y, Seto E, Chang LS, Shenk T (1991) Transcriptional repression by YY1, a human GLI-Kruppel-related protein, and relief of repression by adenovirus E1A protein. Cell 67: 377–388

Simon MC, Kitchener K, Kao HT, Hickey E, Weber L, Voellmy R, Heintz N, Nevins JR (1987) Selective induction of human heat shock gene transcription by the adenovirus E1A gene products, including the 12S E1A product. Mol Cell Biol 7: 2884–2890

Simon MC, Fisch TM, Benecke BJ, Nevins JR, Heintz N (1988) Definition of multiple, functionally distinct TATA elements, one of which is a target in the hsp70 promoter for E1A regulation. Cell 52: 723–729

Simon MC, Rooney RJ, Fisch TM, Heintz N, Nevins JR (1990) E1A-dependent trans-activation of the c-fos promoter requires the TATAA sequence. Proc Natl Acad Sci USA 87: 513–517

Smeal T, Binetruy B, Mercola DA, Birrer M, Karin M (1991) Oncogenic and transcriptional cooperation with Ha-Ras requires phosphorylation of c-Jun on serines 63 and 73. Nature 354: 494–496

Smeal T, Binetruy B, Mercola D, Grover-Bardwick A, Heidecker G, Rapp UR, Karin M (1992) Oncoprotein-mediated signalling cascade stimulates c-Jun activity by phosphorylation of serines 63 and 73. Mol Cell Biol 12: 3507–3513

Spitkovsky D, Steiner P, Lukas J, Lees E, Pagano M, Schulze A, Joswig S, Picard D, Tommasino M, Eilers M, Jansen-Durr P (1994) Modulation of cyclin gene expression by adenovirus E1A in a cell line with E1A-dependent conditional proliferation. J Virol 68: 2206–2214

Stein R, Ziff EB (1984) HeLa cell beta-tubulin gene transcription is stimulated by adenovirus 5 in parallel with viral early genes by an E1a-dependent mechanism. Mol Cell Biol 4: 2792–2801
Subramanian T, Tarodi B, Govindarajan R, Boyd JM, Yoshida K, Chinnadurai G (1993) Mutational analysis of the transforming and apoptosis suppression activities of the adenovirus E1B-175R protein. Gene 124: 173–181
Taylor DA, Kraus VB, Schwarz JJ, Olson EN, Kraus WE (1993) E1A-mediated inhibition of myogenesis correlates with a direct physical interaction of E1A12S and basic Helix-Loop-Helix protein. Mol Cell Biol 13: 4714–4727
Thalmeier K, Synovzik H, Mertz R, Winnacker EL, Lipp M (1989) Nuclear factor E2F mediates basic transcription and trans-activation by E1a of the human MYC promoter. Genes Dev 3: 527–536
Timmers HT, De Wit D, Bos JL, van der Eb AJ (1988) E1A products of adenoviruses reduce the expression of cellular proliferation-associated genes. Oncogene Res 3: 67–76
Timmers HT, van Dam H, Pronk GJ, Bos JL, van der Eb AJ (1989) Adenovirus E1A represses transcription of the cellular JE gene. J Virol 63: 1470–1473
Usheva A, Horikoshi N, Lewis B, Weinmann R, Shenk T (1994) Transcriptional activation by the adenovirus E1A protein: anti-repression. J Cell Biochem [Suppl] 18C: 206
van Dam H, Offringa R, Smits AM, Bos JL, Jones NC, van der Eb AJ (1989) The repression of the growth factor-inducible genes JE, c-myc and stromelysin by adenovirus E1A is mediated by conserved region1. Oncogene 4: 1207–1212
van Dam H, Offringa R, Meijer I, Stein B, Smits AM, Herrlich P, Bos JL, van der Eb AJ (1990) Differential effects of the adenovirus E1A oncogene on members of the AP-1 transcription factor family. Mol Cell Biol 10: 5857–5864
van Dam H, Duyndam M, Rottier R, Bosch A, de Vries-Smits L, Herrlich P, Zantema A, Angel P, van der Eb AJ (1993) Heterodimer formation of cJun and ATF-2 is responsible for induction of c-jun by the 243-amino acid adenovirus E1A protein. EMBO J 12: 479–487
Velcich A, Ziff EB (1989) The adenovirus-5 12S E1a protein, but not the 13S induces expression of the endoA differentiation marker in F9 cells. Oncogene 4: 707–713
Wagner S, Green MR (1993) HTLV-I Tax protein stimulation of DNA binding of bZIP proteins by enhancing dimerization. Science 262: 395–399
Wang CY, Petryniak B, Thompson CB, Kaelin WG, Leiden JM (1993) Regulation of the ets-related transcription factor Elf-1 by binding to the retinoblastoma protein. Science 260: 1330–1335
Webster KA, Muscat GE, Kedes L (1988) Adenovirus E1A products suppress myogenic differentiation and inhibit transcription from muscle-specific promoters. Nature 332: 553–557
Weigel RJ, Nevins JR (1990) Adenovirus infection of differentiated F9 cells results in a global shut-off of differentiation-induced gene expression. Nucleic Acids Res 18: 6107–6112
Weigel RJ, Devoto SH, Nevins JR (1990) Adenovirus 12S E1A gene represses differentiation of F9 teratocarcinoma cells. Proc Natl Acad Sci USA 87: 9878–9882
White E, Cipriani R, Sabbatini P, Denton A (1991) Adenovirus E1B 19-kilodalton protein overcomes the cytotoxicity of E1A proteins. J Virol 65: 2968–2978
White E, Sabbatini P, Debbas M, Wold WSM, Kusher DI, Gooding LR (1992) The 19-kilodalton adenovirus E1B transforming protein inhibits programmed cell death and prevents cytolysis by tumor necrosis factor-alpha. Mol Cell Biol 12: 2570–2580
Whittaker JL, Byrd PJ, Grand RJ, Gallimore PH (1984) Isolation and characterization of four adenovirus type 12-transformed human embryo kidney cell lines. Mol Cell Biol 4: 110–116
Williams GT, McClanahan TK, Morimoto RI (1989) E1a transactivation of the human HSP70 promoter is mediated through the basal transcriptional complex. Mol Cell Biol 9: 2574–2587
Wu BJ, Hurst HC, Jones NC, Morimoto RI (1986) The E1A 13S product of adenovirus 5 activates transcription of the cellular human HSP70 gene. Mol Cell Biol 6: 2994–2999
Yamaguchi M, Hayashi Y, Hirose F, Shiroki K, Matsukage A (1992) Activation of the mouse DNA polymerase-beta gene promoter by adenovirus type-12 E1A proteins. Nucleic Acids Res 20: 2321–2325
Yew PR, Berk AJ (1992) Inhibition of p53 transactivation required for transformation by adenovirus early 1B-protein. Nature 357: 82–85
Yew PR, Liu X, Berk AJ (1994) Adenovirus E1B oncoprotein tethers a transcriptional repression domain to p53. Gene Dev 8: 190–202
Yu D, Suen TC, Yan DH, Chang LS, Hung MC (1990) Transcriptional repression of the neu protooncogene by the adenovirus 5 E1A gene products. Proc Natl Acad Sci USA 87: 4499–4503
Yu DH, Scorsone K, Hung MC (1991) Adenovirus type 5 E1A gene products act as transformation suppressors of the neu oncogene. Mol Cell Biol 11: 1745–1750

Yu DH, Hamada JI, Zhang H, Nicolson GL, Hung MC (1992) Mechanisms of c-erbB2/neu oncogene-induced metastasis and repression of metastatic properties by adenovirus 5 E1A gene products. Oncogene 7: 2263–2270

Yu DH, Shi DR, Scanlon M, Hung MC (1993a) Reexpression of neu-encoded oncoprotein counteracts the tumor-suppressing but not the metastasis-suppressing function of E1A. Cancer Res 53: 5784–5790

Yu DH, Wolf JK, Scanlon M, Price JE, Hung MC (1993b) Enhanced c-erbB-2/neu expression in human ovarian cancer cells correlates with more severe malignancy that can be suppressed by E1A. Cancer Res 53: 891–898

Zantema A, Schrier PI, Davis Olivier A, van Laar T, Vaessen RT, van der Eb (1985) Adenovirus serotype determines association and localization of the large E1B tumor antigen with cellular tumor antigen p53 in transformed cells. Mol Cell Biol 5: 3084–3091

Zu YL, Maekawa T, Matsuda S, Ishii S (1991) Complete putative metal finger and leucine zipper structures of CRE-BP1 are required for the E1A-induced trans-activation. J Biol Chem 266: 24134–24139

Adenovirus E1A: Transcription Regulation and Alteration of Cell Growth Control

J.R. Nevins

1 Introduction

The small DNA tumor viruses, which include the adenovirus, polyomavirus, and papillomavirus groups, have provided enormously valuable model systems with which to study complex events such as RNA processing, transcriptional regulation, and DNA replication, to name but a few examples. In addition, given the fact that these viruses also possess oncogenic activity under certain circumstances and indeed that the papillomaviruses are a significant cause of human cancer, these viruses have also provided insights into the mechanisms of cell growth control. A major advance in the understanding of viral oncogenesis came with the realization that a protein previously identified as an E1A-binding protein was actually the product of the retinoblastoma tumor suppressor gene (Whyte et al. 1988). The rapid subsequent finding that each of the DNA tumor viruses encoded proteins that bound to the Rb protein demonstrated the general nature of this event. The importance of these findings was immediately obvious: the physical interaction of oncogenic viral proteins with a key tumor suppressor protein provided a mechanistic view for the oncogenic capacity of these viruses. Nevertheless, the precise molecular consequence of these events were left unclear.

A series of revelations over the past several years have focused attention on the transcription factor E2F as a target for the growth-suppressing action of the

Department of Genetics, Howard Hughes Medical Institute, Duke University Medical Center, Box 3054, Durham, NC 27710, USA

retinoblastoma gene product Rb. These studies have provided a conceptual picture for how the targeting of Rb might contribute to oncogenesis as well as contribute to the replication of the virus. Moreover, the ability to understand the strategies of the DNA tumor virus oncoproteins with respect to E2F has provided important insights into the role of E2F in cellular growth control events, even in the midst of this growing complexity.

2 E2F Transcription Factor: A Target for Viral Oncoproteins

The E2F transcription factor was identified as a cellular protein involved in the E1A-mediated activation of adenovirus E2 transcription (Kovesdi et al. 1986). Subsequent experiments revealed that in most cell types, E2F was found complexed to cellular proteins that prevented the utilization of E2F for E2 transcription (Bagchi et al. 1990). It was found that the adenovirus E1A protein possessed the ability to disrupt these E2F complexes, releasing E2F in a form that could activate transcription (Bagchi et al. 1990). This activity was found to be dependent on the CR1 and CR2 domains of E1A (Raychaudhuri et al. 1991), and subsequent experiments have provided a mechanistic view of how E1A might accomplish the disruption (Ikeda and Nevins 1993; Fattaey et al. 1993). As depicted in Fig. 1, it appears that the CR2 domain of E1A, by associating with a region of the Rb of p107 protein distinct from the site of E2F interaction, can bring the viral protein to the E2F complex, then allowing the CR1 domain to compete for Rb or p107 binding to E2F. As a consequence of this interaction, E1A is left firmly bound to the protein that was originally bound to E2F and the equilibrium state would be shifted towards the dissociated form of the factor.

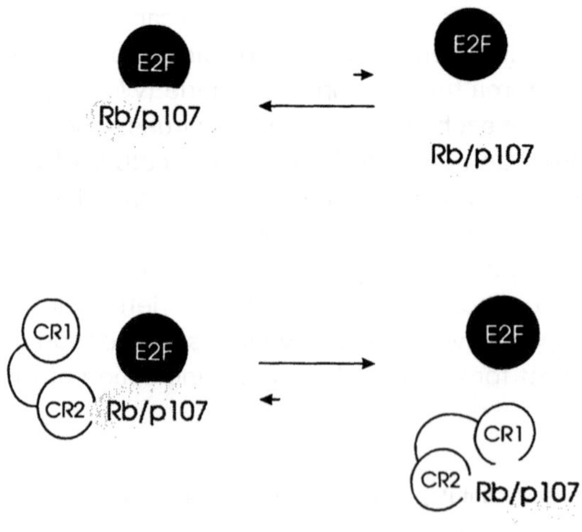

Fig. 1. Disruption of E2F complexes by the adenovirus E1A protein. *Top*, equilibrium between the E2F complex and the dissociated components. *Below*, the interaction of the E1A CR1 and CR2 domains with the complex. The CR2 domains appears to interact with a site on Rb or p107 distinct from the E2F binding site, whereas the CR1 domain competes with E2F for binding to Rb or p107 (Ikeda and Nevins 1993, Fattaey et al. 1993)

Most attention has focused on the ability of Rb and related proteins to inhibit the transcriptional activating capacity of E2F as well as the ability of the DNA tumor virus oncoproteins to relieve this inhibition (see NEVINS 1992 for review). Although these events are generally viewed in the context of oncogenesis, it is nevertheless clear that the normal function of these viral proteins is not to transform cells, but rather to facilitate viral growth. Clearly, these otherwise unrelated viruses have evolved a common need to activate E2F, by inactivating Rb, and in so doing facilitate lytic growth. Nevertheless, although the adenovirus E2 gene does benefit directly from the activation of E2F, there are no E2F sites in either the SV40 genome or the papillomovirus genome. Thus, what is the common need of these otherwise distinct viruses that requires the activation of E2F?

Adenovirus and SV40 infections are known to induce quiescent cells to enter S phase, dependent on E1A and T antigen, respectively, resulting in the activation of cellular DNA synthesis and induction of activities such as dihydrofolate reductase (DHFR), thymidine kinase, and DNA polymerase (see TOOZE 1981). Because in most cases these viruses infect a quiescent cell, it is generally thought that induction of S phase activities creates an environment for efficient viral DNA synthesis. Each virus encodes proteins involved in viral DNA replication but, even with these activities, substrates for DNA synthesis must be produced by the host cell. Thus, the DNA tumor viruses share a common goal to activate S phase events to allow efficient viral DNA replication and it would appear that E2F plays a role in this process.

Recent experiments reveal that E2F elements located at the DHFR transcription initiation site are essential for late G_1 regulated expression of DHFR; indeed, a single E2F element is sufficient for regulated transcription at G_1/S, thus providing direct evidence that E2F controls DHFR transcription during the growth response (SLANSKY et al. 1993). Given that E2F regulates DHFR transcription, and may regulate many of the S phase genes, together with the fact that E2F activation is mediated by E1A, T antigen, and E7, it seems quite likely that these viruses induce S phase, at least in part, by freeing E2F from inhibitory complexes with proteins such as Rb. Moreover, the likelihood that E1A, T antigen, and E7 represent the independent evolution of viral activities targeting E2F strongly suggests an important role for E2F in controlling the entry of cells into S phase.

In principle, any virus that must replicate DNA at a high rate would be dependent on the activities that create the S phase environment. Nevertheless, there is no evidence that infection by members of the herpesvirus family or the poxvirus family results in a stimulation of S phase in the manner achieved by the small DNA tumor viruses, nor is there evidence that either of these virus groups encode proteins that inactivate the Rb protein. However, a parallel can be found, since genes encoding many of the activities suspected to be induced by E2F are found in the herpesvirus and poxvirus genomes (ALBRECHT et al. 1992). Although the entire group of genes is not found in every one of these viruses, it is true that each does contain a ribonucleotide reductase activity, the rate-limiting enzyme in deoxynucleotide biosynthesis (THELANDER and REICHARD 1979). Thus, it is possible that the same end result, the creation of appropriate pools of deoxynucleotides and thus an environment for DNA synthesis, has been achieved by these different types of viruses but through two distinct mechanisms.

3 A Role for E2F in Cell Growth Control

The timing of E2F accumulation during the cell cycle, the role of E2F in activating the S phase gene DHFR, and the targeting of E2F by the viral oncoproteins all underscore an important role for this transcription factor in cell growth control. Recent experiments utilizing the E2F1 cDNA lend support to this view. In one assay, expression of the E2F1 cDNA product in rat embryo fibroblasts was shown to induce DNA synthesis in cells that would otherwise enter quiescence due to serum starvation, with an efficiency similar to that achieved by E1A expression (JOHNSON et al. 1993). In a second assay, the E2F1 product was shown to induce quiescent cells to enter S phase following microinjection of the E2F1 cDNA into serum-starved REF cells (JOHNSON et al. 1993).

These results, together with other revelations of the past 2 years, suggest that the E2F transcription factor resides at a critical juncture in a regulatory pathway that controls cell growth (Fig. 2) such that overexpression of E2F1 can disturb this control. The ability of Rb to arrest cells in G_1, apparently through the regulation of E2F activity (QIN et al. 1992; QIAN et al. 1992), is consistent with other experiments implicating E2F as an Rb target (HIEBERT et al. 1992); indeed, recent assays demonstrate that the Rb-mediated G_1 arrest can be overcome by expression of the E2F1 product (ZHU et al. 1993).

Rb activity is known to be controlled by phosphorylation, and recent experiments implicate the cyclin E/cdk2 kinase or a cyclin D/cdk4 kinase complex, both of which are activated as cells progress through G_1, as mediators of this Rb phosphorylation since they too can overcome Rb growth suppression (see SHERR 1993 for review). It is also possible that the negative growth action of regulatory proteins such as transforming growth factor (TGF)-β (HOWE et al. 1991), whose action results in the accumulation of unphosphorylated Rb (PIETENPOL et al. 1990; LAIHO et al. 1990), may be mediated through an inhibition of G_1 cyclin cdk kinase activity (KOFF et al. 1993). Since phosphorylation of Rb prevents the formation of the Rb–E2F complex (CHELLAPPAN et al. 1991; SCHWARZ et al. 1993; HELIN et al. 1992; KATO et al. 1993), regulatory events that govern the G_1 to S phase transition, either positively (G_1 cyclins/cdk kinases) or negatively (Rb, TGF-β), may ultimately converge on E2F1.

These new revelations provide strong evidence for a role for E2F in cell growth control and in so doing have provided a better understanding of regulatory interactions that control cell growth. Nevertheless, two additional lines of investigation indicate that the situation is undoubtedly more complex.

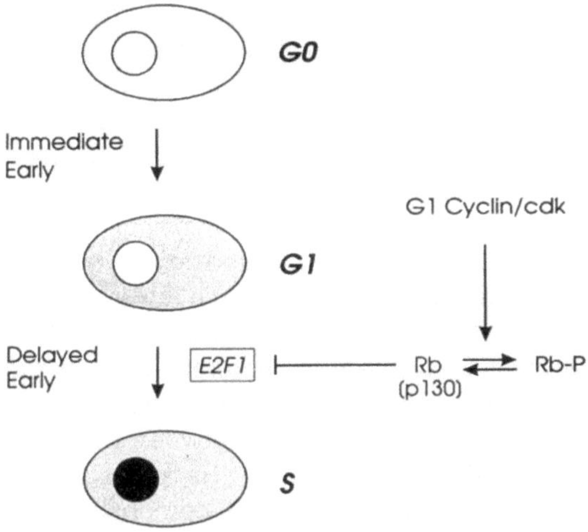

Fig. 2. Gene activation and regulatory events during the progression of cells from G_0 to S phase. *Immediate-early* and *delayed-early* refer to the kinetics of activation of specific sets of genes as well as the dependency on previous protein synthesis. The E2F1 gene is activated with delayed-early kinetics and dependent on protein synthesis (SLANSKY et al. 1993). Only the under-phosphorylated form of the Rb protein is capable of binding to E2F (CHELLAPPAN et al. 1991)

4 Complexity of E2F Control

Whereas much of the initial efforts have focused on the role of Rb in controlling cell growth via E2F, recent studies suggest a role for another member of the Rb family of proteins in controlling E2F in G_0 and G_1. Proteins of 107 kDa and 130 kDa were originally identified along with Rb as E1A-associated proteins. The gene encoding p130 has now been isolated by three independent procedures, and the sequence of p130 identifies it as a relative of Rb and the Rb-related p107 protein (HANNON et al. 1993; LI et al. 1993; MAYOL et al. 1993). Moreover, recent biochemical assays demonstrate an interaction of p130 with E2F, primarily during G_0 and G_1 (COBRINIK et al. 1993). Although previous experiments have provided evidence for the participation of the p107 protein in G_1-specific E2F complexes (SCHWARZ et al. 1993), as well as the formation of an E2F–p107 complex containing cyclin E and the cdk2 kinase (SHIRODAKAR et al. 1992; DEVOTO et al. 1992; LEES et al. 1992), it appears that the p130–E2F interaction could be a major element of E2F control in G_0 and G_1 cells, although the effect of p130 on E2F function is yet to be determined. Further experiments will surely focus on the relationship of Rb and p130 control of E2F to cell cycle regulation and differentiation as well as the possibility that p130 represents yet another tumor suppressor activity.

5 Complexity of E2F

Although originally defined simply as a sequence-specific DNA-binding activity, it has now become clear that E2F activity is made up of a family of DNA-binding polypeptides, of which the E2F1 product is but one. A mouse gene termed DP-1 was isolated based on amino acid sequence information obtained from a purified preparation of E2F/DRTF1 (GIRLING et al. 1993). Sequence analysis demonstrates that the DP-1 gene shares only limited homology with the E2F1 product, primarily within sequences specifying DNA-binding capacity. At least two additional gene products that contribute to the E2F DNA binding activity, termed E2F2 and E2F3, have been identified through low stringency screening with the original E2F1 cDNA (LEES et al. 1993; IVEY-HOYLE et al. 1993). As expected based on their method of isolation, these products exhibit considerable homology to the E2F1 product.

Biochemical assays using components of affinity-purified E2F activity have shown that the functional E2F activity very likely exists as a heterodimer, with the E2F1 product being one participant of the dimer (HUBER et al. 1993). Indeed, recent experiments have now demonstrated a physical interaction between E2F1 and DP-1 that markedly enhances the E2F DNA-binding capacity (HELIN et al. 1993; BANDARA et al. 1993). In addition, these experiments have demonstrated a functional synergy in this interaction, since transactivation mediated by E2F1 is greatly enhanced by cotransfection of DP-1 (HELIN et al. 1993; BANDARA et al. 1993). It is also apparent that the ability of Rb to inhibit E2F1-dependent transcription is greatly facilitated by the expression of DP-1 (HELIN et al. 1993).

This complexity of polypeptide species that constitute the E2F activity may be reflected in the observation of distinct species of E2F binding activity that have kinetically different properties with respect to the cell growth cycle (CHITTENDEN et al. 1993). Thus, it is possible that the polypeptides that constitute the E2F activity found in G_1 may differ from those found in S phase with respect to the interaction with regulatory proteins such as Rb and p107. For example, although both Rb and p107 can suppress growth of Saos-2 cells, only E2F-1 appears to reverse the effect of Rb (ZHU et al. 1993). Moreover, although p107 can inhibit the E2F1-mediated activation of transcription in cotransfection assays (CRESS et al. 1993), other experiments suggest that the E2F target for p107 may not yet be identified, since neither E2F1, E2F2, nor E2F3 could be recovered in a p107 immunoprecipitate (LEES et al. 1993). In short, it is likely that the complexity of E2F will continue to grow and we may find that distinct species of E2F DNA-binding activity perform distinct functional roles in conjunction with other proteins.

6 DNA Tumor Viruses and the Cell Cycle

Much of our present understanding of cell cycle and cell growth control events derives from an elegant combination of genetics and biochemistry that the budding and fission yeast have provided. The ability to identify critical regulatory activities, identify the genes that specify these activities, and then understand the biochemical mechanisms of action of these gene products has been an enormously successful approach. Although the study of cell cycle and cell growth control in mammalian cells suffers from the lack of genetic analysis, a rather different combination of disciplines has nevertheless provided important insights into control of cell growth. The elucidation of strategies of DNA virus replication as they pertain to the host cell, coupled with the alterations in cell growth control that occur when these viruses fail to undergo a productive infection and result in cellular transformation, has now shed new light on the process of cellular proliferation control. Although the story will likely become even more complex as time passes, the ability to explain these events in a context that makes sense for both viral replication as well as oncogenic transformation once again underscores the value of these viruses in helping to provide an understanding of complex cellular events.

References

Albrecht J-C, Nicholas J, Biller D, Cameron KR, Biesinger B, Newman C, Wittmann S, Craxton MA, Coleman H, Fleckenstein B, Honess RW (1992) Primary structure of the herpesvirus saimiri genome. J Virol 66: 5047–5058

Bagchi S, Raychaudhuri P, Nevins JR (1990) Adenovirus E1A proteins can dissociate cellular complexes containing the E2F transcription factor: a novel mechanism for E1A trans-activation.Cell 62: 659–669

Bandara LR, Buck VM, Zamanian M, Johnston LH, La Thangue NB (1993) Functional synergy between DP-1 and E2F-1 in the cell cycle-regulating transcription factor DRTF1/E2F. EMBO J 12: 4317–4324

Chellappan SP, Hiebert SW, Mudryj M, Horowitz JM, Nevins JR (1991) The E2F transcription factor is a cellular target for the RB protein. Cell 65: 1053–1061

Chittenden T, Livingston DM, DeCaprio JA (1993) Cell cycle analysis of E2F in primary human T cells reveals novel E2F complexes and biochemically distinct forms of free E2F. Mol Cell Biol 13: 3975–3983

Cobrinik D, Whyte P, Peeper DS, Jacks T, Weinberg RA (1993) Cell cycle specific association of E2F with the p130 E1A-binding protein. Genes Dev 7: 2392–2404

Cress WD, Johnson DG, Nevins JR (1993) A genetic analysis of the E2F1 gene distinguishes regulation by Rb, p107, and adenovirus E4. Mol Cell Biol 13: 6314–6325

Devoto SH, Mudryj M, Pines J, Hunter T, Nevins JR (1992) A cyclin A-cdc2 kinase complex possesses sequence-specific DNA binding activity: p33 cdk2 is a component of the E2F-cyclin A complex. Cell 68: 167–176

Fattaey A, Harlow E, Helin K (1993) Independent regions of adenovirus E1A are required for binding to and dissociation of E2F-protein complexes. Mol Cell Biol 13: 7267–7277

Girling R, Partridge JF, Bandara LR, Burden N, Totty NF, Hsuan JJ, LaThangue NB (1993) A new component of the transcription factor DRTF1/E2F. Nature 362: 83–87

Hannon GJ, Demetrick D, Beach D (1993) Isolation of the RB-related p130 through its interaction with CDK2 and cyclins. Genes Dev 7: 2378–2391

Helin K, Lees JA, Vidal M, Dyson N, Harlow E, Fattaey A (1992) A cDNA encoding a pRB-binding protein with properties of the transcription factor E2F. Cell 70: 337–350

Hiebert SW, Chellappan SP, Horwitz JM, Nevins JR (1992) The interaction of RB with E2F coincides with an inhibition of the transcriptional activity of E2F. Genes Dev 6: 177–185

Howe PH, Draetta G, Leof EB (1991) Transforming growth factor B1 inhibition of p34-cdc2 phosphorylation and histone H1 kinase activity is associated with G1/S phase-growth arrest. Mol Cell Biol 11: 1185–1194

Huber HE, Edwards G, Goodhart PJ, Patrick DR, Huang PS, Ivey-Hoyle M, Barnett SF, Oliff A, Heimbrook DC (1993) Transcription factor E2F binds DNA as a heterodimer. Proc Natl Acad Sci USA 90: 3525–3529

Ikeda M-A, Nevins JR (1993) Identification of distinct roles for separate E1A domains in the disruption of E2F complexes. Mol Cell Biol 13: 7029–7035
Ivey-Hoyle M, Conroy R, Huber H, Goodhart P, Oliff A, Heimbrook DC (1993) Cloning and characterization of E2F-2, a novel protein with the biochemical properties of transcription factor E2F. Mol Cell Biol 13: 7802–7812
Johnson DG, Schwarz JK, Cress WD, Nevins JR (1993) Expression of E2F1 induces quiescent cells to enter S phase. Nature 365: 349–352
Kato J, Matsushime H, Hiebert SW, Ewen ME, Sherr CJ (1993) Direct binding of cyclin D to the retinoblastoma gene product (pRb) and pRb phosphorylation by the cyclin D-dependent kinase CDK4. Genes Dev 7: 331–342
Koff A, Ohtsuki M, Polyak K, Roberts JM, Massague J (1993) Negative regulation of G1 in mammalian cells: inhibition of cyclin E-dependent kinase by TGF-beta. Science 260: 536–539
Kovesdi I, Reichel R, Nevins JR (1986) Identification of a cellular transcription factor involved in E1A trans-activation. Cell 45: 219–228
Laiho M, DeCaprio JA, Ludlow JW, Livingston DM, Massague J (1990) Growth inhibition by TGF-β linked to suppression of retinoblastoma protein phosphorylation. Cell 62: 175–185
Lees E, Faha B, Dulic V, Reed SI, Harlow E (1992) Cyclin E/cdk2 and cyclin A/cdk2 kinases associate with p107 and E2F in a temporally distinct manner. Genes Dev 6: 1874–1885
Lees JA, Saito M, Vidal M, Valentine M, Look T, Harlow E, Dyson N, Helin K (1993) The retinoblastoma protein binds to a family of E2F transcription factors. Mol Cell Biol 13: 7813–7825
Li Y, Graham C, Lacy S, Duncan AMV, Whyte P (1993) The adenovirus E1A-associated 130 kd protein is encoded by a member of the retinoblastoma gene family and physically interacts with cyclins A and E. Genes Dev 7: 2366–2377
Mayol X, Grana X, Baldi A, Sang N, Hu Q, Giordano A (1993) Cloning of a new member of the retinoblastoma gene family (pRb2) which binds to the E1A transforming domain. Oncogene 8: 2561–2566
Nevins JR (1992) E2F: a link between the Rb tumor suppressor protein and viral oncoproteins. Science 258: 424–429
Peitenpol JA, Stein RW, Moran E, Yaciuk P, Schlegel R, Lyons RM, Pittelkow MR, Munger K, Howley PM, Moses HL (1990) TGF-β1 inhibition of c-myc transcription and growth in keratinocytes is abrogated by viral transforming proteins with pRB binding domains. Cell 61: 777–785
Qian Y, Luckey C, Horton L, Esser M, Templeton DJ (1992) Biological function of the retinoblastoma protein requires distinct domains for hyperphosphorylation and transcription factor binding. Mol Cell Biol 12: 5363–5372
Qin XQ, Chittenden T, Livingston DM, Kaelin WG (1992) Identification of a growth suppression domain within the retinoblastoma gene product. Genes Dev 6: 953–964
Raychaudhuri P, Bagchi S, Devoto SH, Kraus VB, Moran E, Nevins JR (1991)_ Domains of the adenovirus E1A protein that are required for oncogenic activity are also required for dissociation of cellular transcription factor complexes. Genes Dev 5: 1200–1211
Schwarz J, Devoto SH, Smith EJ, Chellappan S, Jakoi L, Nevins JR (1993) Interactions of the p107 and Rb proteins with E2F during the cell proliferative response. EMBO J 12: 1013–1020
Sherr CJ (1993) Mammalian G1 cyclins. Cell 73: 1059–1065
Shirodkar S, Ewen M, DeCaprio JA, Morgan J, Livingston DM, Chittenden T (1992) The transcription factor E2F interacts with the retinoblastoma product and a p107-cyclin A complex in a cell cycle-regulated manner. Cell 68: 157–166
Slansky JE, Li Y, Kaelin WG, Farnham PJ (1993) A protein synthesis-dependent increase in E2F1 mRNA correlates with growth regulation of the dihydrofolate reductase promoter. Mol Cell Biol 13: 1610–1618
Thelander L, Reichard P (1979) Reduction of ribonucleotides. Annu Rev Biochem 48: 133–158
Tooze J (1981) The DNA tumor viruses. Cold Spring Harbor Laboratory Press, Cold Spring Harbor
Whyte P, Buchkovich KJ, Horowitz JM, Friend SH, Raybuck M, Weinberg RA, Harlow E (1988) Association between an oncogene and an anti-oncogene: the adenovirus E1A proteins bind to the retinoblastoma gene product. Nature 334: 124–129
Zhu L, van den Heuvel, Helin K, Fattaey A, Ewen M, Livingston D, Dyson N, Harlow E (1993) Inhibition of cell proliferation by p107, a relative of the retinoblastoma protein. Genes Dev 7: 1111–1125

Regulation of p53-Dependent Apoptosis by E1A and E1B

E. WHITE

Center for Advanced Biotechnology and Medicine, Department of Biological Sciences, Rutgers University, 679 Hoes Lane, Piscataway, NJ 08854, USA

1 Introduction

Regulation of apoptosis (programmed cell death) is a fundamentally important process in development, is required for maintaining homeostasis (reviewed in RAFF 1992), and is an important cellular defense against cancer and viral infection (reviewed in WHITE and GOODING 1994; WHITE 1993; WHITE et al. 1994). The product of the p53 tumor suppressor gene is the most frequently mutated gene in human tumors (HOLLSTEIN et al. 1991; MALKIN et al. 1990), and loss of p53 function in animal models dramatically increases the incidence of cancer (DONEHOWER et al. 1992). Expression of the p53 protein will induce either growth arrest (DILLER et al. 1990; GINSBERG et al. 1991; KUERBITZ et al. 1992; MARTINEZ et al. 1991) or apoptosis (YONISH-ROUACH et al. 1991), depending on the physiological circumstances. It is intriguing that p53 may function as a tumor suppressor by inducing cell death, as this would be the most effective means for irreversibly insuring the elimination of abnormal, emerging cancer cells. We have discovered that the transforming gene products of adenovirus encode activities that both activate (E1A) and repress (E1B) p53-dependent apoptosis in the normal course of productive infection in human cells and during the transformation of primary rodent cells (WHITE et al. 1991, 1992, 1993; RAO et al. 1992; DEBBAS and WHITE 1993; WHITE and GOODING 1994; WHITE 1993; CHIOU et al. 1994a). These discoveries have led to the following conclusions:

1. E1A expression stimulates both p53 accumulation and apoptosis.
2. E1A-induced apoptosis likely derives from perturbation of cell growth control by E1A.
3. The E1B gene encodes two proteins, 19K and 55K, that inhibit apoptosis.
4. Inhibition of apoptosis by E1B 19K is important for sustaining viability of the infected host cell.
5. Induction of cell proliferation by E1A must be coupled to inhibition of apoptosis by E1B for transformation.
6. Induction of apoptosis by E1A is mediated by the product of the p53 tumor suppressor gene.
7. E1B 19K and the human Bcl-2 protein prevent p53-dependent apoptosis but not growth arrest.
8. E1B 19K is likely the adenoviral equivalent of the human *bcl-2* proto-oncogene.

Induction of apoptosis by p53 in response to viral infection may explain why DNA tumor viruses such as adenovirus have evolved overlapping mechanisms for disabling p53 function, i.e., E1B 19K and 55K. While it is well established that E1B 55K complexes with p53, thereby inhibiting its function (SARNOW et al. 1982; YEW and BERK 1992), the mechanism by which E1B 19K affects p53 function is not yet known. Defining the mechanism by which E1A upregulates p53 and how this leads to apoptosis will provide insight into the cellular pathway to control apoptosis. Establishing how E1B 19K and Bcl-2 modulate p53 function away from apoptosis and toward growth arrest will be important for understanding a fundamental aspect of normal growth and development that, when lost, leads to the development of cancer.

2 E1A Expression Induces Apoptosis

2.1 Induction of Apoptosis by E1A During Productive Infection

The focus of the initial work was to identify the role of E1B 19K in adenovirus infection and transformation, which was approached through the chracterization of adenovirus mutants that fail to synthesize a functional E1B 19K protein. Infection of human cells with E1B 19K mutant adenoviruses induced the degradation of host cell and viral DNA (*deg* phenotype), produced enhanced cytocidal effects (*cyt* phenotype), and premature death of the host cell, which greatly impaired virus production (PILDER et al. 1984; TAKEMORI et al. 1984; SUBRAMANIAN et al. 1984; WHITE et al. 1984a, b). Induction of DNA fragmentation and pronounced morphological changes are most commonly associated with the induction of apoptosis (reviewed in WYLLIE 1980a) and were indistinguishable from the consequence of infection without functional E1B 19K (WHITE et al. 1991).

Although pronounced morphological changes can occur under circumstances having nothing to do with apoptosis, the induction of DNA fragmentation was more compelling. One of the characteristics distinguishing apoptotic from necrotic cell death is the digestion of cellular DNA into nucelosome-sized fragments, indicative of DNA cleavage between nucleosomes in chromatin (WYLLIE 1980). The DNA degradation induced upon infection with an E1B 19K mutant adenovirus, however, did not display the typical nucleosomal banding pattern (see PILDER et al. 1984; WHITE et al. 1984a). Southern blotting of the degraded DNA indicated that both cellular and viral DNA was degraded (WHITE et al. 1984a). As adenovirus DNA is not packaged into nucleosomes in the same way as cellular DNA, the degraded viral DNA may have prevented detection of the nucleosomal-sized DNA fragments in degraded cellular DNA. To test this possibility, a recombinant adenovirus was constructed that contained a point mutation in the gene encoding the 72K DNA-binding protein (DBP) from Ad5*ts*125 that rendered the virus temperature sensitive for virus replication in conjunction with a mutation in the E1B 19K gene from Ad2*cyt*106. Cellular DNA fragmentation in the nucleosome degradation pattern could thereby be examined at the restrictive temperature for viral DNA replication unobscured by the degraded viral DNA.

Under conditions in which viral DNA replication is inhibited at 39.5°C, the nucleosomal DNA fragmentation pattern of cellular chromosomal DNA typical of cell death by apoptosis became apparent (Fig.1). Condensation of chromatin into multiple aggeregates typical of apoptotic death was also observed in 19K mutant virus-infected cells (data not shown) and in transformed cells (RAO et al. 1992).

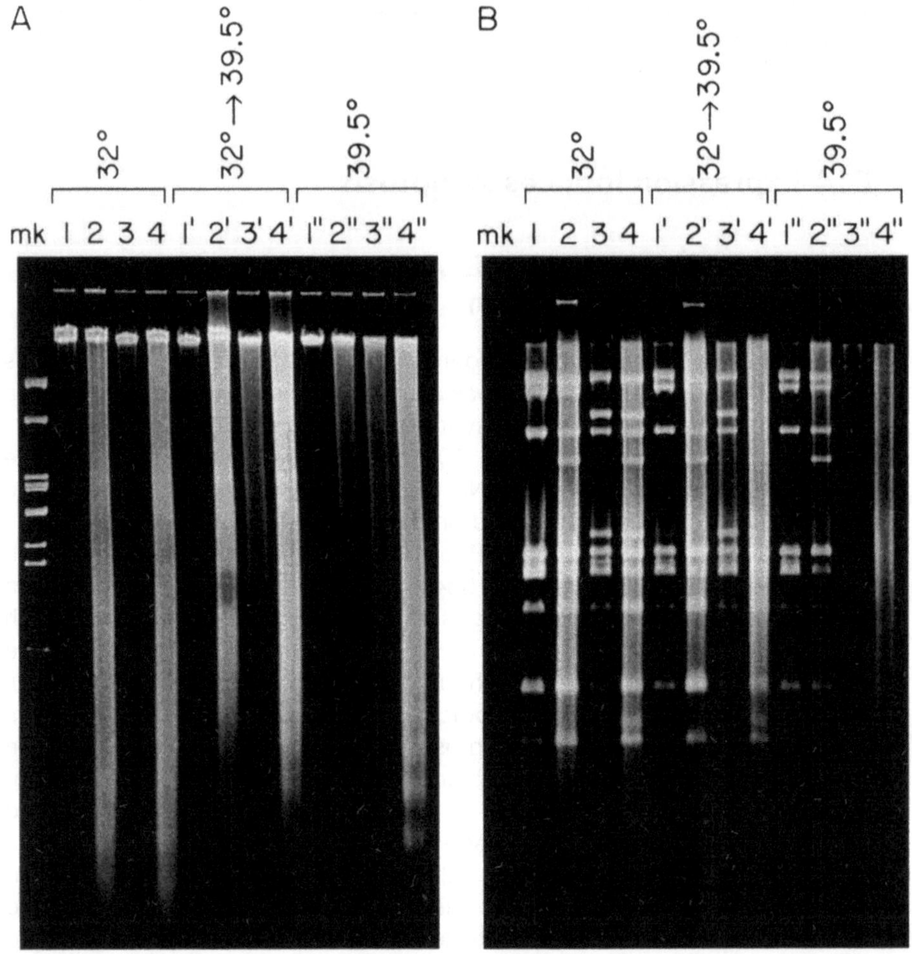

Fig. 1A, B. Induction of DNA fragmentation by infection with E1B 19K gene mutant viruses. HeLa cells were infected with the wild-type adenovirus type 2 (Ad2; *lane 1*), the E1B 19K mutant Ad2*cyt*106 (*lane 2*), the 72K DNA-binding protein (DBP) mutant Ad5*ts*125 (*lane 3*), and the Ad5*ts*125 plus Ad2*cyt*106 double mutant (*lane 4*). HeLa cells were infected with the indicated viruses at a multiplicity of 50 plaque-forming units (pfu) per cells, with the exception of the double mutant, which was at 25 pfu. Infected cells were incubated at the permissive (32°C) or restrictive (39.5°C) temperature for the mutant 72K DBP of Ad5*ts*125 as indicated. One set of samples was incubated for 48 h at 32°C and then shifted to 39.5°C. Low molecular weight Hirt DNA was isolated at 51 h after infection as previously described (WHITE et al. 1984b). Hirt DNA samples were analyzed by agarose gel electrophoresis followed by ethidium bromide staining either uncut (**A**) or digested with the restriction enzyme *SacI* to visualize the viral DNA (**B**). As expected, no detectable viral DNA replication took place with mutant viruses containing the DBP mutation at 39.5°C (**B**, *lanes 3"* and *4"*). DNA degradation is observed in cells infected with mutant viruses containing the E1B 19K mutation (*lanes 2* and *4"*). A small amount of DNA fragmentation was observed in Ad5*ts*125-infected cells at 39.5°C (*lanes 3'* and *3"*), which may be due to insufficient 19K expression when infection is blocked in the early phase by the Ad5*ts*125 mutation. Infection with the Ad2*cyt*106 plus Ad5*ts*125 double mutant virus at 39.5°C produced massive DNA fragmentation in the nucleosome "ladder" pattern characteristic of cell death by apoptosis (**A**, *lane 4"*), which was not otherwise detectable when viral DNA synthesis and degradation also take place (*lane 2*). *mk*, Ad2 DNA digested with *Hin*dIII as molecular weight markers

Two important conclusions were derived from these observations. First, E1B 19K may function as an inhibitor of apoptosis, which is an important activity required to maximize virus production (see below). Second, adenovirus infection and/or expression of viral gene products triggered apoptosis.

To identify the adenovirus gene products responsible for the induction of apoptosis (also known as the *cyt* and *deg* phenotypes), cells were infected with an E1B 19K gene mutant in the presence of hydroxyurea to block viral DNA replication and maintain cells in the early phase of infection. Induction of apoptosis still occurred in the presence of hydroxyurea, establishing that a viral early gene product was responsible for triggering apoptosis (White et al. 1984b). To map the induction of apoptosis, recombinant adenoviruses were constructed that contained mutations in each of the viral early genes in the background of an E1B 19K gene mutation. Based on these studies, the E3 and E2 72K DBP and E1B genes had no role in the induction of apoptosis (White and Stillman 1987). The role of E4 could not be addressed, because E4-E1B 19K double mutant viruses could not be obtained (White and Stillman 1987). Significantly, failure to express a functional E1A gene abolished any indication of apoptosis, which implicated E1A expression as the causative factor triggering apoptosis (White and Stillman 1987). No apoptosis was observed following infection at high multiplicities, i.e., 500 plaque-forming units (pfu)/cell, which was sufficient to permit efficient viral gene expression and replication without E1A expression (White and Stillman 1987).

Induction of apoptosis by E1A could be the direct consequence of the action of the E1A proteins on cell metabolism or could result indirectly from transcriptional activation of another viral gene product. To eliminate the latter possibility, HeLa cells were transfected with a plasmid expression vector where E1A expression was driven by the cytomegalovirus (CMV) promoter/enhancer to determine whether apoptosis could occur following exclusive expression of E1A. Transient expression of the E1A proteins was sufficient to drive cells into apoptosis, implicating a direct role for the E1A proteins (White et al. 1991).

2.2 Influence of Host Cell Type on Induction of Apoptosis

Another parameter established as influencing the induction of apoptosis by viral infection of human cells is the host cell. While HeLa and KB cells, for example, are susceptible to the induction of apoptosis by infection with E1B 19K viral mutants, WI38 cells are not (White and Stillman 1987). Many differences exist between these cell types that could account for these observations such as differing tissue of origin, growth state, or accumulated mutations that permit growth in culture. Nonetheless, host cell factors apparently exist that can influence the ability of E1A to activate an apoptotic pathway, and this is an important consideration when evaluating the requirement for E1B function in different cell lines, particularly those used in transformation assays.

2.3 Induction of Apoptosis by E1A Is Linked to Stimulation of Cell Proliferation

The E1A proteins promote cell proliferation through the interaction with cellular proteins that act as key negative regulators of cell growth (reviewed in DYSON and HARLOW 1992). There are two functional regions of E1A as defined by mutations, one at the amino terminus of E1A associated with binding to the p300 protein and a more central region that interacts with Rb and the related proteins p107 and p130 (reviewed in MORAN 1993). Mutations in either of these two regions that abolish binding to these cellular proteins will profoundly inhibit transformation, although each region alone is sufficient to stimulate cellular DNA synthesis. Thus, two independent activities within E1A, one associated with p300 and the other associated with the Rb family members, are involved in deregulating cell cycle progression. Although the biological function of p300 is not known, insight into how cell cycle control is usurped by E1A has been provided by the observation that Rb forms complexes with the transcription factor E2F and inhibits its activity (reviewed in NEVINS 1992). E2F activation occurs upon phosphorylation of Rb near the G_1/S boundary of the cell cycle, where E2F may be required for the transcription of cellular genes necessary for cellular DNA synthesis. Overexpression of E2F in quiescent cells will stimulate progression into DNA synthesis, substantiating the role for E2F in establishing S phase (JOHNSON et al. 1993).

The E1A function associated with the induction of apoptosis was mapped using E1A mutants defective for transformation and binding to specific cellular proteins. Mutations in the 13S unique region (transactivation domain) or elsewhere in E1A that do not affect the biological activity of E1A did not influence apoptosis (WHITE and STILLMAN 1987; WHITE et al. 1991). Mutations in E1A that impair Rb binding (which are transformation defective but still capable of stimulating cellular DNA synthesis because the p300 binding activity is intact) also did not prevent the induction of apoptosis. Interestingly, an E1A mutant that impaired binding of both Rb relatives and p300 (which is transformation defective and defective for induction of cellular DNA synthesis) was incapable of triggering apoptosis (WHITE et al. 1991). Stimulation of cellular DNA synthesis by E1A is thereby implicated in eliciting the apoptotic response (WHITE et al. 1991). It is possible that both the Rb and p300 binding regions are responsible for apoptosis, since loss of both functions is the only situation where apoptosis is suppressed. As the genetic analysis of E1A function has continued to be refined, more elegant E1A point mutants are now available to permit fine genetic dissection of the E1A functions and cellular protein interactions associated with stimulation of apoptosis.

2.4 Induction of Apoptosis by E1A in Transformation

The first indication that induction of apoptosis by E1A may play a role in the transformation process was that E1A expression in primary baby rat kidney (BRK) cells stimulated proliferation and transformed focus formation, but foci subsequently degenerated (WHITE et al. 1992). Transfection of the E1A gene alone into primary BRK cells stimulates cell proliferation to the point of microfocus formation by 2 weeks in culture (Fig. 2; WHITE et al. 1992). Between 2 and 3 weeks following transfection, focus degeneration begins, and by 4 weeks very few of the original foci remain viable (Fig. 2; WHITE et al. 1992; RAO et al. 1992). E1A expression appears to drive primary BRK cells into proliferation sufficient to initiate transformation, but is not sufficient to sustain the growth of transformed cells. This induction of cell death in primary BRK cells following E1A expression was reminiscent of the induction of cell death by apoptosis following E1A expression in productively infected human cells (see above).

In the process of focus degeneration subsequent to E1A expression, cell morphology is greatly altered, similar to the appearance of the *cyt* phenotype, and cells fail to exclude trypan blue (see TAYLOR 1993). Morphological indications of apoptosis, such as clumped chromatin, were also apparent at the level of the electron microscope (RAO et al. 1992). E1A expression was apparently sufficient to initiate the transformation process by recruiting normally quiescent cells into

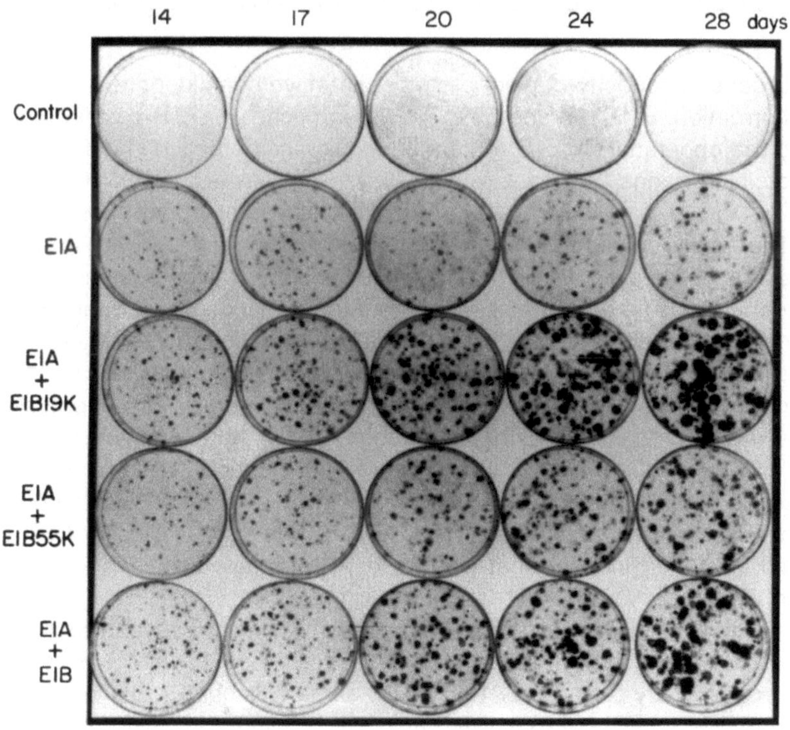

Fig. 2. Role of the E1A and E1B proteins in the transformation of primary baby rat kidney (BRK) cells. Primary BRK cells were transfected with expression vectors designed to express the individual adenovirus-transforming proteins as indicated. At the indicated times, representative plates were fixed and stained with Giemsa to reveal the transformed foci (from RAO et al. 1992)

repeated rounds of cell divison, but was insufficient to sustain the growth of the transformed cells which gradually die. With an equal balance between cell proliferation and cell death, no net accumulation in cell number can occur. Cotransfection of E1A with a second cooperating oncogene such as E1B, which does not have any measurable capacity to promote focus formation in this assay on its own, suppressed all indications of cell death and permitted transformation (Fig. 2; WHITE et al. 1992; RAO et al. 1992). These observations suggested that the inability of E1A to transform cells efficiently may arise from stimulation of cell proliferation and in conjunction with death which would antagonize the transformation process. The implications from these observations were two-fold. First, the transformation may require that induction of cell proliferation by E1A be coupled to inhibition of apoptosis. Second, inhibitors of apoptosis may cooperate with E1A to transform cells. This was confirmed with cooperation between E1A and Bcl-2 (see below).

3 E1B Encodes Two Independent Transforming Activities

The E1B gene encodes two major proteins that have been identified to function in transformation, E1B 19K and 55K. Initial approaches to E1B protein function involved the generation of E1B mutant adenoviruses defective in either or both E1B protein functions (BARKER and BERK 1987; BERNARDS et al. 1986; CHINNADURAI 1983; TAKEMORI et al. 1984; PILDER et al. 1984; WHITE et al. 1984b; EDBAUER et al. 1988). The prevailing observations were that both E1B proteins were important; however, the overlapping reading frames of 19K and 55K, the potential contribution of viral gene products other that E1A to the transformation process, and the use of different cell types (various cell lines and primary cells) complicated interpretation of the results. Analysis of E1B 19K mutant viruses in transformation was particularly problematic because of the cytotoxicity (apoptosis) caused by E1A expression in some assays (WHITE et al. 1984b). Generation of plasmid expression vectors to specifically express the individual E1B proteins under a strong constitutive promoter (CMV) permitted the transforming activity of 19K and 55K in cooperation with E1A to be unambiguously addressed. The use of primary BRK cells alleviated any complications that may arise when stable cell lines are used in transformation assays.

It has long been known that cotransfection of E1A with a second cooperating oncogene such as an activated *ras* gene or the E1B gene was required for transformation of primary cells (RULEY 1983). E1B alone was unable to produce focus formation in the absence of E1A (WHITE and CIPRIANI 1990) and did not increase the frequency with which E1A stimulated focus formation (Fig. 2; WHITE et al. 1992; RAO et al. 1992). Initiation of focus formation was, therefore, clearly an E1A-dependent phenomenon, whereas E1B was required to sustain transformed cell growth once focus formation occurred.

To address the contribution of the individual E1B proteins to the transformation process, E1A was contransfected with plasmid expression vectors that express either E1B 19K or 55K. Expression of either E1B protein dramatically increased the transformation frequency when cotransfected with E1A, indicating that both E1B proteins independently possess transforming activity (Fig. 2; WHITE and CIPRIANI 1990; McLORIE et al. 1991; RAO et al. 1992). This supports evidence for a role for both E1B proteins in transformation from earlier work with E1B mutant adenoviruses (see above) and explains why under some circumstances eliminating expression of only one E1B protein still permits transformation to occur (TELLING and WILLIAMS 1993; ZHANG et al. 1992). More importantly, expression of either E1B protein, but particularly 19K, abrogated all indications of cell death and focus degeneration that would occur by singular expression of E1A (WHITE et al. 1992; RAO et al. 1992). Thus, the prevailing function of the E1B gene in transformation appeared to be inhibition of apoptosis (see below).

4 E1B 55K Inhibits p53 Function

4.1 Complex Formation Between E1B 55K and p53

Immunoprecipitation of adenovirus type 2 (Ad2) E1B 55K from adenovirus-transformed cells resulted in the coprecipitation of the cellular protein p53 (SARNOW et al. 1982). SV40 large T antigen had also been observed to enter into a complex with p53, an observation later to be extended to include the E6-transforming protein from human papilloma virus (reviewed in LEVINE 1990). The p53 protein, which is normally predomionantly localized in the nucleus, translocates to a perinuclear aggregate when complexed with E1B 55K, suggesting that direct physical sequestration of p53 by 55K may contribute to inhibition of p53 function (ZANTEMA et al. 1985a). The equivalent E1B protein from Ad12, E1B 54K, does not detectably bind to p53, but the p53 protein is highly stabilized in Ad12-transformed cells, suggesting that it affects p53 by a mechanism different from Ad5 55K (ZANTEMA et al. 1985b). The binding sites required for interaction between Ad2 E1B 55K and p53 in vitro have been mapped to the activation domain of p53 (amino-terminal 123 amino acids) and to several regions of 55K, but predominantly residues 224–354 (KAO et al. 1990). With one exception (see below), E1B 55K mutants that fail to bind p53 also fail to transform cells, suggesting that complex formation with p53 is necessary but not sufficient for transformation (YEW and BERK 1992).

4.2 Inhibition of p53 Transactivation by E1B 55K

The protein p53 appears to function as a transcription factor (FIELDS and JUNG 1990; RAYCROFT et al. 1990; FARMER et al. 1992), which prompted investigation of

E1B 55K effects on transcriptional activation by p53. Expression of Ad2 E1B 55K will specifically inhibit p53-dependent transcriptional transactivation of a recorder gene (YEW and BERK 1992). E1B 55K mutants that fail to bind p53, and one mutant still capable of complex formation with p53, also fail to inhibit the transcriptional activation activity of p53 (YEW and BERK 1992). All 55K mutants that loose that ability to inhibit the function of p53 as a transcriptional activator are defective for transformation (YEW and BERK 1992). Ad12 E1B 54K, which does not enter into a stable complex with p53, will also inhibit transcriptional activation by p53, providing a strong argument that the tumor suppressor activity of p53 requires regulation of transcription and that E1B 55K/54K function by blocking this activity (YEW and BERK 1992).

4.3 Inhibition of Apoptosis by E1B 55K

Expression of E1B 55K with E1A in a BRK cell transformation assay suppresses the degree of apoptosis and may thereby enhance focus formation, but is substantially less effective than E1B 19K (RAO et al. 1992). Furthermore, transformed cell lines derived from E1A and E1B 55K transfections remain susceptible to apoptosis upon factor withdrawal or growth to high density, whereas the E1A plus E1B 19K transformants do not (RAO et al. 1992). Perhaps alterations in cell growth conditions interfere with the ability of E1B 55K to block p53 function or there exists a p53-independent pathway.

5 E1B 19K Inhibits Apoptosis

5.1 Inhibition of Apoptosis by E1B 19K in Infected Cells

The first indication that E1B 19K functions as an inhibitor of apoptosis derived from the observation that loss-of-function mutations in the E1B 19K gene from Ad2 and Ad5, as well as the highly oncogenic Ad12, accelerated death of the infected host cell and caused the induction of host cell and viral DNA fragmentation and enhanced cytopathic effects (PILDER et al. 1984; TAKEMORI et al. 1984; WHITE et al. 1984b) which strongly resemble cell death by apoptosis (WHITE et al. 1991). Overexpression of the inhibitor of apoptosis, Bcl-2, will complement absence of E1B 19K by suppressing these phenotypes during infection, further substantiating that E1B 19K was functioning to inhibition apoptosis (see below). The prevailing consequence of the induction of apoptosis by E1A during infection is the premature death of the host cell, which can substantially impair virus yield by two orders of magnitude (PILDER et al. 1984; SUBRAMANIAN et al. 1984). Prematurely infected cell lysis is most likely responsible for the large plaque phenotype of E1B 19K mutant viruses (CHINNADURAI 1983; TAKEMORI et al. 1968). Large plaques may form because of accelerated cell death-induced enhanced cell-to-cell spread of virus particles necessary for plaque formation.

Other pleiotropic effects of the E1B 19K gene mutant viruses on viral gene expression have been observed that are likely related, directly or indirectly, to failure to suppress apoptosis. E1B 19K viral mutants display elevated levels of viral early gene expression at the level of transcription relative to the wild-type virus (WHITE et al. 1986, 1988). This could result from modulation of transcription by E1B 19K (HERRMANN and MATHEWS 1989; SHIROKI et al. 1990) or may be an indirect consequence of perturbation of transcriptional regulation caused by cellular destruction brought on by induction of apoptosis. A significant obstacle to obtaining the mechanism of E1B 19K action has been the inability to efficiently utilize comparisons between infections with wild-type and E1B 19K mutant viruses. The profound changes in cell structure and metabolism between non-apoptotic wild-type versus apoptotic E1B 19K mutant virus-infected cells has produced difficulties in discriminating causation from the consequence of apoptosis.

E1B 19K expression has been observed to disrupt the organization of components of the intermediate filament cytoskeleton. Transient expression of E1B 19K in HeLa or COS cells produces collapse of the cytoplasmic vimentin intermediate filaments and disorganization of the lamin-type intermediate filaments within the nuclear envelope (WHITE and CIPRIANI 1989). Vimentin and lamin organization is also disrupted by E1B 19K expression in stably transformed BRK cells and during adenovirus infection (WHITE and CIPRIANI 1990). 19K colocalizes with the disrupted intermediate filaments in the cytoplasm and nuclear envelope, suggesting that 19K expression may be directly related to intermediate filament disorganization (WHITE and CIPRIANI 1989, 1990). It is unclear whether intermediate filament disruption is related to inhibition of apoptosis by E1B 19K or is merely a secondary consequence of 19K expression in the same intracellular locale. To address this point, a dominant-negative mutant form of vimentin which effectively disrupts vimentin intermediate filament organization was tested for the capacity to inhibit apoptosis in several assays. Although vimentin filament organization was effectively disrupted by expression of the dominant-negative vimentin mutant, no ability to suppress apoptosis was detected (E. White and P. Sabbatini, unpublished observations). Disruption of vimentin filaments is apparently not sufficient to inhibit apoptosis; however, the involvement of the lamin-type intermediate filaments in apoptosis has yet to be addressed.

5.2 Inhibition of Apoptosis by E1B 19K During Transformation

Once it had been determined that the E1A proteins produced apoptosis in conjunction with cell proliferation in the course of transformation of primary BRK cells, a possible role for inhibitors of apoptosis such as E1B 19K in the transformation process became apparent (RAO et al. 1992; WHITE et al. 1992). Macroscopic examination of foci from BRK cell transformation assays established that E1B 19K expression prevented the appearance of apoptosis induced by E1A expression (Fig. 2; RAO et al. 1992; WHITE et al. 1992; also see

TAYLOR 1993). Growth of foci, which would otherwise degenerate by apoptosis, was sustained by coexpression of E1B 19K with E1A, increasing the frequency of transformation greater than tenfold (RAO et al. 1992; WHITE et al. 1992). Unlike foci derived from transfection of E1A alone, stable cell lines could readily be derived from E1A plus E1B 19K transfectants, and all were found to express both E1A and E1B 19K (WHITE and CIPRIANI 1990). Thus, it appears that the same phenomenon of induction of apoptosis by E1A and inhibition by E1B 19K observed in productively infected human cells is also observed in the transformation process.

5.3 E1B 19K Expression Blocks Apoptosis by Tumor Necrosis Factor-α and Fas Antigen

E1B 19K can inhibit apoptosis in situations where E1A is expressed, but if it acted at a central regulatory point in the apoptotic pathway, inhibition of E1A-independent pathways would be expected. Tumor necrosis factor-α (TNF-α) and antibodies directed against cell surface antigen Fas will induce apoptosis in many cell types that do not express E1A, and E1B 19K expression will effectively prevent their ability to induce apoptosis (reviewed in WHITE and GOODING 1994). Inhibition of TNF-α and Fas antigen-mediated apoptosis by E1B 19K was achieved in adenovirus-infected cells, stable cell lines, and through transient expression with 19K plasmid expression vectors (CHIOU et al. 1994b; GOODING et al. 1991; HASHIMOTO et al. 1991; WHITE et al. 1992). E1B 19K expression will also suppress apoptotic death brought on by treatment with the anticancer drug cisplatin (SUBRAMANIAN et al. 1993). These studies demonstrate that E1B 19K is capable of blocking apoptosis induced by very divergent signals, indicating that it acts downstream at a step common to multiple pathways for the induction of apoptosis. The ramifications of E1B 19K bypassing apoptosis now extend beyond merely alleviating cytotoxicity of E1A expression. By blocking these E1A-independent apoptotic pathways, E1B 19K may impart a survival advantage to transformed cells in vivo and may contribute to the ability of adenovirus to evade immune surveillance mechanisms of an infected host.

5.4 Cosegregation of 19K Functions

E1B 19K possesses multiple activities (i.e., transformation, TNF-α resistance, and inhibition of Fas antibody killing), which raises the question as to whether 19K is multifunctional or whether all activities derive from the singular function of inhibiting apoptosis. To address this issue, E1B 19K mutants were assayed for transforming activity and the ability to block apoptosis induced by TNF-α and Fas antigen. The anticipation was that these activities would be separable if 19K was multifunctional and inseparable if a single function was responsible for all activities. In the case of all twelve E1B 19K missense mutants, transforming activity (E1A cooperation in a BRK cell transformation assay) cosegregated with resistance to TNF-α and Fas antigen-mediated apoptosis (CHIOU et al. 1994b;

WHITE et al. 1992). Although definitive determination awaits elucidation of the biochemical mechanism, these results are consistent with E1B 19K having a single function to inhibit multiple forms of apoptosis.

There is, however, evidence for multiple events in the cell death process; most notable in the adenovirus system are the cytoplasmic events associated with *cyt* phenotype and the nuclear events associated with the *deg* phenotype. E1B 19K mutants have been identified that retain the capacity to block *deg* but not *cyt* (SUBRAMANIAN and CHINNADURAI 1986; WHITE et al. 1984a), suggesting that these events in the cell death process are separable. DNA fragmentation does not always accompany apoptosis in other systems as well. Initiation of either the *cyt* activity may be sufficient to induce cell death, as E1B 19K mutants that retain wild-type *deg* function but have the *cyt* phenotype still remain defective for transformation (WHITE et al. 1984a; CHINNADURAI 1983; SUBRAMANIAN and CHINNADURAI 1986). Further study of these events will be required to substantiate the existence of two parallel pathways or multiple steps for the induction of apoptosis.

6 Induction of Apoptosis by E1A Is p53 Dependent

E1A expression during adenovirus infection or transformation produces apoptosis, which prevents transformation (WHITE et al. 1992; RAO et al. 1992) and severely impairs virus yield (reviewed in WHITE and GOODING 1994; WHITE 1993; WHITE et al. 1994). The induction of apoptosis by E1A cosegregates genetically with induction of cellular DNA synthesis and transformation, indicating that apoptotic response may be instigated by some aspect of subversion of cell growth control by E1A (WHITE et al. 1991). E1A expression produces p53 accumulation (LOWE and RULEY 1993; S.-K. CHIOU, L. RAO, and E. WHITE, in preparation) similarly to DNA damage (MALTZMAN and CZYZYK 1984), which indicates that one cellular response to E1A is to increase p53 levels. Interference in p53 function with either dominant-negative p53 mutants or E1B 55K prevents the induction of apoptosis by E1A, indicating that wild-type p53 function is required for cell death (RAO et al. 1992; DEBBAS and WHITE 1993). Stable transformed cell lines derived from transfection of primary BRK cells with E1A and the temperature-sensitive mutant *ts*p53 (val135) are transformed when p53 is in the mutant conformation and undergo apoptosis when p53 is returned to the wild-type conformation (DEBBAS and WHITE 1993). Thus, functional wild-type p53 is required for the induction of apoptosis by E1A, suggesting that p53 acts as a molecular switch for the induction of apoptosis in E1A expressing cells.

7 Inhibition of p53-Dependent Apoptosis by E1B 19K

As E1A-associated apoptosis is exclusively dependent on p53, E1B 19K was tested directly for inhibition of p53-dependent apoptosis. Introduction of the E1B

19K gene into BRK cell lines transformed by E1A and *ts*p53 (val135) completely prevented all indications of apoptotic death when p53 assumed the wild-type conformation (DEBBAS and WHITE 1993). Thus, E1B 19K was identified as a modifier of p53 function.

8 Apoptosis May Result from Conflicting Growth Signals

How E1A and p53 interact to regulate apoptosis may arise from perturbation of cell cycle control. E1A can drive quiescent cell from G_0 into S phase, while p53 is known to effectively inhibit cell cycle progression at G_1/S (DILLER et al. 1990; GINSBERG et al. 1991; KASTAN et al. 1991; MARTINEZ et al. 1991). These two opposing growth signals from E1A and p53 may be incompatible and result in apoptosis (DEBBAS and WHITE 1993). Conclusions derived from independent experiments addressing the apoptotic activity of deregulated *c-myc* are remarkably similar. Amplified *c-myc* expression is associated with growth-promoting activities, as is E1A, and will cause apoptosis when cells are deprived of serum (EVAN et al. 1992). Furthermore, Bcl-2 will inhibit apoptosis in response to deregulated *c-myc* (BISSONNETTE et al. 1992; FANIDI et al. 1992; VAUX et al. 1988). Analogous to E1A and p53, amplified *c-myc* provides the stimulus to proliferate, which is diametrically opposed to the signal to arrest from factor withdrawal, creating a situation of conflicting growth signals. If this reasoning is correct, aberrant cell cycle control may be the actual cause of apoptosis.

9 Adenovirus May Require Two Mechanisms to Disable Apoptosis

Expression of E1B 19K and 55K provides two overlapping, redundant mechanisms for disabling p53 function. E1B 55K can enter into a protein complex with p53 and inhibit its activity via a direct mechanism (Fig. 3). E1B 19K does not detectably bind p53 (see below), but prevents the induction of apoptosis by p53 (Fig. 3). As E1B 19K blocks p53-independent apoptosis (TNF-α-mediated, for example; WHITE et al. 1992), it likely acts indirectly and downstream of p53 to inhibit p53-dependent and -independent apoptotic pathways (Fig. 3).

There are several explanations as to why adenovirus encodes two independent mechanisms for disabling p53 function. First, the induction of apoptosis by E1A during infection can drastically compromise virus production (PILDER et al. 1984; SUBRAMANIAN et al. 1984; WHITE et al. 1984b). Thus, inhibition of p53-dependent apoptosis is a very important function required to sustain a

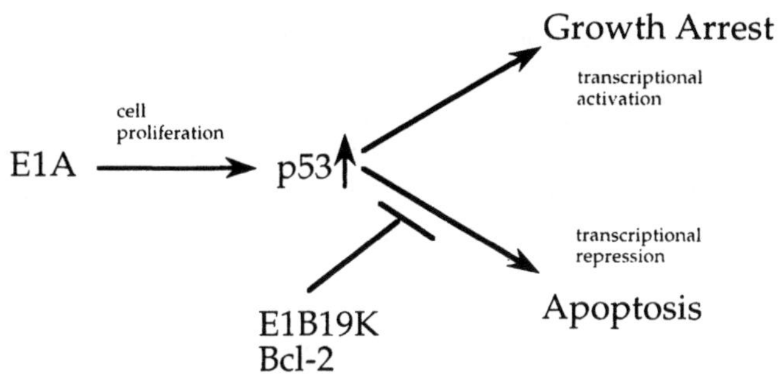

Fig. 3. Regulation of apoptosis by oncogenes and tumor suppressor genes (see text for explanation)

productive adenovirus infection. Having two redundant mechanisms for blocking the activity of p53 would ensure the prevention of apoptosis. Second, E1B 55K is multifunctional and is required for processing and transport of viral late mRNA and the shutoff of host cell protein synthesis in productively infected human cells (Babiss et al. 1983; Pilder et al. 1986; Leppard and Shenk 1989). Presumably, as a consequence of these other functional activities, E1B 55K is found predominantly in a complex with an E4 protein and not p53 in productively infected human cells (Sarnow et al. 1984). Furthermore, E1B 55K expression in the absence of E1B 19K is not sufficient to inhibit apoptopsis in productive infection, since adenovirus mutants that have lost E1B 19K function but retain the function of E1B 55K still induce apoptosis (Pilder et al. 1984; White et al. 1984b). Thus, adenovirus requires a second mechanism other than E1B 55K to prevent the induction of apoptosis by p53 encoded by E1B 19K. Third, E1B 19K can additionally block presumably p53-independent apoptotic pathways where it may convey an additional survival advantage to adenovirus in vivo.

10 Functional Equivalency Between E1B 19K and the Human Bcl-2 Oncoproteins

The human Bcl-2 protein was independently identified as an inhibitor of apoptosis in other systems (Vaux et al. 1988; Hockenbery et al. 1990); this is likely related to its translocation in human B cell follicular lymphoma (Bakhshi et al. 1985; Cleary et al. 1986; Tsujimoto et al. 1985) and its transforming activity in animal models (McDonnell and Korsmeyer 1991). *bcl-2* knockout mice postnatally display polycystic kidney disease, massive apoptosis of T and B cells in the thymus and spleen, and hypopigmented hair (Veis et al. 1993). While Bcl-2 may not initially be essential for inhibition of apoptosis in embryonic development, it is apparently required at later times in specific tissues, perhaps by functioning in an antioxidant pathway (Veis et al. 1993; Hockenbery et al. 1993).

Bcl-2, like E1B 19K, will suppress the induction of apoptosis by E1A and cooperate with E1A to transform primary BRK cells (RAO et al. 1992). As the induction of apoptosis by E1A is p53 dependent (DEBBAS and WHITE 1993), this suggested that Bcl-2 may block the apoptotic activity of p53. Introduction of the human *bcl-2* gene into the E1A plus *ts*p53(val135)-transformed BRK cells that undergo apoptosis when p53 assumes the wild-type conformation effectively prevented all indications of apoptotic death by wild-type p53 (CHIOU et al. 1994a).

In addition to functionally substituting for E1B 19K in transformation, Bcl-2 will also complement for the absence of E1B 19K during adenovirus infection and will confer resistance to TNF-α- and Fas antigen-induced apoptosis (CHIOU et al. 1994b). Currently, E1B 19K and Bcl-2 protein are qualitatively, functionally indistinguishable, which strongly indicates that both possess a related, if not identical mode of action in inhibiting apoptosis.

10.1 Genetic Mapping of E1B 19K Function

E1B 19K sequence is available for 12 different serotypes of adenovirus, from the highly oncogenic Ad12 to tree shrew adenovirus. Comparison of the sequence conservation indicated that the central region (amino acids 81–116) of the 19K is highly conserved, the N terminus (amino acids 1–80) is moderately conserved, and the C terminus (amino acids 117–175) is poorly conserved (WHITE et al. 1992; CHIOU et al. 1994b). Based on this conservation information, a mutagenesis strategy was designed to create missense mutations at predicted key conserved residues. The functional activity of 12 19K mutants bearing single amino acid substitutions (named for the amino acid number from the amino terminus) have been characterized in detail—*pm*7, *pm*28, *pm*44, *pm*44R, *pm*51, *pm*55, *pm*71, *pm*87, *pm*91, *pm*102, *pm*113, and *pm*123— relative to a control plasmid containing a frame shift mutation, *pm*7fs (CHIOU et al. 1994b; WHITE et al. 1992). Introduction of missense mutations in conserved residues in the central and amino terminal regions produced a loss of 19K function and stability, whereas those in the poorly conserved C terminus did not (WHITE et al. 1992; CHIOU et al. 1994b). Based on these studies, the central region and part of the N-terminal region of 19K between residues 44 and 113 are important for maintaining structure and/or function. Conserved functional motifs or interaction sites with cellular proteins might be expected to fall within this region (see below).

10.2 Sequence Similarities Between E1B 19K and Bcl-2 Family Members

It is now apparent that Bcl-2 is not a novel entity but a member of a family of related proteins that function to regulate apoptosis. Proteins that share limited sequence homology with Bcl-2 have been identified in eukaryotic cells and in viruses, and some have been demonstrated to also function to regulate apoptosis.

Programmed cell death in the nematode *Caenorhabditis elegans* resembles apoptosis and some of the regulatory genes have been identified. Of interest is the *ced-9* gene, which acts to inhibit death in the worm in a fashion analogous to Bcl-2 (HENGARTNER et al. 1992). Comparison of the amino acid sequence of Ced-9 and Bcl-2 indicates 23% identity between the two (M. HENGARTNER and R. HORVITZ, personal communication). Inhibitors of cell death are, therefore, likely to contain functional and structural similarities that are highly conserved throughout evolution.

The human Bcl-2 protein heterodimerizes with a protein designated Bax which displays extensive amino acid homology with Bcl-2 (OLTVAI et al. 1993). Bax homodimerizes and forms heterodimers with Bcl-2. When overexpressed, Bax accelerates apoptosis induced by factor withdrawl in dependent cell lines. (OLTVAI et al.1993). The prevalence of Bax homodimers is associated with cell death, whereas the predominance of Bcl-2 homodimers favors cell survival (OLTVAI et al. 1993). A model whereupon apoptosis is regulated by the ration of Bcl-2 Bax has been proposed analogous to the relationship between Myc and Max (OLTVAI et al. 1993). Thus, Bcl-2 family members may function as either activators or repressors of apoptosis.

In a similar vein, the human *bcl-x* gene encodes two products, *bcl-x*$_L$ and *bcl-x*$_{S,}$, generated by alternative splicing (BOISE et al. 1993). While the product of *bcl-x*$_L$ shares both structural and functional homology to Bcl-2, the product of *bcl-x*$_S$ prevents suppression of apoptosis by Bcl-2 (BOISE et al. 1993). The product of *bcl-x*$_S$ is a 170-amino acid protein which has the region of highest homology to Bcl-2 deleted. Unlike Bcl-2 and Bax, however, *bcl-x*$_L$ and *bcl-x*$_S$ do not heterodimerize either with each other or with Bcl-2, indicating that *bcl-x*$_S$ is not a dominant-negative inhibitor of Bcl-2 or *bcl-x*$_L$ but may compete for interaction with the same cellular proteins (C. Thompson, personal communication).

Two other mammalian genes that display homology to Bcl-2, the human *MCL-1* gene (KOZOPAS et al. 1993) and the murine A1 gene (LIN et al. 1993), have been reported. *MCL-1* is a myeloid cell differentiation marker, whereas A1 is an early-reponse gene. Although no functional studies on the regulation of apoptosis are available, the sequence homology with Bcl-2 places *MCL-1* and A1 as members of the Bcl-2 family.

Induction of apoptosis may be a common feature of viral infection and not a phenomenon restricted only to adenovirus (reviewed in WHITE 1993; WHITE and GOODING 1994). It is therefore, not surprising that viruses would encode their own inhibitors of apoptosis as adenovirus does in the E1B gene. Viral homologues of the human *bcl-2* gene have been identified in Epstein-Barr virus (the BHRF1 gene; CLEARY et al. 1986) and in African swine fever virus (the LMW-HL gene; NEILAN et al. 1993). Although no functional information is available for LMW5-HL, BHRF1 can effectively act to inhibit apoptosis (HENDERSON et al. 1993). Thus, the limited sequence homologies between these viral proteins and Bcl-2 are likely to be significant.

The functional similarity between E1B 19K and Bcl-2 prompted a comparision between the amino acid sequences of the Bcl-2 family members *bcl-x*$_L$ Bcl-2, Bax, MCL1, A1, BHRF1, LMW-HL. Although the degree of homology between

E1B 19K and Bcl-2 family members is limited, the region of highest homology is the region between amino acids 39 and 123 of E1B 19K, which is also the region most homologous among the other Bcl-2 family members (CHIOU et al. 1994b). We know from conservation data and mutational analysis that this overlaps the region of 19K that is required for functional and/or structural integrity, residues 44–113 (see above). This region also overlaps with the region deleted in *bcl-x*$_S$ (corresponding to residues 77–133 of 19K). Two conserved regions have been observed among Bcl-2 family members, designated I and II, and of these, region I displays high homology with the E1B 19K sequence. Glycine-87 within region I in particular is one of only four residues of 19K that is conserved among the 19K sequence from 12 adenovirus serotypes. Substitution of glycine-87 with alanine produces a complete loss of function of E1B 19K (WHITE et al. 1992). Because the highest degree of conservation among the 19-kDa proteins occurs in this region and this is the region which displays the highest homology among the Bcl-2 family members, it is anticipated that important functional or structural information is provided by these sequences. In contrast, conserved region II of the Bcl-2 family members is not appreciably conserved in the 19K sequence (CHIOU et al. 1994b). Mutational analysis of Bcl-2 conserved region II may shed some light on the functional significance to this region. Although the sequence homology between 19K and the Bcl-2 family members is limited, the particular region that is conserved may be significant for regulation of apoptosis.

The ultimate determination of the significance of any of these observations awaits determination of the biochemical mechanism of action of Bcl-2 family members. These limited sequence homologies in conjunction with the functional homology suggest possible interaction with similar, if not the same cellular proteins. It is also interesting to note that E1B 19K and Bcl-2 family members appear to have a very similar intracellular distribution within cytoplasmic and nuclear envelope membranes where the information is available (WHITE et al. 1984a; JACOBSON et al. 1993; HOCKENBERY et al. 1990; MONAGAN et al. 1992), and all are relatively low molecular weight proteins ranging from 18 to 37 kDa. Of the Bcl-2 family members, the viral proteins E1B 19K, LMW5-HL, and BHRF1 appear least similar to the cellular Bcl-2 family members. Interestingly, E1B 19K and LMW5-HL do not possess the putative hydrophobic transmembrane region shared by other Bcl-2 family members. E1B 19K is, however, associated with membranes, most likely through fatty acid acylation (MCGLADE et al. 1987; GRAND et al. 1985).

11 Modulation of p53 Function by E1B 19K

11.1 E1B 19K and Bcl-2 Protein Block Apoptosis But Not Growth Arrest by p53

The regulation of p53 function by E1B 19K and Bcl-2 is of extreme interest, since they potentially represent a class of regulators that act indirectly to modify p53 function. p53 is known to induce both growth arrest or apoptosis, depending on the physiological circumstances. Although 19K and Bcl-2 block the apoptotic activity of p53, they do not overcome the growth arrest function (Debbas and White 1993; Chiou et al. 1994a). Introduction of E1B 19K or Bcl-2 gene expression into E1A plus tsp53(val135)-transformed cells, which would otherwise undergo apoptosis, prevented cell death and caused cells to remain in a predominantly growth-arrested state (Debbas and White 1993; Chiou et al. 1994a). Thus, 19K and Bcl-2 can block the apoptotic but not the growth arrest function of p53, suggesting that these two activities of p53 are discrete and separable. Cell cycle analysis of the arrested cells indicated that cell cycle progression was blocked at multiple points and was reversible (Chiou et al. 1994a). The ability of 19K and Bcl-2 to divert the activity of p53 from the induction of apoptosis to growth arrest is extremely significant. Until now, what controls the decision to induce either growth arrest or apoptosis by p53 has not been clear. Our data suggest that Bcl-2 may be one cellular mechanism to regulate this process.

Induction of growth arrest by p53 must normally be overcome during transformation of primary BRK cells, as E1A and E1B 19K or E1A and Bcl-2 are sufficient for transformation. The E1A plus *ts*p53(val135)-transformed cells express much higher levels of p53 than what is seen in E1A-expressing BRK cells. The growth-arrest phenotype of the *ts*p53(val135)-expressing cell lines may be due to insufficient levels of E1A, E1B 19K, or Bcl-2 proteins relative to the high levels of wild-type p53. We expect and are currently testing the more likely possibility that E1A may overcome growth arrest by p53 if expressed at sufficient levels.

11.2 E1B 19K Does Not Interact with p53 Directly

One frequently used mechanism for affecting p53 function is through direct protein complex formation with p53. Although we did not believe that this was a viable possibility as 19K is membrane bound and predominantly in the cytoplasm, whereas p53 functions in the nucleus, a transient interaction between19K and p53 was still possible. In vitro transcription, cotranslation, and immunoprecipitation of 19K and human p53 was utilized to discern intermolecular complex formation. While 19K (and missense mutants *pm*44 and *pm*102) and human p53 could be efficiently immunoprecipitated from in vitro translations with antibodies against 19K and p53, respectively, coprecipitation of 19K and p53 was not observed (P. Verwaerde and E. White unpublished observation). Coprecipitation of the HPV E6 protein, which is known to complex with p53, however, did take place. Attempts to detect 19K and p53 association in lysates from transformed or infected cells have also failed. Neither 19K or Bcl-2 will affect p53 levels or localization, attributes ascribed to other viral modifiers of p53 function (Chiou et al. 1994a; Debbas and White 1993). Although it is still possible that 19K and p53 directly interact, we think that indirect interactions are more probable. Consistent with indirect modification of p53 function are the observations that E1B 19K and Bcl-2 block not only p53-dependent, but also p53-independent apoptotic pathways (CLARKE et al. 1993; CHIOU et al. 1994a).

11.3 E1B 19K Modifies the Transcriptional Activity of p53

The involvement of p53 in apoptosis provides a unique opportunity to investigate the molecular mechanism of apoptosis, since so much is known about the biochemical properties of p53. Although there is evidence for a role for p53 in controlling DNA synthesis, most data suggest that p53 is a transcription factor (reviewed in ZAMBETTI and LEVINE 1993). p53 binds to DNA in a sequence-specific fashion (EL-DEIRY et al. 1992) and will specifically activate the transcription of recorder genes that contain a p53-binding element within the promoter (reviewed in ZAMBETTI and LEVINE 1993). Promoters that lack a p53-binding element can be transcriptionally repressed by p53 (SETO et al. 1992; MACK et al. 1993). p53 will bind to the TATA-binding protein (TBP), which may enhance transcription when p53 is bound to DNA or squelch transcription where p53 cannot bind to DNA. The relationship between the transcriptional activity of p53 on the one hand and the growth arrest and apoptotic functions on the other was investigated by taking advantage of the ability of E1B 19K to block apoptosis but not growth arrest by p53.

A promoter reporter construct with two copies of a 50-base pair p53-binding element from the MCK promoter upstream of the chloramphenicol acetyltransferase (CAT) gene (p50-2) that are specifically transactivated by wild-type p53 (ZAMBETTI et al. 1992) was utilized to measure the effect of E1B 19K on the transactivating ability of p53. Wild-type but not mutant p53 could specifically transactivate the expression of CAT in Saos2 and HeLa cells, as previously reported (ZAMBETTI et al. 1992). E1B 19K expression, while having no effect on basal CAT expression in the absence of wild-type p53, greatly stimulated the ability of wild-type p53 to transactivate expression of CAT (P. SABBATINI and E. WHITE, in preparation). The enhancement of p53 transactivation by E1B 19K was specific for wild-type p53, as no effect of 19K was observed with mutant p53. These results differ greatly from the observations with other DNA tumor virus-transforming proteins that interact with p53, which appear to ablate all p53 function including transactivation (LECHNER et al. 1992; YEW and BERK 1992). Since E1B 19K blocks only the apoptotic function of p53, the transcriptional transactivation function may be specifically associated with the growth arrest.

Transcriptional repression by p53 was approached similarly using minimal promoters that lack a p53-binding site and are specifically repressed by wild-type p53 (MACK et al. 1993). E1B 19K expression had no effect on the basal level of expression, but completely inhibited transcriptional repression by wild-type p53 (P. SABBATINI and E. WHITE, in preparation). Mutant p53 had no effect on the level of CAT expression with or without the presence of 19K. Furthermore, derepression by 19K occurred in Saos2 cells, which do not undergo apoptosis upon expression of wild-type p53, indicating that repression is not the result of nonspecific toxicity of p53. Thus, repression of transcription by p53 appears to be associated with apoptosis, since when apoptosis is inhibited, so is transcriptional repression. E1B 19K is uniquely capable of driving cells away from p53-dependent apoptosis and toward p53-dependent growth arrest and may do so by modulating the transcriptional activity of the p53 protein. These results are consistent with p53 activating the transcription of genes required to induce cell cycle arrest, while repressing transcription of genes required to maintain cell viability. Presumably, the physiological state of the cell or expression of E1B 19K will dictate whether a cell will growth arrest or die byp53. How this regulation takes place has not yet been established, but could be related to post-translational modification of the p53 protein. The ability of the human Bcl-2 protein to similarly modify p53 transcriptional activity is currently under investigation.

11.4 Model for Regulation of p53-Dependent Apoptosis

The model for regulation of p53-dependent apoptosis most consistent with our results is that E1A expression initiates cellular DNA synthesis, and perhaps through the accumulation of DNA damage or perturbation of cell cycle control p53 responds by accumulating to high levels in the nucleus (Fig. 3). Normally p53 accumulation is associated with either growth arrest or apoptosis, but in the presence of E1A, and perhaps due to the creation of conflicting growth signals, apoptosis results (DEBBAS and WHITE 1993). E1B 55K can block p53 function in its entirety by entering into a protein complex with p53 and sequestering its activity. Any activity of p53 would be expected to be inhibited by E1B 55K. E1B 19K and Bcl-2 probably act downstream and indirectly to block only the apoptotic function of p53 (Fig. 3). Since transcriptional transactivation py p53 is enhanced, and only transcriptional repression by p53 is blocked by E1B 19K, transcriptional activation may be associated with growth arrest and transcriptional repression may be either a symptom or cause of apoptosis (Fig. 3). E1B 19K will be an invaluable probe to elucidate the regulation of p53 growth arrest versus apoptosis functions, since cellular mechanisms equivalent to E1B 19K that control p53 (i.e., Bcl-2) appear to exist.

12 Conclusions and Future Outlook

Apoptosis is now recognized as an important biological process essential for normal growth and development and for maintaining homeostasis. Loss of negative control of apoptosis can lead to degenerative diseases and impaired development (VEIS et al. 1993), whereas loss of positive control can produce cancer and autoimmune disease (MCDONNELL and KORSMEYER 1991; DONEHOWER et al. 1992; WATANABE-FUKUNAGA et al. 1992). The identification of E1A and E1B as regulators of apoptosis has provided an opportunity to define the molecular processes by which apoptosis is regulated.

The biochemical mechanisms involved in regulating apoptosis are just beginning to be identified. Endonucleases have long been known to be a part of the process (Wylle 1980a), although it is not yet clear whether they are a causative or downstream event. The identification of the product of the *ced-3* gene, which is a positive effector of apoptosis, as a protease has indicated a role for protein cleavage in the activation of apoptosis (Miura et al. 1993; Yuan et al. 1993). Bcl-2 has been implicated in the function of an antioxidant pathway, suggesting a role for free radical generation in the cell death process (Hockenbery et al. 1993; Kane et al. 1993). Activated forms of H-*ras* will cooperate with E1A (Ruley 1983) and thereby presumably regulate apoptosis. Recently R-*ras* has been identified as a Bcl-2-associated protein, further implicating *Ras* proteins in the apoptotic process (Fernandez-Sarabia and Bischoff 1994). Apoptosis is often brought about in situations where conflicting growth signals occur, suggesting that aberrant cell cycle regulation may elicit apoptosis. E1B 19K and Bcl-2 will restore the ability of p53 to enforce proper growth arrest without apoptosis (Chiou et al. 1994a; Debbas and White 1993). The p53 protein potently suppresses cell cycle progression through the transcriptional activation of an inhibitor of cyclin kinases (El-Deiry et al. 1993; Harper et al. 1993; Xiong et al. 1993) or activates apoptosis (Yonish-Rouach et al. 1991), suggesting that apoptosis may be an abortive attempt at growth arrest. How E1A and E1B fit into these various schemes will be of great interest to determine.

Acknowledgments. I wish to thank S.-K. Chiuo, H.J. Lin, D. Perez, L. Rao, and P. Sabbatini for reading the manuscript and for helpful discussions. This work was supported by grants from the NIH (CA53370), the American Cancer Society (MV71975), and the New Jersey Commission on Science and Technology to E.W.

References

Babiss LE, Ginsberg HS, Darnell JE (1983) Adenovirus E1B proteins are required for accumulation of late viral mRNA and for effects on cellular mRNA translation and transport. Mol Cell Biol 5: 2552–2558

Bakhshi A, Jensen JP, Goldman P, Wright JJ, McBride OW, Epstein AL, Korsmeyer SJ (1985) Cloning the chromosomal break-point of the t(14:18) human lymphomas: clustering around JH on chromosome 14 and near a transcriptional unit on 18. Cell 41: 889–906

Barker DD, Berk AJ (1987) Adenovirus proteins from both E1B reading frames are required for transformation of rodent cells by viral infection and DNA transfection. Virology 156: 107–121

Bernards R, deLeeuw MGW, Houweling A, van der Eb AJ (1986) Role of the adenovirus 1B tumor antigens in oncogenic transformation and lytic infection. Virology 150: 126–139

Bissonnette RP, Echeverri F, Mahboubi A, Green D (1992) Apoptotic cell death induced by c-myc is inhibited by bcl-2. Nature 359: 552–554

Boise LH, Gonzalez-Garcia M, Postema CE, Ding L, Lindsten T, Turka LA, Mao X, Nuñez G, Thompson C (1993) bcl-x, a bcl-2-related gene that functions as a dominant regulator of apoptotic death. Cell 74: 597–608

Chinnadurai G (1983) Adenovirus 2 lp+locus codes for a 19K tumor antigen that plays an essential role in cell transformation. Cell 33: 759–766
Chiou S-K, Rao L, White E (1994a) Bcl-2 blocks p53-dependent apoptosis. Mol Cell Biol 14: 2586–2563
Chiou SK, Tseng CC, Rao L, White E (1994b) Functional complemenation of the adenovirus E1B 19K protein with Bcl-2 in the inhibition of apoptosis in infected cells. J Virol 68: 6553–6566
Clarke AR, Purdie CA, Harrison DJ, Morris RG, Bird CC, Hooper ML, Wyllie AH (1993) Thymocyte apoptosis induced by p53-dependent and independent pathways. Nature 362: 849–852
Cleary ML, Smith SD, Sklar J (1986) Cloning and structural analysis of cDNAs for bcl-2 and a hybrid bcl-2/immunoglobulin transcript resulting from the t(14;18) translocation. Cell 47: 19–28
Debbas M, White E (1993) Wild-type p53 mediates apoptosis by E1A which is inhibited by E1B. Genes Dev 7: 546–554
Diller L, Kassel J, Nelson CE, Gryka MA, Litwak G, Geghardt M, Bressac B (1990) p53 functions as a cell cycle control protein in osteosarcomas. Mol Cell Biol 10: 5772–5781
Donehower LA, Harvey M, Slagle BL, McArthur MJ, Montgomery JCA, Butel JS, Bradley A (1992) Mice deficient for p53 are developmentally normal but susceptible to spontaneous tumors. Nature 356: 215–221
Dyson N, Harlow E (1992) Adenovirus E1A targets key regulators of cell proliferation. Cancer Surv 12: 161–195
Edbauer C, Lamberti C, Tong J, Williams J (1988) Adenovirus type 12 E1B 19-kilodalton protein is not required for oncogenic transformation in rats. J Virol 62: 3265–3273
El-Deiry WS, Kern SE, Pietenpol JA, Kinzler KW, Vogelstein B (1992) Definition of a consensus binding site for p53. Nature Gen 1: 45–49
El-Deiry WS, Tokino T, Velculescu VE, Levy DB, Parsons R, Trent JM, Lin D, Mercer E, Kinzler KW, Vogelstein B (1993) WAF1, a potential mediator of p53 tumor suppression. Cell 75: 817–825
Evan GI, Wyllie AH, Gilbert CS, Littlewood TD, Land H, Brooks M, Waters CM, Penn LZ, Hancock DC (1992) Induction of apoptosis in fibroblasts by c-myc protein. Cell 69: 119–128
Fanidi A, Harrington EA, Evan G (1992) Cooperative interaction between c-myc and bcl-2 proto-oncogenes. Nature 359: 554–556
Farmer G, Bargonetti J, Zhu H, Friedman P, Prywes R, Prives C (1992) Wild-type p53 activates transcription in vitro. Nature 358: 83–86
Fernandez-Sarabia MJ, Bischoff JR (1994) Bcl-2 associates with the ras-related protein R-ras p23. Nature 366: 274–275
Fields S, Jung SK (1990) Presence of a potent transcription activating sequence in the p53 protein. Science 249: 1046–1049
Ginsberg D, Michael-Michalovitz D, Ginsberg D, Oren M (1991) Induction of growth arrest by a temperature-sensitive p53 mutant is correlated with increased nuclear localization and decreased stability of the protein. Mol Cell Biol 11: 582–585
Gooding LR, Aquino L, Duerksen-Hughes PJ, Day D, Horton TM, Yei S, Wold WSM (1991) The E1B-19K protein of group C adenoviruses prevents cytolysis by tumor necrosis factor of human cells but not mouse cells. J Virol 65: 3083–3094
Grand RJ, Roberts C, Gallimore PH (1985) Acylation of adenovirus type 12 early region 1b-18kDa protein. FEBS Lett 181: 229–235
Harper JW, Adami GR, Wei N, Keyomarsi K, Elledge SJ (1993) The p21 cdk-interacting protein cip1 is a potent inhibitor of G1 cyclin-dependent kinases. Cell 75: 805–816
Hashimoto S, Ishii A, Yonehara S (1991) The E1B oncogene of adenovirus confers cellular resistance to cytotoxicity of tumor necrosis factor and monoclonal anti-Fas antibody. Int Immunol 3: 343–351
Henderson S, Huen D, Rowe M, Dawson C, Johnson G, Rickinson A (1993) Epstein-Barr virus-coded BHRF1 protein, a viral homologue of Bcl-2, protects human B cells from programmed cell death. Proc Natl Acad Sci USA 90: 8479–8483
Hengartner MO, Ellis RE, Hovitz HR (1992) Caenorhabditis elegans gene ced-9 protect cells from programmed cell death. Nature 356: 494–499
Hermann CH, Mathews MB (1989) The adenovirus E1B 19-kilodalton protein stimulates gene expression by increasing DNA levels. Mol Cell Biol 9: 5412–5423
Hockenbery D, Nuñez G, Milliman C, Schreiber RD, Korsmeyer S (1990) Bcl-2 is an inner mitochondrial membrane protein that blocks programmed cell death. Nature 348: 334–336
Hockenbery DM, Oltvai ZN, Yin X-M, Milliman CL, Korsmeyer SJ (1993) Bcl-2 functions in an antioxidant pathway to prevent apoptosis. Cell 75: 241–251
Hollstein M, Sidransky D, Vogelstein B, Harris C (1991) p53 mutations in human cancers. Science 253: 49–53

Jacobson MD, Burne JF, King MP, Miyashita T, Reed JC, Raff MC (1993) Bcl-2 blocks apoptosis in cells lacking mitochondrial DNA. Nature 361: 365–369
Johnson DJ, Schwarz JK, Cress WD, Nevins JR (1993) Expression of transcription factor E2F1 induces quiescent cells to enter S phase. Nature 365: 349–352
Kane DJ, Sarafian TA, Anton R, Hahn R, Butler Gralla E, Selverstone Valentine J, Ord T, Bredsen DE (1993) Bcl-2 inhibition of neuronal death: decreased generation of reactive oxygen radical species. Science 262: 1274–1277
Kao CC, Yew PR, Berk AJ (1990) Domains required for in vitro association between the cellular p53 and the adenovirus 2 E1B 55K proteins. Virology 179: 806–814
Kastan MB, Onyekwere O, Sidransky D, Vogelstein B, Craig RW (1991) Participation of p53 protein in the cellular response to DNA damage. Cancer Res 51: 6304–6311
Kozopas KM, Yang T, Buchan HL, Zhou P, Craig RW (1993) MCL1, a gene expressed in programmed myeloid cell differentiation, has sequence similarity to BCL-2. Proc Natl Acad Sci USA 90: 3516–3520
Kuerbitz SJ, Plunkett BS, Walsh WV, Kastan MB (1992) Wild-type p53 is a cell cycle checkpoint determinant following irradiation. Proc Natl Acad Sci USA 89: 7491–7495
Lechner MS, Mack DH, Finicle AB, Crook T, Vousden KH, Laimins LA (1992) Human papillomavirus E6 proteins bind p53 in vivo and abrogate p53-mediated repression of transcription. EMBO J 11: 3045–3052
Leppard KN, Shenk T (1989) The adenovirus E1B 55Kd protein influences mRNA transport via an intranuclear effect on RNA metabolism. EMBO J 8: 2329–2336
Levine AJ (1990) The p53 protein and its interactions with the oncogene products of the small DNA tumor viruses. Virology 177: 419–426
Lin EY, Orlofsky A, Berger MS, Prystowsky MB (1993) Characterization of A1, a novel hemopoietic-specific early response gene with sequence similarity to bcl-2. J Immunol 151: 1979–1988
Lowe S, Ruley HE (1993) Stabilization of the p53 tumor suppressor is induced by adenovirus–5 E1A and accompanies apoptosis. Genes Dev 7: 535–545
Mack DH, Vartikar J, Pipas JM, Laimins L (1993) Specific repression of TATA-mediated but not initiator-mediated transcription by wild-type p53. Nature 363: 281–283
Malkin D, Li FP, Strong LC, Fraumeni JFJ, Nelson CE, Kim DH, Kassel J, Gryka MA, Bischoff FZ, Tainsky MA, Friend SH (1990) Germ line p53 mutations in a familial syndrome of breast cancer, sarcomas, and other neoplasms. Science 250: 1233–1238
Maltzman W, Czyzyk L (1984) UV irradiation stimulates levels of p53 cellular tumor antigen in nontransformed mouse cells. Mol Cell Biol 4: 1689–1694
Martinez J, Georgoff I, Martinez J, Levine AJ (1991) Cellular localization and cell cycle regulation by a temperature-sensitive p53 protein. Genes Dev 5: 151–159
McDonnell TJ, Korsmeyer SJ (1991) Progression from lymphoid hyperplasia to high-grade malignant lymphoma in mice transgenic for the t(14:18). Nature 349: 254–256
McGlade CJ, Tremblay ML, Yee S-P, Ross R, Branton PE (1987) Acylation of the 176R (19-kilodalton) early region 1B protein of human adenovirus type 5. J Virol 61: 3227–3234
McLorie W, McGlade CJ, Takayesu D, Branton PE (1991) Individual adenovirus E1B proteins induce transformation independently but by additive pathways. J Gen Virol 72: 1467–1471
Miura M, Zhu H, Rotello R, Hartwieg EA, Yuan J (1993) Induction of apoptosis in fibroblasts by IL-1β-converting enzyme, a mammalian homolog of the C. elegans cell death gene ced-3. Cell 75: 653–660
Monagan P, Robertson D, Amos TA, Dyer MJ, Mason DY, Greaves MF (1992) Ultrasturctural localization of Bcl-2 protein. J Histochem Cytochem 40: 1819–1825
Moran E (1993) E1A/T antigen/E7 and the cell cycle. Curr Opin Gen Dev 3: 63–70
Neilan JG, Lu Z, Afonzo CL, Kutish GF, Sussman MD, Rock DL (1993) An african swine fever virus gene with similarity to the proto-oncogene bcl-2 and the Epstein-Barr virus gene BHRF1. J Virol 67: 4391–4394
Nevins JR (1992) E2F: a link between the Rb tumor suppressor protein and viral oncoproteins. Science 258: 424–428
Oltvai ZN, Millman CL, Korsmeyer SJ (1993) Bcl-2 heterodimerizes in vivo with a conserved homolog, Bax, that accelerates programmed cell death. Cell 74: 609–619
Pilder S, Logan J, Shenk T (1984) Deletion of the gene encoding the adenovirus 5 early region 1B—21,000-molecular weight polypeptide leads to degradation of viral and cellular DNA. J Virol 52: 664–671
Pilder S, Moore M, Logan J, Shenk T (1986) The adenovirus E1B 55K transforming polypeptide modulates transport or cytoplasmic stabilization of viral and host cell mRNAs. Mol Cell Biol 6: 470–476
Raff MC (1992) Social controls on cell survival and cell death. Nature 356: 398–400

Rao L, Debbas M, Sabbatini P, Hockenberry D, Korsmeyer S, White E (1992) The adenovirus E1A proteins induce apoptosis which is inhibited by the E1B 19K and Bcl-2 proteins. Proc Natl Acad Sci USA 89: 7742–7746

Raycroft L, Wu H, Lozano G (1990) Transcriptional activation by wild-type but not transforming mutants of the p53 anti-oncogene. Science 249: 1049–1051

Ruley HE (1983) Adenovirus early region 1A enables viral and cellular transforming genes to transform primary cells in culture. Nature 304: 602–606

Sarnow P, Ho YS, Williams J, Levine AJ (1982) Adenovirus E1b-58 kd tumor antigen and SV40 large tumor antigen are physically associated with the same 54 kd celluylar protein in transformed cells. Cell 28: 387–394

Sarnow P, Hearing P, Anderson CW, Halbert DN, Shenk T, Levine AJ (1984) Adenovirus early region 1B 58,000-dalton tumor antigen is physically associated with an early region 25,000-dalton protein in productively infected cells. J Virol 49: 692–700

Seto E, Usheva A, Zambetti GP, Momand J, Horikoshi N, Weinmann R, Levine AJ, Shenk T (1992) Wild-type p53 binds to the TATA-binding protein and represses transcription. Proc Natl Acad Sci USA 89: 12028–12032

Shiroki K, Kato H, Kawai S (1990) Tandemly repeated hexamer sequences within the beta interferon promoter can function as an inducible regulatory element in activation by the adenovirus E1B 19-kilodalton protein. J Virol 64: 3063–3068

Subramanian T, Chinnadurai G (1986) Separation of the functions controlled by the adenovirus 2 lp$^+$ locus. Virology 150: 381–389

Subramanian T, Kuppuswamy M, Gysbers J, Mak S, Chinnadurai G (1984) 19-kDa tumor antigen coded by early region E1b of adenovirus 2 is required for efficient synthesis and for protection of viral DNA. J Biol Chem 259: 11777–11783

Subramanian T, Tarodi B, Govindarajan R, Boyd JM, Yoshida K, Chinnadurai G (1993) Mutational analysis of the transforming and apoptosis suppression activities of the adenovirus E1B 175R protein. Gene 124: 173–181

Takemori N, Riggs JL, Aldrich C (1968) Genetic studies with tumorigenic adenoviruses. I. Isolation of cytocidal (cyt) mutants of adenovirus type 12. Virology 36: 575–586

Takemori N, Cladaras C, Bhat B, Conley AJ, Wold WSM (1984) cyt gene of adenovirus 2 and 5 is an oncogene for transforming function in early region E1B and encodes the E1B 19,000-molecular-weight polypeptide. J Virol 52: 793–805

Taylor R (1993) Cell death makes for lively research. J Natl Inst Health Res 5: 59–62

Telling GC, Williams J (1993) The E1B 19-kilodalton protein is not essential for transformation of rodent cells in vitro by adenovirus type 5. J Virol 67: 1600–1611

Tsujimoto Y, Gorham J, Cossman J, Jaffe E, Croce CM (1985) The t(14;18) chromosome translocations involved in B cell neoplasms result from mistakes in VDJ joining. Science 229: 1390–1393

Vaux DL, Cory S, Adams TM (1988) Bcl-2 promotes the survival of haemopoietic cells and cooperates with c-myc to immortalize pre-b cells. Nature 335: 440–442

Veis DJ, Sorenson CM, Shutter JR, Korsmeyer SJ (1993) Bcl-2-deficient mice demonstrate fulminant lymphoid apoptosis, polycyctic kidneys, and hypopigmented hair. Cell 75: 229–240

Watanabe-Fukunaga R, Brannan CI, Copeland NG, Jenkins NA, Nagata S (1992) Lymphoproliferation disorder in mice explained by defects in Fas antigen that mediates apoptosis. Nature 356: 314–317

White E (1993) Regulation of apoptosis by the transforming genes of the DNA tumor virus adenovirus. Proc Soc Exp Biol Med 204: 30–39

White E, Cipriani R (1989) Specific disruption of intermediate filaments and the nuclear lamina by the 19-kDa product of the adenovirus E1B oncogene. Proc Natl Acad Sci USA 86: 9886–9890

White E, Cipriani R (1990) Role of adenovirus E1B proteins in transformation: altered organization of intermediate filaments in transformed cells that express the 19-kilodalton protein. Mol Cell Biol 10: 120–130

White E, Gooding LR (1994) Regulation of apoptosis by human adenoviruses. In: Tomei D, Cope F (eds) Apoptosis: the molecular basis for cell delath II. Cold Spring Harbor Laboratory Press, Cold Spring Harbor, pp 111–141

White E, Stillman B (1987) Expression of the adenovirus E1B mutant phenotypes is dependent on the host cell and on synthesis of E1A proteins. J Virol 61: 426–435

White E, Blose SH, Stillman B (1984a) Nuclear envelope localization of an adenovirus tumor antigen maintains the integrity of cellular DNA. Mol Cell Biol 4: 2865–2875

White E, Grodzicker T, Stillman BW (1984b) Mutations in the gene encoding the adenovirus E1B 19K tumor antigen cause degradation of chromosomal DNA. J Virol 52: 410–419

White E, Faha B, Stillman B (1986) Regulation of adenovirus gene expression in human WI38 cells by an E1B-encoded tumor antigen. Mol Cell Biol 6: 3763–3773

White E, Denton A, Stillman B (1988) Role of the adenovirus E1B 19,000-dalton tumor antigen in regulating early gene expression. J Virol 62: 3445–3454

White E, Cipriani R, Sabbatini P, Denton A (1991) The adenovirus E1B 19-kilodalton protein overcomes the cytotoxicity of E1A proteins. J Virol 65: 2968–2978

White E, Sabbatini P, Debbas M, Wold WSM, Kusher DI, Gooding L (1992) The 19-kilodalton adenovirus E1B transforming protein inhibits programmed cell death and prevents cytolysis by tumor necrosis factor α. Mol Cell Biol 12: 2570–2580

White E, Rao L, Chiou S-K, Tseng C-C, Sabbatini P, Gonzalez M, Verwaerde P (1994) Regulation of apoptosis by the transforming gene products of adenovirus. In: Mihich E (ed) Apoptosis. Plenum, New York, pp 47–62

Wyllie AH (1980a) Cell death: the significance of apoptosis. Int Rev Cytol 68: 251–306

Wyllie AH (1980b) Glucocorticoid-induced thymocyte apoptosis is associated with endogenous endonuclease activation. Nature 284: 555–556

Xiong Y, Hannon G, Zhang H, Casso D, Kobayashi R, Beach D (1993) p21 is a universal inhibitor of cyclin kinases. Nature 366: 701–704

Yew PR, Berk AJ (1992) Inhibition of p53 transactivation required for transformation by adenovirus early 1B protein. Nature 357: 82–85

Yonish-Rouach E, Resnitzky D, Lotem J, Sachs L, Kimchi A, Oren M (1991) Wild-type p53 induces apoptosis of myeloid leukaemic cells that is inhibited by interleukin-6. Nature 352: 345–347

Yuan J, Shaham S, Leudox S, Ellis HM, Horvitz HR (1993) The C. elegans cell death gene ced-3 encodes a protein similar to mammalian interleukin-1β-converting enzyme. Cell 75: 641–652

Zambetti GP, Levine AJ (1993) A comparison of the biological activities of wild-type and mutant p53. FASEB J 7: 855–865

Zambetti GP, Baronetti J, Walker K, Prives C, Levine AJ (1992) Wild-type p53 mediates positive regulation of gene expression through a specific DNA sequence element. Gene Dev 6: 1143–1152

Zantema A, Frasen JAM, Davis-Olivier A, Ramaekers FCS, Vooijs GP, DeLeys B, Van Der Eb AJ (1985a) Localization of the E1B proteins of adenovirus 5 in transformed cells, as revealed by interaction with monoclonal antibodies. Virology 142: 44–58

Zantema A, Schrier PI, Davis-Olivier A, Van Laar T, Vaessen RT, Van der Eb AJ (1985b) Adenovirus serotype determines association and localization of the large E1B tumor antigen with cellular tumor antigen p53 in transformed cells. Mol Cell Biol 5: 3084–3091

Zhang S, Mak S, Branton PE (1992) Overexpression of the E1B 55-kilodalton (482R) protein of human adenovirus type 12 appears to permitt efficient transformation of primary baby rat kidney cells in the absence of the E1B 19K protein. J Virol 66: 2303–2309

Transcriptional Modulation by the Adenovirus E1A Gene

N. JONES

1 Introduction

The DNA tumour viruses are remarkably efficient in replicating themselves in the host mammalian cell. Investigations over the last 10–20 years aimed at a detailed understanding of how this is accomplished have led to many advances in our understanding of how the processes of cell proliferation and gene expression are regulated. These studies have proved to be a relative gold-mine, leading to the identification of key cellular regulatory proteins whose importance is becoming increasingly evident. Why is this the case? It stems from the fundamental needs of the virus for efficient DNA replication, which involves the participation of viral-encoded and cellular proteins. Both sets of proteins are required in abundance. In the case of the cellular proteins, this normally occurs during the S phase of the cell, during which they are required for cellular DNA synthesis. The problem these viruses face, however, is that the majority of the target cells they infect are quiescent where the DNA synthesis machinery is not in operation, and many of the components of the machinery are present in low amounts. Clearly the virus has to overcome this obstacle, and this is accomplished by a viral-mediated stimulation of cell proliferation resulting in the infected cells proceeding into S phase, accompanied by the up-regulation of necessary DNA synthesis components. The virus therefore has the capacity to overcome proliferation arrest, and it does so by interfering with the activity of regulatory proteins that control cell cycle progression. Thus the study of these viruses has been of fundamental

---

Laboratory of Gene Regulation, Imperial Cancer Research Fund, PO Box 123, Lincoln's Inn Fields, London WC2A 3PX, UK

importance in identifying and understanding the function of these regulatory proteins. The second important requirement for successful viral DNA replication is efficient expression of viral genes that encode essential products. The viruses have evolved mechanisms to guarantee such expression; investigating these mechanisms has led to many advances in understanding gene transcription, such as the identification of transcriptional enhancers and *trans*-acting transcriptional regulators.

In the case of the DNA tumour virus, adenovirus, both these requirements are fulfilled through the action of one viral gene, the E1A gene. E1A products deregulate cell proliferation and transcriptionally activate viral as well as cellular genes (for reviews see MORAN and MATHEWS 1987; BOULANGER and BLAIR 1991; SHENK and FLINT 1991; DYSON and HARLOW 1992; JONES 1990; BAYLEY and MYMRYK 1994). Occasionally, the E1A gene integrates into the cellular genome and is constiutively expressed. If this occurs in a primary cell, it becomes immortalised. Given its interesting functional properties, therefore, it is not surprising that the E1A proteins have been the focus of a great deal of attention. What has emerged are two common themes: (1) E1A directly or indirectly modulates the expression of a wide array of viral and cellular genes, and it is this modulation that underlies its multifunctional characteristics; (2) E1A functions by interacting with a variety of cellular proteins, activating or repressing their activities. It is these themes that will be expanded upon in this review.

During a productive infection, the E1A gene produces five mRNAs through differential splicing (BOULANGER and BLAIR 1991; BAYLEY and MYMRYK 1994). The two most abundant and important mRNAs, however, are the 13s and 12s mRNAs encoding proteins of 289 and 243 amino acid residues, respectively (referred to as 289R and 243R). These are the products synthesised at early times of infection when quiescent cells need to be stimulated and viral genes transactivated; experimentally, all the functions attributed to E1A can be carried out by one or both of these products. Functional characterisation of an extensive array of E1A mutants carried out in a number of different laboratories has identified regions required for individual activities as well as regions required for interaction with specific cellular proteins (for reviews see MORAN and MATHEWS 1987; BOULANGER and BLAIR 1991; BAYLEY and MYMRYK 1994). Thus most of E1A's activities were found to correlate with binding to a particular cellular protein. This has proved to be very useful in understanding the role of these cellular proteins in regulating activities within the cell. A summary of such studies is shown in Fig. 1. It is also clear that these activities reside in the regions of E1A that are conserved in a variety of human as well as simian adenovirus (KIMELMAN et al. 1985; VAN ORMONDT et al. 1980). These regions are referred to as conserved regions 1–3 (CR1–3).

Of particular relevance to this review are the following correlations:

1. The activation of viral transcription units requires the CR3 region present in the 289R, but not the 243R protein, and is likely to be accomplished through a direct involvement of E1A in the activation process. This region interacts with

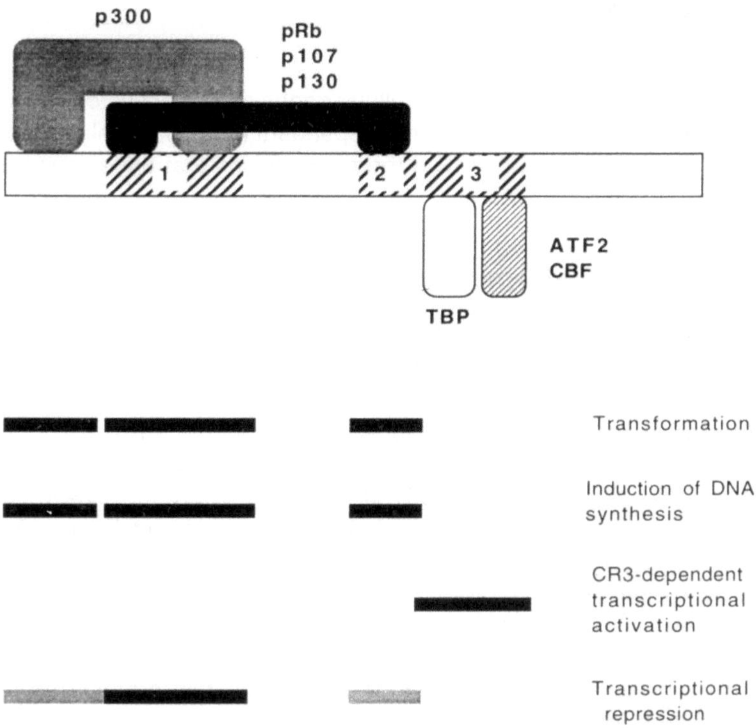

Fig. 1. Map of the regions of E1A required for E1A function and binding to cellular proteins. The 289R protein of E1A is shown with the approximate positions of the three conserved regions, CR1–3. These regions are highly conserved between the E1A proteins of different serotypes. The approximate sites of binding of p300, the pRb family, TATA box-binding protein (TBP) and certain transcription factors are shown. At the *bottom* of the figure, the *solid lines* indicate the regions of E1A that have been shown to be important for the indicated function. The *stippled lines* represent regions shown in some, but not all experiments, to be essential. The diagram depicts those interactions and activities that are discussed in this review. In some reports, sequences within the C-terminal half of E1A have been implicated in the activation and repression of gene expression and tumourigenesis (BAYLEY and MYMRYK 1994); these results are not discussed. *ATF*, activating transcription factor

specific transcription factors and at least one component of the basic transcription machinery.

2. Stimulation of cell proliferation correlates with the binding to two families of proteins, namely the pRb family, which includes the product of the retinoblastoma gene (pRb) itself as well as pRb-like proteins p107 and p130, and the p300 protein, which also appears to constitute a family of proteins rather than being unique (see later).

In all cases these interactions result in the modification of transcription—either the activation of specific genes or the repression of specific genes. This modulation of transcription must underlie the profound changes to the cell that result from E1A expression.

2 Transcriptional Activation by E1A

As described above, a number of viral and cellular genes can be activated by the E1A protein(s) either in a CR3-dependent or -independent manner. I shall discuss the current state of knowledge concerning the mechanisms that underlie these different activation processes.

2.1 Conserved Region 3-Dependent Activation

It is the CR3-dependent form of E1A activation that has received most of the attention over the last few years. All the adenoviral early genes are activated in a CR3-dependent manner, as first shown by comparing the pattern of early gene expression in adenoviral mutants that had alterations in the CR3 region or mutants that could only synthesise the 289R and 243R proteins (MORAN et al. 1986; LILLIE et al. 1986; SCHNEIDER et al. 1987; JELSMA et al. 1988; for review see MORAN and MATHEWS 1987; BOULANGER and BLAIR 1991; SHENK and FLINT 1991; BAYLEY and MYMRYK 1994). Interestingly, however, the kinetics of activation of different early genes was found to differ; in particular, the activation of the E2A gene peaked substantially later than the activation of the E3 and E4 genes (NEVINS 1981). The reason for this is now clear, since efficient expression of E2A requires the synthesis of a protein encoded by E4. The nature of this protein and its function in E2A expression will be discussed in a subsequent section.

As a first step towards understanding CR3-dependent activation, the promoters and the various early genes were analysed in detail to identify *cis*-acting elements required for efficient transcription in the presence of E1A (for reviews see SHENK and FLINT 1991; BERK 1986). Surprisingly, however, the promoters were found to be significantly different, although some elements were identified as being important in more than one case. The E4 promoter was found to be the simplest in that it appeared to consist of three copies of the same element upstream of a standard TATA box (LEE and GREEN 1987; LEE et al. 1989). The element in question was an activating transcription factor (ATF)/cyclic adenosine monophosphate (cAMP) response element (CRE) element which had previously been described as an element that could confer cAMP inducibility on a number of cellular genes (MONTMINY et al. 1983; ROESLER et al. 1988). This element was also found in the E2A and E3 promoters; in these cases, however, it was found in conjunction with other elements rather than being present in multiple copies (MURTHY et al. 1985; ZAJCHOWSKI et al. 1985; MONOHAR et al. 1990; LEFF et al. 1984; WEEKS and JONES 1985; LEE et al. 1987). The analysis of the E4 promoter suggested that this element could mediate an E1A response. The focus was therefore directed at the protein(s) that specifically interacted with it. A very complex picture emerged: a large number of proteins were identified that could bind to CRE/ATF sites (for reviews see (ZIFF 1990; HABENER 1990; HOEFFLER 1992)). All of these proteins, of which there are at least 20, can bind to DNA as homodimers and have one structural feature in common, namely that they contain a bZIP domain. This domain is found in a number of different transcription factors and constitutes a basic region which contacts DNA, adjacent to a leucine zipper, the main feature of which is a heptad repeat of leucine residues; the zipper mediates dimerisation of the protein. Many of the ATF family members can heterodimerise with each other or even with members of the Jun/Fos family of transcription factors (HAI et al. 1989; BENBROOK and JONES 1990; IVANSHKIV et al. 1990; MACGREGOR et al. 1990; HAI and CURRAN 1991; HSU et al. 1991). The total number of dimeric complexes capable of binding CRE/ATF sites is therefore vast. A major challenge in the future is to understand the need for such complexity and the parameters that dictate which complexes bind to particular promoters or are regulated by particular stimuli.

An initial approach involves investigating the characteristics of individual family members, and this approach has led to the identification of ATF factors that can positively respond to E1A in CR3-dependent manner. Since cells contain multiple ATF/CRE species, studying the activity of one member is problematical. A common method for circumventing this problem is to construct and analyse chimera between the protein under analysis and a heterologous DNA-binding domain. So, for example, the coding region of different ATF species could be fused in frame to the DNA-binding domain of GAL4, a yeast transcription factor of which there is no mammalian equivalent. The hybrid protein would bind to GAL4-binding sites in promoters, regulating the expression of an easily assayable product. In this way the ability of the hybrid to activate transcription could be measured. Using such an approach, two very closely related members of the family have been identified, ATF-2 (CRE-BP1, CREB2) and ATFa, which can be activated by E1A (LIU and GREEN 1990; MAEKAWA et al. 1991; FLINT and JONES 1991; CHATTON et al. 1993). Importantly, other members, although activated by other stimuli, are not sensitive to E1A. The regions of these two proteins required for mediating E1A activation have been identified. I will mainly refer to work on ATF-2, although, where tested, the results with ATFa are qualitatively identical, in keeping with their close similarity.

Apart from the bZIP region, the other region of ATF-2 that is critical is that containing the N-terminal 96 amino acids (LIU and GREEN 1990; FLINT and JONES 1991; MATSUDA et al. 1991; LIU and GREEN 1994). Deletion of the region destroys E1A activation, whereas a chimeric protein containing the ATF-2 N terminus fused to the GAL4 DNA-binding domain can be activated by E1A. What does this region do? Recent evidence shows that it constitutes a transcriptional activation domain; thus a GAL4-ATF-2 N terminus chimeric protein can stimulate the expression of an appropriate GAL4-dependent reporter (FLINT and JONES 1991; LIU and GREEN 1994; LIVINGSTONE et al. 1994). In contrast, however, the full-length ATF-2 protein in the context of a GAL4 chimera is inactive. This has led to the notion that the N-terminal activation domain is somehow masked in full-length ATF-2. The functional requirements of this N-terminal region have been analysed in detail. Firstly, a Cys-His type zinc finger motif which has been shown to bind zinc (ABDEL-HAFIZ et al. 1993) is essential for function (FLINT and JONES 1991; ABDEL-HAFIZ et al. 1993). Its destruction by mutation not only lowers the ability of this region to activate transcription, but also affects the ability of the entire ATF-2 protein to mediate E1A stimulation. However, the exact role of the zinc-binding motif is unknown.

A second feature essential for function consists of two Thr residues, Thr-68 and Thr-71, which are predicted substrates for the mitogen-activated and stress-activated (MAPK and SAPK) protein kinases. It is now clear that the ATF-2 N-terminal domain is regulated by particular environmental signals that trigger specific signal transduction pathways (LIVINGSTONE et al. 1994) and provides an important clue as to the normal cellular role of ATF-2. It is regulated by E1A and by the Rb protein (at least in some cells; KIM et al. 1992). It also participates in the regulation of transcription of the c-*Jun* gene, which is activated by a number of diverse stimuli including ultraviolet (UV) irradiation, phorbol esters, cytokines and growth factors (ANGEL et al. 1987; VAN DAM et al. 1990; DEVARY et al. 1991; STEIN et al. 1992; MUEGGE et al. 1992; HERR et al. 1994). In all cases this activation is mediated through two binding sites (jun1 and jun2), which have been shown to preferentially bind ATF-2/c-Jun heterodimers or ATF-2 homodimers (VAN DAM et al. 1993). This suggested that at least some of these stimuli may activate ATF-2 activity, a possibility now shown to be the case. All of these stimuli can activate the activity of the MAPK family, a large group of kinases that are characterised as being proline-directed Ser/Thr kinases and which are themselves activated through phosphorylation of specific tyrosine and threonine residues (for reviews see PELECH and SANGHERA 1992; MARSHALL 1994). The prototypes of this family are Erk1 and Erk2, which are highly related and are strongly activated by a variety of mitogens. In contrast, the SAPKs, which are approximately 40% related to Erk1 and -2, are activated by different stress conditions such as UV (KYRIAKIS et al. 1994). This group of kinases contains the predominant c-Jun N-terminal kinase activity (JNK) (KYRIAKIS et al. 1994; DÉRIJARD et al. 1994). Thr-69 and -71 of ATF-2 are efficiently phosphorylated in vitro by JNK/SAPK and are phosphorylated in vivo dependent upon signals that activate the JNK/SAPK pathway (LIVINGSTONE et al. 1994). Importantly, both Thr-69 and Thr-71 are essential for the N terminus of ATF-2 to activate transcription and also for the intact ATF-2 protein to respond to E1A. A third functional requirement is a region (amino acids 47–66) adjacent to the phosphorylation sites that is required for the interaction of SAPK with ATF-2 (LIVINGSTONE et al. 1994). Without such an interaction, phosphorylation fails to occur. Before turning to the requirements of E1A for activation of ATF-2, it is worth emphasising that the functional requirements of the N-terminal activation domain so far appear indistinguishable from the requirements for an E1A response.

The CR3 region of E1A has also been under intense scrutiny, and a number of important characteristics have emerged, foremost amongst these being that the CR3 region itself can function as a very potent activation domain in the context of a fusion with a heterologous DNA-binding region (LILLIE and GREEN 1989; WEBSTER and RICCIARDI 1991). Furthermore, there is excellent correlation between the requirements for this activation and the requirements for E1A *trans*-activation of sensitive promoters such as the E4 promoter. This finding has underpinned a popular model for E1A activation, namely that E1A is tethered to a promoter by specific protein–protein interaction, whereby it then interacts with, and stimulates, the basic transcription machinery. The model is strengthened by the observation that E1A can indeed bind to a number of specific DNA-binding proteins including ATF-2 (see below) and that this interaction is mediated through a distinct region of CR3, the "promoter targeting" region at its carboxy terminus.

The activation domain of CR3 has been dissected in great detail—every amino acid has been mutated and the effect of such a change on the function of this region measured (WEBSTER and RICCIARDI 1991). An important feature of this region is the presence of a C4-type zinc figure whose integrity is essential for CR3 function (WEBSTER and RICCIARDI 1991; MARTIN et al. 1990; CULP et al.1988). Thus mutation of any of the Cys residues destroys *trans*-activation concomitant with a loss of zinc binding. The role of the zinc finger is unclear, but may either be involved in organising the structure of this domain or may represent an actual interface of a specific protein–protein interaction (or of course both). The interaction of this region with a critical component of the transcriptional machinery has been described. The component in question is TATA box-binding protein (TBP), the DNA-binding subunit of transcription factor IID (TFIID; HORIKOSHI et al. 1991; LEE et al. 1991). Thus in vitro and in vivo interaction between E1A and TBP has been demonstrated, the interaction requiring the activation domain of CR3 but not the targeting region. Furthermore, E1A can interact with TBP even when it is complexed to the TBP-associated factors (TAF) that together constitute TFIID (BOYER and BERK 1993). It appears, therefore, that TBP is not embedded in the centre of the TFIID complex, but rather is exposed on its perimeter and is therefore accessible for a number of interactions, including that with E1A. For the most part, individual amino acid residues within CR3 required for transactivation were also found to be required for TBP interaction (GEISBERG et al. 1994). However, some exceptions were found where particular mutants could bind to TBP normally but were nevertheless incapable of transactivation. This has led to the suggestion that CR3 may interact with other components of the basic machinery besides TBP. Possibilities would be one of the TAF proteins or the general transcription factor TFIIB.

The characteristics of the activation domain of CR3 have therefore been well studied. What about the targeting region? E1A does physically interact with a number of transcription factors in a CR3-dependent manner (CHATTON et al. 1993; LIU and GREEN 1994; LIVINGSTONE et al. 1994; ABDEL-HAFIZ et al. 1993; SCHOLER et al. 1991; LUM et al. 1992; AGOFF and WU 1994). Furthermore, in some cases these interactions have been shown to require the C-terminal CR3 targeting sequences (LIU and GREEN 1994; AGOFF and WU 1994). These interactions appear to involve the DNA-binding regions of the different factors, but surprisingly, the type of DNA-binding domain can vary (LIU and GREEN 1994). Thus in the case of ATF-2 and c-Jun a bZip domain is involved, in the case of Sp1 a zinc finger-containing domain and in the case of USF a basic helix–loop–helix (bHLH) domain. A major question that arises is how the targeting region can recognise such a diverse set of DNA-binding domains. At present there is no answer, but these results do at least present a rational basis for explaining how E1A can activate a wide variety of different promoters, i.e. it is due to promiscuity in interaction.

Considering the wealth of information that has emerged concerning not only the CR3 region of E1A but also ATF-2, a factor that can interact with and mediate E1A activation, can a rational model be drawn up to explain the mechanism of E1A *trans*-activation? The bones of any such model must involve the features of CR3 described above (Fig. 2). The simplest model would view E1A as a "bridging" protein or "co-activator" supplying a pathway between factors that bind to upstream promoter elements and the transcriptional machinery. However, although such bridging has to be important, it cannot be quite that simple, since a number of observations suggest additional complexity. A significant observation that does not fit into the simple model concerns the role of the ATF-2 N terminus, since it is clearly important for E1A stimulation but not for direct

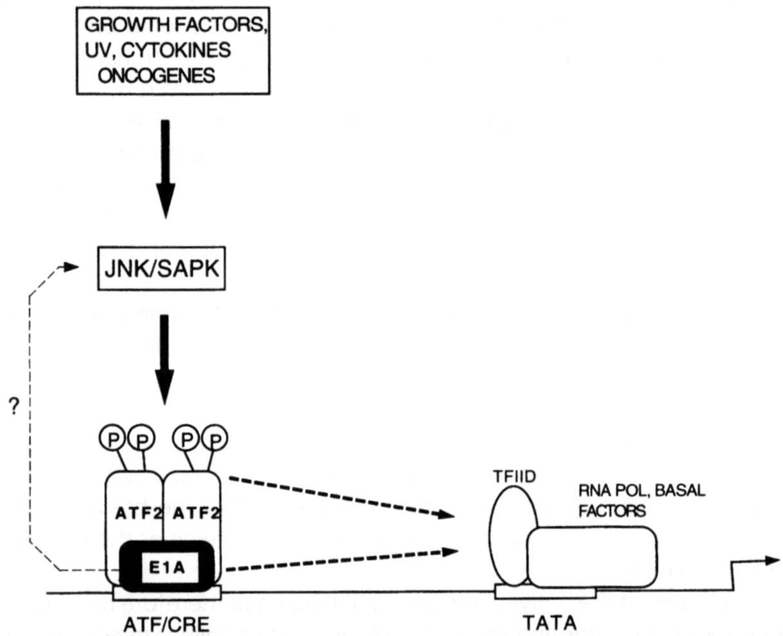

Fig. 2. Proposed model for the E1A-mediated activation of transcription via activating transcription factor (*ATF*) cyclic adenosine monophosphate (cAMP) response element (*CRE*). The ATF-2 factor has been shown to be capable of mediating E1A activation and to bind directly to E1A protein via sequences within or close to its DNA-binding domain. Both E1A and the N terminus of ATF-2 have been shown to harbour transcriptional activation domains, the latter being dependent upon phosphorylation mediated by members of the c-Jun N-terminal Kinase(*JNK*)/Stress-activated protein kinase(*SAPK*) family of kinases; these kinases are activated by an ill-defined signal transduction pathway stimulated by a variety of stress signals. The model proposes that both activation domains synergise by interacting with components of the basic transcriptional machinery (in the case of E1A by direct interaction, but in the case of ATF-2 the interaction could be direct or indirect involving a co-activator). E1A could have multiple effects on ATF-2: (a) as a result of E1A binding the conformation of ATF-2 may be altered to expose the cryptic N-terminal activation domain; (b) E1A may stimulate the activity of the JNK/SAPK pathway, resulting in increased phosphorylation of the critical sites within the activation domain of ATF-2. *UV* ultraviolet; *TFIID*, transcription factor IID

interaction with E1A. Thus E1A can be tethered to ATF/CRE site-containing promoters by mutant ATF-2 proteins that lack the N terminus, but nevertheless transactivation does not occur. Thus a central feature of the E1A response must be some form of co-operation between the activation domains of E1A and ATF-2. Understanding what form this co-operation takes is a major goal for the future. It may involve the interaction of ATF-2 and E1A with different or the same components of the basic machinery, resulting in a synergistic effect on transcription stimulation. Alternatively, it may involve a direct or indirect interaction between the two, resulting in a physical change more conducive to function. For example, the activation domain of ATF-2 is masked in the intact protein, presumably because of its conformation. Upon binding to E1A, a conformational change resulting in unmasking is likely to occur. It has also been suggested that ATF-2 has a similar effect on the CR3 activation domain so that there is mutual unmasking (LIU and GREEN 1994). It also appears that, in some cases, activation can occur in the absence of direct interaction (FLINT and JONES 1991; LIVINGSTONE et al. 1994). For example, the N terminus of ATF-2, when fused to the GAL4 DNA-binding domain, is E1A sensitive. Perhaps in this case, the activation is more indirect and may, for example, involve an E1A-stimulated increase in phosphorylation of the N-terminal activation domain.

2.2 Conserved Region 3-Independent Activation

Although most *trans*-activation by E1A is due to CR3, there are clearly examples where expression of genes is activated in a CR3-independent manner. In most of these cases, the activation is dependent on other conserved sequences of E1A namely CR2, CR1 or the extreme N terminus of E1A. These same regions are required for the transformation and cell cycle modulation activities of E1A and indeed it would appear that these latter activities result, at least in part, from the gene activation functions. The best characterised example of activation in a CR3-independent manner concerns the transcription factor E2F, which was originally described as a factor necessary for efficient expression of the adenovirus E2A gene (for reviews see (NEVINS 1992; LA THANGUE 1994; NEVINS 1994; HELIN and HARLOW 1993)). Since then, however, E2F has been recognised as a key factor controlling the expression of a number of cellular genes in a cell cycle-dependent manner and specifically at the end of G_1. Many of these cellular genes encode products required for efficient DNA replication such as dihydrofolate reductase (DHFR), thymidine kinase, ribonucleotide reductase, B-*myb* and DNA polymerase alpha (BLAKE and AZIZKHAN 1989; MEANS et al. 1992; OGRIS et al. 1993; PEARSON et al. 1991; DALTON 1992; LAM and WATSON 1993). The fact that E2F controls the expression of such genes suggests a key role in cell cycle progression. More convincing evidence has come from understanding how E2F activity is regulated. In cells, E2F is complexed to different cellular proteins, many of which have previously been shown to be key regulators of the cell cycle. These include pRb, p107, cyclin E and cyclin A (for reviews see NEVINS 1992; LA THANGUE

1994; Nevins 1994; Helin and Harlow 1993). The interaction of E2F with pRb has served as a paradigm of how E2F activity is regulated by such interactions. Binding of pRb blocks the ability of E2F to activate transcription but not to bind to DNA (Flemington et al. 1993; Helin et al. 1993). A rationalisation of why this is the case has come from the localisation of the pRb-binding domain within the C-terminal activation domain of E2F-1, the best characterised member of the E2F family of factors (Helin et al. 1992; Kaelin et al. 1992; Shan et al. 1992). Thus by binding to pRb, other productive interactions (for example with TFIID) may be prevented. At times when E2F activity is off, E2F–Rb complexes predominate, for example during the early G_1 phase of the cycle (Mudryj et al. 1991; Schwarz et al. 1993; Shirodkar 1992). However, in late G_1 "free" E2F accumulates, i.e. E2F that is uncomplexed. It is proposed that this free E2F is the active form of E2F and therefore this accumulation explains the increase in E2F transcriptional activity at this time. How does free E2F accumulate? The evidence suggest that it is due to specific phosphorylation of pRb (Schwarz et al. 1993; Chellappan et al. 1991) and E2F-1 (Fagan et al. 1994), interaction being precluded when either partner is phosphorylated on particular residues. The phosphorylations are brought about by particular cyclin/cdks which are activated in a cell cycle-dependent manner (Fagan et al. 1994; Buchkovich et al. 1989; Chen et al. 1989; De Caprio et al. 1989; Mihara et al. 1989). What has this got to do with E1A? It is well established that E1A can bind strongly to pRb or relatives such as p107 (Yee and Branton 1985; Harlow et al. 1986; Egan et al. 1987; Egan et al. 1988; Ewen et al. 1991). As a result, E2F–Rb complexes are disrupted (Fig 3). Thus even in the absence of phosphorylation, E1A can stimulate the appearance of free E2F. Accordingly, genes that are controlled by E2F can be activated. This is likely to be an important feature of the mechanism adenovirus employs to stimulate quiescent cells to progress into S phase. As described earlier, this ensures an ideal environment for viral replication.

E2F plays one other important role in adenovirus infection namely activation of the E2A promoter. The E2 gene encodes products essential for viral replication, including a DNA polymerase, and therefore good expression of this gene is necessary for efficient viral production. The E2A promoter contains an ATF/CRE site and two binding sites for E2F that are inverted with respect to each other. Mutagenesis experiments have indicated that both E2F sites are important for high-level expression and that they appear to act synergistically (for review see Shenk and Flint 1991). It is now known that E2F binds to these sites in conjugation with a 19-kDa product of the viral E4 gene (Hardy et al. 1989; Hardy and Shenk 1989; Huang and Hearing 1989; Reichel et al. 1989). This interaction produces a stable complex on the promoter in contrast to the binding of E2F to a single site, which has a very short half-life. By stabilising the complex, E4 enhances the E2F-driven transcription of this promoter. How this stabilisation is accomplished remains unknown, but what is clear is that the arrangement of the two E2F binding sites is crucial; if they are present as tandem rather than inverted repeats or if the spacing between the repeats is altered, then stable complex formation and hence activation is diminished (Hardy and Shenk 1989). The effect of E1A on the E2A promoter, therefore, is very complex, involving more than one E1A activity:

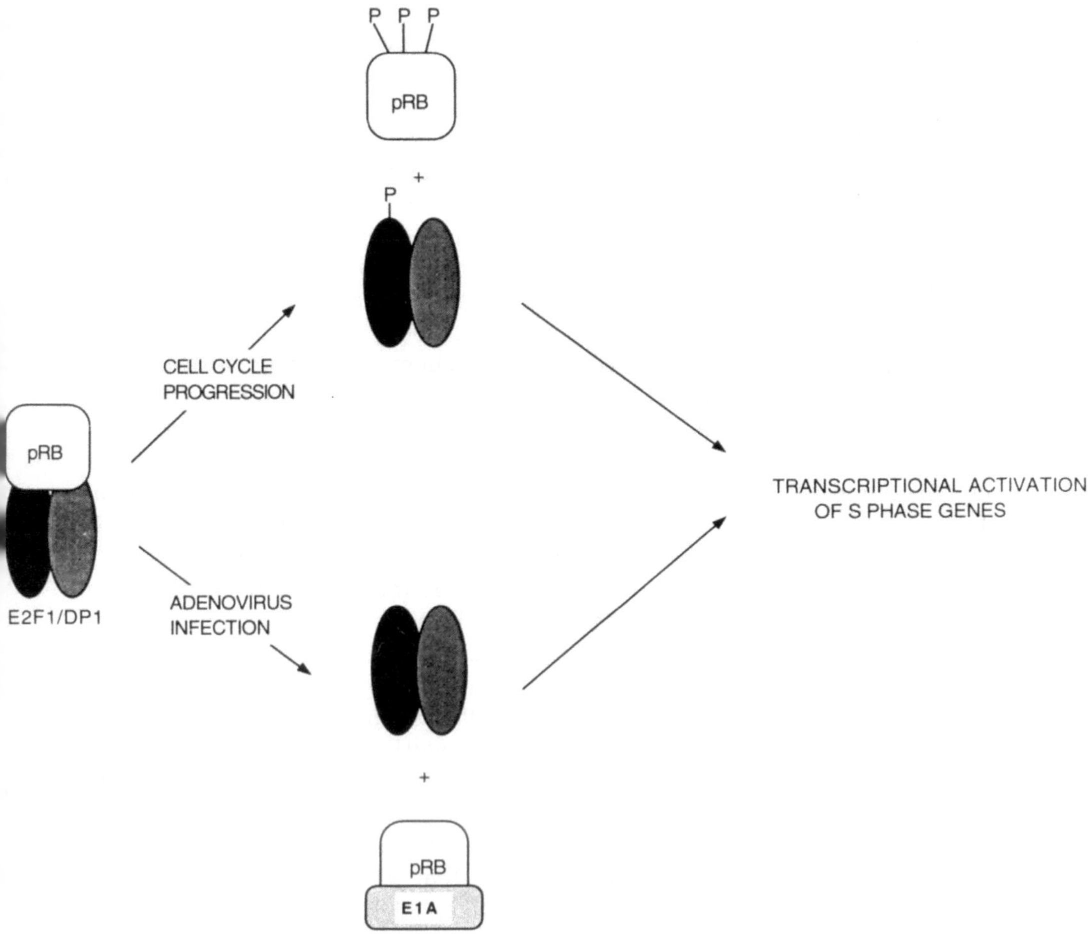

Fig. 3. Regulation of the E2F transcription factor. The E2F factor, which consists of E2F-1 or related proteins heterodimerised to DP-1, is regulated by its interaction with the pRb protein; the E2F–pRb complex is inactive. During the progression of cells through the G_1 phase of the cell cycle, free E2F accumulates due to the phosphorylation of pRb and E2F-1 by cyclin-dependent kinases. The free E2F is the active form and can stimulate the transcription of genes required for S phase. In adenovirus-infected cells, the E2F–pRb complex can be disrupted by the interaction of E1A with pRb, which displaces the E2F factor

1. It is likely that E1A can activate the E2A promoter at least to a limited extent through CR3-dependent activation via the ATF/CRE element.
2. E1A is necessary to activate the E4 gene and hence the synthesis of the E4 19-kDa product which interacts with E2F.
3. E1A can free E2F from complexes with pRb or its relatives. This is necessary for E4 interaction; E4 itself cannot displace E2F from an E2F–pRb complex.

This latter role of E2F in an adenovirus infection, namely the activation of the E2A promoter, appears to be unique for the adenovirus situation. In contrast, however, the freeing of E2F as a result of E1A interaction with pRb or its relatives is a theme that is common to infection by a number of small DNA tumour viruses such as SV40 virus and human papillomavirus (for review see NEVINS 1992)). They direct the synthesis of proteins that, like E1A, can strongly interact with pRb, and indeed they all share a common amino acid motif that is crucial for this interaction. In none of the other cases, however, is E2F involved in the activation of a viral gene. All of these viruses have a common need, which is to stimulate quiescent cells, and they have evolved the same mechanism to fulfil this need, i.e. they disrupt pRb-mediated protein–protein interactions.

Although activation via the E2F factor is the best understood mechanism of CR3-independent activation of transcription, it by no means represents the only mechanism that has been described. The E1A 243R product has been shown to activate certain genes in TATA-dependent manner, and the exact sequence and context of the TATA element determines the degree of sensitivity (SIMON 1988). This activation was shown to be dependent upon the N-terminal sequence of E1A rather than the CR1 and CR2 regions necessary for E2F activation (KRAUS et al. 1991). Kraus et al. (1994) have recently presented a mechanism for the activation that involves a transcriptional repressor Dr1 which binds to TBP and prevents it from interacting with TFIIA and TFIIB; hence initiation complex formation is inhibited (Fig. 4). Importantly, the TBP–Dr1 interaction is disrupted by E1A and requires the E1A N-terminal domain. How this disruption takes place is not entirely clear, but some interaction between Dr1 and E1A can be detected in vitro. This suggests the possibility that, as with E2F, activation by E1A entails the disruption of inhibitory complexes, in this case involving TBP. Future studies should clarify how important this mode of activation is and what relationship, if any, exists between Dr1 and p300, a protein that also binds to the N-terminal region. It is also interesting to note that Dr1 can inhibit not only pol II-directed transcription but also pol III transcription (WHITE et al. 1994). Since E1A can activate both these transcriptional processes, it is possible that activation of both can be influenced by the E1A–Dr1 interaction.

Yet another case of CR3-independent activation that has been reported concerns the c-Jun promoter, which contains two sites for a heterodimer between c-Jun protein and ATF-2 (VAN DAM et al. 1993). This promoter is activated by E1A dependent upon the N-terminal region and appears to be caused, at least in part, by hyperphosphorylation of the activation domain of c-Jun (HAGMEYER et al. 1993).

Recent experiments have identified a second activation domain within the E1A protein, distinct from the one within CR3 (BONDESSON et al. 1994; G. MICHLEM and N.C. JONES, unpublished results). In the context of a fusion with a heterologous DNA-binding domain, the N terminus of E1A, including the CR1 but not the CR2 region, can strongly activate transcription. The exact significance of this finding is presently unclear. It may be linked to the repression activity of E1A, which also maps to the N-terminal region. As will be described below, repression

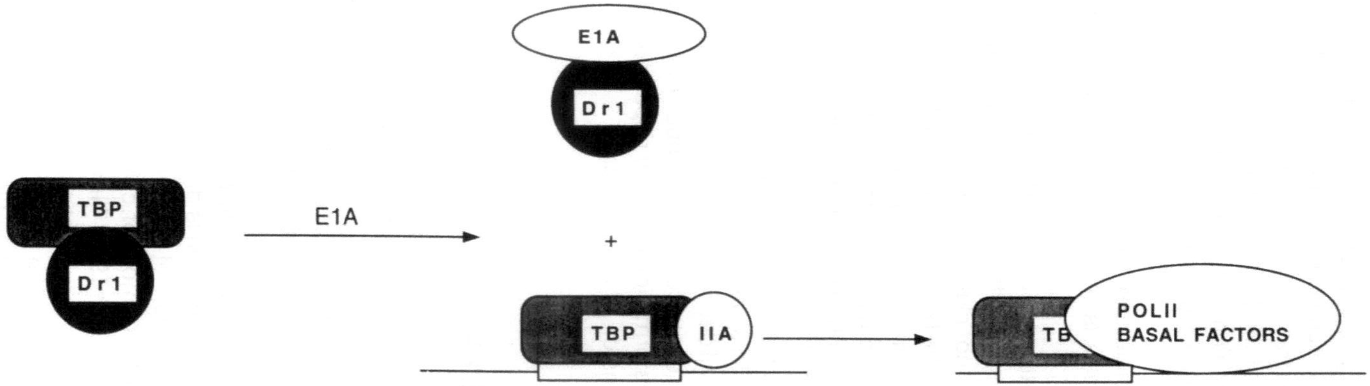

Fig. 4. Possible role of E1A in TATA-dependent transcriptional activation. The inhibitor Dr1 binds to TATA box-binding protein(*TBP*) and prevents interaction with the general transcriptional factor TFIIA. Accordingly, an initiation complex cannot form. In the presence of E1A, the TBP–Dr1 complex is disrupted by the interaction of E1A with Dr1, resulting in TBP displacement. The TBP factor is thus free to participate in initiation complex formation

may result from E1A binding to and preventing co-activator proteins from functioning. However, if E1A is tethered to a promoter by virtue of a fused heterologous DNA-binding domain, the co-activator would be promoter bound and free to activate. Therefore, in the context of the 243R protein the N-terminal activation domain may not play any role in activation, but rather be involved in the repression process. However, it cannot be ignored that the 289R protein has the ability to be tethered to specific promoters via interaction with DNA-binding factors; under these circumstances it is possible that the N-terminal activation domain is important. Obviously, further analysis is required before the significance of this finding becomes evident.

It should be clear from the above sections that activation by E1A can occur via numerous different pathways and can involve many different *cis*-acting elements and DNA-binding transcription factors. What is not clear, however, is the relative importance of the different activation events to the biological function of E1A. The CR3-dependent activation of the viral early genes is obviously crucial to efficient virus replication. Activation of the E2F factor through the action of the CR1 and CR2 regions of E1A is likewise crucial both to ensuring efficient expression of the E2A promoter and, in all probability, stimulation of S phase in quiescent cells. The importance of the other activation events, however, particularly those involving the extreme N terminus of E1A, is less clear. This same region is necessary for E1A to transform primary cells. However, as I will describe below, it has also been shown to be the region that mediates the repression of a number of target promoters. Thus the role of this region in transformation could be dependent on either the activation or the repression functions or, of course, both.

3 Transcriptional Repression by E1A

Over the years, E1A has been demonstrated to repress the efficient transcription of a variety of viral and cellular genes (for review see Bayley and Mymryk 1994). This was initially demonstrated for the viral enhancers of SV40 and polyomavirus and the enhancer of the E1A gene itself. Subsequently, it was extended to a large number of cellular enhancers and promoters, including those that control the expression of differentiation-specific genes. It is not clear in all cases, particularly those connected to differentiation, whether the repression is direct or whether it is an indirect consequence of some other function of E1A. It is known, for instance, that E1A blocks the differentiation of a number of cell lines, including rat PC12 pheochromocytoma cells (Maruyama et al. 1987; Heasley et al. 1991; Boulukos and Ziff 1993; Kalman et al. 1993) and a variety of myoblast cell lines (Webster et al. 1988; Enkemann et al. 1990). The loss of expression of differentiation genes may result from such a differentiation block. On the other hand, the differentiation block may result, at least in part, from the repression of differentiation-specific genes. This latter possibility, which is clearly of great importance and interest, is supported by studies demonstrating that E1A can affect the activity of certain regulatory bHLH proteins implicated in the induction and control of muscle differentiation (BRAUN et al. 1992; CARUSO et al. 1993).

Until recently, the mechanism of E1A repression has been largely neglected, particularly in comparison to the efforts devoted to understanding E1A activation. This was somewhat surprising, since it was clear from mutagenesis studies that the regions of E1A involved in repression closely correlated with regions required for transformation (MORAN and MATHEWS 1987; BOULANGER and BLAIR 1991; SHENK and FLINT 1991; DYSON and HARLOW 1992; JONES 1990; BAYLEY and MYMRYK 1994). In particular, the extreme N-terminal region of E1A was consistently found to be critical, either alone or in conjunction with CR2 region. In no case was repression found to require CR3. The N-terminal region was known to bind to a cellular 300-kDa protein called p300, and in some cases a very good correlation was found between p300 binding and repression. The p300 protein, therefore, was a strong candidate for mediating E1A repression. A cDNA encoding this protein has recently been described (ECKNER et al. 1994) and its structural and functional analysis has provided valuable clues to the possible mechanism of repression.

The p300 protein sequence shows features associated with transcriptional adaptors or co-activators, i.e. proteins implicated in providing a link between DNA-binding transcription factors and the basic transcription machinery. In particular it has a bromodomain, a conserved motif implicated in mediating protein–protein interactions. The most striking finding, however, is that p300 is remarkably similar to CBP, a transcriptional activator which binds to the transcription factor CREB (CHRIVIA et al. 1993; ARANY et al. 1994). Indeed, the similarity is such that it is highly unlikely that p300 will not have a very similar function to CBP.

CBP was originally isolated as a protein that interacted specifically with the *trans*-activation domain of CREB (CHRIVIA et al. 1993). What was particularly exciting about this interaction was that it was dependent upon phosphorylation of CREB on Ser-133, a phosphorylation mediated by protein kinase A. The phosphorylation is critical for CREB function, thus explaining how CREB activity is regulated by intracellular cAMP levels. Therefore, if CBP represented a co-activator, a plausible picture would emerge whereby phosphorylation of CREB regulated its interaction with CBP and hence its link to the basic transcription machinery. More recent studies have supported such a picture. CBP augments transcriptional activation by CREB, and antibodies against CBP repress transcription via the CRE element (KWOK et al. 1994; ARIAS et al. 1994). Clearly, therefore, CREB-mediated stimulation requires CBP, consistent with its role as a co-activator. Furthermore, CBP directly binds to TFIIB (KWOK et al. 1994) through a C-terminal region that has the ability to activate transcription when fused to the GAL4 DNA-binding domain. In all probability, p300 is also a co-activator. This being the case, it suggests a model for E1A-mediated repression (Fig. 5). By binding to p300 it might prevent it from participating in interactions with activator proteins, thereby repressing its coactivator function. Promoters or enhancers regulated by DNA-binding proteins requiring this co-activator would therefore be repressed. One possible problem with such a model is that E1A represses a wide

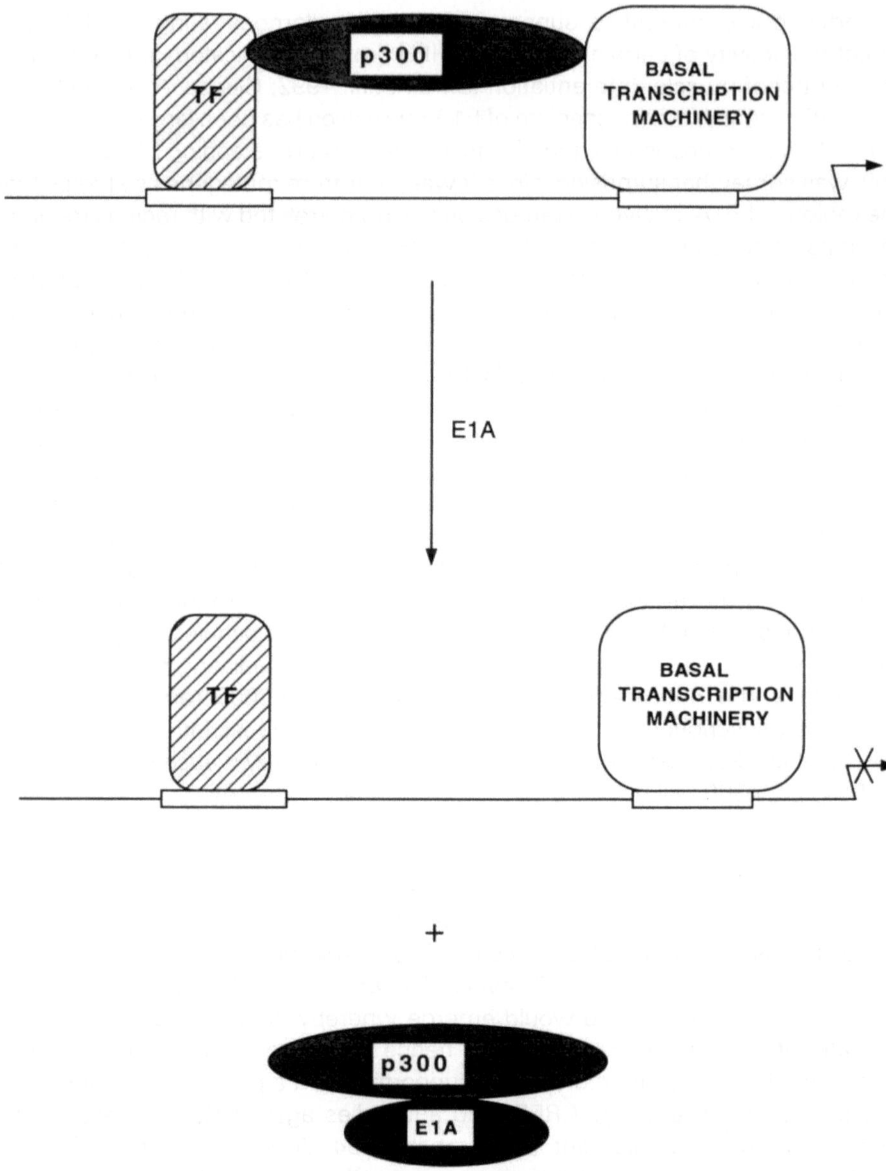

Fig. 5. Model for E1A-mediated repression of transcription. According to the model, the p300 protein functions as a co-activator protein, bridging the gap between certain DNA-binding transcription factors and the basic transcription machinery. This possible role for p300 stems from its considerable similarity to CBP, which has been demonstrated to be co-activator. In the presence of E1A, a p300–E1A complex is formed; as a result p300 is no longer available to bind to target transcription factors *TF* and thus unable to function as a co-activator. Promoters that rely on p300 activity will therefore be repressed

variety of different promoters or enhancers. Could they all be dependent upon the p300 coactivator? Again, studies with CBP provided some comfort. Antibodies against CBP not only prevented transcriptional activation via the CRE element, but also the SRE and TRE elements as well (ARIAS et al. 1994). Thus CBP may not be a specific co-activator of CREB and may indeed be required for the activity of a variety of different activator proteins. In support of this possibility, CBP was found to directly bind to the c-Jun protein (ARIAS et al. 1994). Furthermore, the similarity between CBP and p300 indicates the existence of a family, the extent of which is unknown. Different family members may interact with different activator proteins, but they may all be able to interact with, and therefore be antagonised by, E1A. It may therefore be possible to accommodate the promiscuity of E1A repression within the simple model outlined above. The cloning of p300 represents an important step in understanding E1A repression and should ensure rapid progress over the next few years.

The p300 protein may not be the only cellular protein involved in repression. It is generally agreed that the first exon of the 12S mRNA encodes the repression activities of E1A. However, some discrepancies exist as to the exact domains within this region that are critical. As described above, in some studies there is a good correlation with p300 binding. However, other studies have described mutants that can no longer associate with p300 but nevertheless can still repress certain transcriptional units (BONDESSON et al. 1994; WANG et al. 1993). It is not clear at this time whether these discrepancies represent differences in the experimental systems being employed or genuinely represent different and multiple modes of repression. If different modes are operating it can almost be guaranteed that they also involve interaction between the E1A protein and specific cellular proteins; after all, every function of E1A that has been examined in detail is mediated by protein–protein interactions. There may therefore be novel proteins to be discovered that interact with E1A. Indeed, a p400 protein has been described whose identity is, as yet, unknown (HOWE and BAYLEY 1992). A likely prediction would be that other proteins interacting with the N terminus and CR1 will have properties similar to p300 and may be family members. Different members may regulate the transcription of different genes and may have subtle differences in their binding characteristics with E1A. Precedence for this is seen with the Rb family, since specific point mutations in E1A have differential effects on pRb and p107 binding (WANG et al. 1993).

References

Abdel-Hafiz HAM, Chen C-Y, Marcell T, Kroll DJ, Hoeffler JP (1993) Structural determinants outside of the leucine zipper influence the interactions of CREB and ATF2—interaction of CREB with ATF-2 blocks E1a–ATF-2 complex formation. Oncogene 8: 1161–1174

Agoff SN, Wu B (1994) CBF mediates adenovirus E1a trans-activation by interaction at the C-terminal promoter targeting domain of conserved region 3. Oncogene 9: 3707–3712

Angel P, Imagawa M, Chiu R, Stein B, Imbra RJ, Rahmsdorf HJ, Jonat C, Herrlich P, Karin M (1987) Phorbol ester-inducible genes contain a common cis element recognized by a TPA-modulated trans-acting factor. Cell 49: 729–739

Arany Z, Sellers WR, Livingston DM, Eckner R (1994) E1A associated p300 and CREB-associated CBP belong to a conserved family of coactivators. Cell 77: 799–800

Arias J, Alberts AS, Brindle P, Claret FX, Smeal T, Karin M, Feramisco J, Montminy M (1994) Activation of cAMP and mitogen responsive genes relies on a common nuclear factor. Nature 370: 226–229

Bayley ST, Mymryk JS (1994) Adenovirus E1A proteins and transformation. Int J Onco 5: 425–444

Benbrook DM, Jones NC (1990) Heterodimer formation between CREB and Jun proteins. Oncogene 5: 295–302

Berk AJ (1986) Adenovirus promoters and E1A transactivation. Annu Rev Genet 20: 45–79

Blake MC, Azizkhan JC (1989) Transcription factor E2F is required for efficient expression of the hamster dihydrofolate reductase gene in vitro and in vivo. Mol Cell Biol 9: 4994–5002

Bondesson M, Mannervik M, Akusjärvi G, Svensson C (1994) An adenovirus E1A transcriptional repressor domain functions as an activator when tethered to a promoter. Nucleic Acids Res 22: 3053–3060

Boulanger PA, Blair GE (1991) Expression and interactions of human adenovirus oncoproteins. Biochem J 275: 281–299

Boulukos KE, Ziff EB (1993) Adenovirus 5 E1A disrupts the neuronal phenotype and growth factor responsiveness of PC12 cells by a conserved region 1-dependent mechanism. Oncogene 8: 237–248

Boyer TG, Berk AJ (1993) Functional interaction of adenovirus E1A with holo-TFIID. Genes Dev 7: 1810–1823

Braun T, Bober E, Arnold HH (1992) Inhibition of muscle differentiation by the adenovirus E1a-protein—repression of the transcriptional activating function of the HLH protein Myf-5. Genes Dev 6: 888–902

Buchkovich K, Duffy LA, Harlow E (1989) The retinoblastoma protein is phosphorylated during specific phases of the cell cycle. Cell 58: 1097–1105

Caruso M, Martelli F, Giordano A, Felsani A (1993) Transcription and protein function by the transforming domains of the adenovirus E1A oncoprotein. Oncogene 8: 267–278

Chatton B, Bocco JL, Gaire M, Hauss C, Reimund B, Goetz J, Kedinger C (1993) Transcriptional activation by the adenovirus larger E1a product is mediated by members of the cellular transcription factor ATF family which can directly associate with E1a. Mol Cell Biol 13: 561–570

Chellappan SP, Hiebert S, Mudryj M, Horowitz JM, Nevins JR (1991) The E2F transcription factor is a cellular target for the RB protein. Cell 65: 1053–1061

Chen P-L, Scully P, Shew J-Y, Wang JYJ, Lee W-H (1989) Phosphorylation of the retinoblastoma gene product is modulated during the cell cycle and cellular differentiation. Cell 58: 1193–1198

Chrivia JC, Kwok RPS, Lamb N, Hagiwara M, Montminy MR, Goodman RH (1993) Phosphorylated CREB binds specifically to the nuclear protein CBP. Nature 365: 859–866

Culp JS, Webster LC, Friedman DJ, Smith CL, Huang W-J, Wu FY-H, Rosenberg M, Ricciardi RP (1988) The 289-amino acid E1A protein of adenovirus binds zinc in a region that is important for trans-activation. Proc Natl Acad Sci USA 85: 6450–6454

Dalton S (1992) Cell cycle regulation of the human cdc2 gene. EMBO J 11: 1797–1804

DeCaprio JA, Ludlow JW, Lynch D, Furukawa Y, Griffin J, Piwnica-Worms H, Huang C-M, Livingston DM (1989) The product of the retinoblastoma susceptibility gene has properties of an cell cycle regulatory element. Cell 58: 1085–1095

Dérijard B, Hibi M, Wu I-H, Barrett T, Su B, Deng T, Karin M, Davies RJ (1994) JNK1: a protein kinase stimulated by UV light and Ha-Ras that binds and phosphorylates the c-Jun activation domain. Cell 76: 1025–1037

Devary Y, Gottlieb RA, Lau LF, Karin M (1991) Rapid and preferential activation of the c-jun gene during the mammalian UV response. Mol Cell Biol 11: 2804–2811

Dyson N, Harlow E (1992) Adenovirus E1A targets key regulators of cell proliferation. Cancer Surv 12: 161–195

Eckner R, Ewen ME, Newsome D, Gerdes M, DeCaprio JA, Lawrence JB, Livingston DM (1994) Molecular cloning and functional analysis of the adenovirus E1A-associated 300 kD protein (p300) reveals a protein with properties of a transcriptional adaptor. Genes Devel 8: 869–884

Egan C, Yee S-P, Ferguson B, Rosenberg M, Branton PE (1987) Binding of cellular polypeptides to human adenovirus type 5 E1A proteins produced in Escherichia coli. Virology 160: 292–296

Egan C, Jelsma TN, Howe JA, Bayley ST, Ferguson B, Branton PE (1988) Mapping of cellular protein-binding sites on the products of early-region 1A of human adenovirus. Mol Cell Biol 8: 3955–3959

Enkemann SA, Konieczny SF, Taparowsky EJ (1990) Adenovirus 5 E1A represses muscle-specific enhancers and inhibits expression of the myogenic regulatory factor genes, MyoD1 and myogenin. Cell Growth Differ 1: 375–382

Ewen ME, Xing Y, Lawrence JB, Livingston DM (1991) Molecular cloning, chromosomal mapping, and expression of the cDNA for p107, a retinoblastoma gene product-related protein. Cell 66: 1155–1164

Fagan R, Flint KJ, Jones N (1994) Phosphorylation of E2F-1 modulates its interaction with the retinoblastoma gene product and the adenoviral E4 19kDa protein. Cell 78: 799–811

Flemington EK, Speck SH, Kaelin WG Jr (1993) E2F-1-mediated transactivation is inhibited by complex formation with the retinoblastoma susceptibility gene product. Proc Natl Acad Sci USA 90: 6914–6918

Flint KJ, Jones NC (1991) Differential regulation of three members of the ATC/CREB family of DNA-binding proteins. Oncogene 6: 2019–2026

Geisberg JV, Lee WS, Berk AJ, Ricciardi RP (1994) The zinc finger region of the adenovirus E1A transactivating domain complex with the TATA box binding protein. Proc Natl Acad Sci USA 91: 2488-2492

Habener JF (1990) Cyclic AMP response element binding proteins: a cornucopia of transcription factors. Mol Endocrinol 4: 1087–1094

Hagmeyer BM, Konig H, Herr I, Offringa R, Zantema A, van der Eb AJ, Herrlich P, Angel P (1993) Adenovirus E1A negatively and positively modulates transcription of AP-1 dependent genes by dimer-specific regulation of the DNA binding and transactivation activities of Jun. EMBO J 12: 3559–3572

Hai T, Curran T (1991) Cross-family dimerization of transcription factors Fos/Jun and ATF/CREB alters DNA binding specificity. Proc Natl Acad Sci USA 88: 3720–3724

Hai T, Liu F, Coukos WJ, Green MR (1989) Transcription factor ATF cDNA clones: an extensive family of leucine zipper proteins able to selectively form DNA-binding heterodimers. Genes Dev 3: 2083–2090

Hardy S, Shenk T (1989) E2F from adenovirus-infected cells binds co-operatively to DNA containing two properly orientated and spaced recognition sites. Mol Cell Biol 9: 4495–4506

Hardy S, Engel DA, Shenk T (1989) An adenovirus early region 4 gene product is required for induction of the infection-specific form of cellular E2F activity. Genes Dev 3: 1062–1074

Harlow E, Whyte P, Franza BR Jr, Schley C (1986) Association of adenovirus early-region 1A proteins with cellular polypeptides. Mol Cell Biol 6: 1579–1589

Heasley LE, Benedict S, Gleavy J, Johnson GL (1991) Requirement of the adenovirus E1A transformation domain 1 for inhibition of PC12 cell neuronal differentiation. Cell Regul 2: 479–489

Helin K, Harlow E (1993) The retinoblastoma protein as a transcriptional repressor. Trends Cell Biol 3: 43–46

Helin K, Lees JA, Vidal M, Dyson N, Harlow E, Fattaey A (1992) A cDNA encoding a pRB-binding protein with properties of the transcription factor E2F. Cell 70: 337–350

Helin K, Harlow E, Fattaey A (1993) Inhibition of E2F-1 transactivation by direct binding of the retinoblastoma protein. Mol Cell Biol 13: 6501–6508

Herr I, van Dam H, Angel P (1994) Binding of promoter-associated AP-1 is not altered during induction and subsequent repression of the c-jun promoter by TPA and UV irradiation. Carcinogenesis 15: 1105–1113

Hoeffler JP (1992) Structure/function relationships of CREB/ATF proteins. Invest Dermatol 98: 21S–28S

Horikoshi N, Maguire KJ, Kralli A, Maldonado E, Reinberg D, Weinmann R (1991) Direct interaction between adenovirus E1A protein and the TATA box binding transcription factor IID. Proc Natl Acad Sci USA 88: 5124–5128

Howe JA, Bayley ST (1992) Effects of Ad5 E1A mutant viruses on the cell cycle in relation to the binding of cellular proteins including the retinoblastoma protein and cyclin A. Virology 186: 15–24

Hsu J-C, Laz T, Thomas L, Mohn KL, Taub R (1991) Identification of LRF-1, a leucine-zipper protein that is rapidly and highly induced in regenerating liver. Proc Natl Acad Sci USA 88: 3511–3515

Huang MM, Hearing P (1989) Adenovirus early region 4 encodes two gene products with redundant effects in lytic infection. J Virol 63: 2605–2615

Ivashkiv LB, Liou H-C, Kara CJ, Lamph WW, Verma IM, Glimcher LH (1990) mXBP CRE-BP2 and c-Jun form a complex which binds to the cyclic AMP, but not to the 12-otetradecanoylphorbol-13-acetate, response element. Mol Cell Biol 10: 1609–1621

Jelsma TN, Howe JA, Evelegh CM, Cunniff NF, Skiadopoulos MH, Floroff MR, Denman JE, Bayley ST (1988) Use of deletion and point mutants spanning the coding region of the adenovirus 5 E1A gene to define a domain that is essential for transcriptional activation. Virology 163: 494–502

Jones NC (1990) Transformation by the human adenoviruses. Semin Cancer Biol 1: 425–435

Kaelin WG Jr, Krek W, Sellers WR, De Caprio JA, Ajchenbaum F, Fuchs CS, Chittenden T, Li Y, Farnham PJ, Blanar MA, Livingston DM, Flemington EK (1992) Expression cloning of a cDNA encoding a retinoblastoma-binding protein with E2F-like properties. Cell 70: 351–364

Kalman D, Whittaker K, Bishop JM, O'Lague PH (1993) Domains of E1A that bind p105Rb, p130 and p300 are required to block nerve growth factor-induced neurite growth in PC12 cells. Mol Cell Biol 4: 353–361

Kim S-J, Wagner S, Liu F, O´ Reilly MA, Robbins PD, Green MR (1992) Retinoblastoma gene product activates expression of the human TGF-β2 gene through transcription factor ATF-2. Nature 358: 331–334

Kimelman D, Miller JS, Porter D, Roberts BE (1985) E1A regions of the human adenoviruses and of the highly oncogenic simian adenovirus 7 are closely related. J Virol 53: 399–409

Kraus VB, Moran E, Nevins JR (1991) Promoter-specific trans-activation by the adenovirus-e1a12s product involves separate E1a domains. Mol Cell Biol 12: 4391–4399

Kraus VB, Inostroza JA, Yeung K, Reinberg D, Nevins JR (1994) Interaction of the Dr1 inhibitory factor with the TATA binding protein is disrupted by adenovirus E1A. Proc Natl Acad Sci USA 91: 6279–6282

Kwok RPS, Lundblad JR, Chrivia JC, Richards JP, Bächinger HP, Brennan RG, Roberts SGE, Green MR, Goodman RH (1994) Nuclear protein CBP is a coactivator for the transcription factor CREB. Nature 370: 223–226

Kyriakis JM, Banerjee P, Nikolakaki E, Dai T, Rubie EA, Ahmad MF, Avruch J, Woodgett JR (1994) The stress-activated protein kinase subfamily of c-Jun kinases. Nature 369: 156–160

Lam E W-F, Watson RJ (1993) An E2F-binding site mediates cell-cycle regulated repression of mouse B-myb transcription. EMBO J 12: 2705–2713

La Thangue NB (1994) DRTF1/E2F: and expanding family of heterodimeric transcription factors implicated in cell-cycle control. TIBS 19, 108–114

Lee KAW, Green MR (1987) A cellular transcription factor E4F1 interacts with an E1a inducible enhancer and mediates constitutive enhancer function in vitro. EMBO J 6: 1345–1353

Lee KAW, Hai TY, SivaRaman L, Thimmappaya B, Hurst HC, Jones NC, Green MR (1987) A cellular protein, activating transcription factor, activates transcription of multiple E1A-inducible adenovirus early promoters. Proc Natl Acad Sci USA 84: 8355–8359

Lee KAW, Fink JS, Goodman RH, Green MR (1989) Distinguishable promoter elements are involved in transcriptional activation by E1a and cyclic AMP. Mol Cell Biol 9: 4390–4397

Lee WS, Kao CC, Bryant GO, Liu X, Berk AJ (1991) Adenovirus E1A activation domain binds the basic repeat in the TATA box transcription factor. Cell 67: 365–376

Leff T, Elkaim R, Goding CR, Jalinot O, Sassone-Corsi P, Perricaudet M, Kedinger C, Chambon P (1984) Individual products of the adenovirus 12S and 13S Ela mRNAs stimulate viral EIIA and EIII expression at the transcriptional level. Proc Natl Acad Sci USA 81: 4381–4385

Lillie JW, Green MR (1989) Transcription activation by the adenovirus E1a protein. Nature 338: 39–44

Lillie JW, Green M, Green MR (1986) An adenovirus E1A protein region required for transformation and transcriptional repression. Cell 46: 1043–1051

Liu F, Green MR (1990) A specific member of the ATF transcription factor family can mediate transcription activation by the adenovirus E1A protein. Cell 61: 1217–1224

Liu F, Green MR (1994) Promoter targeting by adenovirus E1a through interaction with different cellular DNA-binding domains. Nature 368: 520–525

Livingstone C, Patel G, Jones NC (1995) ATF2 contains a phosphorylation dependent transcriptional activation domain. EMBO J (in press)

Lum LSY, Hsu S, Vaewhongs M, Wu B (1992) The hsp70 gene CCAAT-binding factor mediates transcriptional activation by the adenovirus E1a protein. Mol Cell Biol 12: 2599–2605

MacGregor PF, Abate C, Curran T (1990) Direct cloning of leucine zipper proteins: Jun binds co-operatively to the CRE with CRE-BP1. Oncogene 5: 451–458

Maekawa T, Matsuda S, Fujisawa J, Yoshida M, Ishii S (1991) Cyclic AMP response element-binding protein, CRE-BP1, mediates the E1A-induced but not the Tyr-induced trans-activation. Oncogene 6: 627–632

Manohar CF, Kratochvil J, Thimmappaya B (1990) The adenovirus EII early promoter has multiple EIA-sensitive elements, two of which function cooperatively in basal and virus-induced transcription. J Virol 64: 2457–2466

Marshall CJ (1994) MAP kinase kinase kinase, MAP kinase kinase and MAP kinase. Curr Opin Genet Dev 4: 82–89

Martin KJ, Lillie JW, Green MR (1990) Evidence for interaction of different eukaryotic transcriptional activators with distinct cellular targets. Nature 346: 147–152

Maruyama K, Schiavi SC, Huse W, Johnson GL, Ruley HE (1987) myc and E1A oncogenes alter the responses of PC12 cells to nerve growth factor and block differentiation. Oncogene 1: 361–367
Matsuda S, Maekawa T, Ishii S (1991) Identification of the functional domains of the transcriptional regulator CRE-BP1. J Biol Chem 266: 18188–18193
Means AL, Slansky JE, McMahon SL, Knuth MW, Farnham PJ (1992) The HIp1 binding site is required for growth regulation of the dihydrofolate reductase gene promoter. Mol Cell Biol 12: 1054–1063
Mihara K, Cao X-R, Yen A, Chandler S, Driscoll B, Murphree AL, T'Ang A, Fung Y-KT (1989) Cell cycle-dependent regulation of phosphorylation of the human retinoblastoma gene product. Science 246: 1300–1303
Montminy MR, Sevarino KA, Wagner JA, Mandel G, Goodman RH (1983) Identification of a cyclic-AMP-responsive element within the rat somatostatin gene. Proc Natl Acad Sci USA 86: 6682–6686
Moran E, Mathews MB (1987) Multiple functional domains in the adenovirus E1A gene. Cell 48: 177–178
Moran E, Zerler B, Harrison TM, Mathews MB (1986) Identification of separate domains in the adenovirus E1A gene for immortalization activity and the activation of early virus genes. Mol Cell Biol 6: 3470–3480
Mudryj M, Devoto SH, Hiebert SW, Hunter T, Pines J, Nevins JR (1991) Cell cycle regulation of the E2F transcription factor involves an interaction with cyclin. Cell 65: 1243–1253
Muegge K, Vila M, Gusella GL, Musso T, Herrlich P, van der Eb A, Rahmsdorf HJ (1992) Photochem Photobiol 55: 409–415
Murthy SCS, Bhat GP, Thimmappaya B (1985) Adenovirus EIIA early promoter: transcriptional control elements and induction by the pre-early EIA gene, which appears to be sequence independent. Proc Natl Acad Sci USA 82: 2230–2234
Nevins JR (1994) Cell cycle targets of the DNA tumor viruses. Curr Opin Genet Dev 4: 130–134
Nevins JR (1981) Mechanism of activation of early viral transcription by the adenovirus E1A gene product. Cell 26: 213–220
Nevins JR (1992) E2F: a link between the Rb tumor suppressor protein and viral oncoproteins. Science 258: 424–429
Ogris E, Rotheneder H, Mudrak I, Pichler A, Wintersberger E (1993) A binding site for transcription of murine thymidine kinase by polyomavirus large T antigen and plays an important role in growth regulation of the gene. J Virol 67: 1765–1771
Pearson BE, Nasheuer H-P, Wang TS-F (1991) Human DNA polymerase α gene: sequences controlling expression in cyclic and serum-stimulated cells. Mol Cell Biol 11: 2081–2095
Pelech SL, Sanghera JS (1992) Mitogen-activated protein kinases: versatile transducers for cell signaling. Trends Biochem Sci 17: 223–238
Reichel R, Neill SD, Kovesdi I, Simon MC, Raychaudhuri P, Nevins JR (1989) The adenovirus E4 gene, in addition to the E1A gene, is important for trans-activation of E2 transcription and for E2F activation. J Virol 63: 3643–3650
Roesler WJ, Vandenback GR, Hanson JR (1988) Cyclic AMP and the induction of eukaryotic gene transcription. J Biol Chem 263: 9063–9066
Schneider JF, Fisher F, Goding CR, Jones NC (1987) Mutational analysis of the adenovirus E1A gene: the role of transcriptional regulation in transformation. EMBO J 6: 2053–2060
Scholer HR, Ciesiolka T, Gruss P (1991) A nexus between Oct-4 and E1A: implications for gene regulation in embryonic stem cells. Cell 6: 291–304
Schwarz JK, Devoto SH, Smith EJ, Chellappan SP, Jakoi L, Nevins JR (1993) Interactions of the p107 and Rb proteins with E2F during the cell proliferative response. EMBO J 12: 1013–1020
Shan B, Zhu X, Chen P-L, Durfee T, Yang Y, Sharp D, Lee W-H (1992) Molecular cloning of cellular genes encoding retinoblastoma-associated proteins: identification of a gene with properties of the transcription factor E2F. Mol Cell Biol 12: 5620–5631
Shenk T, Flint J (1991) Transcriptional and transforming activities of the adenovirus E1A proteins. Adv Cancer Res 57: 47–85
Shirodkar S, Ewen M, DeCaprio JA, Morgan J, Livingston DM, Chittenden T (1992) The transcription factor E2F interacts with the retinoblastoma product and a p107-cyclin A complex in a cell cycle-regulated manner. Cell 68: 157–166
Simon MC, Fisch TM, Benecke BJ, Nevins JR, Heintz N (1988) Definition of multiple, functionally distinct TATA elements, one of which is a target in the hsp70 promoter for E1A regulation. Cell 52: 723–729
Stein B, Angel P, van Dam H, Ponta H, Herrlich P, van der Eb A, Rahmsdorf HJ (1992) Photochem Photobiol 55: 409–415
Van Dam H, Offringa R, Meijer I, Stein B, Smits AM, Herrlich P, Bos JL, van der Eb AJ (1990) Differential effects of the adenovirus E1A oncogene on members of the AP-1 transcription factor family. Mol Cell Biol 10: 5857–5864

Van Dam H, Duyndam M, Rottier R, Bosch A, de Vries-Smits L, Herrlich P, Zantema A, Angel P, van der Eb AJ (1993) Heterodimer formation of c-Jun and ATF-2 is responsible for induction of c-jun by the 243-amino acid adenovirus E1A protein. EMBO J 12: 479–487

Van Ormondt H, Maat J, Dijkema R (1980) Comparison of nucleotide sequences of the early E1a regions for subgroups A, B and C of human adenoviruses. Gene 12: 63–76

Wang H-GH, Rikitake Y, Carter MC, Yaciuk P, Abraham SE, Zerler B, Moran E (1993) Identification of specific adenovirus E1A N-terminal residues critical to the binding of cellular proteins and to the control of cell growth. J Virol 67: 476–488

Webster KA, Muscat GEO, Kedes L (1988) Adenovirus E1A products suppress myogenic differentiation and inhibit transcription from muscle-specific promoter. Nature 332: 553–557

Webster LC, Ricciardi RP (1991) Trans-dominant mutants of E1A provide genetic evidence that the zinc finger of the trans-activating domain binds a transcription factor. Mol Cell Biol 11: 4287–4296

Weeks DL, Jones NC (1985) Adenovirus E3-early promoter: sequences required for activation by E1A. Nucleic Acids Res 13: 5389–5402

White RJ, Khoo BC-E, Inostroza JA, Reinberg D, Jackson SP (1994) Differential regulation of RNA polymerase I, II, and III by the TBP- binding repressor Dr1. Science 266: 448–450

Yee S-P, Branton PE (1985) Detection of cellular proteins associated with human adenovirus type 5 early region 1A polypeptides. Virology 147: 142–153

Zajchowski DA, Boeuf H, Kedinger C (1985) The adenovirus-2 early EIIA transcription unit possesses two overlapping promoters with different sequence requirements for Ela-dependent stimulation. EMBO J 4: 1293–1300

Ziff EB (1990) Transcription factors: a new family gathers at the cAMP response site. Trends Genet 6: 69–72

Regulation of Viral and Cellular Gene Expression by E1A Proteins Encoded by the Oncogenic Adenovirus Type 12

D. BROCKMANN and H. ESCHE

1 Introduction

For more than 15 years human adenoviruses (Ad) have been a powerful tool for studying cellular processes such as regulation of gene expression, alternative splicing, polyadenylation, and replication. Especially the analysis of viral proteins encoded by early region 1A (E1A), which was shown to regulate transcription, has given many insights into how DNA viruses regulate their own and cellular gene expression. Moreover, E1A proteins have been the subject of extensive studies because of their ability to act as oncoproteins that cooperate with the adenovirus E1B gene products to transform rodent cells in culture and, in case of the

Institute of Molecular Biology (Cancer Research), University of Essen Medical School, Hufelandstr. 55, 45147 Essen, Germany

oncogenic adenoviruses (e.g., Ad12), to induce tumors in animals (GALLIMORE et al. 1974; GRAHAM et al. 1974a, b; HOUWELING 1980; JOCHEMSEN et al. 1982). The ability to promote oncogenic transformation and transcriptional regulation appear to be distinct activities of the E1A polypeptides (for review see MORAN and MATHEWS 1987). In oncogenic transformation, protein functions of region E1A are necessary to immortalize primary cells, whereas functions of region E1B are essential to obtain a fully transformed phenotype. The functions of region E1B can be substituted by specific cellular gene products, e.g., activated Ha-ras (BYRD et al. 1988; RULEY 1983). The reasons for the difference in oncogenicity of variant adenovirus serotypes are not yet understood.

The E1A proteins are by far the most extensively studied viral transcriptional regulators. All adenoviral promoters that have been examined respond to E1A proteins, but also specific cellular promoters are transcriptionally regulated. The most striking property to emerge from subsequent studies has been the wide range of promoters on which the E1A proteins can act. Although they do not exhibit sequence-specific DNA-binding activity, they do bind to DNA apparently independent of sequence (FERGUSON et al. 1985). The regulated promoters share no common sequence elements, implying that the *trans*-regulation is not mediated through interaction of E1A proteins with a specific E1A response promoter element.

In this review we will focus on the functions and on gene regulatory mechanisms of E1A gene products encoded by the oncogenic adenovirus serotype 12 and compare these functions with those of non-oncogenic serotypes.

2 Structural and General Properties of the Adenovirus Type 12 E1A Proteins

E1A, located at the leftmost end (0%–11.5%) of the adenovirus genome, is the first transcription unit to be expressed after infection of recipient cells (LEWIS and MATHEWS 1980; NEVINS et al. 1979). By alternative splicing of a common Ad12 E1A RNA precursor, at least six different mRNA are generated (13S, 12S, 11S, 10S, 9.5S, and 9S, according to their sedimentation coefficients; Fig. 1; PERRICAUDET et al. 1980; SAWADA and FUJINAGA 1980; BROCKMANN et al. 1990). All mRNA share the same 5'and 3' ends, but differ in the amount of internal sequences removed by splicing events. At least five distinct proteins of 266 amino acid residues (R) (13S mRNA), 235R (12S mRNA), 106R (11S and 10S mRNA), 52R (9.5S mRNA), and 53R (9S mRNA) are translated from the respective mRNA. The protein products translated from the 13S and 12S mRNA share the same N- and C-terminal sequences and differ only in a 31R internal, cysteine-rich segment, unique to the larger E1A protein. This segment represents one out of three E1A domains which are highly conserved among distantly related human adenoviruses (KIMELMANN et al. 1985), termed conserved regions 1, 2, and 3 (CR1, CR2, CR3).

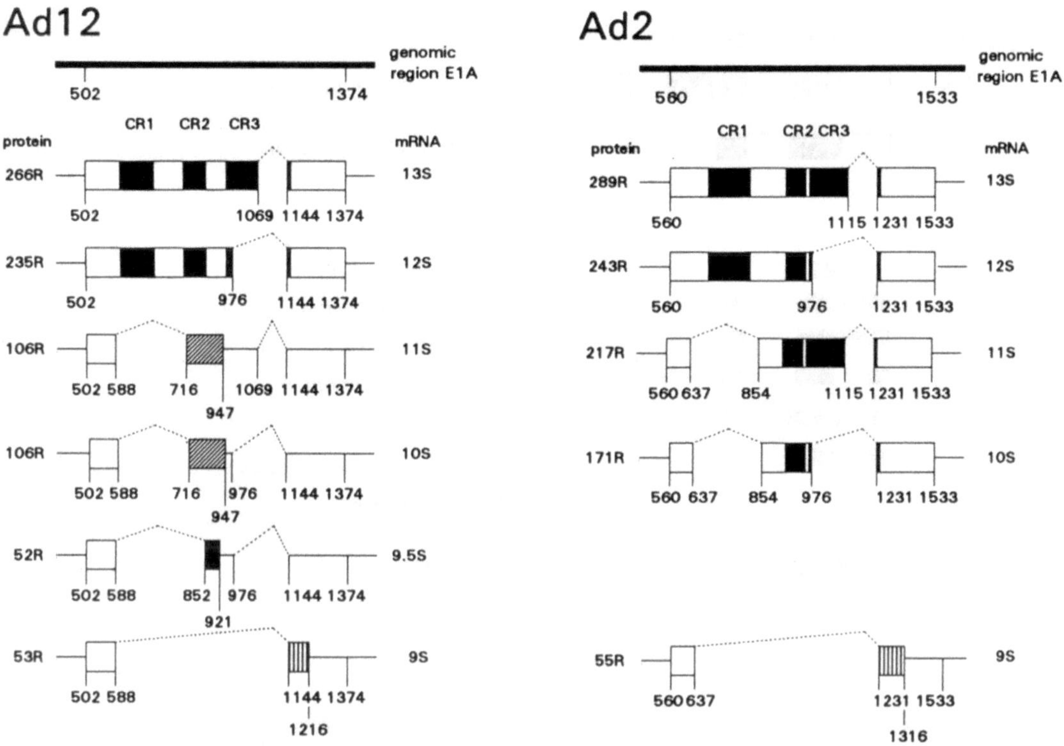

Fig. 1. Structure of adenovirus (*Ad*) type 12 and type 2 E1A mRNA and proteins. *Boxes* represent the proteins translated from the different mRNA. The *numbers below the lines* refer to the positions of nucleotides in the respective adenovirus genomes, indicating the common start codons, the splice junctions, and the first nucleotide of the respective stop codons. Frameshifts are indicated by different *hatching* of the boxes. *Black boxes* represent the conserved regions 1–3 (CR1,CR2,CR3)

As will be discussed later, CR1–3 express most *trans*-regulatory functions of both larger E1A proteins. Historically, it was believed that CR3, which is unique for the 266R protein, is responsible for *trans*-activation functions of the E1A polypeptides, whereas CR1 and CR2 are involved in *trans*-repression functions. Recent data, however, have demonstrated that CR1 and CR2 are also involved in the process of *trans*-activation of specific target genes (IKEDA and NEVINS 1993). In this case *trans*-activation occurs through a cellular transcription factor (E2F), which is activated by dissociating repressor proteins from the inactive transcription factor complex (see Sect. 3.1.2). In addition to *trans*-regulatory functions, the conserved regions are involved in several other properties of the adenoviral E1A proteins (Fig. 2).

In contrast to the proteins encoded by the 13S (266R), 12S (235R), and 9S mRNA (53R), whose structure is more or less conserved among all adenovirus serotypes (PERRICAUDET et al. 1980), protein products of the 11S (106R), 10S (106R), and 9.5S (52R) mRNA are not related between the oncogenic and nononcogenic adenoviruses (BROCKMANN et al. 1990). The smaller E1A proteins of

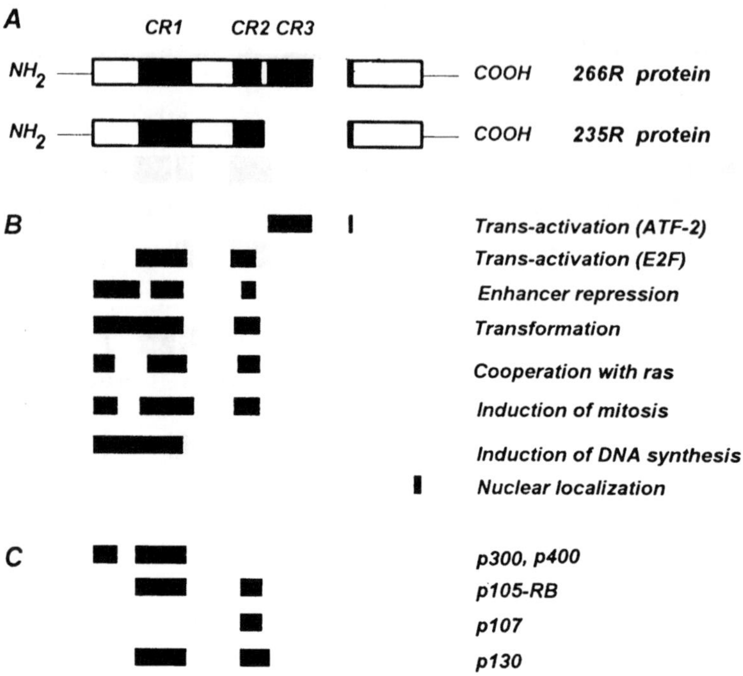

Fig. 2A–C. E1A polypeptides, functional domains, and protein regions essential for physical interactions with specific cellular proteins. **A** The 266R and 235R proteins encoded by the 13S and 12S mRNA of Ad12 E1A are schematically drawn as *rectangles*. Protein domains that are conserved between different adenovirus serotypes are indicated as CR1, CR2, and CR3. **B** Regions encoding functional protein domains, determined by mutation analyses. **C** Protein domains essential for physical interactions with specific cellular proteins

Ad2/Ad5 are translated in the same reading frame as the protein products of the 13S and 12S mRNA (Stephens and Harlow 1987; Ulfendahl et al. 1987). They only differ from the larger proteins in that the functional domain CR1 is removed by splicing events. In the case of Ad12, however, differential splicing leads to smaller proteins (106R, 52R) which are unrelated to the 266R and 235R proteins except for the first 29 amino acids. Therefore, in comparison with the non-oncogenic adenovirus serotypes, all smaller Ad12 E1A proteins differ in their carboxy termini due to the use of varied frame shifts after splicing. The 11S and 10S mRNA give rise to the same protein, which is due to a stop codon in front of the second splice junctions (Fig. 1).

As analyzed so far, the non-oncogenic adenovirus serotypes do not have the coding capacity for a protein equivalent to 52R (Brockmann et al. 1994). Ad5 (the non-oncogenic serotype) lacks nearly all the DNA sequences necessary for encoding the carboxy terminal end of a 52R-like polypeptide. The DNA sequence of Ad7 (a weakly-oncogenic serotype) has little homology (37%) to that of Ad12. In addition, a hypothetical 9.5S mRNA-equivalent splicing of the Ad5 or Ad7 E1A mRNA precursors would result in shifts of the reading frame, generating stop codons right at the beginning of the putative second exon.

The proteins translated from the 13S and 12S mRNA are post-translationally modified by phosphorylation at several sites (LUCHER et al. 1984, 1985, 1986; RICHTER et al. 1988; SPINDLER et al. 1984; TSUKAMOTO et al. 1986; YEE et al. 1983). Phosphorylation, however, appears not to be essential for the functions of *trans*-regulation or oncogenic transformation (DUMONT et al. 1989; RICHTER et al. 1988). In contrast to both larger E1A protein products, nothing is known about the post-translational modifications of the smaller Ad12 E1A proteins. Although they carry consensus sequences for several phosphokinases (e.g., protein kinase C, casein kinase II), which might suggest that these proteins are phosphorylated in vivo, no experimental data have confirmed this assumption up to now. All E1A proteins of the non-oncogenic adenoviruses (except the polypeptide translated from the 9S mRNA) and the two larger ones of oncogenic adenovirus 12 carry a nuclear localization signal at the carboxy terminal end (KRIPPLE et al. 1985; MORAN and MATHEWS 1987) and have been localized in the nucleus of infected as well as transformed cells by immunofluorescense (KRIPPLE et al. 1986; MORAN and MATHEWS 1987). There is at present no direct proof for a nuclear localization of the smaller proteins of Ad12 due to the lack of functional specific antibodies.

Analysis of the expression of E1A during the lytic cycle has shown that 13S and 12S mRNA and their protein products are expressed both early and late after infection, whereas the 9S mRNA, which is almost completely absent early after infection, becomes the most abundant E1A mRNA species at late times (BERK and SHARP 1978; CHOW et al. 1979; KITCHINGMAN and WESTPHAL 1980; PERRICAUDET et al. 1980; SAWADA and FUJINAGA 1980; SPECTOR et al. 1978). The expression kinetics of the smaller mRNA and their respective proteins seem to be dependent on different adenovirus serotypes. STEPHENS and HARLOW (1987) and ULFENDAHL and colleagues (1987) have shown that the 11S and 10S mRNA of the non-oncogenic serotypes Ad2/Ad5 are expressed only at late times in the infectious cycle. We have shown for Ad12 that the 10S mRNA is transcribed early as well as late after infection, although the amount of 10S mRNA increases from early to late phase (BROCKMANN et al. 1990). The 9.5S mRNA was also found early and late after infection.

All Ad12 E1A mRNA (13S, 12S, 11S, 10S, and 9S) except the 9.5S mRNA have been detected in Ad12-transformed cells (BROCKMANN et al. 1990), such as HEK-12 cells (a cell line derived from primary human embryo kidney cells by transformation with the *Eco*RI-C fragment of Ad12; WHITTAKER et al. 1984) and HA 12/7 cells (a cell line derived from primary Syrian hamster embryo cells by transformation with Ad12 virus; FANNING and DOERFLER 1976; ZUR HAUSEN 1973). The reason why we were unable to detect the 9.5S mRNA species in Ad12-transformed cells is unclear; it might be due to its low abundancy in these cells.

3 Transcriptional Regulation by Adenovirus Type 12 and Type 2/5 E1A Proteins

Proteins encoded by E1A of adenoviruses are potent *trans*-regulators. In the lytic cycle they are essential for *trans*-activation of all other adenoviral transcription units (Berk et al. 1979; Jones and Shenk 1979; Nevins 1981) and induce the expression of specific cellular genes (for review see Shenk and Flint 1991). In addition, E1A proteins can repress enhancer-mediated gene expression (for review see Rochette-Egly et al. 1990). These *trans*-regulatory functions are mainly attributed to the protein products of the 13S and 12S mRNA (Flint and Shenk 1989; Lillie et al. 1986, 1987; Schneider et al. 1987). Several reports have demonstrated that CR3, which is unique for the 266R polypeptide encoded by the 13S mRNA, is essential for *trans*-activating functions (for review see Nevins 1993), whereas CR1, CR2 and the amino terminal end are involved in *trans*-repression functions (Jelsma et al. 1989; Lillie et al. 1986; Rochette-Egly et al. 1990; Schneider et al. 1987; Stein et al. 1990; Wang et al. 1993a). Recent experiments have shown, however, that E1A proteins can also act as *trans*-activators through functions of CR1 and CR2 (for review see Nevins 1993).

Functional analyses of the smaller proteins, translated from the 11S mRNA of Ad2 as well as the 9.5S mRNA of Ad12 have shown that they are also involved in *trans*-regulation of viral and specific cellular genes, at least in transient expression assays (Ulfendahl et al. 1987; Brockmann et al. 1994). No *trans*-regulatory functions have yet been described for the protein product of the 11S/10S mRNA of Ad12 and the 9S mRNA of both serotypes.

3.1 Cellular Transcription Factors: Targets for E1A-Mediated *Trans*-activation

The central question of the *trans*-activating activity of E1A proteins concerns the mechanism by which they can act to stimulate transcription from a diverse array of promoters. E1A-responsive promoters neither share common nucleotide sequence elements, nor do E1A polypeptides bind sequence specifically on their own to DNA, at least those translated from the 13S and 12S mRNA (Ferguson et al. 1985). Thus, *trans*-activation of specific target genes has to be an indirect effect, which might be due to alterations or modifications of one or more cellular transcription factors by E1A proteins. Such cellular transcription factors have been recognized by virtue of the presence of their binding sites in the E1A-responsive adenoviral promoters. Two of these factors are the activating transcription factor 2, ATF-2 (Lee et al. 1987), interacting with CR3, and the transcription factor E2F (Koveshdi et al. 1987), which is activated in a CR1- and CR2-dependent manner. Both examples will be discussed in more detail in the following.

3.1.1 Activating Transcription Factor-2, an Example of Transcriptional Activation Through CR3

ATF-2 belongs to a group of transcription factors called the ATF/CREB family (CHATTON et al. 1993; HAI et al. 1989), which consists of at least eight closely related proteins immunologically related to members of the Jun transcription factor family. They form homodimers, although several ATF proteins, but not all, form heterodimers with other members of this family. They bind as dimers to the ATF/CREB consensus sequence (5'-$T^G{}_T$ACGTCA-3') in promoter regions of specific target genes (e.g., adenoviral E3, E4, and E2e, specific cellular genes) and regulate their expression (for review see SHENK and FLINT 1991). Recent experiments have indicated that not all ATF transcription factors mediate *trans*-activation by E1A proteins (NOMURA et al. 1993). ATF-2, for example, but not ATF-1 is a target for E1A protein functions (FLINT and JONES 1991; LIU and GREEN 1990).

As mentioned above, ATF-binding sites are located in E1A protein-dependent E2e, E3, and E4 promoters. The consensus sequence for ATF-binding sites in these promoters (5'-$T^G{}_T$ACGTCA-3') is identical to that of a cellular factor called CREB, which mediates cyclic adenosine monophosphate (cAMP)-dependent induction of genes such as somatostatin, c-fos, and fibronectin (MIYAMOTO et al. 1985; MONTMINY and BILEZIKJIAN 1987; SILVER et al. 1987). The ATF elements are important for both basal and E1A-inducible transcription (for review see SHENK and FLINT 1991). Deletion of the ATF-binding site from the E3 promoter, for example, reduces the efficiency of E3 transcription tenfold (LEFF et al. 1985). Deletion of CR3 of the protein product translated from the 13S mRNA also results in the loss of its ability to *trans*-activate the E3 promoter (for review see MORAN and MATHEWS 1987), indicating that CR3 in cooperation with ATF is responsible for the *trans*-activation function. This assumption was confirmed by experiments, in which it was shown that a synthetic 49-amino acid peptide corresponding exactly to CR3 of Ad5 E1A was sufficient to stimulate expression from adenovirus promoters in vitro and in vivo (GREEN et al. 1988; LILLIE et al. 1987).

CR3 of Ad5 E1A consists of two distinct activities (LILLIE and GREEN 1989; LIU and GREEN 1990): (1) the amino terminal part (Ad5 E1A: amino acids 139–153) containing a transcriptional activating region, which is functionally analogous to that of a typical cellular activator; (2) the carboxy terminal portion (Ad5 E1A: amino acids 175–189) containing a promoter binding function which is required to direct the E1A protein to its natural targets. These regions are separated by a metal-binding domain of the type $CX_2CX_{13}CX_2C$ (C = cysteine; X = any amino acid), which is essential for both activities. LEE and colleagues (1991) have shown that the amino terminal part of CR3 interacts with the basic amino acid repeat in the carboxy terminus of the TATA box-binding factor TFIID, while others have demonstrated that the carboxy terminal domain of CR3 binds to the amino terminus of ATF-2 (FLINT and JONES 1991; LILLIE and GREEN 1989; LIU and GREEN 1990; ZU et al. 1991). From these results a model has been suggested in which CR3 forms a molecular bridge between ATF-2 and TFIID (Fig. 3). In this model CR3 functions as an adapter or coactivator mediating an interaction between the sequence-specific DNA-binding transcription factor ATF-2 and the TATA box-binding protein (TBP; LEE et al. 1991).

The model described above was set up for CR3 of the non-oncogenic serotypes Ad2/Ad5. Although there are no detailed analyses available for CR3 of the oncogenic serotype Ad12, it can be assumed that the mechanism of CR3-dependent gene expression is identical, due to the high homology between the CR3 of the oncogenic and non-oncogenic adenovirus serotypes (Fig. 4).

3.1.2 E2F, an Example of Transcriptional Activation Through CR1 and CR2

The E2F transcription factor complex was originally identified as a sequence-specific DNA-binding protein bound to the adenovirus E2 early (E2e) promoter (KOVESDI et al. 1986). Later on, E2F-binding sites were also found in the promoter regions of several cellular proliferation-related genes such as c-myc, N-myc, the dihydrofolate reductase gene (DHFR), the thymidine kinase gene, cdc2, or the Rb gene (BLAKE and AZIZKHAN 1989; DALTON 1992; HAMEL et al. 1992; HIEBERT et al. 1989; MUDRYJ et al. 1990; SLANSKY et al. 1993; THALMEIER et al. 1989).

E2F, also termed DRTF1 (SHIVJI and LA THANGUE 1991), is a multicomponent transcription factor (GIRLING et al. 1993; SHIVJI and LA THANGUE 1991), consisting of at least two DNA-binding proteins, e.g. E2F-1 and DP-1. They bind after heterodimerization sequence specifically to E2F sites in promoter regions of respective target genes and induce their expression (for review see LA THANGUE and JOHNSTON 1993). In cells in which E2F-dependent promoters are silent, the E2F transcription factor is bound in a complex with the 105-kDa protein product of the retinoblastoma (RB) gene (for review see NEVINS 1992). Bound in this protein complex, E2F is still able to bind to its recognition sequence but is transcriptionally inactive (HIEBERT et al. 1992; WEINTRAUB et al. 1992; ZAMANIAN and LA THANGUE

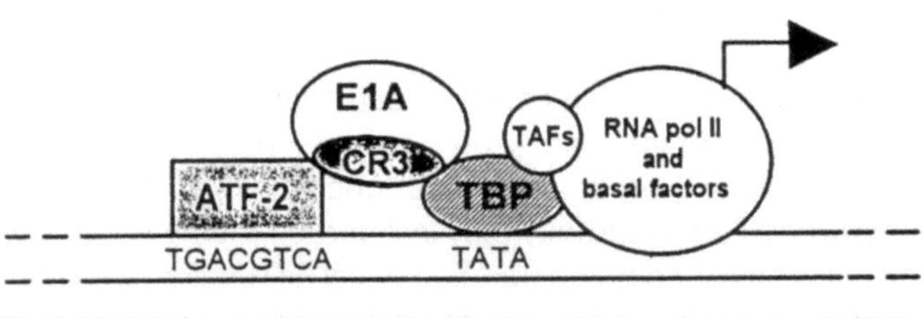

Fig. 3. Model of transcriptional activation through conserved region 3 (CR3) of the E1A protein encoded by the 13S mRNA (LEE et al. 1991). *ATF*, activating transcription factor II; *TBP*, TATA box-binding protein; *TAF*, TBP-associated factor; *RNA pol II*, RNA polymerase II

```
Ad5    140 - EEFVLDYVEHPGHGCRSCHYHRRNTGDPDIMCSLCYMRTCGMFVYSPVS - 188
             *** **  * *** * **  **  **  * ****** *   ** *****
Ad12   145 - EEFQLDHPELPGHNCKSCEHHRNSTGNTDLMCSLCYLRAYNMFIYSPVS - 193
```

Fig. 4. Comparison of the amino acid sequence of CR3 of the E1A 289R protein of adenovirus (*Ad*) type 5 and the E1A 266R polypeptide of Ad12. Amino acids (single letter code) are numbered according to their positions in the protein translated from the respective 13S mRNA. Identical residues are marked by *asterisks*. The Ad5 sequence originates from GREEN et al. (1988), and the Ad12 sequence from VAN ORMONDT and GALIBERT (1984)

1992). In addition, E2F interacts with the pRB-related protein p107, which itself is also bound—depending on the cell cycle — by different cyclins as well as protein kinase cdk2 (CAO et al. 1992; DEVOTO et al. 1992; LEES et al. 1992; MUDRYJ et al. 1991; SHIRODKAR et al. 1992). Several experiments have demonstrated that adenovirus E1A proteins translated from the 13S and 12S mRNA are able to disrupt the E2F/pRB and E2F/p107 complexes, releasing uncomplexed transcriptionally active E2F (BAGCHI et al. 1990; BANDARA and LA THANGUE 1991; CHELLAPPAN et al. 1991, 1992; MUDRYJ et al. 1991). The active E2F then *trans*-activates viral and cellular genes mentioned above (Fig. 5), which result in entering the S phase of the cell cycle (for review see NEVINS 1992). Dissociation of the E2F/pRB and E2F/p107 complexes is dependent on functional CR1 and CR2 of the adenovirus E1A proteins. For these E1A domains it has been shown to interact physically with the pRB and p107 proteins (DYSON et al. 1992; EGAN et al. 1988; RAYCHAUDHURI et al. 1991).

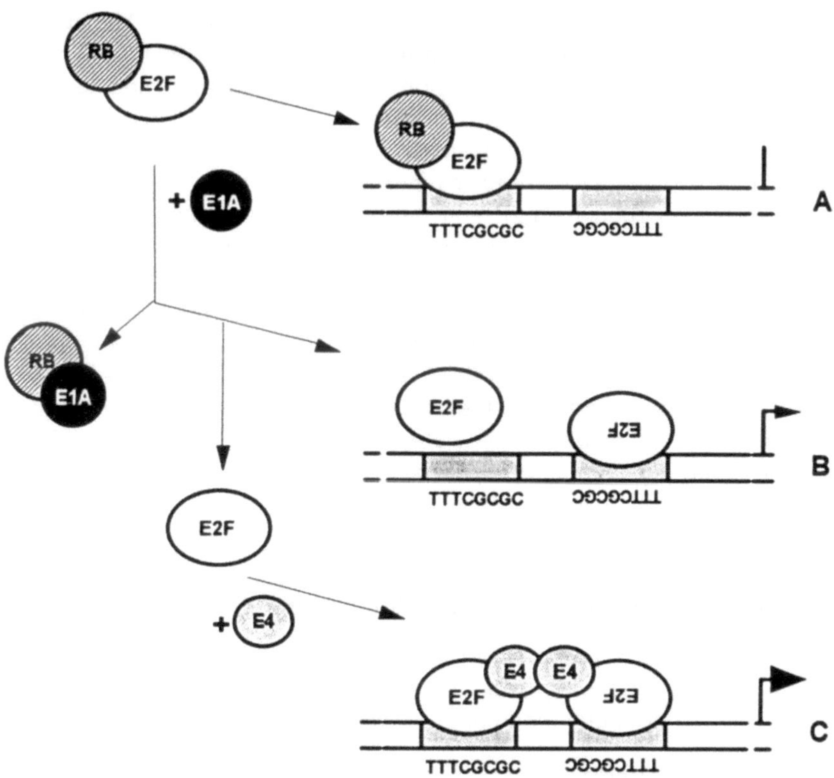

Fig. 5A–C. Model of transcriptional activation of the adenovirus type 2 E2 early promoter by region E1A and region E4 polypeptides. E1A proteins translated from the 13S and 12S mRNA release active E2F transcription factors (**B**) from a transcriptionally inactive E2F/pRB complex (**A**) by interacting with the pRB protein; 19-kDa polypeptides encoded by viral region E4 then enable two E2F transcription factors to bind cooperatively and stably to the two adjacent E2F recognition sequences, increasing the rate of transcription from this promoter

Interestingly, the E2e promoter of several adenoviruses (e.g., Ad2/Ad5) carry two inverted E2F recognition sequences, located 16 bp apart from each other (VAN ORMONDT and GALIBERT 1984; YEE et al. 1987), which can function independently from each other. It has been shown that these E2e promoters are *trans*-activated in a cooperative manner by E1A and adenoviral E4 protein (Fig. 5C; MARTON et al. 1990; RAYCHAUDHURI et al. 1990). While the E1A proteins release active E2F from an inactive, pRB- or p107-bound complex, 19-kDa proteins encoded by ORF6/7 of the viral region E4 enable two E2F transcription factors to bind cooperatively to the two adjacent E2F sites (NEILL et al. 1990). Stabilizing the E2F–DNA complex in this way, the E4 19-kDa polypetides increase the rate of transcription from the E2 gene (NEILL and NEVINS 1991; REICHEL et al. 1989). Not all adenovirus serotypes carry two cooperatively functioning E2F sites in their E2e promoter regions. One exception, for example, is the E2e promoter of the weakly oncogenic subtype Ad3 (HEYSEN et al. 1991).

For both the non-oncogenic adenovirus serotypes 2 and 5 and the oncogenic serotype 12 it has been shown that the proteins translated from the 13S and 12S mRNA are able to interact with the RB protein (BARBEAU et al. 1994; GRAND et al. 1989; PEEPER et al. 1992; H.-C. KIRCH, personal communication). In addition KIRCH and co-workers have demonstrated that both the 266R and 235R proteins of region E1A of Ad12 are able to release p107 from E2F in extracts prepared from HeLa or HL60 cells (personal communication), indicating that non-oncogenic and oncogenic serotypes regulate E2F activity in the same manner. In addition, YAMAGUCHI and colleagues (1992) have shown that the proteins translated from the 13S and 12S mRNA of Ad12 E1A *trans*-activate the promoter of the mouse proliferating cell nuclear antigen (PCNA) gene. The target sequence for *trans*-activation of this gene is located between nucleotides –21 and + 47, a region containing an E2F-binding site as well as other transcription factor recognition sequences. As *trans*-activation of the mouse PCNA gene is independent of functions of CR3, it has been speculated that the E2F transcription factor mediates this *trans*-activation. Further experiments, however, are necessary to prove this hypothesis.

3.1.3 Other Cellular Transcription Factors Involved in E1A-Responsive Gene Regulation

Although studied in less detail, a number of other sequence-specific cellular transcription factors have been reported to be involved in E1A protein *trans*-activation. They will be mentioned here, but without discussing all known data.

3.1.3.1 Activating Protein-1

Cellular AP-1 activity, first detected as transcription and DNA-binding activities, consists of several proteins belonging to the Fos and Jun families which interact with AP-1-binding sites as homo- and heterodimers (for review see ANGEL and KARIN 1991; CURRAN and FRANZA 1988). The AP-1 complex, which binds to the consensus sequence (5'-T$^{G}_{T}$A$^{C}_{G}$TCA-3') that differs by only one nucleotide from the consensus ATF/CREB site (5'-T$^{G}{}_{T}$ACGTCA-3'; ANGEL et al. 1987; LEE et al. 1987), can bind ATF/CREB sequences in vitro, although with reduced affinity (HAI et al. 1988; MÜLLER et al. 1989). This property, as well as the presence of an AP-1 element (E3/AP-1) important to both basal and E1A protein-induced transcription in the adenovirus E3 promoter (HURST and JONES 1987), suggests that AP-1 might be one component in *trans*-activation of viral early promoters. MÜLLER and colleagues (1989) have shown that cAMP treatment of Ad5-infected S49 cells induces the DNA-binding activity of a specific form of AP-1 and that a maximal induction of this activity depends directly on E1A proteins.

As a part of a series of experiments to identify cellular proteins that can bind to ATF/CREB sequences, BUCKBINDER et al. (1989) have identified a HeLa cell factor that stimulates in vitro transcription from the E3 promoter and binds to both its AP-1 and ATF/CREB sites. Moreover, an E3 promoter in which the normal ATF/CREB and AP-1 recognition sequences have been replaced by four copies of the E3/AP-1 sequence was *trans*-activated by E1A proteins even more efficiently than the wild-type E3 promoter. E3 promoter constructs containing two or four copies of the ATF/CREB consensus sequence were stimulated only twofold. These observations indicate, that under appropriate experimental conditions, E3/AP-1-binding sites are sufficient to confer induction of transcription by E1A proteins. The mechanism by which E1A proteins stimulate this transcription is not clear yet.

Another line of evidence comes from experiments in which it has been shown that AP-1 seems to be involved in the pathway resulting in the induction of transcription by the phorbol ester 12-O-tetradecanoylphorbol-13-acetate (TPA), which depends on protein kinase C (ANGEL et al. 1987). TPA treatment of HeLa cells cotransfected with an E3 CAT reporter gene and an E1A-expressing plasmid results in a significantly greater stimulation of CAT activity than is accomplished by either E1A proteins or TPA alone (BUCKBINDER et al. 1989). These findings indicate that synergistic induction of E3 expression by E1A proteins and TPA requires both the E3/AP-1 and ATF/CREB sequences of the E3 promoter, which are presumably recognized by different factors.

3.1.3.2 Major Late Transcription Factor

The adenovirus late promoter, also active during the early phase of infection and responsive to E1A *trans*-activation in vivo (FLINT and LEONG 1986), contains two principal elements, the TATA box and the major late promoter upstream element (MLP-UE) (HEN et al. 1982). The latter one is also found in several mammalian genes (CARTHEW et al. 1987; CHODOSH et al. 1987) and is the target of the major late transcription factor (MLTF; CARTHEW et al. 1985), also called upstream element factor (UEF) (MIYAMOTO et al. 1985) or USF (SAWADOGO and ROEDER 1985). Binding of the USF/MLTF to the MLP-UE leads to a three- to fivefold stimulation of the MLP in an in vitro transcription system prepared from HeLa cells, but an up to ten- to 20-fold stimulation by E1A proteins present in infected cell extracts (LEONG and BERK 1986), by bacterially synthesized E1A proteins (SPANGLER et al. 1987), or by a 49R synthetic peptide containing CR3 (GREEN et al. 1988; LILLIE et al. 1987).

Deletion of the MLP-UE sequence diminishes *trans*-activitation of the MLP in transfected cells (LEWIS and MANLEY 1985). An MLP-UE sequence was also found to be required for the inducibility of the adenovirus-associated virus P5 promoter by E1A proteins (CHANG et al. 1989).

3.1.3.3 E4F and EivF

The E4 promoter of adenoviruses is also bound by cellular factors which were reported to be involved in E1A-responsive *trans*-activation of this transcription unit. One of these is the cellular factor E4F, which binds to a sequence within the E4 promoter overlapping two ATF/CREB-binding sites (RAYCHAUDHURI et al. 1987; REICHEL et al. 1987). No E4F binding sequence has been found in other viral promoters, indicating that E4F and ATF/CREB are distinct factors. Another independently isolated factor not related to E4F is the factor EivF (CORTES et al. 1988). EivF binds to a sequence closely related to the ATF/CREB consensus sequence. For both factors have been shown to stimulate E4 transcription in a reconstituted system (CORTES et al. 1988; MERINO et al. 1989).

3.1.3.4 Transcription Factor Complex IID

Experiments in which regulatory elements upstream of the TATA box in the adenovirus promoters E1B (the SP-1 recognition site) and MLP (the MLP-UE) have been deleted have shown that the TATA element appears to be essential for *trans*-activation of these promoters (reviewed in FLINT and SHENK 1989; LEONG et al. 1988; SHENK and FLINK 1991; WU et al. 1987). The TATA element, present in the majority of eukaryotic RNA polymerase II promoters, has also been shown to be important in E1A protein *trans*-activation of specific cellular genes (GREEN et al. 1988; SIMON et al. 1988). It is recognized by the general transcription factor complex TFIID, which binding appears to be the first, and committing, step in the formation of a functional initiation complex (BURATOWSKI et al. 1989; NAKAJIMA et al. 1988). TFIID itself is the target of cellular proteins that specifically bind to upstream promoter elements.

Surprisingly, TFIID does not play a more general role in *trans*-activation by E1A proteins, as it was shown that certain TATA elements, including that of the SV40 early promoter, are refractory to stimulation by E1A proteins (SIMON et al. 1988), whereas others, such as that of the human hsp 70 gene, are E1A responsive. Similarly, substitution of the E1B TATA box with that from the viral E4 promoter significantly increases basal activity and virtually eliminates E1A protein-dependent stimulation of E1B transcription in infected cells (PEI and BERK 1989). The variance in the response to promoters carrying different TATA sequences is probably due to functional variations among the TATA elements and different additional factors (TAF) associated with different multiprotein transcription factor complexes. In vivo *trans*-activation experiments with E1A polypeptides mutated within CR3 as well as activation of in vitro transcription from the MLP by a CR3 peptide have demonstrated that CR3 is involved in *trans*-activation through the TATA element (GREEN et al 1988; LOEWENSTEIN and GREEN 1989).

3.1.3.5 Transcription Factor Complex IIIC

Nuclear extracts from adenovirus-infected cells harvested early in infection and from adenovirus-transformed cells have demonstrated a markedly increased capacity for transcription by RNA polymerase III of exogenous adenovirus virus-associated (VA) RNA genes, as well as cloned tRNA and 5S RNA (HOEFFLER et al. 1988; YOSHINAGA et al. 1986). HOEFFLER and colleagues have shown that two distinct forms of TFIIIC bound to VA RNA promoter DNA can be distinguished by gel shift assays. The same authors have also demonstrated that the concentration of the complex containing active TFIIIC increased in response to E1A expression, although the overall TFIIIC concentration remained unchanged during viral infection. From these studies it has been concluded that adenovirus infection induces an E1A protein-dependent mechanism that results in increased concentrations of transcriptionally active TFIIIC, leading to a greater number of transcription complexes containing the active form of this factor.

3.2 Differential Regulation of the Expression of the Activating Protein-1 Constituents Jun and Fos by E1A Proteins of Adenovirus Types 12 and 2/5

The transcription factor complex AP-1 represents the prototype of a regulatory protein that converts extracellular signals into changes in the program of gene expression (for reviews see ABATE and CURRAN 1990; ANGEL and KARIN 1991). AP-1 consists mainly of two subunits: (1) members of the Jun family (c-jun, junB, junD) and (2) members of the fos family (c-fos, fosB) including the fos-related antigens fra1 and fra2. To fulfil their gene regulatory functions, these proteins have to dimerize. Jun proteins can form homodimers either with themselves or with other members of the Jun family. In addition, they can heterodimerize with different Fos or Fra proteins. Fos, however, can only form dimers with members of the jun family. As it is known that jun can heterodimerize with specific members of the ATF/CREB transcription factor family (BENBROOK and JONES 1990; HAI and CURRAN 1991; IVASHKIV et al. 1990), numerous different transcription factor complexes are conceivable, all with distinct activities and promoter specificities.

The expression of genes which are under the control of AP-1 (c-Jun:c-Fos) is induced within a few minutes in response to serum growth factors or phorbol esters (for review see KARIN 1991). This induction occurs also in the absence of protein synthesis, indicating two levels of regulation of AP-1 activity: (1) post-translational modifications of preexisting AP-1 complexes and (2) induction of the synthesis of its constituents Jun and Fos.

3.2.1 Jun

All members of the c-jun family (c-jun, junB, junD) can be regulated in their expression by several E1A proteins of the non-oncogenic and oncogenic adenovirus serotypes. The expression of c-jun is well analyzed in respect to different tissues and cultured cells. c-Jun seems to be expressed at constant levels during embryo genesis and is distributed ubiquitously in all tissues, even if at low levels (with some exceptions). Only low basal levels of c-jun mRNA and c-Jun protein are found in serum-starved cultured cells (for review see KARIN 1991). Expression can be induced by the addition of serum, growth factors (e.g., epidermal growth factor, EGF; transforming growth factor-β, TGF-β), tumor necrosis factor (TNF-α), phorbol esters such as TPA, and several oncogenes (e.g., src, activated c-Ha-ras, and mos). The promoter of the c-jun gene contains binding sequences for several transcription factors (SP-1, CTF) including AP-1-binding sites, which are called c-jun1-TRE (5'-TGACATCA-3') and c-jun2-TRE (5'-TTACCTCA-3') (VAN DAM et al. 1993). c-jun1-TRE appears to be necessary and sufficient for c-jun autoregulation and inducibility by TPA (ANGEL et al.1988). The functions of the SP-1 and CTF sites in the regulation of c-jun expression are not yet clearly understood (ANGEL et al. 1988).

In contrast to the collagenase gene, whose expression is repressed through the AP-1-binding site (see Sect. 3.3.2), the expression of the c-jun gene is strongly induced through the distal of the two TPA-responsive elements in the c-jun promoter (VAN DAM et al. 1993). The proximal AP-1 binding site also contributes to E1A inducibility, but does not respond significantly to E1A functions if c-jun 1-TRE is deleted. Thus, adenovirus E1A proteins can differentially regulate AP-1-responsive genes, probably due to the difference in the nucleotide sequence of the AP-1 sites. The collTRE element consists of 5'-TGAGTCA-3', whereas the distal AP-1-binding site in the c-jun promoter is made up of 5'-TTACCTCA-3'. The different sequences give rise to the speculation that the two promoter elements might be bound by different members of the AP-1 transcription factor family. Using in vitro binding studies, VAN DAM and colleagues (1993) have shown that in contrast to the AP-1-binding site of the collagenase promoter, which is preferentially targeted by dimers composed of the Jun and Fos families, the distal jun-TRE element is bound by dimers consisting of c-Jun and ATF-2 or ATF-2 like proteins. Activation of c-Jun/ATF-2 is dependent on CR1 and is most probably due to hyperphosphorylation of the *trans*-activation domain of c-Jun (HAGMEYER et al. 1993).

There is at least one more TRE element which can be induced by the 13S E1A protein of Ad2. DE GROOT and colleagues (1991a) have shown that the human metallothionein IIA TRE element (5'-TGACTCA-3') can mediate *trans*-activation of a metTRE/tk-CAT reporter construct in transient expression assays in different cell types (HeLa cells, P19 cells, JEG-3 cells). In addition, these authors observed an increased binding of an AP-1 transcription factor complex consisting most probably of Jun:Fos heterodimers in Ad5-infected HeLa cells, indicating a mechanism by which the 289R protein of Ad5 might *trans*-activate this promoter.

While c-jun expression is strongly induced by E1A functions of both serotypes in various cell types (DE GROOT et al. 1991b; KITABAYASHI et al. 1991; VAN DAM et al. 1990, 1993), the regulation of junB gene expression is dependent on both the cell type and the adenovirus serotype used for infection (DE GROOT et al. 1991b; MEIJER et al. 1991; VAN DAM et al. 1990). In Ad5 E1A-transformed rat kidney cells, for example, E1A functions stimulate junB expression, whereas those of Ad12 E1A do not. In BRK cells transformed with region E1A of Ad5, however, junB RNA levels are not altered compared with those in untransformed BRK cells, while BRK cells transformed with region E1A of Ad12 express reduced levels of junB mRNA. Moreover, in P19 EC cells, Ad5 E1A functions induce the expression of all three cellular jun genes, while in those cells transformed by Ad12 E1A only c-jun expression is strongly induced; the expression of junB and junD is not altered (DE GROOT et al. 1991b). The activation of the respective jun promoters by E1A functions is strongly dependent on the protein encoded by the E1A 13S mRNA (DE GROOT et al. 1991a). Most interestingly, the expression of Ad12 E1A polypeptides leads to a complete differentiation of P19 EC cells, comparable to the effect of retinoic acid, while expression of Ad5 E1A proteins leads only to partial differentiation of these cells (DE GROOT et al. 1991b). This might be explained by the differentially regulated expression of the different jun genes in P19 EC cells by Ad5 and Ad12 E1A proteins. It has been shown that over-expression of c-jun leads to the differentiation of PC19 EC cells (DE GROOT et al. 1990; YAMAGUCHI-IWAI et al. 1990); junB expression, however, blocks neuronal differentiation of these embryonic carcinoma cells (DE GROOT et al. 1991b). Since junB seems to be an antagonist of c-jun (CHIU et al. 1989; SCHÜTTE et al. 1989), in a coordinated induction of both genes by Ad5 E1A functions JunB might repress the differentiation-inducing capacity of c-jun. This is an attractive hypothesis, which can also be used to explain the different oncogenic potential of both adenovirus serotypes (DE GROOT et al. 1991b). As junB has been shown to repress the transforming capacity of c-jun, the lack of induction of junB expression might be one of several possible reasons for the oncogenicity of Ad12 viruses.

3.2.2 Fos

At least four proteins belong to the fos family of transcription factors: c-fos, fosB, and the fos-related antigens fra1 and fra2 (COHEN and CURRAN 1988; CURRAN et al. 1983; MATSUI et al. 1990; NISHINA et al. 1990; ZERIAL et al. 1989). The expression of c-fos, the most prominent member of this family, is very low in most cell types, but can be transiently induced by several orders of magnitude as a result of mitogenic, differentiation, or neural depolarizing stimuli (for review see MORGAN and CURRAN. 1991). Depending on the cell line these stimuli include: growth factors and cytokines, e.g., platelet-derived growth factor, (PDGF), FGF, EGF, nerve growth factor (NGF), TNF-α, TNF-β; thyrotropic hormone; phorbol esters; Ca ionophore; metal ions, e.g., K^+, Ba^{2+}, neurotransmitters; heat shock; cAMP; and ultraviolet (UV) irradiation (for review see ANGEL and KARIN 1991). Most of these factors transduce the extracellular signal through the protein kinase C or the cAMP-dependent protein kinase A pathway (ANGEL and KARIN 1991; HIPSKIND and NORDHEIM 1991; MORGAN and CURRAN 1991). The target sequence of the protein kinase C pathway in the c-fos promoter is the serum response element (SRE, nt –317 to –298), which is bound by a protein called serum response factor (SRF). A cAMP-responsive element (CRE) located around nt –60 is also involved in *trans*-activating the basal and induced c-fos expression. In addition, several other *cis*-acting elements have been described in the c-fos promoter: a glucocorticoid and progesterone responsive element (GRE/PRE), several SP-1 binding sites, a sis-factor element, a nuclear factor-1 (NF-1)- and an AP-1-binding site (FAP-1; 5'-TGACGCA-3'), and a retinoblastoma control element (Robbins et al. 1990).

Sassone-Corsi and Borrelli (1987) have shown that the c-fos promoter can be activated by the E1A protein product of the 13S mRNA of Ad2. *Trans*-activation has been measured in NIH3T3 and HeLa cells after cotransfection of a c-fos promoter–reporter construct with the respective E1A expression plasmid. No *trans*-activation of c-fos has been observed expressing the protein from the 12S mRNA, indicating that functions of the CR3, unique for the 289R E1A polypeptide, are involved in *trans*-activation of the c-fos gene. Mutant analysis of the c-fos promoter has suggested that a sequence located between nt –402 to –240, harbouring among others the SRE element, is the target of the 13S *trans*-activation.

The fos/AP-1 (FAP) element has also been discussed as a target for *trans*-activation mediated by the 13S gene product of Ad2 (De Groot et al. 1991a). Using transient expression assays in the human choriocarcinoma cell line JEG-3, De Groot and colleagues have shown that the expression of a chimeric CAT construct containing the fos promoter-specific AP-1-binding site in combination with the thymidine kinase (tk) promoter is upregulated by the 289R adenovirus protein. These data, however, remain to be verified using other promoter constructs, as the CAT expression from the basal tk-CAT construct can be already induced, at least by the E1A protein encoded by the 13S mRNA of Ad12 (Brockmann et al. 1994).

Experiments performed by Simon and colleagues (1990) have indicated that the TATA box of the c-fos promoter is most probably one of the important targets. These authors replaced the TATA box of the c-fos promoter (TATAA) with the TATA box of the SV40 early promoter (TATTTAT), which is not affected by E1A functions. The chimeric construct could no longer be *trans*-activated by E1A gene products.

Furthermore, there seems to be a second target for the *trans*-activation function of E1A in the c-fos promoter in addition to the TATA box (Engel et al. 1991; Gedrich et al. 1992). The protein products translated from the 13S and the 12S mRNA can act synergistically with cAMP to induce the transcription of c-fos in mouse S49 cells. *Trans*-activation requires the presence of cAMP, as no effect was observed in these cells if E1A was expressed in the absence of cAMP. For this cAMP-dependent *trans*-activation two domains in the E1A proteins have been identified: the amino-terminal end (amino acids 4–25) and the amino acid residues 36–81, which correspond to CR1, indicating that the latter mechanism of *trans*-activation of c-fos is different from that described earlier. However, as in these experiments the concentration of the endogenous mouse c-fos mRNA was measured by RNase protection assays after infection of S49 cells with adenoviruses, no target sequence(s) could be identified in the mouse c-fos promoter.

3.3 E1A-Mediated *Trans*-repression of Viral and Cellular Genes

Adenovirus E1A proteins are also able to *trans*-repress the expression of viral and specific cellular genes. This repression of gene expression by E1A functions is mediated primarily through enhancer sequences that are associated with the target genes. E1A proteins can repress, for example, the transcriptional activity directed by the SV40 early promoter (BORELLI et al. 1984; VELCICH and ZIFF 1985), the polyomavirus early promoter (VELCICH et al. 1986), the immunoglobulin H (IgH) gene enhancer (HEN et al. 1985), the insulin promoter region (STEIN and ZIFF 1987), the MyoD1 and myogenin enhancers (ENKEMANN et al. 1990), and the cytochrome P-450c control region (SOGAWA et al. 1989). The wide variety of cellular and viral enhancer sequences negatively affected by E1A functions, coupled with the observation that E1A proteins do not bind directly to these sequences, suggests that E1A polypeptides negatively regulate transcription by interfering with cellular transcription factors (e.g., AP-1, NF-κB). Most interestingly, the activity of some of these factors is differently regulated by the E1A polypeptides of different adenovirus serotypes. NF-κB-responsive transcription of the major histocompatibility complex (MHC) class I genes, for example, is repressed, at least in rats and rat cell lines, by Ad12 E1A but not Ad2/Ad5 E1A proteins (see Sect. 4.2).

3.3.1 p300, a Model of Transcriptional Repression by the Larger E1A Proteins

The cellular nuclear phosphoprotein p300 (YACIUK and MORAN 1991), first identified by co-immunoprecipitation with E1A proteins encoded from the 13S and 12S mRNA of Ad5 from infected (KB, HeLa) and transformed cells (293, human embryonic kidney cell line transformed with the *Eco*RI-C fragment of Ad5) (HARLOW et al. 1986; YEE and BRANTON 1985), binds sequence specifically to double-stranded DNA (consensus sequence 5'-GGGAGTG-3'; RIKITAKE and MORAN 1992). This and related sequences were found in the promoter/enhancer region of several genes whose expression is known to be regulated by E1A functions, such as the MHC class I gene H-2K^b (regulated only by Ad12 E1A proteins), the κ immunoglobulin gene enhancer, and the enhancers of SV40 and the human immunodeficiency virus (HIV). UV cross-linking experiments using immunoaffinity-purified p300 protein have documented that p300 can bind specifically to the respective sequences at least within the SV40 and H-2K^b enhancer (RIKITAKE and MORAN 1992). Although there are as yet no published data demonstrating the biological significance of the p300 DNA binding to these target sequences, it has been speculated that at least in some cases repression by adenovirus E1A proteins might be expressed in cooperation with the sequence-specific DNA-binding activity of p300.

Two domains of the larger E1A proteins are essential for binding to p300: amino acids 1–25 and 68–85; the latter domain is located at the carboxy terminal end of CR1 (EGAN et al. 1988; WANG et al. 1993a). Deletion of the basic arginine residue within the amino-terminal domain, which is conserved among all adenoviruses, totally abolishes p300 binding, whereas mutation of amino acids 3 and 20 reduces but does not eliminate p300 binding. The binding domain located within CR1 does not overlap with the binding site for the pRB protein, which is located at the N-terminal part of CR1 (DYSON et al. 1992; EGAN et al. 1988; WANG et al. 1993a). Although the amino-terminal part of the E1A proteins is not conserved among the different serotypes with the exception of the first two amino acids, which are the same in all subtypes, it has been shown that the E1A protein products encoded by the 13S and 12S mRNA of Ad12 bind p300 with an affinity similar to that of the Ad5 E1A polypeptides (WANG et al. 1993b). Interestingly, WANG and coworkers have detected E1A polypeptides in Ad12- but not in Ad5-infected cells, which seem to be modified at the amino-terminal end and are not able to bind p300. The nature of this amino-terminal modification is not known; however, the authors speculate on the basis of preliminary experimental data that at least three amino acids of the amino-terminal end of the E1A polypeptides are deleted, probably by specific proteolytic cleavage. An alternative explanation might be that a second translation start signal, located four amino acids apart from the first ATG and not present in the first 14 amino-terminal amino acids of Ad5, might sometimes be used in translation of the Ad12 E1A proteins. E1A mutants defective for p300 binding are also defective for an E1A-mediated repression function (JELSMA et al. 1989; RIKITAKE and MORAN 1992; STEIN et al. 1990), suggesting that p300 may play a crucial role in enhancer-stimulated expression of tissue-specific genes.

The mechanism of E1A-mediated repression of transcription through the p300 protein is still unclear. ABRAHAM and coworkers have shown that p300 can be co-immunoprecipitated with the TATA box-binding protein (TBP), which is required by all three eukaryotic polymerases as a component of different multi-subunit complexes (for reviews see PUGH and TJIAN 1992; SHARP 1992) as well as with at least two other cellular phosphoproteins of 64 kDa and 59 kDa (p64 and p59) (ABRAHAM et al. 1993). Their results also suggest that p300 is associated with a distinct subpopulation of TBP. This interaction might be either direct or indirect through additional proteins. The fact that p300 is implicated in the regulation of RNA polymerase II-transcribed promoters implies that p300-associated TBP is most likely part of TFIID complexes. Several models have been set up in an attempt to explain the repression of cellular genes by E1A products by p300. A simple one would be that E1A dissociates p300 (and additional proteins) from TBP complexes, thus altering the expression of cellular promoters affected by this association. Another hypothesis is that binding of E1A proteins to p300 may exclude or promote the binding of additional factors necessary for transcriptional activity of the respective genes. A third assumption is that E1A proteins alter the conformation of the TBP–p300 complex, leading to the repression of the enhancer/promoter activity, without changing its protein composition. Recently, WANG and colleagues (1993b) have shown that additional proteins which bind to CR2 of the E1A polypeptides, e.g., the p105 retinoblastoma gene product, can be found within the same Ad12 E1A protein–p300 complex. Therefore, it can be speculated that E1A proteins can stimulate cell proliferation only if p300 and CR2-associated proteins are present in the same multiprotein complex (BARBEAU et al. 1994).

3.3.2 Repression of the Activity of Different Activating Protein-1 Transcription Factor Complexes by E1A Proteins

It has been shown by several research groups that E1A represses at least three members of the metalloprotease gene family: stromelysin, collagenase I, and collagenase IV (GARBISA et al. 1987; OFFRINGA et al. 1988). The inhibition of the collagenase gene expression is mediated through the phorbol ester responsive element (collTRE; OFFRINGA et al. 1990), which was first identified as a promoter element responsible for the induction of gene expression by TPA, a number of mitogens, and by the oncogenes ras, mos, src, and trk (ANGEL et al. 1987; BRENNER et al. 1989; SCHÖNTHAL et al. 1988). The regulatory functions of these substances are mediated by the transcription factor complex AP-1 (Jun:Jun homodimers and Jun:Fos heterodimers), which directly binds to the collTRE element (ABATE and CURRAN 1990; ANGEL et al. 1987).

Proteins encoded by E1A of the non-oncogenic and oncogenic adenovirus serotypes repress AP-1-mediated induction of collagenase gene expression (OFFRINGA et al. 1990; BROCKMANN et al. 1994). In vivo footprinting analyses of the collagenase promoter in Ad5 E1-transformed 293 cells, Ad5-transformed human embryonic retinoblasts (HER), and nontransformed primary HER cells have demonstrated that the binding of Jun:Jun homodimers and Jun:Fos heterodimers to the collTRE element is totally blocked by E1A (HAGMEYER et al. 1993). Although E1A functions cause hyperphosphorylation of the *trans*-activating domain of Jun, the phosphorylation pattern of the DNA-binding domain is not altered. In addition, repression of DNA binding by E1A functions is not due to redox regulation of the Jun DNA-binding domain. These observations indicate that E1A does not prevent binding of AP-1 to its target sequence through post-translational modifications of the Jun DNA-binding domain. As E1A inhibits DNA binding of Jun:Fos heterodimers in vivo without affecting its formation, it can be postulated that E1A influences the interaction of Jun with other cellular proteins which are somehow involved in its DNA binding (HAGMEYER et al. 1993). Inhibition of the DNA-binding activity of AP-1 is dependent on CR1, indicating that both proteins, the protein products of the 13S and 12S mRNA, are able to repress Jun:Jun- and Jun:Fos-induced collagenase gene expression (OFFRINGA et al. 1990).

As described earlier in Sect. 2, Ad12 E1A expresses a 52R protein which cannot be expressed by the non-oncogenic serotypes (Fig. 1). We have shown in transient expression assays that this 52R protein selectively represses the *trans*-activating activity of Jun:Jun homodimers, while the activity of Jun:Fos heterodimers is most probably not downregulated (BROCKMANN et al. 1994). As the 52R protein does not contain CR1, the mechanism of repression, which is not yet understood, is most likely different from that observed in the 13S and 12S gene products. This ability to selectively repress the activity of a particular subset of transcription factor complexes might be beneficial for Ad12 in specific phases of the lytic cycle or during the process of oncogenic transformation.

4 Transforming Activity of E1A Proteins

E1A gene products are able to immortalize primary rodent cells on their own (HOUWELING et al. 1980; RULEY 1983; ZERLER et al. 1986). In order to obtain a fully transformed phenotype, they have to cooperate with functions of the adenovirus E1B gene, the polyomavirus middle T antigen gene, or the activated Ha-*ras* oncogene (RULEY 1983). Using cDNA constructs it has been shown that E1A proteins expressed either by the 13S or 12S mRNA can immortalize rodent cells (HALEY et al. 1984; ROBERTS et al. 1985; ZERLER et al. 1986). The product of either cDNA can also cooperate with the Ha-ras protein to transform primary rat cells (ULFENDAHL et al. 1987; ZERLER et al. 1986). As the protein encoded by the E1A 12S mRNA does not contain CR3, it has been concluded that this domain is not essential for either immortalization or transformation of primary cells. As will be discussed later, however, CR3 of the Ad12 E1A 266R protein might be involved in oncogenicity, at least in rat cells.

Adenovirus E1A oncogene-encoded sequences essential for transformation and cell growth regulation activities are localized at the N terminus and in CR1 and CR2 (Fig. 2). Mutant studies and the observation that the protein products of the 11S and 10S mRNA of Ad2/Ad5, which lack CR1, support this view (EGAN et al. 1988, 1989; HOWE and BAYLEY 1992; HOWE et al. 1990; LILLIE et al. 1986; MORAN et al. 1986; SCHNEIDER et al. 1987; STEIN et al. 1990; STEPHENS and HARLOW 1987; SVENSSON et al. 1991; ULFENDAHL et al. 1987; WHYTE et al. 1988a). Interestingly, mutant E1A proteins lacking either CR1 or CR2 can complement to transform cells (MORAN and ZERLER 1988), indicating that both domains can function independently and are required for transformation.

4.1 Binding of Specific Cellular Proteins Correlates with Transformation

One role of the E1A regions described above (CR1, CR2, amino terminus) in transformation appears to be to form complexes with two classes of cellular proteins involved in regulation of the cell cycle and cell proliferation, e.g., the pocket protein p105 RB tumor suppressor (EGAN et al. 1989; WHYTE et al. 1988b), the pRB-related p107 (EWEN et al. 1991), and a 130-kDa polypeptide. Their association with the E1A proteins is specifically dependent on CR2, although mutation analysis has suggested that sequences of CR1 also play a role in binding the p105 RB protein (DYSON et al. 1990, 1992). The other class known so far comprises only the large cellular DNA-binding protein p300 and a p300-related p400 polypeptide (BARBEAU et al. 1994; EGAN et al. 1988; WHYTE et al. 1989), whose association with the E1A proteins is specifically dependent on amino-terminal residues and sequences within CR1. In fact, the binding of these cellular proteins strongly correlates with the transforming activity of E1A proteins (BARBEAU et al. 1994; EGAN et al. 1989; WHYTE et al. 1989). Furthermore, the domains of the E1A proteins involved in this binding are precisely the regions required for their transformation activity (Fig. 2).

In addition to the cellular proteins discussed above, at least two other E1A-binding proteins have been described. One is cyclin A ($p60^{cycA}$), a cell cycle control protein (GIORDANO et al. 1989; PINES and HUNTER 1990) which appears to interact indirectly through association with p107 (EWEN et al. 1992; FAHA et al. 1992; HOWE and BAYLEY 1992). The other is the protein kinase $p33^{cdk2}$ (GIORDANO et al. 1991; HERRMANN et al. 1991; HU et al. 1992; KLEINBERGER and SHENK 1991), which is involved in the regulation of the cell cycle in a wide variety of eukaryotes. However, it is not yet clear how interactions with the latter proteins might contribute to the transforming activity of E1A proteins.

4.2 Repression of the Binding of NF-κB to the Major Histocompatibility Complex Class I Promoter by the Adenovirus Type 12 E1A 266R Gene Product

Although both the non-oncogenic and the oncogenic adenovirus serotypes can transform primary rodent cells in vitro, only the oncogenic adenovirus-transformed cells are able to form tumors when injected into syngeneic immunocompetent animals (HUEBNER et al. 1962; MCALLISTER and MACPHERSON 1968; MCBRIDE and WIENER 1964; MUKAI and KOBAYASHI 1972; MUKAI and MURAO 1975; POPE and ROWE 1964; TRENTIN et al. 1962). The mechanisms underlying this observation are not yet fully understood.

A key role in the oncogenicity of transformed cells may be played by the MHC class I cell surface antigens, because their presence on the cell surface in connection with antigenic peptides is essential for the recognition of tumor cells by cytotoxic T lymphocytes. Therefore, it was speculated that reduction of class I expression might facilitate the escape of Ad12 E1-transformed cells form the T cell immune response of the host (BERNARDS et al. 1983). Analyses of MHC class I gene expression in human and mouse cells have shown that, in comparison to either normal or Ad5 E1-transformed cells, the Ad12 E1-transformed cells exhibit reduced levels of MHC class I antigens (FRIEDMAN and RICCIARDI 1988; VASAVADA et al. 1986). The downregulation of MHC class I genes by Ad12, which is dependent on the presence of the E1A protein encoded by the 13S mRNA (MEIJER et al. 1991), occurs at the transcriptional level (ACKRILL and BLAIR 1988; FRIEDMAN and RICCIARDI 1988; MEIJER et al. 1989; VASAVADA et al. 1986).

In various cell types, transcription of the MHC class I genes is controlled mainly by the class I regulatory element (CRE) (BALDWIN and SHARP 1987; KIMURA et al. 1986). This *cis*-element consists of two distinct transcription factor-binding sites, H2TF1 and CREII (SHIRAYOSHI et al. 1987). CREII is recognized by a factor called H-2RIIBP. This factor shows a high homology to the nuclear hormone receptor family (HAMADA et al. 1989). H2TF1 is the target for several proteins, including NF-κB, a heterodimer consisting of two proteins of 65 kDa (p65) and 50 kDa (p50), and KBF1 (MEIJER et al. 1992). In contrast of NF-κB, KBF1 consists of p50 homodimers. Both p65 and p50 are members of the re-oncogene family. The H2TF1 sequence seems to be the target of the Ad12 E1-mediated repression of MHC class I gene expression. It was observed by MEIJER and colleagues (1992) that the binding activities of NF-κB and/or KBF1 to this element were reduced in BRK cell lines transformed by the 13S protein product of Ad12 E1A in comparison to cell lines expressing the 12S product of Ad12 E1A or the 13S product of Ad5 E1A. Introduction of the 13S protein product of Ad5 E1A into Ad12 13S-transformed cells counteracts the repression mediated by the Ad12 E1A 266R protein and results in an increase of protein-binding activity to H2TF1. From these results the conclusion can be drawn that oncogenicity of Ad12 in BRK cells is most probably due to the downregulation of MHC class I gene expression by functions encoded by CR3 (MEIJERS et al. 1992).

5 Future Prospects

The E1A proteins are able to form complexes with a variety of cellular proteins. These interactions seem to be essential and critical for all E1A-mediated activities: induction and repression of transcription as well as transformation of primary cells. The best-supported models for the mechanism of transcriptional activation are based on the action of E1A proteins as bridging molecules, interacting with enhancer-bound factors (e.g., ATF-2) as well as components of the transcriptional initiation complex (e.g., TFIID), and the activation of a transcription factor (e.g., E2F) by dissociating this factor from its inhibitors (e.g., pRB). However, less is known about E1A-mediated repression of transcription. The most attractive and logical model for the role of E1A proteins in transformation suggests that they bind to and interfere with the activity of proteins that regulate cellular growth and differentiation. Initial evidence for this model has been obtained by investigating the biochemical activities of cellular growth regulators, such as pRB, p107, p130, and p300, and their interaction with E1A polypeptides, and by demonstrating that E1A functions modify their activities. Assuming these models for E1A functions in gene regulation and transformation are correct, the next steps of research will involve the exploration of single molecular events, mediated by E1A proteins, that alter the activity of transcription complexes and the behavior of differentiation and growth-regulatory proteins.

Acknowledgments. Work in our laboratory was supported by the Deutsche Forschungsgemeinschaft, grant no. SFB354/TP3, and by the Fonds der Chemischen Industrie.

References

Abate C, Curran T (1990) Encounters with Fos and Jun on the road to AP-1. Semin Cancer Biol 1: 19–26

Abraham SE, Lobo S, Yaciuk P, Wang H-GH, Moran E (1993) p300, and p300-associated proteins, are components of TATA-binding protein (TBP) complexes. Oncogene 8: 1639–1647

Ackrill AM, Blair GE (1988) Regulation of major histocompatibility class I gene expression at the level of transcription in highly oncogenic adenovirus-transformed rat cells. Oncogene 3: 483–487

Angel P, Karin M (1991) The role of Jun, Fos and the AP-1 complex in cell-proliferation and transformation. Biochim Biophys Acta 1072: 129–157

Angel P, Imagawa M, Chiu R, Stein B, Imbra RJ, Rahmsdorf HJ, Jonat C, Herrlich P, Karin M (1987) Phorbol ester-inducible genes contain a common cis element recognized by a TPA-modulated trans-acting factor. Cell 49: 729–739

Angel P, Hattori K, Smeal T, Karin M (1988) The Jun proto-oncogene is positively autoregulated by its product, Jun/AP-1. Cell 55: 875–885

Bagchi S, Raychaudhuri P, Nevins JR (1990) Adenovirus E1A proteins can dissociate heteromeric cellular complexes involving the E2F transcription factor: a novel mechanism for E1A trans-activation. Cell 62: 659–669

Baldwin AS, Sharp PA (1987) Binding of a nuclear factor to a regulatory sequence in the promoter of the mouse H-2K^b class I major histocompatibility gene. Mol Cell Biol 7: 305–313

Bandara LR, La Thangue NB (1991) Adenovirus E1A prevents the retinoblastoma gene product from complexing with a cellular transcription factor. Nature 351: 494–497

Barbeau D, Charbonneau R, Whalen SG, Bayley ST (1994) Functional interactions within adenovirus E1A protein complexes. Oncogne 9: 359–273

Benbrook DM, Jones NC (1990) Heterodimer formation between CREB and Jun proteins. Oncogene 5: 295–302

Berk AJ, Sharp PA (1978) Structure of the adenovirus 2 early mRNAs. Cell 14: 695–711

Berk AJ, Lee F, Harrison T, Williams J, Sharp PA (1979) Pre-early adenovirus 5 gene product regulates synthesis of early viral messenger RNAs. Cell 17: 935–944

Bernards R, Schrier PI, Houweling A, Bos JL, Van der Eb AJ, Zijlstra M, Melief CJM (1983) Tumorigenicity of cells transformed by adenovirus type 12 by evasion of T-cell immunity. Nature 305: 776–779

Blake MC, Azizkhan JC (1989) Transcription factor E2F is required for efficient expression of the hamster dihydrofolate reductase gene in vitro and in vivo. Mol Cell Biol 9: 4994–5002

Borrelli E, Hen R, Chambon P (1984) Adenovirus 2 E1A products repress enhancer-induced stimulation of transcription. Nature 312: 608–612

Brenner DA, O'Hara M, Angel P, Chojkier M, Karin M (1989) Prolonged activation of jun and collagenase genes by tumour necrosis factor-α. Nature 337: 661–663

Brockmann D, Tries B, Esche H (1990) Isolation and characterization of novel adenovirus type 12 E1A mRNAs by cDNA PCR technique. Virology 179: 585–590

Brockmann D, Feng L, Kröner G, Tries B, Esche H (1994) Adenovirus type 12 early region 1A express a 52R protein repressing the trans-activating activity of transcription factor c-Jun/AP-1. Virology 198: 717–723

Buckbinder L, Miralles BJ, Reingerg D (1989) TPA can overcome the requirement for E1A and other act synergistically in stimulating expression of adenovirus EIII promoter. EMBO J 8: 4239–4250
Buratowski S, Hahn S, Guarente L, Sharp PA (1989) Five intermediate complexes in transcription initiation by RNA polymerase II. Cell 56: 549–561
Byrd PJ, Grand RJ, Gallimore PH (1988) Differential transformation of primary human embryo retinal cells by adenovirus E1 regions and combinations of E1A + ras. Oncogene 2: 477–484
Cao L, Faha B, Dembski M, Tsai LH, Harlow E, Dyson N (1992) Independent binding of the retinoblastoma protein and p107 to the transcription factor E2F. Nature 355: 176–179
Carthew RW, Chodosh LA, Sharp PA (1985) An RNA polymerase II transcription factor binds to an upstream element in the adenovirus major late promoter. Cell 43: 439–448
Carthew RW, Chodosh LA, Sharp PA (1987) The major late transcription factor binds to and activates the mouse metallothionein I promoter. Gene Dev 1: 973–980
Chang LS, Shi Y, Shenk T (1989) Adeno-associated virus P5 promoter contains an adenovirus-inducible element and a binding site for the major late transcription factor. J Virol 63: 3479–3488
Chatton B, Bocco JL, Gaire M, Hauss C, Reimund B, Goetz J, Kedinger C (1993) Transcriptional activation by the larger E1A product is mediated by members of the cellular transcription factor ATF family which can directly associate with E1A. Mol Cell Biol 13: 561–570
Chellappan SP, Hiebert S, Mudryj M, Horowitz JM, Nevins JR (1991) The E2F transcription factor is a cellular target for the Rb protein. Cell 65: 1053–1061
Chellappan S, Kraus VB, Kroger B, Munger K, Howley PM, Phelps WC, Nevins JR (1992) Adenovirus E1A, simian virus 40 tumor antigen, and human papillomavirus E7 protein share the capacity to disrupt the interaction between transcription factor E2F and the retinoblastoma gene product. Proc Natl Acad Sci USA 89: 1053–1061
Chiu R, Angel P, Karin M (1989) Jun-B differs in its biological properties from, and is a negative regulator of, c-Jun. Cell 59: 979–986
Chodosh LA, Carthew RW, Morgan JG, Crabtree GR, Sharp PA (1987) The adenovirus major late transcription factor activates the rat gamma-fibrinogen promoter. Science 238: 684–688
Chow LT, Brooker TR, Lewis JB (1979) Complex splicing pattern of RNAs from early regions of adenovirus-2. J Mol Biol 124: 265–303
Cohen DR, Curran T (1988) fra-1: a serum inducible, cellular immediate early gene that encodes for a fos-related antigen. Mol Cell Biol 8: 2063–2069
Cortes P, Buckbinder L, Leza MA, Rak N, Hearing P, Merino A, Reinberg D (1988) EivF, a factor required for transcription of the adenovirus EIV promoter, binds to an element, involved in E1A-dependent activation and cAMP induction. Gene Dev 2: 975–990
Curran T, Franza BR Jr (1988) Fos and Jun: the AP-1 connection. Cell 55: 395–397
Curran T, MacConnel WP, Van Straaten F, Verma IM (1983) Structure of the FBJ murine osteosarcoma virus genome: molecular cloning of its associated helper virus and the cellular homolog of the v-fos gene from mouse and human cells. Mol Cell Biol 3: 914–921
Dalton S (1992) Cell cycle regulation of the human cdc2 gene. EMBO J 11: 1797–1804
De Groot RP, Kruyt FAE, Van der Saag PT, Kruijer W (1990) Ectopic expression of c-jun leads to differentiation of P19 embryonal cells. EMBO J 9: 1831–1837
De Groot RP, Foulkes N, Mulder M, Kruijer W, Sassone-Corsi P (1991a) Positive regulation of jun/AP-1 by E1A. Mol Cell Biol 11: 192–201
De Groot RP, Meijer I, Van den Brink S, Mummery C, Kruijer W (1991b) Differential regulation of JunB and JunD by adenovirus type 5 and 12 E1A proteins. Oncogene 6: 2357–2361
Devoto SH, Mudryj M, Pines J, Hunter T, Nevins JR (1992) A cyclin A-cdc2 kinase complex possesses sequence-specific DNA binding activity: p33 cdk2 is a component of the E2F-cyclin A complex. Cell 68: 167–176
Dumont DJ, Tremblay ML, Branton PE (1989) Phosphorylation at serine 89 induces a shift in gel mobility but has little effect on the function of adenovirus type 5 E1A proteins. J Virol 63: 987–991
Dyson N, Bernards R, Friend SH, Gooding LR, Hassell JA, Major EO, Pipas JM, Vandyke T, Harlow E (1990) Large T antigen of many polyomaviruses are able to form complexes with the retinoblastoma protein. J Virol 64: 1353–1356
Dyson N, Guida P, McCall C, Harlow E (1992) Adenovirus E1A makes two distinct contacts with the retinoblastoma protein. J Virol 66: 4606–4611
Egan C, Jelsma TN, Howe JA, Bayley ST, Ferguson B, Branton PE (1988) Mapping of cellular protein-binding sites on the products of early-region 1A of human adenovirus type 5. Mol Cell Biol 8: 3955–3959

Egan C, Bayley ST, Branton PE (1989) Binding of the RBI protein to E1A products is required for adenovirus transformation. Oncogene 4: 383–388

Engel DA, Muller U, Gedrich RW, Eubanks JS, Shenk T (1991) Induction of c-fos mRNA and AP-1 DNA-binding activity by cAMP in cooperation with either the adenovirus 243- or the adenovirus 289-amino acid E1A protein. Proc Natl Acad Sci USA 88: 3957–3961

Enkemann SA, Konieczny SF, Taparowsky EJ (1990) Adenovirus 5 E1A represses muscle-specific enhancers and inhibits expression of myogenic regulatory factor genes, MyoDI and myogenin. Cell Growth Differ 1: 375–382

Ewen ME, Xing Y, Bentley-Lawrence J, Livingston DM (1991) Molecular cloning, chromosomal mapping, and expression of the cDNA for p107, a retinoblastoma gene product-related protein. Cell 66: 1155–1164

Ewen ME, Faha B, Harlow E, Livingston DM (1992) Interaction of p107 with cyclin A independent of complex formation with viral oncoproteins. Science 255: 85–87

Faha B, Ewen ME, Tsai LH, Livingston DM, Harlow E (1992) Interaction between human cyclin A and adenovirus E1A-associated p107 protein. Science 255: 87–90

Fanning E, Doerfler W (1976) Intracellular forms of adenovirus DNA. V. Viral DNA sequences in hamster cells abortively infected and transformed with human adenovirus type 12. J Virol 20: 373–383

Ferguson B, Krippl B, Andrisani O, Jones N, Westphal H, Rosenberg M (1985) E1A 13S and 12S mRNA products made in Escherichia coli both function as nucleus-localized transcription activators but do not directly bind DNA. Mol Cell Biol 5: 2653–2661

Flint SJ, Jones N (1991) Differential regulation of three members of the ATF/CREB family of DNA-binding proteins. Oncogene 6: 2019–2026

Flint SJ, Leong K (1986) Enhanced transcription activity of HeLa cell extracts infected with adenovirus type 2. Cancer Cells 4: 137–146

Flint SJ, Shenk T (1989) Adenovirus E1A protein paradigm viral transactivator. Annu Rev Genet 23: 141–161

Friedman DJ, Ricciardi RP (1988) Adenovirus type 12 E1A gene represses accumulation of MHC class I mRNA at the level of transcription. Virology 165: 303–305

Gallimore PH, Sharp PA, Sambrook J (1974) Viral DNA in transformed cells. II. A Study of the sequence of adenovirus 2 DNA in nine cell lines of transformed rat cells using specific fragments of the viral genome. J Mol Biol 89: 49–72

Garbisa S, Pozzatti R, Muschel RJ, Saffiotti U, Ballin M, Goldfarb RH, Khoury G, Liotta LA (1987) Secretion of type IV collagenolytic protease and metastatic phenotype by transfection with c-Ha-ras but not c-Ha-ras plus Ad2-E1A. Cancer Res 47: 1523–1528

Gedrich RW, Bayley S, Engel DA (1992) Induction of AP-1 DNA-binding activity and c-fos mRNA by the adenovirus 243R E1A protein and cyclic AMP requires domains necessary for transformation. J Virol 66: 5849–5659

Giordana A, Whyte P, Harlow E, Franza BR, Beach D, Draetta G (1989) A 60 kd cdc2-associated polypeptide complexes with E1A proteins in adenovirus-infected cells. Cell 58: 981–990

Giordano A, McCall C, Whyte P, Franza BR (1991) Human cyclin A and the retinoblastoma protein interact with similar but distinguishable sequences in the adenovirus E1A gene product. Oncogene 6: 481–485

Girling R, Patridge JF, Bandara LR, Burden N, Totty NF, Hsuan JJ, La Thangue NB (1993) A new component of the transcription factor DRTF1/E2F. Nature 362: 83–87

Graham FL, Abrahams PJ, Mulder C, Heijneker SO, Warnaar SO, De Vries FAJ, Fiers W, Van der Eb AJ (1974a) Studies on in vitro transformation by DNA and DNA fragments of human adenoviruses and simian virus 40. Cold Spring Harbor Symp Quant Biol 39: 637–650

Graham FL, Van der Eb AJ, Heijneker HL (1974b) Size and location of the transforming region in human adenovirus type 5 DNA. Nature 251: 687–691

Grand RJ, Byrd PJ, Grabham PW, Gregory CD, Huen DS, Merrick RM, Young LS, Gallimore PH (1989) The expression of the retinoblastoma gene product Rb1 in primary and adenovirus-transformed human cells. Oncogen 4: 1291–1298

Green M, Loewenstein PM, Pusztai R, Symington JS (1988) An adenovirus E1A protein domain activates transcription in vivo and in vitro in the absence of protein synthesis. Cell 53: 921–926

Hagmeyer BM, König H, Herr I, Offringa R, Zantema A, Van der Eb AJ, Herrlich P, Angel P (1993) Adenovirus E1A negatively and positively modulates transcription of AP-1 dependent genes by dimer-specific regulation of the DNA binding and transactivation activities of Jun. EMBO J 12: 3559–3572

Hai T, Curran T (1991) Cross-family dimerization of transcription factors Fos/Jun and ATF/CREB alters DNA binding specificity. Proc Natl Acad Sci USA 88: 3720–3724
Hai T, Lui F, Allegretto EA, Karin M, Green MR (1988) A family of immunologically related transcription factors that includes multiple forms of ATF and AP-1. Gene Dev 2: 1216–1226
Hai T, Liu F, Coukos WJ, Green MR (1989) Transcription factor ATF cDNA clones: an extensive family of leucine zipper proteins able to selectively form DNA-binding heterodimers. Gene Dev 3: 2083–2090
Haley KP, Overhauser G, Babiss LE, Ginsberg HS, Jones NC (1984) Transformation properties of type 5 adenovirus mutants that differentially express the E1A products. Proc Natl Acad Sci USA 81: 5734–5738
Hamada K, Gleason SL, Levi B-Z, Hirschfeld S, Appella E, Ozato K (1989) H-2RIIBP, a member of the nuclear hormone receptor superfamily that binds to both the regulatory element of major histo-compatibility class I genes and the estrogen response element. Proc Natl Acad Sci USA 86: 8289–8293
Hamel PA, Gill RM, Phillips RA, Gallie BL (1992) Transcriptional repression of the E2-containing promoters EIIaE, c-myc, and RB1 by the product of the RB1 gene. Mol Cell Biol 12: 3431–3438
Harlow E, Whyte P, Franza BR, Schley C(1986) Association of adenovirus early-region 1A proteins with cellular polypeptides. Mol Cell Biol 6: 1579–1569
Hen R, Sassone-Corsi P, Corden P, Glaub MP, Chambon P (1982) Sequences upstream from the TATA box are required in vivo and in vitro for efficient transcription from the adenovirus serotype 2 major late promoter. Proc Natl Acad Sci USA 79:7132–7136
Hen R, Borrelli E, Chambon P (1985) Repression of the immunoglobulin heavy chain enhancer by the adenovirus-2 E1A products. Science 230: 1391–1394
Herrmann CH, Su LK, Harlow E (1991) Adenovirus E1A is associated with a serine/threonine protein kinase. J Virol 65: 5848–5859
Heysen A, Verwaerde P, D'Halluin JC (1991) Nucleotide sequence and regulation of the adenovirus type 3 E2A early promoter. Virology 181: 241–250
Hiebert SW, Lipp M, Nevins JR (1989) E1A-dependent trans-activation of the human MYC promoter is mediated by the E2F factor. Proc Natl Acad Sci USA 86: 3594–3598
Hiebert SW, Chellappan SP, Horowitz JM, Nevins JR (1992) The interaction of Rb with E2F coincides with an inhibition of the transcriptional activity of E2F. Gene Dev 6: 177–185
Hipskind RA, Nordheim A (1991) Functional dissection in vitro of the human c-fos promoter. J Biol Chem 226: 19583–19592
Hoeffler WK, Kovelman R, Roeder RG (1988) Activation of transcription factor IIIC by the adenovirus E1A protein. Cell 53: 907–920
Houweling A, Van den Elsen PJ, Van der Eb AJ (1980) Partial transformation of primary rat cells by the leftmost 4.5% fragment of adenovirus 5 DNA. Virology 105: 537–550
Howe, JA, Bayley ST (1992) Effects of Ad5 E1A mutant viruses on the cell cycle in relation to the binding of cellular proteins including the retinoblastoma protein and cyclin A. Virology 186: 15–24
Howe JA, Mymryk JS, Egan C, Branton PE, Baley ST (1990) Retinoblastoma growth suppressor and a 300-kDa protein appear to regulate cellular DNA synthesis. Proc Natl Acad Sci USA 87: 5883–5887
Hu Q, Lees J, Buchkovich K, Harlow E (1992) The retinoblastoma protein physically associates with the human cdc2 kinase. Mol Cell Biol 12: 971–980
Huebner RJ, Rowe WP, Lane WT (1962) Oncogenic effects in hamsters of human adenovirus types 12 and 18. Proc Natl Acad Sci USA 48: 2051–2058
Hurst HC, Jones NC (1987) The identification of factors that interact with the E1A-inducible adenovirus E3 promoter. Gene Dev 1: 1132–1146
Ikeda M-A, Nevins JR (1993) Identification of distinct roles for separate E1A domains in disruption of E2F complexes. Mol Cell. Biol 13: 7029–7035
Ivashkiv LB, Liou H-C, Kara CJ, Lamph WW, Verma IM, Glimcher LH (1990) mXBP/CRE-BP2 and c-Jun form a complex which binds to the cyclic AMP, but not to the 12-O-tetradecanoylphorbol-13-acetate, response element. Mol Cell Biol 10: 1609–1621
Jelsma TN, Howe JS, Mymryk JS, Evelegh CM, Cunniff NFA, Bayley ST (1989) Sequences in E1A proteins of human adenovirus 5 required for cell transformation, repression of a transcriptional enhancer, and induction of proliferating cell nuclear antigen. Virology 170: 120–130
Jochemsen H, Daniels GSG, Hertoghs JJL, Schrier PI, Van Elsen PJ, Van der Eb AJ (1982) Identification of adenovirus-type 12 gene products involved in transformation and oncogenesis. Virology 122: 15–28

Jones N, Shenk T (1979) An adenovirus type 5 early gene function regulates expression of other early genes. Proc Natl Acad Sci USA 76: 3665–3669

Karin M (1991) The AP-I complex and its role in transcriptional control by protein kinase C. In: Cohen P, Foulkes JG (eds) The hormonal control regulation of gene transcription. Elsevier Science, Amsterdam, pp 235–252

Kimelmann D, Miller JS, Porter D, Roberts BE (1985) E1A regions of the human adenoviruses and of the highly oncogenic simian adenovirus 7 are closely related. J Virol 53: 399–409

Kimura A, Israel A, Le Bail O, Kourilsky P (1986) Detailed analysis of the mouse H-2K^b promoter: enhancer-like sequences and their role in the regulation of class I gene expression. Cell 44: 261–272

Kitabayashi I, Chiu R, Gachelin G,Yokoyama K (1991) E1A dependent up-regulation of c-jun/AP-1 activity. Nucleic Acids Res 19: 649–655

Kitchingman GR, Westphal H (1980) The structure of adenovirus 2 early nuclear and cytoplasmic RNAs. J Mol Biol 137: 23–48

Kleinberger T, Shenk T (1991) A protein kinase is present in a complex with adenovirus E1a proteins. Proc Natl Acad Sci USA 88: 11143–11147

Koveshdi I, Reichel R, Nevins JR (1986) Identification of a cellular transcription factor involved in E1A trans-activation. Cell 45: 219–228

Koveshdi I, Reichel R, Nevins JR (1987) Role of an adenovirus E2 promoter binding factor in EIA-mediated coordinate gene control. Proc Natl Acad Sci USA 84: 2180–2184

Kripple B, Ferguson B, Jones N, Rosenberg M, Westphal H (1985) Mapping of functional domains in adenovirus E1A proteins. Proc Natl Acad Sci USA 82: 7480–7484

Kripple B, Andrisani O, Jones N, Westphal H, Rosenberg M, Ferguson B (1986) Adenovirus type 12 E1A protein expressed in Escherichia coli is functional upon transfer by microinjection or protoplast fusion into mammalian cells. J Virol 59: 420–427

La Thangue NB, Johnston LH (1993) Transcriptional complexity. Curr Biol 8: 554–557

Lee KA, Hai TY, SivaRaman L, Thimmappaya B, Hurst HC, Jones NC, Green MR (1987) A cellular protein, activating transcription factor, activates transcription of multiple E1A-inducible adenovirus early promoters. Proc Natl Acad Sci USA 84: 8355–8359

Lee WS, Kao CC, Bryant GO, Liu X, Berk AJ (1991) Adenovirus E1A activation domain binds the basic repeat in the TATA box transcription factor. Cell 67: 365–376

Lees E, Faha B, Dulic V, Reed SI, Harlow E (1992) Cyclin E/cdk2 and cyclin A/cdk2 kinases associate with p107 and E2F in a temporally distinct manner. Gene Dev 6: 1874–1885

Leff T, Corden J, Elkaim R, Sassone-Corsi P (1985) Transcriptional analysis of the adenovirus-5 EIII promoter: absence of sequence specificity for stimulation by E1A gene products. Nucleic Acids Res 13: 1209–1221

Leong K, Berk AJ (1986) Adenovirus early region 1A protein increases the number of template molecules transcribed in cell-free extracts. Proc Natl Acad Sci USA 83: 5844–5852

Leong K, Brunet L, Berk AJ (1988) Factors responsible for the higher transcriptional activity of extracts of adenovirus infected cells fractionate with the TATA box transcription factor. Mol Cell Biol 8: 1765–1774

Lewis JB, Mathews MB (1980) Control of adenovirus early gene expression: a class of immediate early products. Cell 21: 303–313

Lewis ED, Manley JL (1985) Control of adenovirus late promoter expression in two human cell lines. Mol Cell Biol 5: 2433–2442

Lillie JW, Green MR (1989) Transcription activation by the adenovirus E1A protein. Nature 338: 39–44

Lillie JW, Green M, Green MR (1986) An adenovirus E1A protein region required for transformation and transcriptional repression. Cell 46: 1043–1051

Lillie JW, Loewenstein PM, Green MR, Green M (1987) Functional domains of adenovirus type 5 E1A proteins. Cell 50: 1091–1100

Liu F, Green MR (1990) A specific member of the ATF transcription factor family can mediate transcription activation by the adenovirus E1A protein. Cell 61: 1217–1224

Loewenstein PM, Green M (1989) In vitro trans-activation by chemically synthesized adenovirus conserved region 3 peptides: reaction properties and mutational analysis. J Biol Chem 264: 21504–21508

Lucher LA, Kimelman D, Symington JS, Brackmann KH, Cartas MA, Thornton H, Green M (1984) Identification of adenovirus 12-encoded E1A tumor antigens synthesized in infected and transformed mammalian cells and in Escherichia coli. J Virol 52: 136–144

Lucher LA, Loewenstein PM, Green M (1985) Phosphorylation in vitro of Escherichia coli-produced 235R and 266R tumor antigens encoded by human adenovirus type 12 early transformation region E1A. J Virol 56: 183–193

Lucher LA, Brackmann KH, Symington JS, Green M (1986) Posttranslational modification at the N terminus of the human adenovirus type 12 E1A 235R tumor antigen. J Virol 58: 592–599

Marton MJ, Baim SB, Ornelles DA, Shenk T (1990) The adenovirus E4 17-kilodalton protein complexes with the cellular transcription factor E2F, altering its DNA-binding properties and stimulating E1A-independent accumulation of E2 mRNA. J Virol 64: 2345–2359

Matsui M, Tokuhara M, Konuma Y, Nomura N, Ishizaki R (1990) Isolation of human fos-related genes and their expression during monocyte macrophage differentiation. Oncogene 5: 249–255

McAllister RM, Macpherson I (1968) Transformation of a hamster cell line by adenovirus type 12. J Gen Virol 2: 99–106

McBride WD, Wiener A (1964) In vitro transformation of hamster kidney cells by human adenovirus type 12. Proc Soc Exp Biol Med 115: 870–874

Meijer I, Jochemsen AG, De Wit CM, Bos JL, Morello D, Van der Eb AJ (1989) Adenovirus type 12 E1A down regulates expression of a transgene under control of a major histocompatibility complex class I promoter: evidence for transcriptional control. J Virol 63: 4039–4042

Meijer I, Van Dam H, Boot AJM, Bos JL, Zantema A, Van der Eb AJ (1991) Co-regulated expression of junB and MHC class I genes in adenovirus-transformed cells. Oncogene 6: 911–916

Meijer I, Boot AJM, Mahabir G, Zantema A, Van der Eb AJ (1992) Reduced binding activity of transcription factor NF-κB accounts for MHC class I repression in adenovirus type 12 E1-transformed cells. Cell Immunol 145: 56–65

Merino A, Buckbinder L, Mermelstein FH, Reinberg DH (1989) Phosphorylation of cellular proteins regulates their binding to the cAMP response element. J Biol Chem 264: 21266–21276

Miyamoto NG, Moncolin V, Egly JM, Chambon P (1985) Specific interaction between a transcription factor and the upstream element of the adenovirus 2 major late promoter. EMBO J 4: 3563–3570

Montminy MR, Bilezikjian LM (1987) Binding of a nuclear protein to the cyclic AMP response element of the somatostatin gene. Nature 328: 175–178

Moran E, Mathews MB (1987) Multiple functional domains in the adenovirus E1A gene. Cell 48: 177–178

Moran E, Zerler B (1988) Interactions between cell growth-regulating domains in the products of the adenovirus E1A oncogene. Mol Cell Biol 8: 1756–1764

Moran E, Grodzicker T, Roberts RJ, Mathews MB, Zerler B (1986) Lytic and transforming functions of individual products of the adenovirus E1A gene. J Virol 57: 765–775

Morgan JI, Curran T (1991) Stimulus-transcription coupling in the nervous system: involvement of the inducible proto-oncogenes fos and jun. Annu Rev Neurosci 14: 421–451

Mudryj M, Hiebert SW, Nevins JR (1990) A role for the adenovirus inducible E2F transcription factor in a proliferation-dependent signal transduction pathway. EMBO J 7: 2179–2184

Mudryj M, Devoto SH, Hiebert SW, Hunter T, Pines J, Nevins JR (1991) Cell cycle regulation of the E2F transcription factor involves an interaction with cyclin A. Cell 65: 1243–1253

Müller U, Roberts MP, Engel AD, Doerfler W, Shenk T (1989) Induction of transcription factor AP-1 by adenovirus E1A protein and cAMP. Gene Dev 3: 1991–2002

Mukai N, Kobayashi S (1972) Undifferentiated intraperitoneal tumors induced by human adenovirus type 12 in hamsters. Am J Pathol 69: 331–348

Muki N, Murao T (1975) Retinal tumor induction by ocular inoculation of human adenovirus in 3 day old rats. J Neuropathol Exp Neurol 34: 28–35

Nakajima H, Horikoshi M, Roeder RG (1988) Factors involved in specific transcription by mammalian RNA polymerase II: purification, genetic specifity, and TATA box-promoter interactions of TFIID. Mol Cell Biol 8: 4028–4040

Neill SD, Nevins JR (1991) Genetic analysis of the adenovirus E4 6/7 transactivator: interaction with E2F and induction of a stable DNA-protein complex are critical for activity. J Virol 65: 5364–5373

Neill SD, Hemstrom C, Virtanen A, Nevins JR (1990) An adenovirus E4 gene product trans-activates E2 transcription and stimulates stable E2F binding through a direct association with E2F. Proc Natl Acad Sci USA 87: 2008–2012

Nevins JR (1981) Mechanism of activation of early viral transcription by the adenovirus E1A gene product. Cell 26: 213–220
Nevins JR (1992) E2F: A link between the Rb tumor suppressor protein and viral oncoproteins. Science 258: 424–429
Nevins JR (1993) Transcriptional activation by the adenovirus E1A proteins. Semin Virol 4: 25–31
Nevins JR, Ginsberg HS, Blanchard J-M, Wilson MC, Darnell JE (1979) Regulation of the primary expression of the early adenovirus transcription units. J Virol 32: 727–733
Nishina H, Sato H, Suzuke T, Sato M, Iba H (1990) Isolation and characterization of fra-2, an additional member of the fos gene family. Proc Natl Acad Sci USA 87: 3619–3623
Nomura N, Zu YL, Maekawa T, Tabata S, Akiyama T, Ishii S (1993) Isolation and characterization of a novel member of the gene family encoding the cAMP response element-binding protein CRE-BP1. J Biol Chem 268: 4259–4266
Offringa R, Smits AMM, Houweling A, Bos JL, Van der Eb AJ (1988) Similar effects of adenovirus E1A and glucocorticoid hormones on the expression of the metalloprotease stromelysin. Nucleic Acids Res 16: 10973–10984
Offringa R, Gebel S, Van Dam H, Timmers M, Smits A, Zwart R, Stein B, Bos JL, Van der Eb AJ, Herrlich P (1990) A novel function of the transforming domain of E1A: repression of AP-1 activity. Cell 62: 527–538
Peeper DS, Zantema A, Dowdy SF, Van der Eb AJ (1992) Expression, purification, and functional characterization of adenovirus 5 and 12 E1A proteins produced in insect cells. Virology 190: 733–745
Pei R, Berk AJ (1989) Multiple transcription factor binding sites mediate adenovirus E1A transactivation. J Virol 63: 3499–3506
Perricaudet M, LeMoullec J-M, Tiollais P, Pettersson U (1980) Structure of two adenovirus type 12 transforming polypeptides and their evolutionary implications. Nature 288: 174–176
Pines J, Hunter T (1990) Human Cyclin A is adenovirus E1A-associated protein p60 and behaves differently from cyclin B. Nature 346: 760–763.
Pope JH, Rowe WP (1964) Immunofluorescent studies of adenovirus 12 tumors and of cells transformed or infected by adenoviruses. J Exp Med 120: 577–587
Pugh BF, Tjian R (1992) TATA-binding protein is a classless factor. J Biol Chem 267: 679–682
Raychaudhuri P, Rooney R, Nevins JF (1987) Identification of an E1A-inducible factor that interacts with sequences within the adenovirus E4 promoter. EMBO J 6: 4073–4081
Raychaudhuri P, Bagchi S, Neill SD, Nevins JR (1990) Activation of the E2F transcription factor in adenovirus-infected cells involves E1A-dependent stimulation of DNA-binding activity and induction of cooperative binding mediated by an E4 gene product. J Virol 64: 2702–2710
Raychaudhuri P, Bagchi S, Devoto SH, Kraus VB, Moran E, Nevins JR (1991) Domains of the adenovirus E1A protein required for oncogenic activity are also required for dissociation of E2F transcription factor complexes. Gene Dev 5: 1200–1211
Reichel R, Kovesdi I, Nevins JR (1987) Developmental control of a promoter-specific factor that is also regulated by the E1A gene product. Cell 48: 501–506
Reichel R, Neill SD, Koveshdi I, Simon C, Raychaudhuri P, Nevins JR (1989) The adenovirus E4 gene, in addition to the E1A gene, is important for trans-activation of E2 transcription and for E2F activation. J Virol 63: 3643–3650
Richter JD, Slavicek JM, Schneider JF, Jones NC (1988) Heterogeneity of adenovirus type 5 E1A proteins: multiple serine phosphorylations induce slow-migrating electrophoretic variants but do not affect E1A-induced transcriptional activation or transformation. J Virol 62: 1948–1955
Rikitake Y, Moran E (1992) DNA-binding properties of the E1A-associated 300-kilodalton protein. Mol Cell Biol 12; 2828–2836
Robbins PD, Horowitz JM, Mulligan RC (1990) Negative regulation of human c-fos expression by the retinoblastoma gene product. Nature 346: 668–671
Roberts BE, Miller SJ, Kimelmann D, Cepko C, Lemischka IR, Mulligan RC (1985) Individual adenovirus type 5 early region 1A gene products elicit distinct alterations of cellular morphology and gene expression. J Virol 56: 404–413
Rochette-Egly C, Fromental C, Chambon P (1990) General repression of enhanson activity by the adenovirus-2 E1A proteins. Gene Dev 4: 137–150
Ruley HE (1983) Adenovirus early region 1A enables viral and cellular transforming genes to transform primary cells in culture. Nature 304: 602–606

Sassone-Corsi P, Borrelli E (1987) Promoter trans-activation of protooncogenes c-fos and c-myc, but not Ha-ras, by products of adenovirus early region 1A. Proc Natl Acad Sci USA 84: 6430–6433
Sawada Y, Fujinaga K (1980) Mapping of adenovirus 12 mRNA's transcribed from the transforming region. J Virol 36: 639–651
Sawadogo M, Roeder RG (1985) Interaction of a gene-specific transcription factor with the adenovirus major late promoter upstream the TATA box region. Cell 43: 165–175
Schneider JF, Fisher F, Goding CR, Jones NC (1987) Mutational analysis of the adenovirus E1A gene: the role of transcriptional regulation in transformation. EMBO J 6: 2053–2060
Schönthal A, Herrlich P, Rahmsdorf HJ, Ponta H (1988) Requirement for fos gene expression in the transcriptional activation of collagenase by other oncogenes and phorbol esters. Cell 54: 325–334
Schütte J, Throughlet J, Nau M, Segal S, Fedorko J, Minna J (1989) junB inhibits and c-fos stimulates the transforming and transactivating activities of c-jun. Cell 59: 987–997
Sharp PA (1992) Diverse transcriptional functions of the multisubunit eucaryotic TFIID complex. Cell 68: 819–821
Shenk T, Flint SJ (1991) Transcriptional and transforming activites of the adenovirus E1A proteins. Adv Cancer Res 57: 47–83
Shirayoshi Y, Miyazaki J-I, Burke PA, Hamada K, Appella E, Ozato K (1987) Binding of multiple nuclear factors to the 5'upstream regulatory element of the murine major histocompatibility class I gene. Mol Cell Biol 7: 4542–4548
Shirodkar S, Ewen M, DeCaprio JA, Morgan J, Livingston DM, Chittenden T (1992) The transcription factor E2F interacts with the retinoblastoma product and a p107-cyclin A complex in a cell cycle regulated manner. Cell 68: 157–166
Shivji MK, La Thangue NB (1991) Multicomponent differentiation-regulated trascription factors in F9 embryonal carcinoma stem cells. Mol Cell Biol 11: 1686–1695
Silver BJ, Bokar JA, Virgin JB, Vallen EA, Milsted A, Nelsen J (1987) Cyclic-AMP regulation of the human glycoprotein hormone α-subunit gene is mediated by an 18 base-pair element. Proc Natl Acad Sci USA 84: 2198–2202
Simon M, Fisch TM, Benecke BJ, Nevins JR, Heintz N (1988) Identification of multiple, functionally distinct TATA elements, one of which is the target in the hsp70 promoter for E1A regulation. Cell 52: 723–729
Simon MC, Rooney RJ, Fisch TM, Heintz N, Nevins JR (1990) E1A-dependent trans-activation of the c-fos promoter requires the TATAA sequence. Proc Natl Acad Sci USA 87: 513–517
Slansky J, Li Y, Kaelin WG, Farnham PJ (1993) A protein synthesis-dependent increase in E2F1 mRNA correlates with growth regulation of the dihydrofolate reductase promoter. Mol Cell Biol 13: 1610–1618
Sogawa K, Handa H, Fujisawa-Sehara A, Hiromasa T, Yamane M, Fujii-Kuriyama Y (1989) Repression of cytochrome P-450c gene expression by co-transfection with adenovirus E1A DNA. Eur J Biochem 181: 539–544
Spangler R, Bruner M, Dalie B, Harter ML (1987) Activation of adenovirus promoters by the adenovirus E1A protein in cell-free extracts. Science 237: 1044–1046
Spector DJ, McGrogan M, Raskas HJ (1978) Regulation of the appearence of cytoplasmatic RNAs from region 1 of the adenovirus 2 genome. J Mol Biol 126: 395–414
Spindler KR, Rosser DSE, Berk AJ (1984) Analysis of adenovirus transforming proteins from early regions 1A and 1B with antisera to inducible fusion antigens produced in Escherichia coli. J Virol 49: 132–141
Stein RW, Ziff EB (1987) Repression of insulin gene expression by adenovirus type 5 E1A proteins. Mol Cell Biol 7: 1164–1170
Stein RW, Corrigan M, Yaciuk P, Whelan J, Moran E (1990) Analysis of E1A-mediated growth regulation functions: binding of the 300-kilodalton cellular product correlates with E1A enhancer repression function and DNA synthesis-inducing activity. J Virol 64: 4421–4427
Stephens C, Harlow E (1987) Differential splicing yields novel adenovirus 5 E1A mRNAs that encode 30 kD and 35 kD proteins. EMBO J 6: 2027–2035
Svensson C, Bondesson M, Nyberg E, Linder S, Jones N, Akusjarvi G (1991) Independent transformation acitivity by adenovirus 5 E1A-conserved regions 1 or 2 mutants. Virology 182: 553–561
Thalmeier K, Synovzik H, Mertz R, Winnacker E-L, Lipp M (1989) Nuclear factor E2F mediates basic transcription and trans-activation by E1A of the human MYC promoter. Gene Dev 3: 527–436
Trentin JJ, Yabe Y, Taylor G (1962) The quest for human cancer viruses. Science 137: 835–841

Tsukamoto A, Ponticelli A, Berk AJ, Gaynor RB (1986) Genetic mapping of a major site of phosphorylation in adenovirus type 2 E1A proteins. J Virol 59: 14–22

Ulfendahl PJ, Linder S, Kreivi J-P, Nordqvist K, Sevensson C, Hultberg H, Akusjärvi G (1987) A novel adenovirus 2 E1A mRNA encoding a protein with transcription activation properties. EMBO J 6: 2037–2044

Van Dam H, Offringa R, Meijer I, Stein B, Smits AM, Herrlich P, Bos JL, Van der Eb AJ (1990) Differential effects of the adenovirus E1A oncogene on members of the AP-1 transcription factor family. Mol Cell Biol 10: 5857–5864

Van Dam H, Duyndam M, Rottier R, Bosch A, De Vries-Smits L, Herrlich P, Zantema A, Angel P, Van der Eb AJ (1993) Heterodimer formation of c-Jun and ATF-2 is responsible for induction of c-jun by the 243 amino acid adenovirus E1A protein. EMBO J 12: 479–487

Van Ormondt H, Galibert F (1984) Nucleotide sequences of adenovirus DNAs. In: Doerfler W (ed) The molecular biology of adenovirus II. Springer, Berlin Heidelberg New York, pp73–143 (Current topics in microbiology and immunology vol 110)

Vasavada R, Eager KB, Barbanti Brodano G, Caputo A, Ricciardi RP (1986) Adenovirus type 12 early region 1A proteins repress class I HLA expression in transformed human cells. Proc Natl Acad Sci USA 83: 5257–5261

Velcich A, Ziff EB (1985) Adenovirus E1A proteins repress transcription from the SV40 early promoter. Cell 40:705–716

Velcich A, Kern FG, Basilico C, Ziff EB (1986) Adenovirus E1A proteins repress expression from polyomavirus early and late promoters. Mol Cell Biol 6: 4019–4025

Wang H-GH, Rikitake Y, Carter MC, Yaciuk P, Abraham SE, Zerler B, Moran E (1993a) Identification of specific adenovirus E1A N-terminal residues critical to the binding of cellular proteins and to the control of cell growth J Virol 67: 476–488

Wang H-GH, Yaciuk P, Ricciardi RP, Green M, Yokoyama K, Moran E (1993b) The E1A products of oncogenic adenovirus serotype 12 include amino-terminally modified forms able to bind the retinoblastoma protein but not p300. J Virol 67: 4804–4813

Weitraub SJ, Prater CA, Dean DC (1992) Retinoblastoma protein switches the E2F site from positive to negative element. Nature 358: 259–261

Whitaker JL, Byrd PJ, Grand RJA, Gallimore PH (1984) Isolation and characterization of four adenovirus type 12-transformed human embryo kidney cell lines. Mol Cell Biol 4: 110–116

Whyte P, Ruely HE, Harlow E (1988a) Two regions of the adenovirus early region 1A proteins are required for transformation. J Virol 62: 257–265

Whyte P, Buchkovich KJ, Horowitz JM, Friend SH, Raybuck M, Weinberg RA, Harlow E (1988b) Association between an oncogene and an anti-oncogene: the adenovirus E1A proteins bind the retinoblastoma gene product. Nature 334: 124–129

Whyte P, Williamson NM, Harlow E (1989) Cellular targets for transformation by the adenovirus E1A proteins. Cell 56: 67–75

Wu L, Rosser DSE, Schmidt M, Berk AJ (1987) A TATA box implicated in E1A transcriptional activation of a simple adenovirus 2 promoter. Nature 326: 512–515

Yaciuk P, Moran E (1991) Analysis with specific polyclonal antiserum indicates that the E1A-associated 300-kDa product is a stable nuclear phosphoprotein that undergoes cell cycle phase-specific modification. Mol Cell Biol 11: 5389–5397

Yamaguchi M, Hayashi Y, Hirose F, Matsuoka S, Shiroki K, Matsukage A (1992) Activation of the mouse proliferating cell nuclear antigen gene promoter by adenovirus type 12 E1A proteins. Jpn J Cancer Res 83: 609–617

Yamaguchi-Iwai Y, Satake M, Murakami Y, Sakai M, Muramatsu M, Ito Y (1990) Differentiation of F9 embryonal carcinoma cells induced by the c-jun and activated c-Ha-ras oncogenes. Proc Natl Acad Sci USA 87: 8670–8674

Yee S-P, Branton PE (1985) Detection of cellular proteins associated with human adenovirus type 5 early region 1A polypeptides. Virology 149: 142–153

Yee S-P, Rowe DT, Tremblay ML, McDermott M, Branton PE (1983) Identification of human adenovirus early region 1 products by using antisera against synthetic peptides corresponding to the predicted carboxy termini. J Virol 46: 1003–1013

Yee AS, Reichel R, Koveshdi I, Nevins JR (1987) Promoter interaction of the E1A-inducible factor E2F and its potential role in the formation of a multi-component complex. EMBO J 6: 2061–2068

Yoshinaga S, Dean N, Han M, Berk AJ (1986) Adenovirus stimulation of transcription by RNA polymerase III: evidence of an E1A-dependent increase in transcription factor IIIC concentration. EMBO J 5: 343–354

Zamanian M, La Thangue NB (1992) Adenovirus E1A prevents the retinoblastoma gene product from repressing the activity of a cellular transcription factor. EMBO J 11: 2603–2610

Zerial M, Toschi L, Ryseck R-P, Schuermann M, Müller R, Bravo R (1989) The product of a novel growth factor activated gene, fosB, interacts with Jun proteins enhancing their DNA binding activity. EMBO J 8: 805–813

Zerler B, Moran B, Maruyama K, Moomaw J, Grodzicker T, Ruley HE (1986) Adenovirus E1A coding sequences which enables ras and pmt oncogenes to transform cultured primary cells. Mol Cell Biol 6: 887–899

Zu YL, Maekawa T, Matsuda S, Ishii S (1991) Complete putative metal finger and leucine zipper structures of CRE-BP1 are required for the E1A-induced trans-activation. J Biol Chem 266: 24134–24139

zur Hausen H (1973) Interactions of adenovirus type 12 with host cell chromosomes. Prog Exp Tumor Res 18: 240–259

Transcriptional Regulation of the Adenovirus E1A Gene

K. YOSHIDA, F. HIGASHINO, and K. FUJINAGA

1 Introduction

Two distinct classes of the regulatory sequences have been identified in eukaryotic transcription systems. Promoter sequences are located close to the transcription initiation site, and enhancer sequences can be located far upstream or downstream from the initiation site. Most of the enhancer consists of multiple DNA-binding sites for proteins which act either synergistically or antagonistically to modulate transcriptional activity (JOHNSON and MCKNIGHT 1989; MITCHELL and TJIAN 1989). The 72-base pair (bp) repeats of the simian virus 40 (SV40) early promoter/enhancer region constitute one of the best-characterized enhancer elements (HERR and CLARKE 1986; ZENKE et al. 1986; ONDEK et al. 1988). The region is composed of a number of modular units, and the combined action of these different units determines the degree of enhancer activity in different cell types. Promoter sequences include the TATA box and initiation sites of transcription, which are involved in the positioning of the transcription initiation site. Promoters of house-keeping genes contain GC-rich sequences and an initiator element instead of the TATA box. These *cis*-acting elements operate by interacting with

---

Department of Molecular Biology, Cancer Research Institute, Sapporo Medical University, School of Medicine, S-1, W-17, Chuo-ku, Sapporo 060, Japan

specific DNA-binding proteins. Therefore, an understanding of the mechanisms operational in transcriptional regulation of eukaryotic genes entails unravelling the mechanisms controlling the activity of these DNA-binding proteins (McKnight and Yamamoto 1992).

The adenovirus early region 1A (E1A) transcription unit is located within the left-terminal 4.5% of the double-stranded, linear viral DNA (Berk and Sharp 1978; Chow et al. 1979; Perricaudet et al. 1979; Kitchingman and Westphal 1980; Virtanen and Petterson 1983). The E1A gene is the first to be transcribed in productive infection (Nevins et al. 1979), and its transcription does not depend on the action of any other viral gene products; however, transcription of the remaining early viral genes requires E1A gene expression (Berk et al. 1979; Jones and Shenk 1979; Nevins 1981). Several enhancer elements which allow transcription from the E1A gene promoter immediately after infection have been localized upstream of and inside the E1A protein-coding region (Hearing and Shenk 1983a, b, 1986; Hen et al. 1983; Imperiale et al. 1983; Weeks and Jones 1983; Osborne et al. 1984; Herbst et al. 1990). These E1A enhancer elements operate by interaction with specific cellular proteins, some of which have been purified to homogeneity and the corresponding cDNA have also been isolated (Barrett et al. 1987; Bolwig et al. 1992; Higashino et al. 1993; Fognani et al. 1993).

Two related nuclear E1A proteins of 289 and 243 amino acids (aa) that differ only in an internal sequence of 46 aa unique to the larger protein are expressed from the E1A gene at early times of infection (Perricaudet et al. 1979; van Ormondt et al. 1980). The E1A proteins, especially the 289-aa protein, augment E1A gene transcription to high levels (Berk et al. 1979; Nevins 1981; Osborne et al. 1984; Hearing and Shenk 1985) and also stimulate promoter activities of other viral genes, including the promoters for the E1B, E2A, E3, and E4 early genes, as well as the major late promoter (Riccardi et al. 1981; Montell et al. 1982; Leff et al. 1984; Svenson and Akusjärvi 1984; Winberg and Shenk 1985). Although the mechanism of how E1A *trans*-activates other genes remains to be clearly defined, a great deal of progress has been made regarding the critical regions or domains of E1A proteins necessary for activation.

2 Transcriptional Control Elements

The transcriptional control region of the adenovirus E1A gene contains multiple *cis*-acting elements, and transcription of the E1A gene is regulated by a variety of cellular and viral protein factors. The upstream region of adenovirus subgroup C (Ad2 and Ad5) E1A genes has been analyzed extensively. More than 20 binding sites for cellular nuclear proteins have been mapped upstream of the E1A gene (Yoshida et al. 1989; Table 1). This region extends from the left terminus of viral genome to the E1A cap site, i.e., nucleotides (nt) 1–499. It contains the replication

Table 1. Binding sites and putative factors in the promoter/enhancer region of the adenovirus 5 E1A gene

Positions of DNase I Footprints	Core Sequences	Factors	Regulatory elements
–25 to +34			Hyphenated symmetry at the
–12 to +9			cap site
–21 to –38	TATTTATA	TFIID	TATA box
–45 to –72			
–36 to –83	TGACGT	ATF	
–91 to –99			
–165 to –178			
–181 to –197	TGAAATCTGAATAA	OCT1	
–195 to –204	AGGAAGT		Enhancer element I
–221 to –293	C/GGCA/TAA		Enhancer element II
	TTTTCGCGG/C	E2F	Two E2F binding sites
–257 to –273	CGGATGT	E1A-F	Degenerate enhancer element I
–289 to –304	AGGAAGT	E1A-F	Enhancer element I
–305 to –319			
–319 to –338	GTGACGT	ATF	24-bp enhancer element
–334 to –346	CGGAAGT	E1A-F	
–363 to –384			
–375 to –394	CGGAAGT	E1A-F	
–391 to –414	GTGACGT	ATF	
–391 to –420	GGGCGG	SP1	
–428 to –450	GTGACGT	ATF	
–442 to –447	GGGTGG	SPI	
–452 to –464	CCAATATATGATAAT	NFIII	DNA replication origin
458 to –498	TGGATTGAAGCCA	NFI	
		ORP-A	

origin of Ad DNA, inverted terminal repeat, signal sequences for the packaging viral DNA into virion, transcriptional enhancer elements for the E1A gene and other viral early genes, and promoter proximal elements of the E1A gene (Fig. 1).

2.1 Inverted Terminal Repeat

The Ad5 inverted terminal repeat (ITR) is 103 bp long and is located at either end of viral genome (Shinagawa and Padmanabhan 1980). The ITR contains the replication origin of viral DNA. The most terminal 51-bp sequence of the ITR is sufficient for maximal initiation of DNA replication (Challberg and Kelly 1989; Stillman 1989). Within this sequence, the most terminal 18 bp sequence can effect low levels of initiation and is required for initiation reaction. The early E2B gene products, the Ad preterminal protein (pTP), and DNA polymerase (Ad pol) bind to the terminal 18-bp sequence to make an initiation complex for DNA replication (Temperley and Hay 1992). The sequence adjacent to the 18-bp sequence is occupied by cellular nuclear factor I (NF I) and III (NF III) and these factors, especially NF I, augment initiation of DNA replication (Chen et al. 1990; Bosher

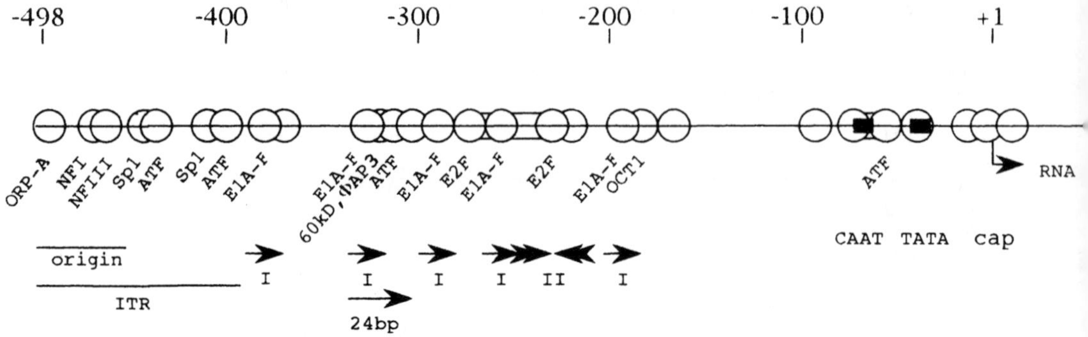

Fig. 1. The upstream promoter/enhancer region of the adenovirus E1A gene. Positions of the binding sites and names of the factors are indicated (Barrett et al. 1987; Jones et al. 1987; Rosenfeld et al. 1987; Kovesdi et al. 1987; Hardy and Shenk 1988; Leza and Hearing 1988; Lin and Green 1988; O'Neil et al. 1988; Bruder and Hearing 1989; Miralles et al. 1989; Prujin et al. 1989; Yoshida et al. 1989; Herbst et al. 1990). The positions relative to the cap site are shown *above the line*. *Arrows* below the line depict enhancer elements I and II and the 24-bp enhancer

et al. 1990; Mul et al. 1990; Mul and van der Vliet 1992; Hatfield and Hearing 1993). NF I and NF III are homologous to members of the CTF transcription factor family (Jones et al. 1987) and OCT-1 transcription factor of the POU domain protein family (O'Neil et al. 1988; Prujin et al. 1989), respectively. It appears that NF-I recruites the pTP–Ad pol complex to the terminal 18-bp sequence and stabilizes it through protein–protein interactions (Chen et al. 1990; Bosher et al. 1990; Mul and van der Vliet 1992). The ITR contains two other conserved sequences in different adenovirus subgroups (Shinagawa and Padmanabhan 1980): GC-rich sequences (GGGTGG: nt 51–56, and GGGCGG: nt 88–94) and the GTGACGT sequence (nt 63–69 and nt 95–101). These sequences have been shown to be binding sites for cellular proteins SP-1 and ATF, respectively (Hardy and Shenk 1988; Lin and Green 1988; Leza and Hearing 1988; Miralles et al. 1989). The ATF site resembles the cyclic adenosine monophosphate (AMP) response element (CRE), which mediates the response to cyclic AMP in cyclic AMP-inducible promoters (Roesler et al. 1988; Hardy and Shenk 1988; Lin and Green 1988).

A number of distinct and related ATF proteins have been reported to recognize the ATF/CRE sequence (Hoeffler et al. 1988; Hai et al. 1989). Several groups have found intrinsic promoter and enhancer activities in the ITR region using in vitro transcription and plasmid transfection assays. When the ITR was placed adjacent to the E1A TATA box, it stimulated E1A transcription in transfected cells (Hatfield and Hearing 1991). When the ITR or 51-bp replication origin were linked to a promoter-removed foreign gene, they were found to initiate transcription from a defined site (Miralles et al. 1989; Yoshida et al. 1990; Hatfield and Hearing 1991). The adenovirus E4 gene which is located close to the right-terminal ITR of the viral genome is activated by the ITR sequence as well (Ooyama et al. 1989).

2.2 E1A Enhancer Region

Several enhancer elements have been mapped upstream of the Ad5 and Ad2 E1A genes using a variety of assays (Fig. 1). The E1A gene should be expressed early after the entry of viral DNA into the cell nucleus, and E1A protein products *trans*-activate promoters of other viral early genes. Therefore, it is logical that the E1A gene is preceded by multiple enhancer elements. Transient expression assays revealed a 24-bp enhancer segment located between positions –320 and –343 relative to the Ad2 E1A cap site (Fig. 1) (HEN et al. 1983). It can stimulate transcription from the Ad major late promoter when inserted in its natural orientation with respect to the direction of transcription.

However, the 24-bp enhancer element has been found to be dispensable for efficient transcription within virus-infected HeLa cells (HEARING and SHENK 1983b). The viral mutants used for such infection experiments contained all the viral genes including the E1A-coding sequences. It is likely that in virus-infected cells after an accumulation of E1A proteins has occurred, E1A gene transcription can be efficiently stimulated by an autoactivation mechanism of E1A proteins, independently of the 24-bp enhancer element. Indeed, the requirement for the E1A enhancer element appears to be less pronounced in the presence of E1A-coding region (SASSONE-CORSI et al. 1983). It is not clear whether the 24-bp enhancer element plays a role in the regulation of E1A gene transcription before the onset of E1A protein synthesis within infected cells. On the other hand, infection of rat embryo fibroblast CREF cells with a virus mutant containing a deletion from positions –393 to –304 including the 24-bp enhancer element resulted in more rapid E1A mRNA expression and by a three- to fivefold greater amount, as compared to that of wild-type virus (HERBST et al. 1990). This indicates another role for the 24-bp enhancer element: suppression of E1A enhancer function in rodent CREF cells.

Two distinct cellular proteins have been reported to bind to the 24-bp enhancer element (BARRETT et al. 1987; FOGNANI et al. 1993). A 60-kDa protein, the binding site of which overlaps a TGTGG sequence in this enhancer, was isolated from HeLa cells (BARRETT et al. 1987). Although a GTGG motif is present at the core C element (GT1 motif) of the SV40 enhancer, this protein appears to be different from a core C-binding protein AP3 (CHIU et al. 1987). In vitro quantitation of the nuclear protein that binds to the E1A upstream sequence demonstrated similar levels of activities for ATF and E2F, but a ten- to 20-fold higher activity for AP3-like factor (ϕAP3) in rat CREF cells than in HeLa cells (HERBST et al. 1990). ϕAP3 binds to an ATGTGGCAAAAG sequence of the 24-bp enhancer element (Fig. 2) and negatively regulate E1A enhancer function in undifferentiated rodent cells through the ϕAP3-binding site (HERBST et al. 1990; FOGNANI et al. 1993). ϕAP3 protein is 100–110 kDa in size (FOGNANI et al. 1993) and is distinct from the 60-KDa HeLa cell protein. The 24-bp enhancer element also contains potential binding sequences for ATF (GTGACGT: positions –330 to –323) and *ets*-related proteins (CGGATGT: positions –344 to –338) (LIN and GREEN 1988; BRUDER and HEARING 1989; YOSHIDA et al. 1989), both of which partially overlap the ϕAP3-binding sequence (Fig. 2).

Deletion analysis with viral mutants defined two separate E1A enhancer elements which regulate transcription of the early genes within infected cells (HEARING and SHENK 1986). Enhancer element I stimulates transcription of the E1A gene, whereas enhancer element II stimulates transcription from other early genes (E2A, E3, and E4 genes) located far from the E1A gene on the viral genome. E1A-specific element I is repeated at positions –301 to –289 and positions –203 to –192 and contains a 10-bp consensus sequence (C/AGGAAGTGAA/C) (HEARING and SHENK 1986; Table 1). This sequence is found at the same location in the distantly related human adenovirus subgroup B (Ad7) and A (Ad12) (VAN ORMONDT and GALIBERT 1984). The element II with a C/GGCCA/TAA consensus sequence is present in four copies between the two repeats of element I and spans a region of about 30 bp (Fig. 1, Table 1). Element II moderately stimulates transcription from all other early genes in infected cells. The factors that interact with element II have not yet been described. Deletion of one copy of repeated element I decreases E1A mRNA levels two- to threefold, but when the entire region, including both copies of element I and four copies of element II (e.g., positions –141 to –305), is removed, E1A mRNA levels drop 15- or 20-fold (HEARING and SHENK 1983b, 1986).

Further analysis of E1A upstream mutants with small internal deletions identified an additional but degenerate element I sequence located between –263 and –272 (Table 1) and revealed an apparent hierarchy of deletion effects on E1A mRNA levels (BRUDER and HEARING 1991). Element I at the –270 site had the most significant effect, whereas the –300 site had an intermediate effect and the element 1 at the –200 site was marginally required for E1A mRNA expression. Analysis of double mutants of these element I sites revealed synergistic interactions between the element 1 sites at –270 and –200 and also those at –270 and –300. Chemical interference assays with the E1A upstream region (BRUDER and HEARING 1989; YOSHIDA et al. 1989) demonstrated in vitro the activity that is specific for the core sequence of the E1A enhancer element I. These binding factors (called EF-1A and E1A-F) also bound to element I-related sequences

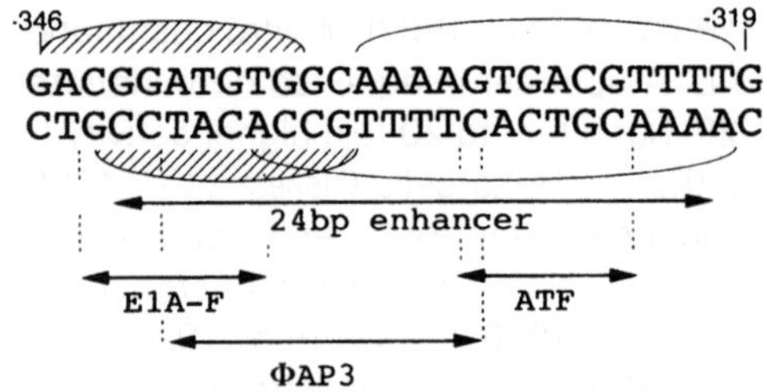

Fig. 2. Protein-binding sites in the 24-bp enhancer element. Sequences of both strands in the 24-bp enhancer element (positions –320 to –343) are shown. Numbers *above* the sequence represent positions relative to E1A cap site (+1). DNase I footprints with crude HeLa cell nuclear extracts (YOSHIDA et al. 1989) are shown by *ovals*. The core-binding sequences for E1A-F, ATF, and ϕAP3 factors (LIN and GREEN 1988; BRUDER and HEARING 1989; YOSHIDA et al. 1989; HERBST et al. 1990) are shown *below* the sequence

located further upstream (positions –337 to –346 and positions –380 to –389). Overall, five element I-related sequences reside separately on the upstream region of the E1A gene (Fig. 1). Comparison of these sequences revealed the consensus motif for the element I-binding factor: A/CGGAA/TGT (Table 2). This resembles the sequence (C/GAGGAAGT/C) recognized by the ets-1 protein, a member of the ets protein family (MacLeod et al. 1992; Wasylyk et al. 1993). Adjacent to element I, the E2F-binding site is repeated at positions –218 to –225 and positions –281 to –288 (Fig. 1, Table 1). The latter E2F site has been shown to stimulate transcription of a heterogeneous, linked gene in an E1A-dependent fashion in a cotransfection assay (Kovesdi et al. 1987). E2F was initially identified as a positively acting transcription factor which mediates transcriptional activation of the E2A gene by E1A gene products (Kovesdi et al. 1986). However, in virus-infected HeLa cells, transcription of the E1A gene is not influenced by deletion of the two E2F sites in the context of the intact E1A promoter/enhancer region (Bruder and Hearing 1989).

2.3 Proximal Promoter Region

Sequences required for transcriptional activation in the promoter proximal region were identified using viral mutants and an in vitro transcription system (Hearing and Shenk 1983a). All of the mutants carrying a small deletion between the E1A cap site (+1) and position –170 grow well in HeLa cells. None of the mutants altered E1A mRNA levels by more than twofold. However, deletions of the TATA box are the most effective (38% reduction of wild-type E1A mRNA level) and generated E1A mRNA with heterogeneous 5′ ends. The hyphenated symmetry at the E1A cap site that is conserved among different adenovirus serotypes (van Ormondt and Galibert 1984) is not essential for efficient transcription and cap site selection of the E1A gene (Hearing and Shenk 1983a). A DNA fragment upstream of the E1A TATA box (SacII to pVUII: positions –45 to –188) can enhance expression of the 72-kDa protein product of the E2A gene in a

Table 2. Comparison of E1A enhancer element I and its related sequences

Promoters	Sequences	Positions
Ad5 E1A	GGCGGAAGTG	–389/–380
Ad5 E1A	GACGGATGTG	–346/–377
Ad5 E1A	ACAGGAAGTG	–302/–293
Ad5 E1A	GGCGGATGTT	–272/–263
Ad5 E1A	AACGGAAGTG	–204/–195
Ad5 E4	AACGGAAGTG	–138/–147
Polyomavirus	GCAGGAAGTG	5106/5115
Consensus	A/CGGAA/TGT	

Element I and its related sequences of the adenovirus (Ad) E1A promoter/enhancer (Hearing and Shenk 1986; Bruder and Hearing 1989, 1991; Yoshida et al. 1989), E4TF1 site of the adenovirus E4 promoter (Watanabe et al. 1993), and PEA3 site of the polyomavirus enhancer (Xin et al. 1992).

transfection assay, when the DNA fragment is linked to the E2A gene (IMPERIALE et al. 1983). However, this DNA fragment does not have any stimulatory effects on the activity of the homologous E1A promoter or heterologous Ad2 major late promoter (SASSONE-CORSI et al. 1983). This region is also dispensable for efficient transcription of the E1A gene within virus-infected cells (HEARING and SHENK 1983a).

2.4 Control Elements in the E1A-Coding Region

Viruses carrying mutations within the E1A-coding region display a five- to 25-fold reduction of E1A mRNA levels within infected cells (BERK et al. 1979; NEVINS 1981; OSBORNE et al. 1984; HEARING and SHENK 1985). This defect can be complemented when E1A proteins is supplied *in trans*. Thus, E1A gene products positively regulate expression of the cognate gene. Since some mutants are not completely complemented, their defects may include *cis*-acting mutations which depress transcription of the E1A gene. Such a critical DNA sequence within the E1A coding region has been identified 399-bp downstream from the E1A cap site (OSBORNE et al. 1984). However, mutations in the E1A-coding region may have this effect post-transcriptionally. The out-of-frame mutation of the E1A protein-coding region may cause premature termination in translation and instability of mutant E1A mRNA and result in a reduction in E1A mRNA levels (HEARING and SHENK 1985).

3 Transcription Factors

Several enhancer and promoter elements are required for E1A gene transcription. These elements operate by interacting with specific cellular transcription factors. Thus far, three distinct cellular proteins which bind to E1A enhancer elements have been characterized and the corresponding cDNA have been cloned (BOLWIG et al. 1992; HIGASHINO et al. 1993; FOGNANI et al. 1993).

3.1 *ets*-Related Proteins

E1A enhancer element contains the core sequence C/AGGAA / TGT (Table 2), the mutation of which abolishes the binding of cellular factors (BRUDER and HEARING 1989; YOSHIDA et al. 1989). By utilizing the ability to bind to E1A enhancer sequences, the cDNA encoding an E1A enhancer-binding protein (called E1A-F) was isolated from a HeLa cell λgt11 expression library (HIGASHINO et al. 1993).

Sequence analysis of the cDNA reveals that E1A-F belongs to the *ets* oncogene family and appears to be a human homologue of mouse *ets*-related protein PEA3 (polyomavirus enhancer activator 3) (HIGASHINO et al. 1993). PEA3 cDNA has been isolated by specific binding to the E1A element I-related PEA3 sequence of polyomavirus enhancer (XIN et al. 1992) (Table 2). Thus far, several *ets* oncogene family members have been cloned (MACLEOD et al. 1992; WASYLYK et al. 1993). The v-*ets* oncogene, the first member of the *ets* oncogene family, was found in the viral fusion protein ($p135^{gag\text{-}myb\text{-}ets}$) of the chicken acutely transforming retrovirus E26 (LEPRINCE et al. 1983; NUNN et al. 1983). All of the *ets* oncogene family members contain the ETS domain, which is approximately 85 aa long and in most cases is located at the carboxy terminus of the protein (Fig. 3). The ETS domain is required for the *ets* site-specific DNA binding, but does not contain the classical structure involved in DNA binding such as homeodomain, leucine zipper, helix-turn-helix, and zinc finger motif. Transcriptional activation domains of *ets*-related proteins are located in the amino-terminal part. E1A-F contains an acidic and glutamine-rich region in the amino-terminal half (Fig. 3) which is found in the activation domains of many transcription factors (PTASHNE 1988). Cotransfection experiments indicate that E1A-F can activate an SV40 early core promoter construct linked to multimeric wild-type E1A-F-binding sites (HIGASHINO et al. 1993).

When a E1A-F expression plasmid was cotransfected with a CAT reporter gene under the control of Ad5 E1A promoter/enhancer (positions –498 to 51) in human oesteosarcoma MG63 cells, E1A-F enhanced CAT gene expression five-fold (our unpublished results). However, E1A gene products can induce E1A promoter activity to much higher levels in a transfection assay (SASSONE-CORSI et al 1983; YAMAZAKI et al. 1992). Cotransfection with the E1A gene expression plasmid resulted in a 20- to 50-fold increase of CAT gene expression (our unpublished results). E1A-F may exert its effect on E1A promoter activity in the absence of E1A proteins within infected cells at very early times of infection.

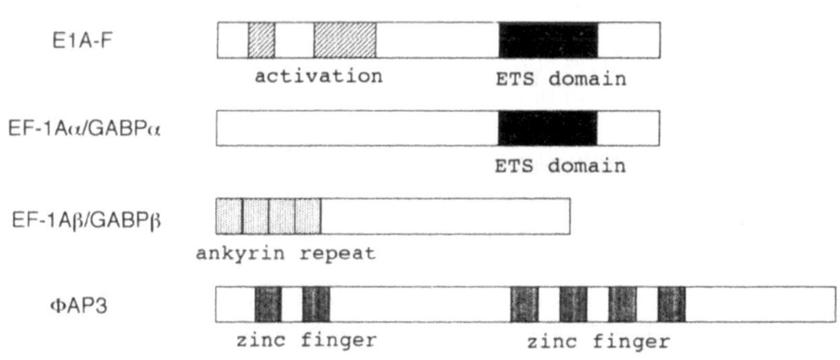

Fig. 3. The *ets*-related proteins E1A-F and EF-1A/GABP and the zinc finger protein ϕAP3. The DNA-binding domain (residues 315–399) and putative activation domain (residues 27–53 and 126–222) of E1A-F are shown (HIGASHINO et al. 1993). The DNA-binding domain of GABPα is from residue 316 to 400 and the ankyrin repeat of GABPβ is from residue 1 to 130 (LAMARCO et al. 1991). Human EF-1A protein is highly related to murine GABP protein, judging from subunit composition, DNA-binding property, and antibody reactivity (BOLWIG et al. 1992). The zinc finger regions of ϕAP3 required for DNA-binding are located at two separate regions (residues 35–84 and 277–409) (FOGNANI et al. 1993)

Another *ets*-related protein, EF-1A, has been purified from HeLa cells using conventional chromatography (Bolwig et al. 1992). EF-1A protein is composed of two subunits, EF-1Aα (approximately 60kDa) and EF-1Aβ (approximately 50kDa), which can make an approximately 200-kDa multimeric complex in solution and on a dimeric EF-1A binding site. EF-1Aα can bind to E1A enhancer element I, whereas EF-1Aβ possesses no apparent DNA-binding activity. The subunit composition and DNA-binding properties of EF-1A highly resemble murine GA-binding protein (GABP), which binds to the purine-rich regulatory sequence (CGGAAG/A) of the ICP4 immediately early gene of human herpes simplex virus type 1 (LaMarco et al. 1991).

Immunochemical assays with the anti-GABP antibody revealed that EF-1Aα is related to the *ets*-related α-subunit of GABP and EF-1Aβ is related to the β-subunit of GABP (Fig. 3). The β-subunit of GABP contains four 32- to 33–aa ankyrin repeats in the amino-terminal region (Fig. 3). The ETS domain of GABPα and ankyrin repeats of GABPβ are required for multiheterodimerization of α- and β-subunits (Thompson et al. 1991). A human homologue of murine GABP, E4TF1, has been purified and the corresponding cDNA has been cloned (Watanabe et al. 1993). E4TF1 is composed of a p53 subunit harboring ankyrin repeats and a p60 subunit harboring the ETS domain. Functional aspects of E4TF1 have been characterized by in vitro transcription assay of the Ad5 E4 gene (Watanabe et al. 1990). Whereas the p53–p60 complex can activate E4 gene transcription, neither the p53 nor the p60 subunits alone can stimulate transcription of the E4 gene. E4TF1 and EF-1A appear to be the same protein. However, the transcriptional properties of E4TF1 and EF-1A have not yet been described in the context of the E1A promoter/enhancer. There is no distinct homology between GABPα/E4TF1p60 and E1A-F/PEA3 except for the ETS domain.

3.2 Negatively Acting Zinc Finger Protein

Expression of the Ad5 E1A gene appears to be suppressed in rodent fibroblast cells by the ϕAP3 protein, which binds to the 24-bp enhancer element located approximately 340bp upstream of the E1A cap site (Herbst et al. 1990). The cDNA encoding ϕAP3 protein has been isolated by screening a λgt11 mouse liver cDNA expression library for ϕAP3 site-specific DNA binding and has been characterized (Fognani et al. 1993). ϕAP3 is 624 aa long and contains six well-conserved zinc finger DNA-binding motifs ($C\text{-}X_2\text{-}C\text{-}X_{12}\text{-}H\text{-}X_3\text{-}H$) found in the GLI-Kruppel family of human zinc finger proteins. Two of the six zinc finger motifs are located at the amino terminus and the other four are located at the middle of the ϕAP3 protein (Fig. 3). Expression of the ϕAP3 is ubiquitous, although levels varied in different mouse tissues. Overexpression of the ϕAP3 protein in human 293 and WI38 cells using a cytomegalovirus-based expression vector induced a fivefold reduction of reporter gene expression under the control of Ad5 E1A promoter/enhancer (positions 45 to −498). When the ϕAP3-binding site of the 24-bp enhancer element was mutated, ϕAP3 did not reduce expression of the reporter gene.

Thus, ϕAP3 appears to be a negative regulator of E1A enhancer activity. When rodent cells are infected with human adenoviruses, the yield of virus is very low compared with that in infected human cells, and cells with characteristically transformed phenotypes appear with low frequency. Negative regulation of the E1A gene expression may provide a mechanism that distinguishes between productive infection and virus-induced cell transformation.

4 Autoactivation

The adenovirus E1A gene encodes nuclear proteins which have pleiotropic effects on viral and cell physiology. These include transcriptional activation or repression, immortalization of primary cells, cooperative transformation with other oncogenes, and modulation of cell cycle progression (BERK 1986; MORAN and MATHEWS 1987; FLINT and SHENK 1989). These functions are carried out by two related E1A proteins of 289 aa and 243 aa which differ in an internal 46-aa sequence unique to the larger 289-aa protein (PERRICAUDET et al. 1979; VAN ORMONDT et al. 1980). During the early phase of productive infection, E1A proteins stimulate transcription from other viral genes including the promoters for the early region E1B, E2A, E3, and E4, as well as the E1A promoter itself and the major late promoter (RICCARDI et al. 1981; MONTELL et al. 1982; LEFF et al. 1984; SVENSON and AKUSJÄRVI 1984; WINBERG and SHENK 1985). Attention has been focused mainly on the larger 289-aa protein, because the 46-aa region conserved among different adenovirus subgroups (conserved region 3, CR3) is important for the transcriptional activation function of the E1A gene (LILLIE et al. 1987). The 243-aa protein has been reported to stimulate gene transcription only in limited cases (KRIPPL et al. 1984; LEFF et al. 1984; WINBERG and SHENK 1985). E1A proteins also activate transcription from a variety of cellular promoters, including the 72-kDa heat shock protein gene (KAO and NEVINS 1983) and β-tubulin gene (STEIN and ZIFF 1984). These viral and cellular genes do not necessarily contain common promoter elements, and E1A proteins appear not to have any particular sequence-specific DNA-binding activities (FERGUSON et al. 1985; CHATTERJEE et al. 1988).

Mutational analysis of E1A-inducible promoters indicates that most of the *cis*-acting promoter elements which are required for basal activities of these promoters also contribute to their E1A-induced activities (BERK 1986; FLINT and SHENK 1989). Thus far, several *cis*-acting elements and cellular factors such as TFIID, ATF-2/CREBP1, E4F, E2F, SP-1, and TFIIIC have been reported to mediate E1A-induced transcriptional activation (BERK 1986; FLINT and SHENK 1989). As described in Sects. 2.4 and 3.1, the adenovirus E1A gene is activated by its own gene products. Nuclear run-on assays of virus-infected cells indicate that E1A proteins increase the rate of transcription of the E1A gene (HEARING and SHENK 1985). Although the mechanism by which E1A autoactivation occurs is not clearly understood, the E1A protein might interrupt the function of a repressor which binds to the regulatory sequence and inhibits E1A gene transcription. Alternatively, the E1A protein might stimulate an interaction of positively acting transcription factors with the regulatory sequence of the E1A gene. However, specific regulatory sequences which mediate E1A autoactivation could not be found upstream of the E1A gene (HEARING and SHENK 1985). Double mutant viruses carrying a variety of deletions in the upstream region of the E1A gene and containing a second mutation within the E1A coding region were used to analyze the roles of E1A upstream sequences on E1A autoactivation (HEARING and SHENK 1985). All double mutants express only low levels of E1A mRNA, but expressed high levels of E1A mRNA when E1A proteins were provided by coinfection with a wild-type virus; however, the restoration of E1A mRNA levels by E1A proteins supplied *in trans* was less pronounced in the E1A TATA box mutants. Furthermore, E1A proteins can activate transcription from a foreign gene transcription unit substituted for the E1A gene on the viral genome. Another study raises the possibility that the DNA region responsible for E1A autoregulation cannot be separated from those sequences required for the basal level of the Ad12 E1A gene in a transfection assay (YAMAZAKI et al. 1992). These studies suggest that sequences necessary for the constitutive expression of the E1A gene also respond to E1A-induced autoactivation.

The 289– but not the 243-aa E1A protein has been shown to be required for E1A autoactivation in virus-infected and plasmid-transfected cells (HEARING and SHENK 1985; YAMAZAKI et al. 1992). This suggests that the CR3 region unique to the 289-aa protein is essential for E1A autoactivation. To identify the E1A protein-coding region required for E1A autoactivation, the CAT reporter gene under the control of E1A promoter/enhancer (positions –498 to 51) was cotransfected with mutants carrying a deletion in the E1A gene (Fig. 4). The requirement for the CR3 region is clear, because deletions of the CR3 region resulted in loss of the activity for autoactivation. The involvement of the carboxyl-terminal region of the E1A protein could be excluded, because the 243-aa protein has no apparent effects on E1A promoter activity (HEARING and SHENK 1985; YAMAZAKI et al. 1992). The CR2 region is also dispensable (Fig. 4). The CR1 and its adjacent amino-terminal region are required for efficient autoactivation (Fig. 4).

E1A proteins can dissociate inactive E2F-cellular protein complexes including Rb (retinoblastoma susceptibility gene-1) and Rb-related p107 and p130 proteins and release transcriptionally active E2F, which can then stimulate transcription of genes, for example the adenovirus E2A gene, whose promoter contains E2F-binding sites (NEVINS 1992; LEVINE 1993). For activation of E2F, both the CR1 and CR2 regions have been shown to be required (BAGCHI et al. 1990; RAYCHAUDHURI et al. 1991). Since the CR2 region is dispensable for E1A autoactivation (Fig. 4), the activity of E1A to dissociate the E2F complex is not associated with autoactivation. The CR3 region has been shown to be required for direct protein–protein interaction with cellular transcription factors, TATA box-binding factor TFIID, and ATF-2 (HORIKOSHI et al. 1991; LEE et al. 1991; LIU and GREEN 1994). It is believed that these cellular factors may serve to recruit the CR3 region of the E1A protein to a target promoter DNA. Interestingly, adenovirus E1A promoter/

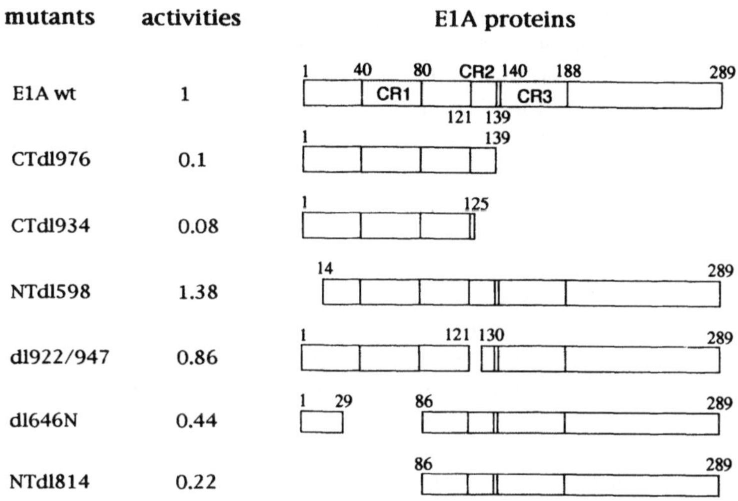

Fig. 4. Effects of E1A mutations on E1A autoactivation. Various mutants of the E1A-coding region used to map the regions of E1A required for E1A autoactivation are shown *below* a representation of the 289-amino acid E1A protein. CR1, CR2, and CR3 refer to the conserved regions 1, 2, and 3 of E1A proteins, respectively (MORAN and MATHEWS 1987). Numbers located *above the bars* refer to the amino acid positions at the boundary of the deletion. These mutants were provided by WHYTE et al. (1988, 1989). Human oesteosarcoma MG63 cells were transfected with 1 µg p5E1A cat plasmid and 0.5µg wild-type or mutant Ad5 E1A genomic plasmids. CAT (chloramphenicol acetyltransferase) gene expression is under the control of the E1A promoter/enhancer (positions –498 to 51). Fold stimulation of the CAT gene expression is shown in values normalized to that of wild-type E1A plasmid. Wild-type E1A plasmid (E1A *wt*) enhanced CAT gene expression 20- to 50-fold

enhancer harbors at least four putative ATF/CREB sites (Fig. 1). However, whether these ATF sites play a role in E1A autoactivation in the context of the E1A promoter/enhancer is not clear.

5 Autorepression

E1A proteins negatively regulate SV40 and polyomavirus early promoters whose activities depend on the enhancer sequence (BORELLI et al. 1984; VELCICH and ZIFF 1985; VELCICH et al. 1986). E1A proteins also repress the expression of their own gene (SMITH et al. 1985, 1986; LARSEN and TIBBETS 1987). Several cellular genes, including the class I major histocompatibility complex (MHC) and the insulin genes, are targets for the negative action of E1A proteins (SCHRIER et al. 1983; VAESSEN et al. 1987; STEIN and ZIFF 1987). In contrast to transcriptional activation by E1A, both 289- and 243-aa proteins appear to be active in repression, and CR1 and CR2 common to both E1A proteins are required for E1A-induced repression (LILLIE et al. 1986; SCHNEIDER et al. 1987). Although the mechanism of E1A repression is not understood, enhancer-dependent repression of the SV40 early promoter by E1A does not occur through a specific enhancer element; however, individual *cis*-acting elements constituting the enhancer can respond to E1A repression (ROCHETTE-EGLY et al. 1990). With respect to E1A-induced repression of the E1A gene, the E1A gene could be down-regulated at late times of infection (SMITH et al. 1985, 1986). E1A mRNA levels increase rapidly and abundantly at early times of infection, but after the onset of adenovirus DNA replication, E1A mRNA levels increase only modestly, compared with the significant increase in late stage-specific mRNA and with the increase in copy number of adenovirus DNA.

An Ad2 E1A expression vector which contains the SV40 origin of DNA replication can replicate to high copy numbers in monkey COS cells and has permitted the examination of E1A gene transcription from newly replicated templates (SMITH et al. 1985). In this system, a mutant expressing the amino-terminal 70 aa of E1A proteins overexpressed E1A mRNA approximately 30-fold. In combination with the fact that E1A proteins are capable of repressing promoters whose activities depends on an enhancer element, E1A proteins could down-regulate E1A mRNA levels through their repression of transcription and thus account for the inhibition of E1A gene expression following DNA replication. Another report also defined autorepression activity of E1A proteins (LARSEN and TIBBETS 1987). Ad3*hr*15, a defective mutant of subgroup B Ad3, duplicates an enhancer sequence located approximately 200 bp upstream of the Ad3 E1A cap site (LARSEN et al. 1986). Ad3*hr*15 yields no detectable E1A mRNA after abortive infection of HeLa cells, although wild-type Ad3 virus expresses abundant E1A mRNA. Revertants of Ad3*hr*15, which retain a duplicated enhancer sequence similar to that present in parental Ad3*hr*15 but carry a short in-frame deletion (residues 128–150 or 134–142) of the E1A protein-coding region, have been isolated (LARSEN and TIBBETS 1987). These revertants restored expression of E1A mRNA to wild-type levels. The in-frame deletions were located immediately downstream of the CR2 domain that is critical for Ad5 E1A function of transcriptional repression. This region may play a role in autorepression of the Ad3 E1A gene expression.

References

Bagchi S, Raychaudhuri P, Nevins JR (1990) Adenovirus E1A proteins, can dissociate heteromeric complexes involving the E2F transcription factor: a novel mechanism for E1A trans-activation. Cell 62: 659–669

Barrett P, Clark L, Hay RT (1987) A cellular protein binds to a conserved sequence in the adenovirus type 2 enhancer. Nucleic Acids Res 15: 2719–2735

Berk AJ (1986) Adenovirus promoters and E1A transactivation. Annu Rev Genet 20: 45–79

Berk AJ, Sharp PA (1978) Structure of the adenovirus 2 early mRNAs. Cell 12: 45–55

Berk AJ, Lee F, Harrison T, Williams J, Sharp PA (1979) Pre-early Ad5 gene product regulates synthesis of early mRNAs. Cell 17: 935–944

Bolwig GM, Bruder JT, Hearing P (1992) Different binding site requirements for binding and activation for the bipartite enhancer factor EF-1A. Nucleic Acids Res 20: 6555–6564

Borelli E, Hen R, Chambon P (1984) Adenovirus-2 E1a products repress enhancer-induced stimulation of transcription. Nature 312: 608–612

Bosher J, Robinson EC, Hay RT (1990) Interactions between the adenovirus type 2 DNA polymerase and the DNA binding domain of nuclear factor 1. New Biol 2: 1083–1090
Bruder JT, Hearing P (1989) Nuclear factor EF-1A binds to the adenovirus E1A core enhancer element and to other transcriptional control regions. Mol Cell Biol 9: 5143–5153
Bruder JT, Hearing P (1991) Cooperative binding of EF-1A to the E1A enhancer region mediates synergistic effects on E1A transcription during adenovirus infection. J Virol 65: 5084–5087
Challberg MD, Kelly JK (1989) Animal virus DNA replication. Annu Rev Biochem 58: 671–717
Chatterjee PK, Bruner M, Flint SJ, Harter ML (1988) DNA-binding properties of an adenovirus 289R E1A proteins. EMBO J 7: 835–841
Chen M, Mermod N, Horwitz MS (1990) Protein–protein interactions between adenovirus DNA polymerase and nuclear factor I mediate formation of the DNA replication preinitiation complex. J Biol Chem 265: 18634–18642
Chiu R, Imagawa M, Imbra RJ, Bockoven JR, Karin M (1987) Multiple cis- and trans-acting elements mediate the transcriptional response to phorbol esters. Nature 329: 648–651
Chow LT, Broker TR, Lewis JB (1979) Complex splicing patterns of RNAs from the early regions of adenovirus-2. J Mol Biol 134: 265–303
Ferguson B, Krippl B, Andrisani O, Jones N, Westphal H, Rosenberg M (1985) E1A 13s and 12s mRNA products made in *Escherichia coli* both function as nucleus-localized transcription activators but do not directly bind DNA. Mol Cell Biol 5: 2653–2661
Flint SJ, Shenk T (1989) Adenovirus E1A protein paradigm viral transactivator. Annu Rev Genet 23: 141–161
Fognani C, DellaValle G, Babiss LE (1993) Repression of adenovirus EIA enhancer activity by a novel zinc finger-containing DNA-binding protein related to the GLI-Kruppel protein. EMBO J 12: 4985–4992
Hai T, Liu F, Coukos WJ, Green MR (1989) Transcription factor ATF cDNA clones: an extensive family of leucine zipper proteins able to selectively form DNA-binding hetero-dimers. Genes Dev 3: 2083–2090
Hardy S, Shenk T (1988) Adenoviral control regions activated by EIA and the cAMP response element bind to the same factor. Proc Natl Acad Sci USA 85: 4171–4175
Hatfield L, Hearing P (1991) Redundant elements in the adenovirus type 5 inverted terminal repeat promote bi-directional transcription in vitro and are important for virus growth in vivo. Virology 184: 265–276
Hatfield L, Hearing P (1993) The NFIII/OCTI binding site stimulates adenovirus DNA replication in vivo and is functionally redundant with adjacent sequences. J Virol 67: 3931–3939
Hearing P, Shenk T (1983a) Functional analysis of the nucleotide sequence surrounding the cap site for adenovirus type 5 region 1A messenger RNAs. J Mol Biol 167: 809–822
Hearing P, Shenk T (1983b) The adenovirus type 5 E1A transcriptional control region contains a duplicated enhancer element. Cell 33: 695–703
Hearing P, Shenk T (1985) Sequence-independent autoregulation of the adenovirus type 5 E1A transcription unit. Mol Cell Biol 5:3214–3221
Hearing P, Shenk T (1986) The adenovirus type 5 E1A enhancer contains two functionally distinct domains: one is specific for E1A and the other modulates all early units *in cis*. Cell 45: 229–236
Hen R, Borrelli E, Sassone-Corsi P, Chambon P (1983) An enhancer element is located 340 base pairs upstream from the adenovirus E1A gene. Nucleic Acids Res 11: 8747–8760
Herbst RS, Pelletier M, Boczko EM, Babiss LE (1990) The state of cellular differentiation determines the activity of the adenovirus E1A enhancer element: evidence for negative regulation of enhancer function. J Virol 64: 161–172
Herr W, Clarke J (1986) The SV40 enhancer is composed of multiple functional elements that can be compensated for one another. Cell 45: 461–470
Higashino F, Yoshida K, Fujinaga Y, Kamio K, Fujinaga K (1993) Isolation of a cDNA encoding the adenovirus E1A enhancer binding protein: a new human member of the *ets* oncogene family. Nucleic Acids Res 21: 547–553
Hoeffler JP, Meyer TE, Yun Y, Jameson JL, Habener JF (1988) Cyclic-AMP-responsive DNA-binding protein: structure based on a cloned placental cDNA. Science 242: 1430–1433
Horikoshi N, Maguire K, Kralli A, Maldonado E, Reinberg D, Weinmann R (1991) Direct interaction between adenovirus E1A protein and the TATA box binding transcription factor II D. Proc Natl Acad Sci USA 88: 5124–5128
Imperiale MJ, Feldman LT, Nevins JR (1983) Activation of gene expression by adenovirus and herpesvirus regulatory genes acting *in trans* and by a *cis*-acting adenovirus enhancer element. Cell 35: 127–136
Johnson PF, McKnight SL (1989) Eukaryotic transcriptional regulatory proteins. Annu Rev Biochem 58: 799–839

Jones KA, Kadonaga TJ, Rosenfeld PJ, Kelly TJ, Tjian R (1987) A cellular DNA-binding protein that activates eukaryotic transcription and DNA replication. Cell 48: 79–89
Jones N, Shenk T (1979) An adenovirus type 5 early gene function regulates expression of other early viral genes. Proc Natl Acad Sci USA 76: 3665–3669
Kao H-T, Nevins JR (1983) Transcriptional activation and subsequent control of the human heat shock gene during adenovirus infection. Mol Cell Biol 3: 2058–2065
Kitchingman GR, Westphal H (1980) The structure of adenovirus 2 early nuclear and cytoplasmic RNAs. J Mol Biol 137: 23–48
Kovesdi I, Reichel R, Nevins JR (1986) Identification of a cellular transcription factor involved in E1A trans-activation. Cell 45: 219–228
Kovesdi I, Reichel R, Nevins JR (1987) Role of an adenovirus E2 promoter binding factor in E1A-mediated coordinate gene control. Proc Natl Acad Sci USA 84: 2180–2184
Krippl B, Ferguson B, Rosenberg M, Westphal H (1984) Functions of purified E1A protein microinjected into mammalian cells. Proc Natl Acad Sci USA 81: 6988–6992
LaMarco K, Thompson CC, Byers BP, Walton EM, McKnight SL (1991) Identification of Ets-and Notch-related subunits in GA binding protein. Science 253: 789–792
Larsen PL, Tibbets C (1987) Adenovirus E1A gene autorepression: revertants of an E1A promoter mutation encode altered E1A proteins. Proc Natl Acad Sci USA 84: 8185–8189
Larsen PL, McGrane MM, Robinson CC, Tibbetts C (1986) An E1A mutant of adenovirus type 3: Ad3hr15 has reiterated DNA sequences 5' to its E1A gene. Virology 155: 149–159
Lee WS, Kao C, Bryant GO, Liu X, Berk AJ (1991) Adenovirus E1A activation domain binds to the basic repeat in the TATA box transcription factor. Cell 67: 365–376
Leff T, Elkaim R, Goding CR, Jalinot P, Sassore-Corsi P, Perricaudet M, Kedinger C, Chambon P (1984) Individual products of the adenovirus 12s and 13s E1a mRNAs stimulate viral E2a and E3 expression at the transcription level. Proc Natl Acad Sci USA 81: 4381–4385
Leprince D, Gegonne A, Coll J, de Taisne C, Schneeberger A, Lagrou C, Stehelin D (1983) A putative second cell-derived oncogene of the avian leukemia retrovirus E26. Nature 306: 395–397
Levine AJ (1993) The tumor suppressor genes. Annu Rev Biochem 62: 623–651
Leza MA, Hearing P (1988) Cellular transcription factor binds to adenovirus early region promoters and to a cyclic AMP response element. J Virol 62: 3003–3013
Lillie JW, Green M, Green MR (1986) An adenovirus E1a protein region required for transformation and transcriptional repression. Cell 46: 1043–1051
Lillie JW, Lowenstein P, Green M, Green MR (1987) Functional domains of adenovirus type 5 E1a protein. Cell 50: 1091–1100
Lin YS, Green MR (1988) Interaction of a common cellular transcription factor, ATF, with regulatory elements in both E1a- and cyclic AMP-inducible promoters. Proc Natl Acad Sci USA 85: 3396–3400
Liu F, Green MR (1994) Promoter targeting by adenovirus E1a through interaction with different cellular DNA binding domains. Nature 368: 520–525
Macleod K, Leprince D, Stehelin D (1992) The ets gene family TIBS 17: 251–256
McKnight SL, Yamamoto KR (eds) (1992) Transcriptional regulation. Cold Spring Harbor Laboratory Press, New York
Miralles VJ, Cortes P, Stone N, Reinberg D (1989) The adenovirus inverted terminal repeat functions as enhancer in a cell-free systems. J Biol Chem 264: 10763–10772
Mitchell PJ, Tjian R (1989) Transcriptional regulation in mammalian cells by sequence-specific DNA binding proteins. Science 245: 371–378
Montell C, Fisher EF, Caruthers MH, Berk AJ (1982) Resolving the functions of overlapping viral genes: site-specific mutagenesis into an mRNA splice site. Nature 295: 380–384
Moran E, Mathews MB (1987) Multiple functional domains in the adenovirus E1A gene. Cell 48: 177–178
Mul YM, van der Vliet PC (1992) Nuclear factor I enhances adenovirus DNA replication by increasing the stability of a preinitiation complex. EMBO J 11: 751–760
Mul YM, Verrijzer CP, van der Vliet PC (1990) Transcription factors NFI and NFIII/OCT-1 function independently, employing different mechanisms to enhance adenovirus DNA replication. J Virol 64: 5510–5518
Nevins JR (1981) Mechanism of activation of early viral transcription by the adenovirus E1A gene product. Cell 26: 213–220
Nevins JR (1992) E2F: a link between the Rb tumor suppressor protein and viral oncoproteins. Science 258: 424–429

Nevins JR, Ginsberg HS, Blanchard JM, Wilson MC, Darnell JE Jr. (1979) Regulation of the primary expression of the early adenovirus transcription units. J Virol 32: 727–733

Nunn MF, Seeburg PH, Moscovici C, Duesberg PH (1983) Tripartite structure of the avian eythroblastosis virus E26 transforming gene. Nature 306: 391–395

Ondek B, Gloss K, Herr W (1988) The SV40 enhancer contains two distinct levels of organization. Nature 333: 40–45

O'Neil EA, Fletcher C, Burrow CR, Heintz N, Roeder RG, Kelly TJ (1988) Transcription factor OTF-1 is functionally identical to the DNA replication factor NF-III. Science 241: 1210–1213

Ooyama S, Imai T, Hanaka S, Handa H (1989) Transcription in the reverse orientation at either terminus of the adenovirus type 5 genome. EMBO J 8: 863–868

Osborne TF, Arvidson DN, Tyau ES, Dunsworth-Browne M, Berk AJ (1984) Transcriptional control region within the protein-coding portion of adenovirus E1A genes. Mol Cell Biol 4: 1293–1305

Perricaudet M, Akusjärvi G, Virtanen A, Petterson U (1979) Structure of two spliced mRNAs from the transforming region of human subgroup C adenoviruses. Nature 281: 694–696

Prujin GJM, van der Vliet P, Dathan NA, Mattaj IW (1989) Anti-OTF-1 antibodies inhibit NFIII stimulation of in vitro adenovirus DNA replication. Nucleic Acids Res 17: 1845–1863

Ptashne M (1988) How eukaryotic transcriptional activator work. Nature 335: 683–689

Raychaudhuri P, Bagchi S, Devoto SH, Kraus VB, Moran E, Nevins JR (1991) Domains of the adenovirus E1A protein required for oncogenic activity are also required for dissociation of E2F transcription complexes. Genes Dev 5: 1200–1211

Riccardi RP, Jones RL, Cepko CL, Sharp PA, Roberts BE (1981) Expression of early adenovirus genes requires a viral encoded acidic polypeptide. Proc Natl Acad Sci USA 78: 6121–6125

Rochette-Egly C, Fromental C, Chambon P (1990) General repression of enhancer activity by the adenovirus-2 E1A proteins. Genes Dev 4: 137–150

Roesler WJ, Vandenback GR, Hanson RW (1988) Cyclic AMP and the induction of eukaryotic gene transcription. J Biol Chem 236: 9063–9066

Rosenfeld PJ, O'Neil E, Wides RJ, Kelly TJ (1987) Sequence-specific interactions between cellular DNA-binding proteins and the adenovirus origin of DNA replication. Mol Cell Biol 7: 875–886

Sassone-Corsi P, Hen R, Borrelli E, Leff T, Chambon P (1983) Far upstream sequences are required for efficient transcription from the adenovirus-2 E1A transcription unit. Nucleic Acids Res 11: 8735–8745

Schneider JF, Fisher F, Goding CR, Jones NC (1987) Mutational analysis of the adenovirus E1a gene: the role of transcriptional regulation in transformation. EMBO J 6: 2053–2060

Schrier PI, Bernards R, Vaessen RTMJ, Houweling A, van der Eb AJ (1983) Expression of class I major histocompatibility antigens switched off by high oncogenic adenovirus 12 in transformed rat cells. Nature 305: 771–776

Shinagawa M, Padmanabhan R (1980) Comparative sequence analysis of the inverted terminal repetitions from different adenovirus serotypes. Pro Natl Acad Sci USA 77: 3831–3835

Smith DH, Kegler DM, Ziff EB (1985) Vector expression of adenovirus type 5 E1a proteins: evidence for E1a autorepression. Mol Cell Biol 5: 2684–2696

Smith DH, Velcich A, Kegler D, Ziff E (1986) Transcriptional control by the adenovirus type 5 E1A proteins. Cancer Cell 4: 217–225

Stein R, Ziff EB (1984) HeLa cell β-tubulin gene transcription is stimulated by adenovirus-5 in parallel with viral early genes by an E1a-dependent mechanism. Mol Cell Biol 4: 2792–2801

Stein R, Ziff EB (1987) Repression of insulin gene expression by adenovirus type 5 E1a protein. Mol Cell Biol 7: 1164–1170

Stillman B (1989) Initiation of eukaryotic DNA replication in vitro. Annu Rev Cell Biol 5: 197–245

Svenson B, Akusjärvi G (1984) Adenovirus 2 early region 1A stimulates expression of both viral and cellular genes. EMBO J 3: 789–794

Temperley SM, Hay RT (1992) Recognition of the adenovirus type 2 origin of DNA replication by the virally encoded DNA polymerase and preterminal proteins. EMBO J 11: 761–768

Thompson CC, Brown TA, McKnight SL (1991) Convergence of ets- and notch-related structural motifs in a heteromeric DNA binding complex. Science 253: 762–768

Vaessen RTMJ, Houweling A, van der Eb AJ (1987) Post-transcriptional control of class I MHC mRNA expression in adenovirus 12-transformed cells. Science 235: 1486–1488

Van Ormondt H, Galibert F (1984) Nucleotide sequences of adenovirus DNAs. In: Doefler W (ed) The molecular biology of adenovirus 2. (Current topics in microbiology and immunology, vol 110) Springer, Berlin Heidelberg New York, pp 73–143

Van Ormondt H, Maat J, Van Beveren CP (1980) The nucleotide sequence of the transforming early region E1 of adenovirus type 5 DNA. Gene 11: 299–309

Velcich A, Ziff E (1985) Adenovirus E1a proteins repress transcription from the SV40 early promoter. Cell 40: 705–716
Velcich A, Kern FG, Basillico C, Ziff EB (1986) Adenovirus E1a proteins repress expression from polyoma-virus early and late promoters. Mol Cell Biol 6: 4019–4025
Virtanen A, Petterson U (1983) The molecular structure of the 9s mRNA from early region 1A of adenovirus sero type 2. J Mol Biol 165: 496–499
Wasylyk B, Hahn S, Giovane A (1993) The Ets family of transcription factors. Eur J Biochem 211: 7–18
Watanabe H, Wada T, Handa H (1990) Transcription factor E4TF1 contains two subunits with different functions. EMBO J 9: 841–847
Watanabe H, Sawada J, Yano K, Yamaguchi K, Goto M, Handa H (1993) cDNA cloning of transcription factor E4TF1 subunits with ETS and Notch motif. Mol Cell Biol 13: 1385–1391
Weeks DL, Jones NC (1983) E1A control of gene expression is mediated by sequence 5' to the transcriptional starts of the early viral genes. Mol Cell Biol 3: 1222–1234
Whyte P, Ruley HE, Harlow E (1988) Two regions of the adenovirus early region 1A proteins are required for transformation. J Virol 62: 257–265
Whyte P, Williamson NM, Harlow E (1989) Cellular targets for transformation by the adenovirus E1A proteins. Cell 56: 67–75
Winberg G, Shenk T (1985) Dissection of overlapping functions within the adenovirus type 5 E1A gene. EMBO J 3: 1907–1912
Xin JH, Cowie A, Lachance P, Hassell AJ (1992) Molecular cloning and characterization of PEA3, a new member of the ETS oncogene family that is differentially expressed in mouse embryogenesis. Genes Dev 6: 481–496
Yamazaki Y, Shimada Y, Sakurai SH, Masamune Y, Nakanishi Y (1992) Multiple cis-acting DNA elements that regulate transcription of the adenovirus 12 E1A gene. Virus Genes 6: 261–271
Yoshida K, Narita M, Fujinaga K (1989) Binding sites of HeLa cell nuclear proteins on the upstream region of adenovirus type 5 E1A gene. Nucleic Acids Res 17: 10015–10034
Yoshida K, Sugawara Y, Higashino F, Fujinaga K (1990) Potential activity of transcriptional promoter in the replication origin region of adenovirus type 5 DNA. Tumor Res 25: 69–84
Zenke M, Grundtstrom T, Matthes H, Wintzereith M, Schatz C, Wildeman A, Chambon P (1986) Multiple sequence motifs are involved in SV40 enhancer function. EMBO J 5: 378–397

Functional Domains of Adenovirus E1A Oncogenes Which Control Interactions with Effectors of Cellular Immunity

K. RAŠKA JR[1,2]

1 Introduction

Group A adenoviruses, such as adenovirus type 12 (Ad12), are oncogenic in newborn rodents, inducing tumors in the majority of animals within 4 months (TRENTIN et al. 1962). The group C adenoviruses, such as adenovirus type 2 (Ad2) or adenovirus type 5 (Ad5), on the other hand, are nononcogenic (HUEBNER 1967). Both groups of viruses transform rodent cells in vitro (BERNARDS et al. 1982). The

[1]Institute for Molecular Diagnostics and Pathology, St. Peter's Medical Center, 254 Easton Avenue, New Brunswick, NJ 08903, USA
[2]Department of Pathology, UMD-Robert Wood Johnson Medical School, 254 Easton Avenue, New Bryunswick, NJ 08903, USA

transforming ability of these two adenovirus groups is encoded within the E1 region. Two viral genes, E1A and E1B, are required for transformation of cultured rodent cells. The E1A gene alone can immortalize primary rat cells (Houweling et al. 1980). Complete, high-efficiency transformation requires, however, that the E1B gene be coexpressed with E1A (Barker and Berk 1987). The E1B gene alone does not show detectable transforming activity even when its expression is controlled by a constitutionally active promoter (van den Elsen et al. 1983). Although the two genes under normal conditions cooperate to transform rodent cells, the E1A gene predominates in determining the phenotype and tumorigenic potential of the transformed cells (Sawada et al. 1985). Mammary tumorigenesis in the rat by group D Ad9 requires functions of the E4 region from the right end of the genome (Javier et al. 1991, 1992). The E4 region is solely responsible for the difference in oncogenicity of Ad9 (tumorigenic) and Ad26 (non-tumorigenic).

Tumorigenicity of Ad-transformed cells usually follows that of the serotype used for transformation (Gallimore and Paraskeva 1980). Rat cells transformed by the E1A and E1B genes of Ad12 are highly tumorigenic in immunocompetent syngeneic animals, while Ad5 E1A/E1B-transformed cells usually fail to induce tumors in these hosts. In immunosuppressed or in T cell-deficient athymic (nude) animals, however, both group A and group C Ad-transformed cells are tumorigenic. Tumorigenicity of highly oncogenic Ad12 also increases in thymectomized mice or after treatment of the host with antilymphocyte serum (Allison et al. 1967).

The difference in oncogenic potential of the group A and group C viruses has not been satisfactorily explained, but various relevant factors controlled by functions of both E1A and E1B genes were identified. They include interactions with cellular p53 protein (van der Heuvel et al. 1990) and specific differences in interaction of transformed cells with matrix (Bober et al. 1987, 1988). It has been implied, however, that most critical to tumorigenicity are differences in their sensitivity to mechanisms of host immune surveillance. Rodent cells transformed by oncogenic and nononcogenic Ad types and their recombinants offer a suitable model not only for the study of tumor growth in vivo, but also for the analysis of the role of different virus genes in determining the phenotype of transformed cells with respect to their interactions with various effectors of host cellular immunity.

In man, Ad have not been associated with any malignancy (Green et al. 1979). Specific analysis of both benign and malignant human breast tumors, analogous to those induced by Ad in the rat, for virus sequences was also negative (R. Javier, T. Shenk, and K. Raška, unpublished observations). Ad are common pathogens causing self-limited infections, except in patients with impaired cellular immunity (Shields et al. 1985; Michaels et al. 1992). Factors involved in defense against Ad tumorigenicity in rodents are likely to control the outcome of Ad infections in man. Dissection of interactions of Ad E1A gene and other Ad regions with effectors of cellular immunity may help to explain not only differences in tumorigenicity of individual types in experimental models, but may also provide information essential for understanding of pathogenesis and resolution of Ad infections in man.

2 Adenovirus Tumor-Specific Transplantation Antigen

Trentin and Bryan (1966) have demonstrated that immunization with Ad12 induces a strong transplantation immunity against Ad12 tumors both in hamsters and in mice. The specificity of this transplant immunity was well documented by isograft experiments and Ad12 tumor-specific transplantation antigen (TSTA) immunity was defined (SJÖGREN et al. 1968). The identity of Ad12 TSTA, however, was unknown. While immunization with group C viruses did not protect against Ad12 tumors, TSTA cross-reactivity was reported between Ad7 and Ad12 both in cytolytic assays and in isograft rejection tests. It was also reported that sera of tumor-bearing mice were cytolytic to the transformed cells in complement-dependent assay in vitro, but that they had "blocking" effects in cellular cytotoxicity assays (ANKERST 1971). With the availability of rat cells transformed with Ad12 virions and the isolated *Eco*RI-C fragment of Ad12 DNA, which were tumorigenic in syngeneic rats, it was shown that cytolytic antibodies were reactive with cellular products dependent on the expression of the E1 gene block (RAŠKA et al. 1980). Antibodies with blocking effects, on the other hand, were directed against the products of the right-hand end of the genome and appeared in tumor-bearing rats significantly earlier than those directed against the products of the E1 region (RAŠKA et al. 1980). Results of the humoral and cellular cytolytic assays in animals bearing tumors transformed with Ad12 virions or with the transforming *Eco*RI-C fragment (0–16.5 map units) of Ad12 were interpreted as evidence that the E1 region codes for membrane-associated antigen active in both humoral and cellular immunity (RAŠKA et al. 1980). Study of the ability of Ad5 mutants to induce syngraft immunity in the rats to a syngeneic, highly tumorigenic clone of Ad2-transformed 50A cells suggested that viruses defective in the E1 region fail to induce TSTA immunity (GALLIMORE and WILLIAMS 1982). These results were in complete agreement with the earlier direct evidence that expression of only the Ad2 E1 region in transformed rat cells is sufficient to elicit strong secondary syngeneic cytolytic T lymphocytes (CTL) against Ad2-transformed cells (RAŠKA et al. 1982).

3 Adenovirus Tumor-Specific Transplantation Antigen Is a Function of the E1A Gene

Transplant immunity, by definition, can only be examined in vivo, by monitoring transplant rejection. Early experiments provided unequivocal evidence that immunization with group C viruses does not protect against Ad12 tumors (SJÖGREN et al. 1968) and vice versa (GALLIMORE and WILLIAMS 1982). This fact offered an ideal opportunity to identify the Ad genes coding for both group A and group C TSTA. A series of viable Ad5 and Ad12 recombinant viruses containing heterologous E1 genes was constructed (SAWADA et al. 1988). These viruses, shown in Fig. 1, are viable and transform rat cells, but are not tumorigenic in immunocompetent rats. Application of these virus constructs established that the serotype origin of the E1A gene alone, irrespective of the rest of the virus chromosome, controls the induction of specific transplant immunity. All virus constructs containing the Ad5 E1A gene induced immunity against tumorigenic Ad2-transformed cells, but not against Ad12 transformants. Conversely, viruses containing the Ad12 E1A gene protected against Ad12, but not against Ad2 tumors (SAWADA et al. 1986). The serotype origin of the E1B gene (SAWADA et al. 1986) or of the rest of the virus chromosome (URBANELLI et al. 1989) was irrelevant to induction of type-specific TSTA immunity. The viruses H5sub370-12E1A and H5sub371-12E1A induce immunity only against Ad12 transformants and not against cells transformed with group C Ad types, despite the fact that the E1B gene and the rest of the virus genome is of type 5 origin. The virus H12sub1004-5E1A, on the other hand, protects only against group C Ad-transformed and not against Ad12-transformed cells.

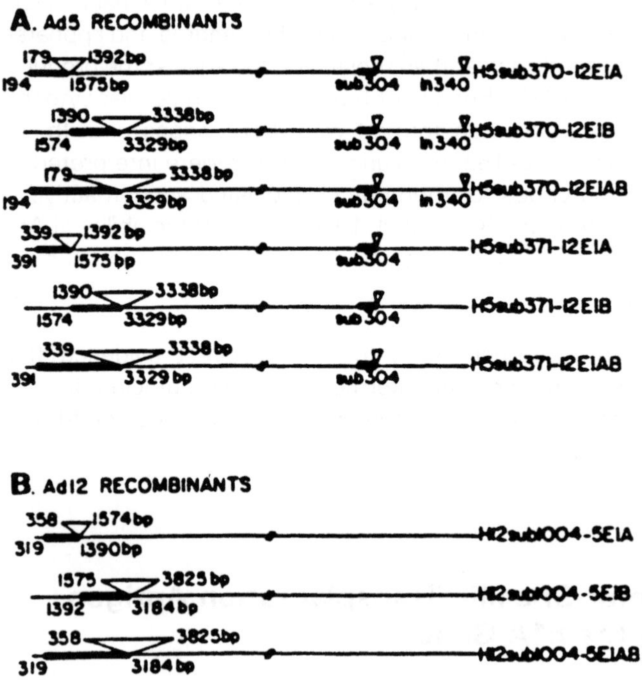

Fig. 1A,B. Chimeric adenoviruses used for demonstration of the tumor-specific transplantation antigen (TSTA) activity of the E1A gene. Recombinants were produced in either an adenovirus (Ad)5 (**A**) or Ad12 (**B**) background. The viral chromosome is represented by a *line*, deletions by *filled rectangles*, and insertions by *triangles*. The *numbers* represent either the Ad5 or Ad12 sequence position of nucleotides present on either side of deletions or the position of nucleotides at the extreme ends of insertions (SAWADA et al. 1988)

4 Functional E1A Domains Which Control Tumor-Specific Transplantation Antigen Activity

Further mapping of the functional domain of group C Ad TSTA was performed by the use of a library of Ad recombinants containing mutations in the E1A gene in LIS rats (URBANELLI et al. 1989). The product encoded by either 13S or by 12S E1A mRNA alone was sufficient for eliciting strong TSTA activity. A series of viruses with mutations within the first exon of the E1A message also induced strong TSTA, while Ad5 mutants with lesions within the second exon failed to induce syngraft immunity.

Definitive positive evidence in this model for the role of the second exon in Ad5 TSTA activity was provided in experiments using viruses containing chimeric Ad5/Ad12 E1A genes (Fig. 2). These viruses are viable and induce a variable transplant immunity against Ad2 and Ad12 tumors (SAWADA et al. 1994) All viruses containing the second E1A exon derived from Ad5 induce a strong Ad group C TSTA, irrespective of the origin of the rest of the E1A gene. Transplant immunity against Ad12 tumors, on the other hand, is exclusively dependent on the Ad12 origin of the N-terminal 68 amino acid residues in the products of chimeric E1A genes, irrespective of the rest of the E1A sequence. The viruses (H5sub1101 and H5sub1102) which carry the 5' E1A terminus of the first exon of Ad12 and the second exon of Ad5 thus induced protective immunity against both group A and group C tumors, while those which carry the 5' terminus of the first exon derived from Ad5 and the second exon from Ad12 (H5sub1107 and H5sub108) did not protect against either tumor. For induction of Ad12 TSTA, viruses expressing only the truncated 5' terminus of the E1A genes (H5sub1104 and H5sub1105) were adequate. These experiments provide unequivocal evidence that in rats of RT-1^{u} haplotype functional domains of Ad5 and Ad12 TSTA, respectively, are encoded by different exons of the E1A gene (SAWADA et al. 1994). Similarly to the case of Ad5, additional experiments have established that small mutations in the functional domain of the relevant Ad12 E1A sequence do not abrogate TSTA activity: the small in-frame insertions or deletions within the N-terminal portion of the first exon of the Ad12 E1A gene did not abrogate TSTA activity of the virus, and either of 12S or 13S Ad12 mRNA alone was sufficient for induction of TSTA (SAWADA et al. 1990).

5 Immunodominant Epitopes Recognized by Major Histocompatibility Complex-Restricted Cytolytic T Lymphocytes Are Encoded by the E1A Gene

The major histocompatibility complex (MHC)-restricted rat CTL directed against syngeneic Ad2-transformed cells were specific for epitopes coded by the E1 region (RAŠKA et al. 1982). The induction of CTL by immunization with Ad5 virions

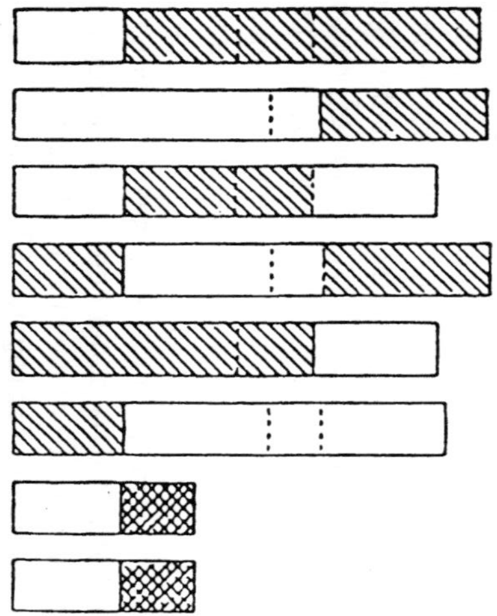

VIRUS	GENE	PRODUCT SIZE	ALTERATION OF A. A. SEQUENCE
H5sub1101	N	(288R) (242R)	^{1}M---^{69}S(Ad5) → ^{1}M---^{68}S(Ad12)
H5sub1102	NI	(294R) (263R)	^{1}M---^{185}S(Ad5) → ^{1}M---^{190}S(Ad12) ^{1}M---^{139}G(Ad5) → ^{1}M---^{159}C(Ad12)
H5sub1103	NC	(260R) (214R)	^{69}L---^{190}S(Ad12) → ^{70}V---^{185}S(Ad5) ^{69}L---^{159}C(Ad12) → ^{70}V---^{139}G(Ad5)
H5sub1106	I	(295R) (264R)	^{69}S---^{185}S(Ad5) → ^{68}S---^{190}S(Ad12) ^{69}S---^{139}G(Ad5) → ^{68}S---^{159}C(Ad12)
H5sub1107	C	(261R) (215R)	^{186}P---^{289}P(Ad5) → ^{191}P---^{266}N(Ad12) ^{140}P---^{243}P(Ad5) → ^{160}P---^{235}N(Ad12)
H5sub1108	IC	(267R) (236R)	^{69}S---^{289}P(Ad5) → ^{68}S---^{266}N(Ad12) ^{69}S---^{243}P(Ad5) → ^{68}S---^{235}N(Ad12)
H5sub1104	N'	(115R)	^{1}M---^{67}E(Ad12) plus 48 a. a.
H5sub1105	NC'	(115R)	^{1}M---^{67}E(Ad12) plus 48 a. a.

Fig. 2. The Chimeric E1A regions of adenoviruses (*Ad*)used for identification of the functional domains of Ad tumor-specific transplantation antigen (TSTA). The splicing difference between the 13S and 12S mRNA products is identified. *Open Boxes* show Ad12 sequence; *hatched boxes* show Ad5 sequence; *cross-hatched boxes* show extra sequence caused by a frame shift. The *product size* indicates products coded by 13S (*upper*) and 12S ((*lower*) type mRNA. Alterations of the predicted amino acid sequences are based on the products coded by 13S (*upper*) and 12S (*lower*) type mRNA (SAWADA et al. 1994)

in mice was studied by MÜLLBACHER et al. (1989). Both primary (in vivo) and secondary (in vitro-boosted) MHC-restricted CTL were elicited in CBA/H (H-2k^k) mice. The response was also directed against products of the E1 region, with some possible involvement of E2. Definitive evidence for the E1A target structure recognized by H-2^d-restricted mouse CTL has been established by the in vitro use of cloned CTL against syngeneic mouse embryo cells sensitized by addition of overlapping peptides spanning the Ad5 E1A gene product as targets (KAST et al. 1989; KAST and MELIEF 1991). Only peptides spanning the sequence residues 232–247 were active, showing the target relevance of the second E1A exon. Induction of secondary Ad5 specific CTL was studied in rats by immunization with Ad5/Ad12 hybrid viruses and recombinant Ad5 viruses containing E1A mutations (URBANELI et al. 1989). Both wild-type Ad5 and recombinants expressing only the 12S or 13S E1A early mRNA elicited MHC-restricted CTL. Cytolytic T cells were not induced with viruses unable to express E1A protein sequences normally encoded within the second exon of the E1A early messages. The results with both mouse H-2D^b-restricted (KAST et al. 1989) and rat RT-1^u-restricted (URBANELLI et al. 1989) CTL thus point to the second E1A exon as a functional domain which encodes the CTL recognition epitopes. The biological relevance of these CTL, specific for epitopes coded by the 3' exon of the E1A message for rejection of Ad tumors, has been documented by the eradication of syngeneic Ad5 tumors in nude mice (KAST et al. 1989) and by rejection of syngeneic tumors in immunocompetent rats (URBANELLI et al. 1989). In Fisher rats, Ad5-specific MHC-restricted CTL are also directed against products encoded by the E1A gene (ROUTES et al. 1991). These effectors lyse only targets infected with Ad of the group C, but not of the viruses belonging to groups A, B, and E. Preliminary mapping studies suggested, however, that the immunodominant epitopes in this rat strain are encoded within the first E1A exon. The difference seen between CTL-immunodominant Ad5 epitopes between LIS (URBANELLI et al. 1989) and Fisher (ROUTES et al. 1991) allogeneic rat strains is not unique to the Ad system. Different nucleoprotein peptides of influenza virus have been recognized in different murine or human MHC haplotypes (BASTIN et al. 1987; TOWNSEND et al. 1986). Within each MHC class I background, however, only one immunodominant antigenic peptide has been detected.

6 Serotype Origin of the E1A Gene Controls Expression of the Class I Major Histocompatibility Complex Antigen

Immune reactions are controlled through surface antigens coded by the MHC genes, and cytotoxic T cells are restricted by class I MHC molecules (ZINKERNAGEL and DOHERTY 1979). The expression of class I MHC antigen in Ad12-transformed rat cells is severely reduced (SCHRIER et al. 1983; MELLOW et al. 1984; VAN DER EB and OOSTRA 1985; SAWADA et al. 1985) as a result of expression of the Ad12 E1A gene. Molecular mechanisms responsible for this effect of Ad12 E1A gene were identified (VAESSEN et al. 1986; NIELSCH et al. 1991; MEIJER et al. 1992). It was postulated that this reduction, seen in Ad12- but not Ad5-transformed cells, is responsible for the observed differences in tumorigenicity of the two serotypes (SCHRIER et al. 1983), which are caused by escape of T cell immune surveillance (BERNARDS et al. 1983). This hypothesis was formulated on the basis of experimental evidence which shows that cells expressing Ad12 E1A gene are usually more resistant to allogeneic CTL than their syngeneic analogues expressing the Ad5 E1A gene (BERNARDS et al. 1983; SAWADA et al. 1985). According to this concept, syngeneic tumor cells which possess unique virus-specific antigens, but express cell surface levels of class I MHC antigens which are below the threshold required for CTL recognition, should be resistant to specific CTL-mediated mechanisms of immune surveillance. This hypothesis to explain the differences im tumorigenicity of cells transformed with group A and C adenoviruses, respectively, has been supported by some experiments (TANAKA et al. 1985), but not by others (MELLOW et al. 1984; HADDADA et al. 1986). Neither tumorigenicity nor the latency period of rat tumors induced by syngeneic cells expressing Ad12 E1A gene reflects the absolute level of the class I MHC antigen (MELLOW et al. 1984; SAWADA et al. 1985), and highly tumorigenic Ad2-transformed rat cells have normal levels of class I MHC antigen (MELLOW et al. 1984). Similarly, Ad5-transformed hamster cells with strikingly low MHC expression are non-tumorigenic (HADDADA et al. 1986). In mice, Ad12-transformed cells also have low levels of class I MHC antigen. Graft rejection studies show, however, that such cells elicit a strong syngraft immunity in syngeneic hosts and are not transplantable across the MHC barrier (SODDU and LEWIS 1992). Experiments have been reported in which tumorigenicity of high-passage Ad12-transformed mouse cells was reversed by coexpression of H-2D and H-2L genes (TANAKA et al. 1985). In more recent studies it was shown that tumorigenicity of Ad12-transformed cells driven to express high levels of class I MHC antigens enhances, rather than abrogates, their tumorigenicity (SODDU and LEWIS 1992). The ability of both oncogenic and nononcogenic Ad serotypes to induce a comparable, strong, and highly specific syngraft immunity in immunologically competent hosts (GALLIMORE and WILLIAMS 1982; SAWADA et al. 1986) also does not support the hypothesis that modulation of the MHC class I protein expression alone accounts for differences in tumorigenicity between groups A and C viruses.

The cell surface expression of class I MHC is also modulated by the Ad E3 19-kDa glycoprotein (PAÄBO et al. 1986) in Ad serotypes belonging to all groups except the highly oncogenic group A. The cells expressing this protein have decreased level of class I MHC antigen and are less sensitive to alloreactive CTL (BURGERT et al. 1987). Affinities of this protein for specific alleles of class I MHC differ in both mouse and human cells. In human cells, endogeneous E1A expression enhances this effect of the E3 region on MHC class I expression (ROUTES et al. 1993). Ad5 E3 gene products were also reported to interfere with the expression of the immunodominant E1A CTL epitope by mechanisms independent of modulation of the level of MHC class I expression (ZHANG et al. 1991).

7 Serotype Origin of the E1A Gene Controls Sensitivity to Natural Killer Cells

Cellular effectors of immune surveillance contain, in addition to specific cytolytic mechanisms mediated by MHC-restricted CTL, various forms of nonspecific immunity including natural killer (NK) cells and activated macrophages (HERBERMAN et al. 1975). The NK cells can be stimulated by T cell-derived cytokines, particularly interleukin (IL)-2 (GRIMM et al. 1982) or interferon (IFN)-α and γ (ELLIS et al. 1989). The resulting effector cell population has been identified as "lymphokine-activated killers" or LAK cells. Resident macrophages can also be activated by various stimuli, including bacille Calmette-Guérin (BCG; LEWIS and COOK 1985). The early experiments designed to study sensitivity of various Ad-transformed cells to NK cells have provided uniform results. An inverse relation was found between the oncogenic potential of Ad-transformed cells and their sensitivity to killing by syngeneic NK cells in the rat (RAŠKA and GALLIMORE 1982). In general, Ad2-transformed cells were sensitive to NK cytolysis in vitro, while Ad12 transformants were resistant. Very similar observations were reported in a hamster model (COOK et al. 1982). Binding and depletion assays established that these differences in sensitivity to NK lysis were not the result of lack of binding of NK cells to syngeneic Ad2- or Ad12-transformed targets (RAŠKA and GALLIMORE 1982). The direct role of the E1A gene in the control of sensitivity of transformed cells to NK cytolysis was established by means of an extensive library of syngeneic rat cells transformed with viruses containing recombinant Ad5/Ad12 E1 regions (SAWADA et al. 1985). All cells which expressed the E1A gene derived from Ad5 were sensitive to NK lysis, irrespective of the serotype origin of the E1B gene, and were nontumorigenic in syngeneic rats. All cells expressing the E1A gene of highly oncogenic Ad12, on the other hand, were resistant to NK lysis and induced tumors. The tumor latency was influenced by the serotype origin of the E1B gene. The cells with Ad12 E1B gene induced tumors in a short time, while in those with Ad5 E1B gene tumors appeared later. The latency period per se did not correlate with NK sensitivity (SAWADA et al. 1985).

Studies of NK cytolysis in various tumor models have implied that an inverse relation exists between NK susceptibility and level of MHC class I expression on tumor cells (KLEIN and KLEIN 1985; PIONTEK et al. 1985; KÄRRE et al. 1986; LJUNDGGREN et al. 1986). Conclusions from these reports clearly do not apply to the Ad system. The cells expressing the E1A gene of Ad12 have both reduced class I MHC level and are resistant to NK cells (SAWADA et al. 1985; KENYON and RAŠKA 1986). The role of the serotype origin of the E1A gene in the induction of sensitivity to NK cells (SAWADA et al. 1985), later confirmed in Ad-infected rat cells (COOK et al. 1987) and in Ad-transformed mouse cells (KAST et al. 1989), thus appears to be a general mechanism in Ad interactions with rodent cells. The difference in sensitivity to NK cytolysis per se is independent of the levels of MHC class I expression in the transformed cells. Modulation of class I MHC expression in rat (KENYON and RAŠKA 1986) or mouse (KAST et al. 1989) transformants, expressing Ad12 E1A, by IFN, which restored the level of its expression to that seen in cells transformed with

Ad5 E1A, did not increase their sensitivity to a highly enriched population of NK cells. On the other hand, expression of Ad5 E1A gene in Ad12-transformed mouse cells increased their NK sensitivity without effects on the level of MHC expression (Soddu and Lewis 1992).

The effects of IFN have to be evaluated in the context of the complex changes it induces in NK cell interactions with their targets. IFN on one hand increase the cytolytic capacity of NK cells, but on the other hand protect cells from lysis by NK cells (Trinchieri et al. 1982). Virus-mediated inhibition of IFN-induced resistance to NK cytolysis makes it possible to distinguish the virus-infected from uninfected cells by NK effectors and to increase effectiveness of their antiviral activity in vivo (Bukowski and Welsh 1985). IFN effects on the interaction of target and NK cell can thus be antagonistic: a severalfold increase in the NK activity was seen when effector cells were pretreated with IFN or when IFN was added directly to cytolytic assays (Trinchieri et al. 1982), but a significant increase of resistance was observed after IFN pretreatment of sensitive targets (Welsh et al. 1981; Trinchieri et al. 1982). In the rat cell system, IFN did not increase NK lysis of either highly sensitive Ad2-transformed cells or resistant cells expressing either the complete E1 region or only the E1A gene of Ad12 (Kenyon and Raška 1986).

Experiments applying the LAK cell effector mechanism to the Ad model have shown that IL-2-stimulated rat LAK cells have a strikingly higher cytolytic effect on sensitive cells expressing the E1A gene of either Ad2 or Ad5 (Kenyon and Raška 1986). In addition, these LAK cells effectively lysed all tested cell lines which expressed the E1A gene of highly oncogenic Ad12 (Kenyon and Raška 1986; Kenyon et al. 1991). IFN pretreatment of both Ad2- and of Ad12-transformed rat cells did not protect them from IL-2-activated LAK cell cytolysis (Kenyon and Raška 1986). Human cell targets, sensitive to lysis by IFN-α-stimulated LAK cells, are often protected by preincubation with IFN-γ (Ellis et al. 1989). This protective mechanism of IFN-induced cytolytic resistance is inhibited by expression of both Ad5 and Ad12 E1A gene (Routes 1993). In the case of Ad5, this inhibition appears to be a function of the first E1A exon.

The cytolytic sensitivity of transformed cells is lost after transformation with the Ad5 mutant H5*hr*1 (Su et al. 1993). Such cell transformants are oncogenic both in nude mice and immunocompetent syngeneic rats (Babiss et al. 1986). Coexpression of wild-type Ad5 E1A, E1B, or both E1A and E1B genes in H5*hr*1 transformants reduced their oncogenic potential, but in most cases did not restore their sensitivity to NK cells (Su et al. 1993). Wild-type Ad5-transforming genes thus function as transdominant suppressors of oncogenesis in mutant Ad5-transformed cells. This suppression of oncogenicity, however, appears to be independent of NK cytolytic mechanisms.

The in vivo relevance of NK cytolytic mechanisms for lack of tumorigenicity of transformed cells expressing Ad2 and Ad5 E1A has been documented by tumorigenicity assays in syngeneic rats with the NK cell pool depleted by anti-asialoGM1 antibody. This treatment effectively eradicated NK cytolytic activity without significantly affecting the in vitro T cell-proliferative responses both after phytohemagglutinin (PHA) stimulation and in the mixed lymphocyte culture. Cells expressing the E1A gene, which are nontumorigenic in unmanipulated rats, have induced tumors after a long latency in rats treated with the anti-asialoGM1 serum (KENYON et al. 1991).

8 E1A Gene Induces Sensitivity to the Cytotoxic Action of Tumor Necrosis Factor-α

Tumor necrosis factor (TNF) is a multifunctional cytokine produced by activated macrophages and lymphocytes which is involved in the regulation of various aspects of immune and inflammatory processes (BEUTLER and CERAMI 1986). TNF also lyses cells infected with some viruses.

CHEN et al. (1987) have shown that Ad E1A gene induced sensitivity of transformed mouse NIH-3T3 cells to cytotoxicity by human TNF-α, while six other oncogenes failed to do so. Further studies have established that either the large 289 E1A or the smaller 243 E1A protein alone can induce TNF susceptibility in infected cells (DUERKSEN-HUGHES et al. 1989). In transformed cells Ad group C E1A gene induction of TNF sensitivity has been demonstrated in both mouse (AMES et al. 1990; VANHAESEBROECK et al. 1990; DUERKSEN-HUGHES et al. 1991) and rat (VANHAESEBROECK et al. 1990; KENYON et al. 1991) but not human (VANHAESEBROECK et al. 1990) lines. Modulation of cellular susceptibility to the cytotoxic/cytostatic action of TNF by the E1A gene is cell type dependent (VANHAESEBROECK et al. 1990). Mapping of the functional domain of the Ad5 E1A gene required for induction of TNF sensitivity in mouse cells revealed that the amino-terminal portion of the conserved region 1 (CR1) is required both in transfected (AMES et al. 1990) and virus-infected (DUERKSEN-HUGHES et al. 1991) cells.

9 Does Serotype Origin of the E1A Gene Affect Induction of Tumor Necrosis Factor Sensitivity?

Several reports link TNF-α to the lytic cycle of NK/LAK cells (OLD 1985; ORTALDO et al. 1986). Based on the inhibitory effects of anti-TNF antibodies, it has been suggested that TNF-α sensitivity may account partially or fully for the induction of sensitivity of mouse cells to NK and macrophage cytolysis by the Ad5 E1A gene (COOK et al. 1989). In an attempt to correlate tumorigenicity and NK/LAK cytolysis with sensitivity to TNF-α, libraries of established rat cell lines expressing the E1A and/or E1B genes of Ad5 and Ad12 have been studied (VANHAESEBROECK et al. 1990; KENYON et al. 1991). Results showed that the sensitivity of these transformed rat cells was variable and did not correlate either with their tumorigenicity or with differences in their sensitivity to NK/LAK cytolysis.

All studies confirm that expression of the Ad5 E1A gene in NIH-3T3 cells enhances their TNF sensitivity (CHEN et al. 1987; VANHAESEBROECK et al. 1990).

Chen et al. (1987) have reported that NIH-3T3 cells transfected not only with Ad5, but also with Ad12 E1A genes are highly sensitive to recombinant TNF-α. Since the expression of the E1B 19-kDa protein prevents TNF cytotoxicity in human, but not in mouse cells (Gooding et al. 1991a), it was possible to examine the effects of Ad5 and of Ad12 E1A genes, respectively, on the induction of sensitivity of NIH-3T3 cells to TNF-α (Zhang et al. 1995). NIH-3T3 cells were cotransfected with plasmid PN 2 (neo) and with seven recombinant plasmids which carry wild-type Ad5 E1A, wild-type Ad12 E1A, or five different chimeric Ad5/Ad12 E1A genes constructed by Hamamatsu and Sawada (1989). Coselected stably transfected cell lines were tested for expression of E1A genes and their sensitivity to TNF-α (Zhang et al. 1995). A schematic representation of E1 mRNA and proteins encoded by recombinant plasmids is shown in Fig. 3. Studies with multiple cloned lines transfected with each plasmid showed that the E1A gene of Ad5, but not that of Ad12 raises TNF-α sensitivity of 3T3 cells to a level comparable to sensitive reference 929 cells. Studies with cells expressing chimeric E1A genes showed that substitution of the N-terminal 68 amino acid residues in the Ad5 E1A gene product by an analogous sequence of Ad12 completely abrogates the TNF sensitivity. On the other hand, cells expressing only the first 68 residues of Ad5 E1A gene product (the rest was derived from Ad12) were TNF sensitive. These results are in complete harmony with earlier reports that the amino-terminal portion of the conserved region 1 of Ad5 makes cells sensitive to TNF-α (Duerksen-Hughes et al. 1991; Ames et al. 1990). These results fail to confirm, however, the initial report that Ad12 E1A gene also renders 3T3 cells sensitive to TNF-α (Chen et al. 1987). This discrepancy could reflect the fact that only a single cell line was used in the earlier study or that significantly different cytolytic assays were used in the two studies. The transfected 3T3 cells lysed by TNF-α were also relatively sensitive to NK cells (Zhang et al. 1995). In mouse cells, the effects of TNF-α thus appear to correlate with cytolytic sensitivity to NK cells (Cook et al. 1989; Zhang et al. 1995). Correlation of the type origin of the E1A gene and induction of TNF sensitivity has not been seen in transformed rat cells (Vanhaesebroeck et al. 1990; Kenyon et al. 1991), demonstrating that effects of E1A gene with respect to TNF-α sensitivity are cell specific.

10 Adenovirus Control of Tumor Necrosis Factor Sensitivity Is Not Limited to the E1A Gene

The sensitivity of Ad-infected and Ad-transformed cells to TNF-α is not controlled by the E1A gene alone. Gooding et al. (1988) have shown that the E3 14.7-kDa protein prevents TNF lysis of infected mouse C3H cells. Analysis of a library of mouse cell lines infected with Ad mutants showed that the majority (14 out of 15) were protected from the cytolytic activity of TNF by the 14.7-kDa protein coded by the E3 region (Gooding et al. 1990). An analogous E3 protein was also

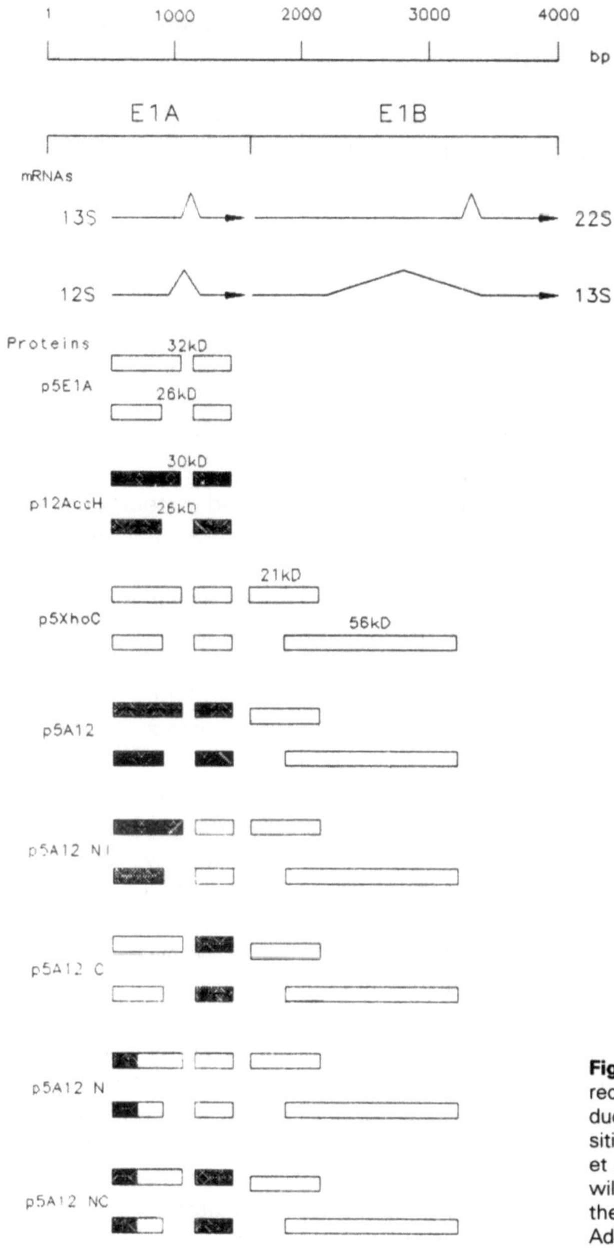

Fig. 3. E1 mRNA and proteins encoded by recombinant plasmids used for study of induction of tumor necrosis factor (TNF)-α sensitivity by E1A genes in NIH-3T3 cells (ZHANG et al. 1995). The predicted structure of the wild-type and chimeric protein is shown by the *open box* for Ad5 and by the *solid box* for Ad12 sequence

documented in groups A, B, D, and E Ad types (HORTON et al. 1990). Proteins in these other Ad groups are not functionally but also serologically related to the 14.7-kDa protein of the group C viruses. In other mouse cell lines, the E3 complex of 10.4-kDa/14.5-kDa proteins also inhibits TNF cytolysis (GOODING et al. 1991b).

Expression of the E3 region in Ad2-transformed rat cells also protected against TNF-α cytolysis (KENYON et al. 1991).

The E1B 19-kDa protein of group C Ad prevents TNF cytotoxicity in infected human, but not in infected mouse cells (GOODING et al. 1991a). The transfected baby rat kidney (BRK) cells coexpressing the Ad5 E1A and E1B genes were also resistant to the cytolytic effects of TNF (WHITE et al. 1992). In other studies, however, some established Ad2- or Ad5-transformed rat cell lines which also coexpress E1A and E1B genes were found to be lysed effectively by TNF-α (VANHAESEBROECK et al. 1990; KENYON et al. 1991).

11 Conclusions

The mechanisms of immune surveillance have profound effects on the tumorigenic potential of Ad and Ad-transformed cells in rodents, and immunosuppressive interventions increase tumorigenicity. The outcome of tumor–host interactions represents a dynamic process, the equilibrium of which depends on a multitude of factors. Evidence accumulated over the last 15 years has shown that there are striking differences in the effects of expression in rodent cells of the E1A genes of oncogenic versus nononcogenic Ad types on cellular interactions with various effectors of host cytolytic mechanisms involved in tumor immunity.

There is a significant difference in the dynamics of the tumorigenic process initiated by virus injection to a newborn host and that after transplantation of a large number of transformed cells to either a newborn or fully competent syngeneic host. Based on the specific conditions in each instance, different defense mechanisms may determine the outcome balance of that specific tumor–host interaction. In general terms, mechanisms of both specific and nonspecific immunity are highly relevant for tumorigenic potential. In immunocompetent animals injected with Ad at birth, the differences in sensitivity to NK-mediated cytolytic mechanism may be critical to eradication of the initially small tumor load, which in itself may be below the threshold required for induction of a specific CTL response. Experiments in athymic (nude) animals show that the relative degree of sensitivity to NK cells is reflected in the tumor cell dose required for tumor induction, but the NK cytolytic mechanism is not sufficient to reject a large tumor load, which can be eradicated by transfer of specific CTL. Differences in relative sensitivity to not only NK, but also CTL, seen between Ad5 and Ad12, may become critical for rejection of tumor transplants in immunocompetent animals. Modulation of MHC class I antigen expression by the E1 gene of tumorigenic Ad12 is not sufficient to abrogate in vivo transplant immunity induced by the TSTA domain of E1A. This MHC antigen downregulation also is not sufficient to evade allogeneic CTL mechanisms, and Ad12-transformed cells are not tumorigenic across the MHC barrier. However, selective manipulations of either T lymphocyte or NK cell pools in immunocompetent hosts may increase tumorigenicity.

A critical review of all the available experimental results provides overwhelming evidence that is not possible to explain differences between oncogenic and nononcogenic phenotypes in the context of a single immunologic parameter. Only the combination of effects that the E1A genes of oncogenic and non-oncogenic Ad types, respectively, have on interactions of transformed cells with both specific and nonspecific immune mechanisms provides a satisfactory explanation of the striking differences in their oncogenic potential in vivo.

References

Allison AC, Berman LD, Levey RH (1967) Increased tumor induction by adenovirus type 12 in thymectomized mice and mice treated with anti-lymphocyte serum. Nature 215: 185–187

Ames RS, Holskin B, Mitcho M, Shalloway D, Chen M-J (1990) Induction of sensitivity to the cytotoxic action of tumor necrosis factor alpha by adenovirus E1A is independent of transformation and transcriptional activation. J Virol 64: 4115–4122

Ankerst J (1971) Demonstration and identification of cytotoxic antibodies and antibodies blocking cell-mediated antitumor immunity against adenovirus 12-induced tumors. Cancer Res 31: 997–1003

Babiss LE, Liew WS, Zimmer SG, Godman GC, Ginsberg HS, Fisher PB (1986) Mutations in the E1A gene of adenovirus type 5 alter the tumorigenic properties of transformed cloned rat embryo fibroblast cells. Proc Natl Acad Sci USA 83: 2167–2171

Barker DD, Berk AJ (1987) Adenovirus proteins from both reading frames are required for transformation of rodent cell by viral infection and DNA transfection. Virology 156: 107–131

Bastin J, Rothbard J, Davey J, Jones I, Townsend A (1987) Use of synthetic peptides of influenza nucleoprotein to define epitopes recognized by class I restricted cytotoxic T lymphocytes. J Exp Med 165: 1508–1523

Bernards R, Houweling A, Schrier PI, Bos JL, van der Eb AJ (1982) Characterization of cells transformed by Ad5/Ad12 hybrid early region 1 plasmids. Virology 120: 422–432

Bernards R, Schrier PI, Houweling A, Bos JL, van der Eb AJ, Ziljstra M, Melief CJM (1983) Tumorigenicity of cells transformed by adenovirus type 12 by evasion of T-cell immunity. Nature 305: 776–779

Beutler B, Cerami A (1986) Cachectin and tumor necrosis factor as two sides of the same biological coin. Nature 320: 584–588

Bober FJ, Birk DE, Raška K (1987) Expression of varying portions of the adenovirus early region 1 in transformed cells affects tumorigenicity and interaction with extracellular matrix components. Lab Invest 56: 37–43

Bober FJ, Birk DE, Shenk T, Raška K Jr (1988) Tumorigenicity of adenovirus-transformed cells; collagen interaction and cell surface laminin are controlled by the serotype origin of the E1A and E1B genes. J Virol 62: 580–585

Bukowski JF, Welsh RM (1985) Inability of interferon to protect virus-infected cells against lysis by natural killer (NK) cells correlates with NK cell-mediated antiviral effects in vivo. J Immunol 135: 3537–3542

Burgert H-G, Maryanski JL, Kvist S (1987) "E3/19K" protein of adenovirus type 2 inhibits lysis of cytolytic T lymphocytes by blocking cell-surface expression of histocompatibility class I antigens. Proc Natl Acad Sci USA 84: 1356–1360

Chen M-J, Holskin B, Strickler J, Gorniak J, Clark MA, Johnson PJ, Mitcho M, Shalloway D (1987) Induction by E1A oncogene expression of cellular susceptibility to lysis by TNF. Nature 330: 581–583

Cook JL, Hibbs JB, Lewis AM Jr (1982) DNA virus transformed hamster cell-host effector cell interactions: level of resistance to cytolysis is correlated with tumorigenicity. Int J Cancer 30: 795–803

Cook JL, May DL, Lewis AM Jr, Walker TA (1987) Adenovirus E1A gene induction of susceptibility to lysis by natural killer cells and activated macrophages in infected rodent cell. J Virol 61: 3510–3520

Cook JL, May DL, Wilson BA, Holskin B, Chen M, Shalloway D, Walker TA (1989) Role of tumor necrosis factor—in E1A oncogene-induced susceptibility of neoplastic cells to lysis by natural killer cells and activated macrophages. J Immunol 142: 4527–4534

Duerksen-Hughes PJ, Wold WSM, Gooding LR (1989) Adenovirus E1A renders infected cells sensitive to cytolysis by tumor necrosis factor. J Immunol 143: 4193–4200

Duerksen-Hughes PJ, Hermiston TW, Wold WSM, Gooding LR (1991) The amino-terminal portion of CD1 of the adenovirus E1A proteins is required to induce susceptibility to tumor necrosis factor cytolysis in adenovirus-infected mouse cells. J Virol 65: 1236–1244

Ellis T, McKenzie R, Simms P, Helfrich B, Fisher R (1989) Induction of human lymphokine-activated killer cells by IFN-α and IFN-γ. J Immunol 143: 4282–4290

Gallimore PH, Paraskeva C (1980) A study to determine the reasons for differences in the tumorigenicity of rat cell lines transformed by adenovirus type 2 and adenovirus type 12. Cold Spring Harbor Symp Quant Biol 44: 703–713

Gallimore PH, Williams J (1982) An examination of adenovirus 5 mutants for their ability to induce group C adenovirus tumor-specific transplantation antigenicity in rats. Virology 120: 146–156

Gooding LR, Elmore LW, Tollefson AE, Brady HA, Wold WSM (1988) A 14,700 MW protein from the E3 region of adenovirus inhibits cytolysis by tumor necrosis factor. Cell 53: 341–346

Gooding LR, Sofola IO, Tollefson, AE, Duerksen-Hughes P, Wold WSM (1990) The adenovirus E3-14.7k protein is a general inhibitor of tumor necrosis factor-mediated cytolysis. J Immunol 145: 3080–3086

Gooding LR, Aquino L, Duerksen-Hughes PJ, Day D, Horton TM, Yei S, Wold WSM (1991a) The E1B 19,000-molecular-weight protein of group C adenoviruses prevents tumor necrosis factor cytolysis of human cells but not of mouse cells. J Virol 65: 3083–3094

Gooding LR, Ranheim TS, Tollefson AE, Aquino L, Duerksen-Hughes P, Horton TM, Wold WSM (1991b) The 10,400- and 14,500-dalton protein encoded by region E3 of adenovirus function together to protect many but not all mouse cell lines against lysis by tumor necrosis factor. J Virol 65: 4114–4123

Green M, Wold W, Mackey J, Rigden P (1979) Analysis of human tonsil and cancer DNAs and RNAs for DNA sequences of group C (serotypes 1, 2, 5 and 6) human adenoviruses. Proc Natl Acad Sci USA 76: 6606–6612

Grimm EA, Mazumder A, Zhang HZ, Rosenberg SA (1982) Lymphokine activated killer phenomenon. Lysis of natural killer-resistant fresh solid tumor cells by interleukin 2-activated autologous human peripheral blood lymphocytes. J Exp Med 155: 1823–1841

Haddada H, Lewis AM Jr, Sogn JA, Coligan JE, Cook JL, Walker TA, Levine AS (1986) Tumorigenicity of hamster and mouse cells transformed by adenovirus types 2 and 5 is not influenced by the level of class I major histocompatibility antigens expressed on the cells. Proc Natl Acad Sci USA 83: 9684–9688

Hamamatsu C, Sawada Y (1989) Functional analyses of human adenovirus types 5 and 12 chimeric early region 1A recombinants. Sapporo Med J 58: 98–109

Herberman RB, Nunn NE, Lavrin DH (1975) Natural cytotoxic reactivity of mouse lymphoid cells against syngeneic and allogeneic tumors. I. Distribution of reactivity and specificity. Int J Cancer 16: 216–229

Horton TM, Tollefson AE, Wold WSM, Gooding LR (1990) A protein serologically and functionally related to group C E3 14,700-kilodalton protein is found in multiple adenovirus serotypes. J Virol 64: 1250–1255

Houweling A, van den Elsen PJ, van der Eb AJ (1980) Partial transformation of primary rat cells by the leftmost 4.5% fragment of adenovirus 5 DNA. Virology 105: 537–550

Huebner RJ (1967) Adenovirus-directed tumor and T-antigens. In: Pollard M (ed) Perspectives in virology. Academic, New York, pp 147–166

Javier R, Raška K Jr, Macdonald GJ, Shenk T (1991) Human adenovirus type 9-induced rat mammary tumors. J Virol 65: 3192–3202

Javier R, Raška K Jr, Shenk T (1992) Requirement for the adenovirus type 9 E4 region in production of mammary tumors. Science 257: 1267–1271

Kärre K, Ljunggren HG, Piontek G, and Kiessling R (1986) Selective rejection of H-2 deficient lymphoma variants suggests alternative immune defence strategy. Nature 319: 675–678

Kast WC and Melief CJ (1991) Fine peptide specificity of cytotoxic T lymphocytes directed against adenovirus-induced tumors and peptide–MHC binding. Int J Cancer [Suppl] 6: 90–94

Kast WC, Offringa R, Peters PJ, Voordouw AC, Meloen RH, van der Eb AJ, Melief JM (1989) Eradication of adenovirus E1-induced tumors by E1A-specific cytotoxic T lymphocytes. Cell 59: 603–614

Kenyon DJ, Raška K Jr (1986) Region E1A of highly oncogenic adenovirus 12 in transformed cells protects against NK but not LAK cytolysis. Virology 155: 644–654

Kenyon DJ, Dougherty J, Raška K Jr (1991) Tumorigenicity of adenovirus-transformed cells and their sensitivity to tumor necrosis factor α and Nk/LAK cell cytolysis. Virology 180: 818–821

Klein G and Klein E (1985) Evolution of tumors and the impact of molecular oncology. Nature 315: 190–195

Lewis AM Jr, Cook JL (1985) A new role for DNA virus early proteins in viral carcinogenesis. Science 227: 15–20

Ljunggren H-G, Ohlen C, Höglund P, Yamasaki T, Klein G, Kärre K (1986) Afferent and efferent cellular interactions in natural resistance directed against MHC class I deficient tumor grafts. J Immunol 140: 671–678

Meijer I, Boot AJ, Mahabir G, Zantema A, van der Eb AJ (1992) Reduced binding activity of transcription factor NF-kappa B accounts for MHC class I repression in adenovirus type 12 E 1-transformed cells. Cell Immunol 145: 56–65

Mellow GH, Föhring B, Dougherty J, Gallimore PH, Raška K Jr (1984) Tumorigenicity of adenovirus rat cells and expression of class I major histocompatibility antigen. Virology 134: 460–465

Michaels MG, Green M, Wald ER, Starzl TE (1992) Adenovirus infection in pediatric liver transplants. J Infect Dis 165: 170–174

Müllbacher A, Bellett AJD, Hla RT (1989) The murine cellular immune response to adenovirus type 5. Immunol Cell Biol 67: 31–39

Nielsch U, Zimmer SG, Babiss LE (1991) Changes in HF-kB and ISGF3 DNA binding activities are responsible for differences in MHC and β-IFN gene expression in Ad5 versus Ad12-transformed cells EMBO J 10: 4169–4175

Old LJ (1985) Tumor necrosis factor (TNF). Science 230: 630–632

Ortaldo JR, Mason LH, Mathieson BJ, Liang SM, Flick DA, Herbeman RB (1986) Mediation of mouse natural cytotoxic activity by tumor necrosis factor. Nature 321: 700–702

Pääbo S, Weber F, Nilsson T, Schaffner W, Peterson PA (1986) Structural and functional dissection of an MHC class I antigen-binding adenovirus glycoprotein. EMBO J 5: 1921–1927

Piontek GE, Taniguchi K, Ljunggren H-G, Grönberg A, Kiessling R, Klein G, Kärre K (1985) YAC-1 MHC class I variants reveal an association between decreased NK sensivity and increased H-2 expression after interferon treatment of in vivo passage. J Immunol 135: 4281–4288

Raška K Jr, Gallimore PH (1982) An inverse relation of the oncogenic potential of adenovirus-transformed cells and their sensitivity to killing by syngeneic natural killer cells. Virology 123: 8–18

Raška K Jr, Morrongiello MP, Föhring B (1980) Adenovirus type 12 tumor antigen III. Tumorigenicity and immune response to syngeneic rat cells transformed with virions and isolated transforming fragment of adenovirus 12 DNA. Int J Cancer 26: 79–86

Raška K Jr, Dougherty J, Gallimore PH (1982) Product of adenovirus type 2 early gene block E_1 in transformed cells elicits cytolytic response in syngeneic rats. Virology 117: 530–535

Routes JM (1993) Adenovirus E1A inhibits IFN-induced resistance to cytolysis by natural killer cells. J Immunol 150: 4315–4322

Routes JM, Bellgrau DL, McGrory WJ, Bautista DS, Graham FL, Cook JL (1991) Anti-adenovirus type 5 cytotoxic T lymphocytes. Immunodominant epitopes are encoded by the E1A gene. J Virol 65: 1450–1457

Routes JM, Metz BA, Cook JL (1993) Endogenous expression of E1A in human cells enhances the effect of adenovirus E3 on class I major histocompatibility complex antigen expression. J Virol 67: 3176–3181

Sawada Y, Föhring B, Shenk TE, Raška K Jr (1985) Tumorigenicity of adenovirus-transformed cells: region E1A of adenovirus 12 confers resistance to natural killer cells. Virology 147: 413–421

Sawada Y, Urbanelli D, Rašková J, Shenk TE, Raška K Jr (1986) Adenovirus tumor-specific transplantation antigen is a functon of the E1A early region. J Exp Med 163: 563–572

Sawada Y, Raška K Jr, Shenk TE (1988) Adenovirus type 5 and adenovirus type 12 recombinant viruses containing heterologous E1 genes are viable, transform rat cells, but are not tumorigenic in rats. Virology 166: 281–284

Sawada Y, Fujinaga K, Urbanelli D, Raška K Jr (1990) Functional analysis of mutant E1A genes of adenovirus type 12. Abstracts of the 8th international congress of virology, Berlin, Germany, p 360

Sawada Y, Rašková J, Fujinaga K, Raška K Jr (1994) Identification of functional domains of adenovirus tumor-specific transplantation antigens in types 5 and 12 by viable viruses carrying chimeric E1A genes. Int J Cancer 57: 598–607

Schrier PI, Bernards R, Vaessen RMTJ, Houweling A, van der Eb AJ (1983) Expression of class I major histocompatibility antigens switched off by highly oncogenic adenovirus 12 in transformed rat cells. Nature 305: 771–775

Shields AF, Hackman RC, Fife KF, Corey L, Meyers JD (1985) Adenovirus infections in patients undergoing bone-marrow transplantation. N Engl J Med 312: 529–533

Sjögren HO, Minowada J, Ankerst J (1968) Specific transplantation antigens of mouse sarcomas induced by adenovirus type 12. J Exp Med 125: 689–701

Soddu S, Lewis AM Jr (1992) Driving adenovirus type 12-transformed BALB/c mouse cells to express high levels of class I major histocompatibility complex protein enhances, rather than abrogates, their tumorigenicity. J Virol 66: 2875–2884
Su Z, Leon JA, Tiang H, Austin VN, Zimmer SG, Fisher PB (1993) Wild type adenovirus type 5 transforming genes function as transdominant suppressors of oncogenesis in mutant adenovirus type 5 transformed rat embryo fibroblast cells. Cancer Res 53: 1929–1938
Tanaka K, Isselbacher KJ, Khoury G, Jay G (1985) Reversal of oncogenesis by the expression of a major histocompatibility complex class I gene. Science 228: 26–30
Townsend ARM, Rothbard J, Gotch FM, Bahadur G, Wraith D, McMichael AJ (1986) The epitope of influenza nucleoprotein recongnized by cytotoxic T lymphocytes can be defined with short synthetic peptides. Cell 44: 959–968
Trentin JJ, Bryan E (1966) Virus induced transplantation immunity to human adenovirus type 12 tumors of the hamster and mouse. Proc Soc Exp Biol Med 121: 1216–1219
Trentin JJ, Yabe Y, Taylor G (1962) Quest for human cancer viruses. Science 137: 835–841
Trinchieri G, Santoli D, Granato D, Perussia B (1982) Antagonistic effects of interferons on the cytoxicity mediated by natural killer cells. Fed Proc 40: 2705–2710
Urbanelli D, Sawada Y, Rašková J, Jones NC, Shenk TE, Raška K Jr (1989) C-terminal domain of the adenovirus E1A oncogene product is required for induction of cytotoxic T lymphocytes and tumor specific transplantation immunity. Virology 173: 607–614
Vaessen RTMJ, Houweling A, Israel A, Kourilsky P, van der Eb AJ (1986) Adenovirus E1A-mediated regulation of class I MHC expression EMBO J 5: 335–341
van der Eb AJ, Oostra BA (1985) Expression of class I major histocompatibility antigens in adenovirus transformed cells. In: Doerfler W, Adenoviruses, 10. Nijhoff, Boston, pp 343–366
van der Elsen P, Houweling A, van der Eb AJ (1983) Morphological transformation of human adenoviruses is determined to a large extent by gene products of region E1A. Virology 131: 242–246
van den Heuvel SJ, van Laar T, Kast WXM, Melief CJ, Zantema A, van der Eb AJ (1990) Association between the cellular p53 and the adenovirus 5 E1B-55kd proteins reduces the oncogenicity of Ad-transformed cells. EMBO J 9: 2621–2629
Vanhaesebroeck B, Timmers HTM, Pronk GJ, van Roy F, van der Eb AJ, Fiers W (1990) Modulation of cellular susceptibility to the cytotoxic/cytostatic action of tumor necrosis factor by adenovirus E1 gene expression is cell type-dependent. Virology 176: 362–368
Welsh RM, Kärre K, Hansson M, Kunkel LA, Kiessling RW (1981) Interferon-mediated protection of normal and tumor target cells against lysis by mouse natural killer cells. J Immunol 128: 219–225
White E, Sabbatini P, Debbas M, Wold WSM, Kusher DI, Gooding L (1992) The 19-kilodalton adenovirus E1B transforming protein inhibits programmed cell death and prevents cytolysis by tumor necrosis factor. Mol Cel Biol 12: 2570–2580
Zhang X, Bellet AD, HLA RT, Braithwaite AW, Müllbacher A(1991) Adenovirus type 5 E2 gene products interfere with the expression of the cytolytic T Cell immunodominant E1A antigen. Virology 180: 199–206
Zhang L, Sawada Y, Raška K Jr (1995) First exon of Ad5 but not Ad12 E1A gene induces sensitivity to TNFα in NIH-3T3 cells. (manuscript in preparation)
Zinkernagel RM, Doherty PC (1979) MHC-restricted cytotoxic T cells: studies on the biological role of polymorphic major transplantation antigens determining T-cell restriction-specificity, function and responsiveness. Adv Immunol 27: 51–177

Assessing the Role of E1A in the Differential Oncogenicity of Group A and Group C Human Adenoviruses

J. Williams[1], M. Williams[1], C. Liu[1], and G. Telling[2]

1 Introduction: Biology of Tumor Induction by Human Adenoviruses

It was first shown in 1962 that human adenovirus (Ad) types 12 and 18 rapidly induce undifferentiated sarcomas in newborn hamsters (Huebner et al. 1962; Trentin et al. 1962). In the decade that followed this initial discovery, numerous investigators confirmed the finding and showed that such tumors could be induced in other rodent species and that, depending upon the route of inoculation, neurogenic, neuroepithelial, medulloblastic, adenocarcinomatous, and retinoblastoma-like tumors were induced. During that same period it was shown that most of the other human adenoviruses either do not induce tumors in rodents or do so slowly and with low efficiency. In fact, it is now known that of the five viral

[1]Department of Biological Sciences, Carnegie Mellon University, 4400 Fifth Avenue, Pittsburgh, PA 15213, USA
[2]Department of Neurology, University of California, San Francisco, CA 94143, USA

groups of human adenoviruses defined on the basis of nucleic acid homology (and most likely also the more recently identified enteric adenoviruses), only those of group A (Ad12, Ad18, and Ad31) are highly oncogenic. Viruses of group B (for example, Ad3 and Ad7) are weakly oncogenic and induce tumors slowly and with low frequency. Viruses of group C (Ad1, Ad2, and Ad5) are considered to be nononcogenic, and those of groups D and E which have been tested are likewise negative for tumor induction, with the signal exception of Ad9. It is of considerable interest that this virus induces mammary fibroadenomas, some of which progress into malignant tumors in female Wistar/Furth rats (Jonsson and Ankerst 1977). Original references to many of the earlier oncogenic studies of the 1960s and 1970s have been chronicled previously in a number of reviews (Gallimore et al 1984a; Graham 1984; Williams 1986).

Despite this remarkable pattern of differential oncogenicity displayed by the human adenoviruses, the members of all subgroups tested (with the possible exception of enteric serotypes 40 and 41; Takiff and Strauss 1982) are capable of transforming a variety of rodent cells in vitro, with broadly similar efficiencies. The transformants generated following infection with different subgroup serotypes possess similar properties in vitro (such as rounded, epithelioid morphology, capacity to grow to high saturation density, ability to grow in agar suspension, etc.), but the tumorigenicity of the different subgroup transformants does differ markedly. For example, rat cells transformed by Ad12 are tumorigenic in newborn and even adult syngeneic rats, while those transformed by Ad2 and Ad5 are not, following the pattern of tumor induction by these viruses in rats. The situation even with these two extreme serotypes is, however, not that straightforward. For example, cells transformed by Ad2, unable to form tumors is syngeneic rats, are capable of tumor formation in rats immunosuppressed by antithymocyte serum (Gallimore 1972), and cells transformed by both Ad5 and Ad12 are able to form tumors in immunodeficient nude mice (Bernards et al. 1983). Further, while Ad12 induces tumors in hamsters and Ad2 and Ad5 do not, hamster cell transformants induced in vitro by all three viruses are tumorigeneic in newborn hamsters, although the capacity of the group C transformants to do so is lost in immunocompetent weanling animals (Lewis and Cook 1982).

Collectively, these various results indicated that the tumorigenicity of transformed cells is very much dependent on the nature of the host used and that thymus-dependent cytotoxic T cell components of the host immune system are involved in rejecting the cells transformed by the nononcogenic group C viruses. The Ad12 transformants are, in fact, the only ones capable of malignant growth in all of the various host systems listed, including adult, fully immunocompetent animals. In part, at least, this may stem from the fact that Ad12, but not Ad5, transformants possess reduced levels of class I major histocompatibility (MHC) antigens (Eager et al. 1985; Schrier et al. 1983). Because foreign antigens on cell surfaces are recognized by cytotoxic T lymphocytes (CTL) in the context of MHC I antigens, the reduced levels of these proteins on Ad12 transformants would allow them to evade the CTL-mediated immune response.

While this particular host response is almost certainly a crucial player in the attack on inoculated cell lines, it is most likely not the only one. There is, for example, considerable evidence that natural defense components, such as natural killer cells (NK), also play a role in determining the tumorigenicity of Ad-transformed cells (Cook et al. 1982; Raska and Gallimore 1982). Furthermore, it is by no means clear that either of these processes prevent the growth of newly induced Ad5 tumors in the animal and that their suppression or modulation is vital for tumor induction by Ad12 virus. In this respect, it is worth noting that even with Ad12, which is extremely oncogenic in newborn and very young animals, the recipients become resistant to tumor induction by the time they are about 3 weeks old; if suppression of MHC and/or resistance to NK cell killing in Ad12 transformants induced in vitro are requirements for high tumorigenic capacity of these cells, these processes, if they do occur at the time of tumor inception, are clearly not sufficient to permit complete tumorigenicity in vivo in some cases. Finally, it should be kept in mind that Ad12, which is considered to be generally highly oncogenic in rodents, is strongly influenced by the genetic make-up of the host. Not all strains of mice and rats are equally susceptible, even as newborns, to Ad12 tumor induction. For example, we have found that inbred Fischer rats are almost totally refractory to induction, while inbred Hooded Lister rats are 100% susceptible to rapid induction at identical input doses (J. Williams, unpublished results). Potentially this could be related to the immunogenetic constitution of the host (eg., RT1 haplotype) or it could result from the influence of some other gene or genes. It goes without saying that the roles of viral gene products in oncogenesis must be evaluated against this background of host-determined factors.

2 Adenovirus Transforming Genes and Their Products

Numerous molecular and genetic analyses carried out largely with group C serotypes 2 and 5, with plasmid-based genes derived from these viruses, and with transformants induced by both, have shown that early region 1 (E1) gene products are necessary and apparently sufficient for complete transformation of rodent cells in vitro (reviewed in Boulanger and Blair 1991; Gallimore et al. 1984a; Graham 1984; van der Eb and Zantema 1992; Williams 1986). While this finding with the group C viruses may well be generally applicable to most human adenoviruses, it has recently been found that at least one serotype does not conform to this generally accepted rule. With group D Ad9, an E4 gene function is required in addition to E1 for complete transformation in vitro and E4 is essential for tumor induction (Javier et al. 1992; see also section 5 below). Other exceptions to this rule may turn up as more adenoviruses are analyzed for transforming characteristics.

The E1 region, located at the extreme left end of the viral genome, has a generally similar organization in all serotypes examined and is comprised of two discrete transcription units, E1A and E1B. The E1A gene is located between map units 1.3 and 4.5 on the genome, and transcripts of that gene are differentially spliced to generate five mRNA size species with common 5' and 3' termini. The two largest species (13S and 12S) are abundant at early times after infection of permissive cells and in rodent cells, while the three smaller species (11S, 10S, and 9S) arise only at much later times during the infection of permissive cells (STEPHENS and HARLOW 1987). In Ad5, translation of the 13S and 12S species, commencing at a common AUG, gives rise to polypeptides of 289 and 243 residues (R), respectively; these are identical in sequence but for the internal residues missing from the smaller protein. In Ad12, the corresponding mRNAs again use a common reading frame to give rise to related proteins of 266R and 235R, respectively.

When the amino acid sequences of the larger E1A proteins of these two serotypes were compared with those of the corresponding E1A proteins of Ad3 and Ad7 (group B) and simian adenovirus type 7 (SA7), it was found that they possess three conserved regions (CR) interspersed with divergent, non-conserved regions (KIMELMANN et al. 1985; VAN ORMONDT and GALIBERT 1984). The CR are termed CR1, CR2, and CR3, and E1A functions influencing transcription repression and activation map to the first two and the third, respectively (JONES 1992; MORAN and MATHEWS 1987). CR1 and CR2 are encoded by the first exon sequence of the 13S and the 12S mRNA, while CR3 is almost entirely excluded (Ad5) or partially excluded (Ad12) from the smaller protein product of the 12S mRNA. The relative positions of the 13S and 12S splice sites and CR in types 5 and 12 are depicted in Fig. 1. As mentioned above, a subset of smaller mRNA (11S, 10S, and 9S) is generated by differential splicing at later times during productive infection of human cells. The 11S and 10S species make use of the same reading frame as the early 13S and 12S mRNA, while the 9S species switches reading frames in the second exon. Thus, with the exception of the protein encoded by the 9S mRNA, each of the other four E1A protein species share common amino and carboxyl-terminal residues and differ only in some of the internal segments present in the largest moiety. The 10S species encodes CR2 only, the 11S species contain CR2 and CR3, but not CR1, while as mentioned above, the 12S mRNA specifies only CR1 and CR2. Because the two larger species appear at very early times in infection and the other three are made at later times, and because each of the four smaller proteins is likely to carry out a different subset of E1A functions, it is possible that the expression of these different products could be regulated transiently to increase and/or decrease the individual E1A functions at different times in the viral replication cycle as the need arises. Further regulation of E1A expression may also occur post-translationally, for as many as 60 species of the primary translation polypeptides, probably resulting from phosphorylation, have been detected using a combination of techniques involving precipitation with monoclonal antibodies and two-dimensional isoelectric focusing gel electrophoresis (HARLOW et al. 1985). The

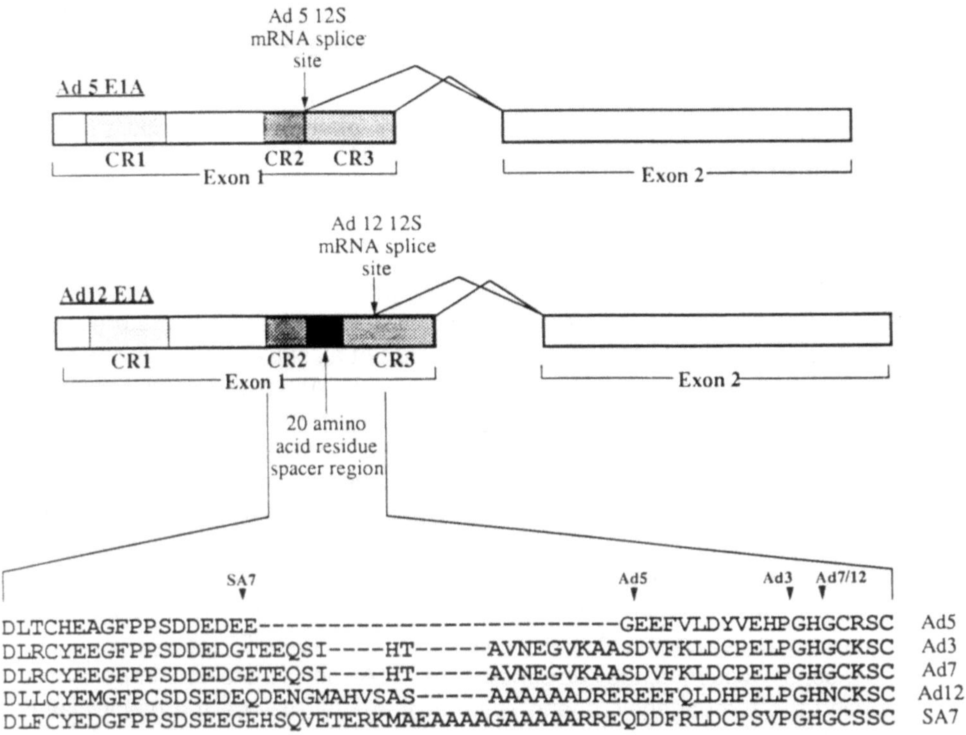

Fig. 1. Comparative physical organization of the adenovirus (*Ad*)5 and Ad12 E1A genes. The *upper* section of the figure compares the general structural features of the larger E1A proteins for each virus; the end of the first exon of the small E1A proteins is indicated by the position of the 12S mRNA donor splice site in each case. The positions of the conserved regions (*CR*) in the first exon are indicated in *shaded blocks*. The position of the unique nonconserved spacer region in Ad12 is shown as a *black block*. The *lower* part of the figure depicts an expanded comparative view of the amino acid sequences of CR2, the spacer region (where present), and part of CR3 for Ad5, Ad3, Ad7, Ad12, and SA7 (adapted from Kimelmann et al. 1985; Larsen and Tibbetts 1987; Moran and Mathews 1986; van Ormondt and Galibert 1984). The *black arrowheads* above these sequences represent the 12S mRNA splice sites of the viruses. (From Telling and Williams 1994, reprinted with permission from the American Society of Microbiology)

functional significance of the various E1A protein forms is not clear; perhaps they play different roles in different subsets of the natural host cells, for example, in target cells existing in different physiological states.

The E1B gene lies adjacent to E1A, between map units 4.6 and 11.2, and transcripts copied from this region are differentially spliced to give rise to two major mRNA species of 22S and13S and two minor species of 14.5S and 14S, all of which possess identical 5' and 3' ends (Saito et al. 1983; van Ormondt and Galibert 1984; Virtanen and Pettersson 1985). The 22S mRNA, which is made in transformed rodent cells, encodes two proteins from discrete, but overlapping reading frames (Bos et al. 1981). The 5' proximal AUG designates the commencement of the first frame, which encodes the 19K protein. This protein is comprised of 176R in Ad5 and 163R in Ad12. The second AUG marks the initiation point for the 55K protein, which is 496R in Ad5 and 482R in Ad12. In cells infected by either Ad5 or Ad12, the E1B 13S mRNA also encodes the 19K protein and a smaller species related to the 55K protein.

2.1 Role of E1A and E1B in Viral Oncogenesis

Phenotypic analysis of baby rat kidney (BRK) transformants induced by Ad5/Ad12 E1A/E1B hybrid plasmids and viruses has revealed that the tumorigeneity of these cells in immunocompetent, syngeneic rats is strongly correlated with the expression of the Ad12 E1A gene, but is increased significantly by the presence and expression of the Ad12 E1B gene (BERNARDS et al. 1982, 1983; SAWADA et al. 1988). Although the transformed cells expressing the Ad5 E1A gene did not form tumors in syngeneic rats, they were tumorigenic in immunodeficient nude mice and rats. In nude mice, however, the expression of Ad12 E1B seems to increase the tumorigeniety of transformants. The results have been interpreted to mean that transformants expressing Ad12 E1A are able to evade the CTL defense force of the host and go on to form tumors, while those which express Ad5 E1A fail to do so. Support for this view is provided by the finding that the expression of Ad12 E1A, but not Ad5 E1A, in rat transformants prevented their lysis by allogeneic CTLs in vitro (BERNARDS et al. 1983). While these results with E1 recombinants are by no means exhaustive, they do suggest that E1A and E1B are both important in endowing the tumorigeneic phenotype, but that E1A may be the major player while E1B exerts a lesser influence. Support for this view stems from the finding that some BRK cells lines produced by transformation with E1A plasmid are tumorigenic in syngeneic rats (GALLIMORE et al. 1984b). It should be pointed out, however, that these studies do not address the problem of tumor induction by virus per se. In this chapter we will evaluate the role of the E1A gene in tumor induction by group A and group C adenoviruses and in the tumorigenic capacity of Ad transformants.

3 Functions of the E1A Proteins

Analyses carried out largely with group C Ad2 and Ad5 have shown that the E1A gene products are made almost immediately after the virus enters host cells, and they play important but not entirely obligatory roles in regulating the expression of other viral early genes and cellular genes and in the establishment of infection in human cells. It has also been shown that these products are necessary for transformation of rodent cells in vitro, are sufficient to immortalize and partially transform BRK cells, and their cooperation with E1B is necessary and perhaps sufficient for complete transformation of these and other rodent primary cell types by most adenoviruses.

3.1 Role of the E1A Proteins in Productive Infection

It was first demonstrated some 15 years ago that E1A protein activates the expression of the viral early genes (Berk et al. 1979; Jones and Shenk 1979). Activation was shown to occur at the level of transcriptional initiation (Nevins 1981) and, through analyses of E1A mutants, to be mediated chiefly via CR3 (see Berk 1986). E1A has also been found to activate some cellular genes, for example, the 70K heat-shock gene (Kao and Nevins 1983). In addition, a range of cellular promoters transfected into cells have been shown via transient assay to be activated, but it appears that most cellular genes, including those activated transiently, are unresponsive to E1A (Berk 1986). Although the CR3 region is the main mediator of E1A-induced transactivation, there is considerable evidence that regions in both the first and second exons (common to the 13S and 12S products) are also capable of activating cellular and/or viral gene expression (Engel et al. 1991; Gedrich et al. 1992; Nevins 1992; Mymryk and Bayley 1993; Winberg and Shenk 1984; Zerler et al. 1987). It has, for example, been shown that the Ad5 E1A 243R protein acts cooperatively with cyclic adenosine monophosphate (cAMP) to induce activating protein-1 (AP-1) DNA binding activity and transcription of the *c-fos* gene, a function which maps to the N terminus and the CR1 domain of E1A (Engel et al. 1991; Gedrich et al. 1992). Numerous analyses have identified and characterized the E1A-responsive, *cis*-acting elements involved in transcription from the promoters of viral early and other genes and the cellular transcription factors which interact with them, and it has been shown that unlike some of these cellular transactivating proteins, E1A protein does not bind specifically to such elements. Most likely E1A protein acts by stimulating the level or the activity of required transcription factors either by interacting directly or indirectly with them as part of the transcriptional initiation complex, by acting indirectly to switch on the expression of cell genes whose products then activate the transcription factor, or in some cases blocking the activity of a cell factor which normally represses a transcription factor. An alternative model proposes that E1A does not enter the initiation complex or switch on cell genes, but instead induces post-translational modifications (perhaps phosphorylation) of cellular transcription factors. Some of these various topics concerning the mechanism(s) of transactivation have been discussed at length in a previous review (Jones 1992).

In addition to discovering that E1A is a transcriptional activator, it was shown that both E1A proteins repress promoters controlled by viral enhancers (Borelli et al. 1984; Velcich and Ziff 1985), and it was determined that the *trans*-repressive activity maps to CR1 and CR2 (Lillie et al. 1987; Schneider et al. 1987). It is now known that these proteins can repress the transcription of a variety of cellular genes, including, in the case of Ad12, the MHC I genes. In this case the capacity to repress maps to CR3 and seems to involve the induction of a repressor to the MHC I enhancer region. Further discussion of this topic appears in Sect. 3.3

below. Based on the fact that E1A mRNAs are found at reduced levels in cells infected with E1A-defective mutants, it has been suggested that the E1A transcriptional unit is autoregulated (BERK et al. 1979; HEARING and SHENK 1985), and negative autoregulation via *trans*-repression apparently also occurs (SMITH et al. 1985). It has been shown in Ad3 (group B) that the transrepression is directed towards E1A enhancer sequences, and using a set of Ad3 enhancer mutants and revertants it was demonstrated that the putative autorepression domain maps to the Ad3 E1A sequences lying between CR2 and CR3 (LARSEN and TIBBETTS 1987). This region is missing from Ad5 and is longer in Ad12 and SA7 (see Fig. 1), and it might also modulate autorepressive activity of Ad12 and SA7. The mechanism of repression in any one case is not well understood, but it is most unlikely that all cases follow a common pathway.

In human cells infected with human adenoviruses the E1A proteins form complexes with a number of cellular proteins (EGAN et al. 1988; GIORDANO et al. 1991; HARLOW et al. 1986; WHYTE et al. 1988a, b; YEE and BRANTON 1985). These include proteins of 300K, 130K, 107K, 105K, and 60K. The p105 species has been identified as the product of the retinoblastoma gene, RB, and it binds to sequences in the N-terminal segment of CR1 (and just adjacent to that) and CR2 (EGAN et al. 1989; WHYTE et al. 1989). RB is a classic antioncogene or tumor suppressor gene, and its product, the Rb protein, imposes cell cycle and proliferative blocks. The fact that Rb binds to sequences of the virus protein which are critical for transformation suggested that inactivation of Rb is involved in that process; this inactivation might also be important for viral growth, particularly in non-proliferating target cells for the virus in the natural host. The Rb protein binds to E2F, a cellular transcriptional activator of certain cell cycle-regulated genes, and when E1A protein binds to Rb, which may have a higher affinity for E1A than for E2F, the latter is released and can then activate target promoters in the cell (NEVINS 1992). This pathway of trans-activation does not go via CR3, but is dependent on the Rb-binding sites in the CR1 and CR2 sequences, which also possess repressor activity.

The p60 protein is known to be cyclin A (PINES and HUNTER 1990), a cell cycle regulator protein which associates with a cdk2 kinase and with p107, a protein closely related to Rb which binds to CR2 of E1A; its association with E1A is most likely indirect, via p107 (FAHA et al. 1992). Cyclin A is also found in complex with E2F. The exact nature and consequences of these various interactions are not entirely clear, but they do implicate regulatory controls of the cell cycle and cellular proliferation, which could influence viral growth in certain cell types (e.g., target cells in the natural host) and transformation. The p300 protein is known to be a nuclear phosphoprotein with DNA-binding activity (YACIUK and MORAN 1991) which interacts with the N terminus and CR1 (EGAN et al. 1988; WANG et al. 1993a; WHYTE et al. 1989). This N-terminal region of E1A is involved in enhancer repression function, DNA synthesis induction, and transformation (JELSMA et al. 1989; STEIN et al. 1990). The functional significance of p300 to any of these activities is not clear at present, but it seems likely that it could be involved in viral growth in certain cell types and/or in transformation. Unlike the other cellular proteins binding to E1A, p300 binds to a nonconserved region which diverges markedly between phenotypes. Comparison of Ad5 and Ad12 sequences in this region of E1A reveals that only two residues are common to both, but in spite of the difference, the Ad12 E1A proteins bind p300 with affinity similar to that of the Ad5 proteins, although a subset of N-terminally modified Ad12 E1A proteins do not (WANG et al. 1993b). Although p300 binding per se seems to be similar with these two E1A species, the functional repertoire of the two complexes could be quite different, with different consequences for repressor, DNA induction, transforming activities, and perhaps even the oncogenicity of these two viruses. The same could also be true for some of the complexes formed between Ad5 and A12 E1A and other cellular proteins such as Rb and p107, although these do, of course, bind to E1A regions conserved between Ad5 and Ad12.

3.2 Role of the E1A Proteins in Transformation of Rodent Cells

Tests in which Ad5 and Ad12 DNA fragments or recombinant plasmids were transfected into primary BRK cells showed that the E1A gene of each of these viruses is capable by itself of transforming those cells (GALLIMORE et al. 1984b; HOUWELING et al. 1980; SHIROKI et al. 1979). The transformation is incomplete in that the transformants are less epithelioid than cells transformed by the viruses (i.e., more fibroblastic) and do not grow to high cell densities; many are, however, immortalized and can be subcultivated continuously and, as mentioned above in Sect. 2.1, some of the Ad12 E1A transformants are tumorigenic (GALLIMORE et al. 1984b). Complete transformation of these cells by adenoviruses requires the cooperation of the E1A and E1B genes (BYRD et al. 1988; GRAHAM et al. 1974, 1978; SHIROKI et al. 1979; VAN DER EB et al. 1977). Fully transformed cells are also produced by cooperation of the E1A gene with the activated Ha-*ras* oncogene (RULEY 1983). Based upon analyses using group C mutants which make either the 13S or the 12S E1A mRNA, but not both, it was determined that both major E1A (289R and 243R proteins) are required for complete transformation of rodent cells (HURWITZ and CHINNADURAI 1985; MONTELL et al. 1984; MORAN et al. 1986; WINBERG and SHENK 1984). Studies with E1A plasmids have shown that both major E1A products are equally effective at immortalizing BRK cells and generating complete transformation in cooperation with the Ha-*ras* oncogene (HALEY et al. 1984; MORAN et al. 1986; ROBERTS et al. 1985; ZERLER et al. 1986). In contrast to the group C viruses, only the larger E1A product of Ad12 (266R) is required for transformation; a mutant which expresses only the 13S mRNA is totally transformation competent in various rodent cell types and induces tumors in rats, while a mutant expressing only the 12S mRNA is totally transformation defective (LAMBERTI and WILLIAMS 1990).

Extensive mutational analysis of group C E1A plasmids has been carried out to determine which regions of the E1A proteins are required for various transforming properties; because, in this case, the 243R protein apparently bestows all of the transforming functions determined by the E1A gene, it is the E1A form most commonly used in these studies. Expression of the 243R Ad5

E1A product is apparently necessary for efficient growth of the virus in growth-arrested WI38 cells, in which it most likely functions by inducing cellular DNA synthesis (SPINDLER et al. 1985). In quiescent rodent epithelial cells it also induces cellular DNA synthesis, cellular proliferation, and production of a growth factor which itself can stimulate these cells to proliferate and may be required for transformation (QUINLAN and GRODZICKER 1987; QUINLAN et al. 1987, 1988). The region of the 243R E1A protein encoded by exon 1 (see Fig. 1) is all that is required for complete transforming capacity of the E1A gene in cooperation with the activated *ras* oncogene, and within that half of the protein the extreme N terminus, CR1, and CR2 are found to be required for this cooperative transformation (JELSMA et al. 1989; LILLIE et al.1986, 1987; MORAN et al. 1986; QUINLAN et al. 1988; SCHNEIDER et al. 1987; SMITH and ZIFF 1988; SUBRAMANIAN et al. 1988; WHYTE et al. 1988b, 1989; ZERLER et al. 1986).

CR1 (and the N-terminal nonconserved region) is considered to be essential for the induction of DNA synthesis in quiescent rodent cells, but this function seems to be separate from the transforming function. CR2 is believed to be required for the progression of cells through the cell cycle. These three regions are also necessary for immortalization, but not sufficient to permit the continuous growth of the stimulated cells; a region at the C terminus of the E1A protein, encoded by the second exon, is required for immortalization and allows them to escape senescence (QUINLAN and DOUGLAS 1992). This region is also required to induce the epithelial growth factor mentioned above. The C-terminal half of the E1A protein, encoded by exon 2, is not required for cooperative transformation by E1A and *ras* (see block of references quoted above), but it does have the effect of modulating that transformation. Mutants bearing deletions located within the 3' end of the second exon transformed more efficiently in cooperation with T24-ras to produce larger colonies of less adherent cells; the transformants were more tumorigenic in syngeneic rats and athymic mice and displayed greater metastasis in the latter hosts (SUBRAMANIAN et al. 1989; BOYD et al. 1993). While this region of E1A may, therefore, play a negative role in determining oncogenicity of the group C viruses, it is not possible at present to assess the potential role(s) of the three N-terminal regions discussed above, as all are required for transformation in vitro, and that capacity is eliminated or reduced by mutation of each region. In this context, it should be added that a similar, extensive characterization of these various E1A domains has not yet been made for the highly oncogenic group A viruses such as Ad12.

Clearly, the E1A proteins are complex, multifunctional entities, and it is likely that a number of their many activities play vital roles in their transforming capacities. CR1 and CR2, two of the regions required for transformation, are, as discussed above in Sect. 3.1, also required for *trans*-repression, which fits with the view that E1A elicits transformation in rodent cells by repressing the expression of certain (as yet undefined) cellular genes. These regions are also the sites to which the Rb protein binds, and this is consistent with an alternative scenario for E1A-mediated transformation, in which the viral proteins complex with the RB gene product to abolish its antioncogenic activity in the host cells.

Similarly, E1A binding of cyclin A, which complexes with the cdk2 protein and is a cell cycle-regulatory component, occurs via binding of p107 to CR2 and could play a role in the proliferative, immortalization, and transformation activities of E1A. While binding of cellular proteins at these sites is likely to be necessary for these E1A-induced processes, it is unlikely that it is sufficient for all of them, and the rules may differ for different viral serotypes. Another E1A function (discussed above in Sect. 3.1) which may have relevance for transformation is the co-operative interaction of the E1A proteins with cAMP to induce AP-1-binding activity and transcription of the c-*fos* gene, which influences the regulation of cell growth (Engel et al. 1991; Gedrich et al. 1992). These authors point out that pRB is known to suppress c-*fos* expression and AP-1 transcriptional activity (Robbins et al. 1990) and suggest that the induction of these processes in virally infected cells treated with cAMP might result from neutralization of Rb protein activity.

3.3 Influence of E1A on the Susceptibility of Transformed Cells to Host Defense Mechanisms

As mentioned briefly in the introduction, expression of the MHC I genes is reduced in Ad12 rat and mouse transformants, but not in the corresponding cells transformed by Ad5 (Eager et al. 1985; Schrier et al. 1983). The MHC I genes encode antigens which reside on the surface of cells (H-2 in mice, RT-1 in rats, HLA in humans) and are necessary for the recognition and elimination of cells bearing foreign antigens by CTL (Cortese-Hassett et al. 1991; Gill et al. 1978; Tanaka et al. 1988). Because Ad12 transformants possess reduced levels of MHC I antigens, they can evade the seek and destroy mission of the CTL and are able to go on to form tumors in the host; the Ad5 transformants possess normal amounts of MHC I antigens and are killed by CTL. Via analyses of antigen levels in rat kidney cells transformed with Ad12/Ad5 E1A/E1B hybrid plasmids, it was shown that downregulation of MHC I gene expression is directed by the E1A gene (Schrier et al. 1983), which has been shown to act at the level of transcription of the MHC I genes (Ackrill and Blair et al. 1988; Friedman and Ricciardi 1988). The MHC-repressing activity of E1A maps to the 5' exon (Jochemsen et al. 1984), and although it might be expected to be determined by the CR1 and CR2 domains like other repressive activities of the E1A proteins, the CR3 region unique to the product of the E1A 13s mRNA is apparently implicated in this function (Meijer et al. 1991). The effect of E1A on MHC expression is mediated via the class I enhancer sequence and seems to involve a concomitant increase in binding of a factor (presumed to be a repressor) to the DNA in the 5' region of the enhancer (Ackrill and Blair 1989; Ge et al. 1992; Meijer et al. 1991). The putative repressor is most likely activated via the CR3 region unique to the Ad12 266R E1A protein, but involvement of other E1A enhancer repression mechanisms has not been ruled out. In this context it may be significant that the CR3 unique region of Ad12 is considerably shorter (31R) than that of Ad5 (46R). It should be pointed out that others report that the enhancer is activated rather than repressed and that a sequence far upstream of the MHC I enhancer is the target for repression by Ad12 E1A (KATOH et al. 1990; OZAWA et al. 1993); these apparent discrepancies might perhaps be explained by use of different transformed cell lines and/or transfection and assay procedures by these various investigators. Clearly, the problem is complex and far from being completely resolved.

The adenovirus gene required for induction of tumor-specific transplantation antigenicity (TSTA), against which the group C-specific CTL react, was mapped to the E1 region via analyses of Ad5 mutants (GALLIMORE and WILLIAMS 1982). While the MHC-repressing activity of E1A in Ad12 has been shown to map to the 5' exon, it has been found that the immunological determinant recognized by these Ad5- or Ad2-specific CTL is encoded by the second exon of the E1A gene (BELLGRAU et al. 1988; KAST et al. 1989; URBANELLI et al. 1989). The dominant epitope presented by the H-2D^b molecules to the immune CTL was determined by using a collection of overlapping 12-mer peptides spanning the entire E1A amino acid sequence to sensitize uninfected mouse cells for use in a cytotoxicity assay with Ad5-specific CTL; it maps to the region encompassing residues 232–247 of the 289R protein (KAST et al. 1989). The Ad5 E1A-specific CTL will cause Ad5-induced tumors formed in nude mice to regress completely when introduced with recombinant isoleukin-2, which adds support to the view that Ad5-induced transformants are nontumorigenic in immunocompetent, syngeneic hosts because of their elimination by specific CTL. The situation with Ad12-induced tumor cells is less clear. Ad12-specific TSTA immunity can be induced in syngeneic hosts and Ad12-specific CTL can kill these transformants in vitro and it has been suggested that the epitope recognized by the CTL is encoded by the Ad12 E1B gene (FÖHRING et al. 1983; SHIROKI et al. 1979); others propose that Ad12 TSTA is specified by E1A (SAWADA et al. 1986). While the level of the MHC I protein available to present this epitope to CTL is likely to be reduced, thus contributing to the increased tumorigenicity of these Ad12 transformants, it also has to be considered that oncogenicity could be influenced by the degree of immunogenicity of these different TSTA. In other words, the Ad12 E1B-related epitope might elicit a weaker immune response than the epitope in Ad5 E1A, aiding the Ad12 tumor cells to evade immune surveillance. Interestingly, the C-terminal sequence identified as an Ad5 E1A immunodominant epitope (R232–247) overlaps the proposed negative determinant of transformation and tumorigenicity (R225–238), discussed above in Sect. 3.2 (BOYD et al. 1993), providing an explanation for the observation that mutants missing this sequence induce transformants both more tumorigenic and more metastatic. This E1A sequence is quite different in Ad12, which could mean that it does not act as a good immunogenetic determinant for that virus, perhaps weak or inefffective in the context of MHC I, or perhaps not reaching the correct location in the cell for association with the MHC I protein. Instead, the dominant epitope that is recognized might be encoded by the Ad12 E1B gene.

Compelling evidence in support of the hypothesis that E1A suppression of MHC I expression results in increased tumorigenicity of transformants by CTL evasion was provided by tests in which the tumorigenicity of an Ad12-

transformed mouse cell line was reduced after introduction of functional H-2 genes resistant to repression by E1A protein (TANAKA et al. 1985). Equally compelling, however, is the report that transfection of MHC I genes into Ad12 transformants actually increases rather than decreases their tumorigenicity (SODDU and LEWIS 1992). In addition, others have been unable to find a clear correlation between tumorigenicity, level of MHC I antigen, and CTL killing in group C and Ad12 transformants (HADDADA et al. 1986; MELLOW et al. 1984). Thus, while these discrepancies could be explained in part by species, strain, haplotype, and other differences, it may be unwise or premature to generalize about the significance of reduction in MHC I gene expression to Ad12 oncogenicity, and it seems probable that other processes are involved in transformed cell growth and/or rejection in syngeneic hosts. As mentioned in the introduction, there is evidence to suggest that natural killer (NK) cells also play an important role in the rejection of such tumor cells. It has been reported that the degree of resistance to NK cells correlates with differences in tumorigenicity of Ad transformants; for example, Ad12 transformants are more resistant to killing by these cells than cells transformed by Ad5 (COOK et al. 1982; HADDADA et al. 1986; MELLOW et al. 1984; KAST et al. 1989; RASKA and GALLIMORE 1982). It has been shown that Ad12 E1A confers resistance to NK action (SAWADA et al. 1985) and that Ad5 E1A confers susceptibility to NK cell activity (KAST et al. 1989). At present, little is known about how NK cells recognize target cells or whether there is any meaningful relationship between this property and the expression of MHC I gene in Ad transformants.

Undoubtedly, E1A gene products play more than one role in shaping the immune response of the host animal to Ad-transformed cells, but the various ways (and their possible interconnections) in which they do so are not clear. In addition, most likely in cooperation with E1B protein(s), E1A protein functions almost certainly influence the inherent capacity of transformed cells to grow in a nonregulated manner in vivo. This may involve phenotypic traits displayed by transformants in vitro, such as growth to high density, anchorage independence, and association of E1A proteins with cellular antioncogene proteins such as pRb. Determining the mechanisms and significance of these various E1A activities in the induction and growth of tumor cells will likely be a complex task.

4 Might There Be Other Ways in Which E1A Affects the Differential Oncogenicity of Adenoviruses?

At this point in time an impressive collection of functions has been ascribed to the E1A proteins required in productive infections of permissive cells and in non-permissive cells undergoing transformation. Some of the roles that these proteins play in infection and transformation in vitro can be broadly defined, but we cannot yet point to a precise functional role of E1A in viral oncogenicity. In considering additional factors which could perhaps contribute to differential oncogenicity of adenoviruses, it seemed rational to us to evaluate the potential roles in that process of the E1A nonconserved regions, which are conspicuously diverse among the different serotypes. In fact, while we were pursuing this approach, it was shown, as discussed above, that the second exon of Ad5 E1A (an extensive nonconserved region) bears a determinant which negatively influences transformation in cooperative assay with activated *ras* and the tumorigenicity of these transformants (Boyd et al. 1993). One nonconserved region which initially attracted us was the unique segment of 20 amino acids (R124 to 143) located between CR2 and CR3 in Ad12. This region, which we call the spacer region, is totally absent from the Ad5 E1A proteins (Kimelman et al. 1985; Larsen and Tibbetts 1987; van Ormondt and Galibert 1984). A slightly larger spacer region than that in Ad12 is also found in the E1A proteins of SA7 , and slightly shorter ones are present in the Ad3 and Ad7 proteins. The various spacer regions and their flanking regions are depicted in Fig.1. Of these five viruses, Ad5 is non-oncogenic, Ad3 and Ad7 are weakly oncogenic, while Ad12 and SA7 are highly oncogenic; there is, therefore, a strong correlation between the spacer length and the degree of oncogenicity, an observation made previously by Larsen and Tibbetts in 1987. In view of this, we postulated that the spacer segment of Ad12 might play a key role in the oncogenicity of that virus.

4.1 Construction and Analysis of Viruses Containing Chimeric E1A Genes; Evidence that the Unique Spacer of Adenovirus Type 12 Plays a Role in Oncogenesis

Our initial approach to assessing the functional significance of the Ad12 E1A spacer was to exchange that segment of Ad12 with the equivalent region in Ad5 (and vice versa) and to determine the phenotypic effects of these rearrangements on the viruses produced. Fortunately, the Ad5 E1A sequences flanking this region contain pre-existing *cla*I and *sma*I restriction sites, and it was possible to create each of these sites by introduction of single point mutations at the equivalent locations in the E1A gene of Ad12; the resulting variant (*var*747) displays a wild-type phenotype for viral growth, transformation, and tumor induction in rats. In this way we could reciprocally exchange small Ad5 and Ad12 cassettes, without disruption of reading frame, to create Ad12-based and Ad5-based, fully functional chimeric E1A genes lacking and containing, respectively, the Ad12 spacer. These chimeric genes were rescued into the appropriate complete virus genome (e.g., the Ad12-based E1A gene into Ad12 genome). The structures of the Ad12 variant containing the new restriction sites and those of selected Ad12/Ad5 chimeric genes are shown in Fig.2. Extensive phenotypic analysis of the Ad12-based chimeric viruses 12*ch*702 and 12*ch*704 shows them to be phenotypically wild type in all respects, except that they display reduced oncogenicity (Telling and Williams 1994). For example, they grow normally in human A549 cells,

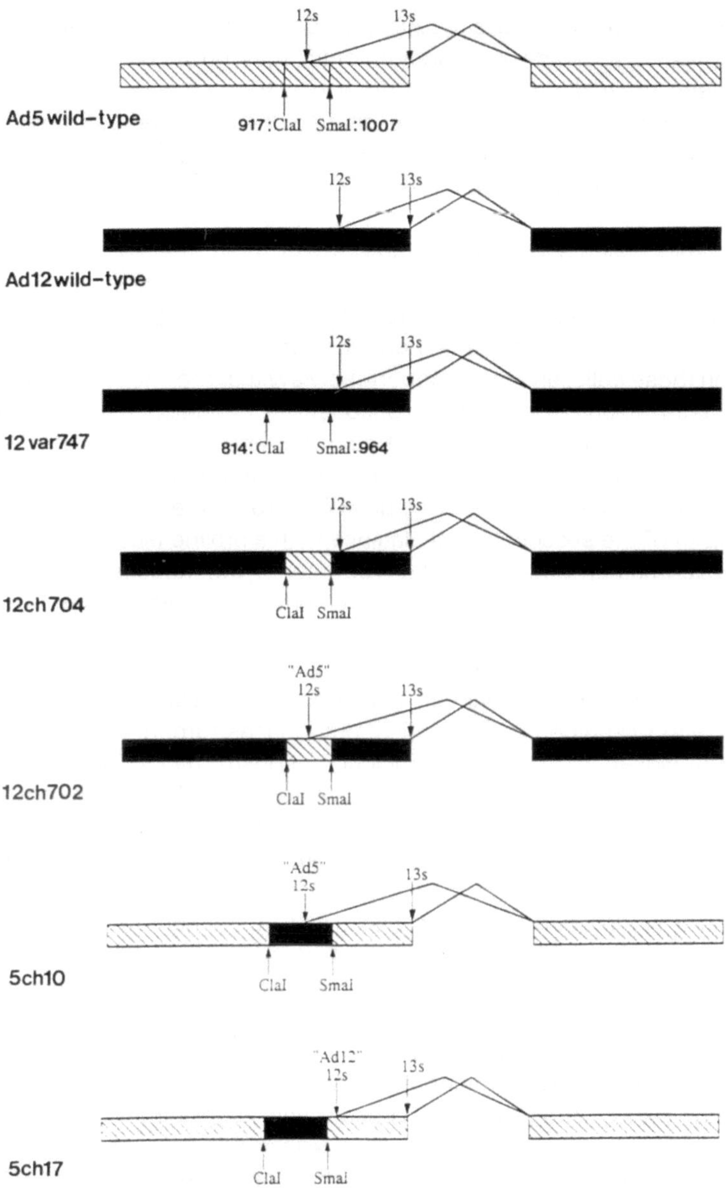

Fig. 2. The structure of the adenovirus (*Ad*)12/Ad5 chimeric E1A genes in Ad12-based chimeric viruses and the Ad5/Ad12 chimeric E1A genes in Ad5-based viruses. Ad12 sequences are depicted as *black blocks*, and Ad5 sequences are shown as *hatched blocks*. (Adapted from a similar figure in TELLING and WILLIAMS 1994)

generating the appropriately shorter E1A RNA and proteins expected as a result of the Ad5 insert, and the E1A proteins transactivate normally in transient assay. They transform BRK and baby mouse kidney cells with the same frequencies induced by Ad12 and *var*747, and although the foci produced on these cells are flatter and larger than wild-type transformants, they grow vigorously and continuously, attain very high densities, and grow in agar suspension, just as the wild-type transformants do. Thus, cassette exchange does not markedly disrupt the broad functional integrity of the chimeric E1A gene products. Their capacity to induce tumors in baby Hooded Lister rats is, however, significantly reduced (Table 1). Further, the tumorigenicity in syngeneic rats of transformants induced in vitro is in almost all cases considerably lower than that of wild-type transformants (Table 2).

These results, taken together, demonstrate for the first time that the capacity of Ad12 to transform cells can be separated from its capacity to confer tumorigenicity on these cells, and they support the view that the spacer region of Ad12 E1A is a determinant of adenovirus oncogenicity. Additional support for this view stems from phenotypic analysis of point mutants *pm*715 and *pm*714 (C. Liu, S. Perera, G.C. Telling, J. Williams, manuscript in preparation). The *pm*715 mutant carries a transversion of the G at position 913 in the spacer to a C, replacing the fourth of the six contiguous alanines with a proline (see Fig.1). The mutant virus, which has an otherwise wild-type phenotype, induces tumors at reduced efficiency, and cells transformed by it are nontumorigenic (Tables 1 and 2, respectively). In mutant *pm* 714, the G at position 932 is also changed to a C, replacing the arginine at the start of CR3 with a glycine, which is the residue found at that relative location in Ad5 (see Fig.1). This mutant, like *pm*715, grows and transforms with wild-type characteristics, but induces tumors in rats more rapidly and with closer to wild-type efficiency; transformants induced by this mutant are highly tumorigenic (Tables 1 and 2, respectively). These results with the point mutants strongly suggest that the spacer region per se, and not one of the flanking regions carried in by cassette exchange, is the oncogenic determinant. They also argue against the view that removal of spacer sequence results simply in a spatial perturbation giving rise to alterations of functional domains elsewhere on the E1A proteins, which in turn bring about reduced oncogenicity. In this context, we have obtained preliminary results which suggest that the capacity to bind pRb (to sites in CR1 and CR2) is not altered in the chimeric E1A (Liu, G.C. Telling, and J. Williams unpublished results). We have also examined the levels of H-2 antigen in a set of mouse cell transformants induced by Ad5, Ad12, and chimeric viruses and determined NK killing susceptibility in some of these and in a set of rat cell lines transformed by these viruses. Loss of the spacer does not seem to alter the pattern of low MHC I levels and low NK susceptibility observed in a variety of Ad12 wild-type transformants; they did not exhibit the high levels of H-2 antigen and NK killing found in parallel sets of Ad5 transformants (C. Liu, M. Wagner, A. Waggoner, and J. Williams, unpublished results). While incomplete at present, these results suggest that there may not be a clear-cut correlation between tumorigenicity and

Table 1. Induction of tumors in Hooded Lister rats inoculated as newborns with adenovirus type 12 (Ad12) E1A chimeric viruses and point mutants

Virus	Tumor-positive rats(*n*)/Rats inoculated [a](*n*)				Latency period[b] (days)			
	1×10^8pfu	3×10^8pfu	6×10^8pfu	1×10^9pfu	1×10^8pfu	3×10^8pfu	6×10^8pfu	1×10^9pfu
*var*747	14/14	13/13	n.d.	n.d.	57±5	53±7	n.d.	n.d.
*ch*702	0/11	3/14	4/5	n.d.	>250	115±5	129±51	n.d.
*ch*704	0/13	3/16	2/12	n.d.	>250	126±18	151+57	n.d.
*pm*715	8/19	3/19	8/19	5/18	118±33	124±19	109±24	105±5
*pm*714	10/17	10/14	12/14	n.d.	103±27	74±19	61±6	n.d.

n.d., not determined, pfu, plaque-forming unit.
[a]In each case 24- to 36-h-old newborns were inoculated intraperitoneally with 0.05 ml of virus suspension.
[b]Average time in days after inoculation at which palpable tumors were first detected ± standard deviation.

Table 2. Tumorigenicity of transformed cell lines derived from baby rat kidney (BRK) cells infected with adenovirus type 12 (Ad12) E1A chimeric viruses and point mutants

Cell line	Tumor incidence by no. of cells inoculated[a]			Latency period[b]			Termination point[c]		
	1×10^7	3×10^6	1×10^6	1×10^7	3×10^6	1×10^6	1×10^7	3×10^6	1×10^6
RK747A	6/6	6/6	6/6	8	9	16	24	32	42
RK747B	6/6	6/6	6/6	9	9	16	33	32	52
RK747C	6/6	6/6	5/6	8	9	16	33	35	48
RK702E	3/4	0/4	0/4	12	14	–	94	>90	>100
RK702F	1/4	0/4	0/4	12	14	–	90	>90	>90
RK702G	4/4	4/4	4/4	14	23	30	49	52	52
RK704B	0/4	0/4	0/4	–	–	–	>90	>90	>90
RK704C	0/4	2/4	2/4	–	44	60	>90	98	90
RK704G	0/4	1/4	1/4	–	30	48	>100	90	90
RK715A	0/6	0/6	0/6	15	22	–	>90	>90	>90
RK715B	0/6	0/6	0/6	15	22	–	>90	>90	>90
RK715C	0/6	0/6	0/6	–	–	–	>90	>90	>90
RK714A	6/6	6/6	6/6	8	8	9	21	21	25
RK714B	6/6	6/6	6/6	10	10	10	28	26	25
RK714C	6/6	6/6	5/6	11	11	14	30	30	34

In some cases, e.g., in animals inoculated with 3×10^6 RK702E, RK702F, RK715A, and RK715B cells, tumors appeared at the time shown, grew slowly for a few weeks, and then regressed completely. These animals were kept for a total of 90 days. In other cases, e.g., with all inocula of RK704B, tumors did not appear at all; these animals were kept for a minimum of 90 days.

[a]No. of rats positive/total no. of rats. The appropriate numbers of cells , contained in Dulbecco's modified Eagle's medium (DMEM), were inoculated subcutaneously into 6–week-old syngeneic Hooded Lister rats. Recipients were examined every 2 days for evidence of tumor growth, and when tumor masses reached about 3 cm in diameter the animals were killed.

[b]Day on which palpable tumors first appeared.

[c]Day on which animals with or without tumors were killed.

level of MHC I and/or degree of susceptibility to NK cell killing of Ad transformants. This implies that other factors may have additional and perhaps more significant influence on tumorigenicity.

4.2 Another E1A Region Also Affects Tumorigenicity of Transformants

These results with the viruses and the point mutants identify the spacer region as an oncogenic determinant of Ad12, but they do not really say that this region alone is sufficient to confer tumorigenic potential on cells. In fact, this seemed unlikely, as *ch*702, *ch*704, and *pm*715 are still partially oncogenic (see Table 1). Other regions of E1A and/or possibly E1B might also influence oncogenicity and the tumorigenicity of transformants; this can be tested by exchanging or mutating these other regions in wild-type or chimeric background. The degree to which the Ad12 spacer can act autonomously in this respect can be tested by transferring it into an Ad5-based background. Chimeric viruses 5*ch*10 and 5*ch*17 carry the Ad12 spacer cassette in Ad5 genome (see Fig. 2). Although apparently wild type for viral growth and transformation of rat cells, these viruses are not oncogenic in Hooded Lister baby rats (J. WILLIAMS, unpublished results), and the transformants we have tested to date are not tumorigenic in syngeneic animals. This suggests that the Ad12 spacer region may be required, but is not sufficient, for oncogenic activity. It is perhaps not too surprising, however, that the Ad5-based virus fails to induce tumors directly, for it is well known that the group C viruses replicate to some extent and are much more cytotoxic for many types of rodent cells by comparison with Ad12. In fact, it has been shown that viruses which have the entire E1 region of Ad12 built into an Ad5 background fail to induce tumors in hamsters or rats (BERNARDS et al. 1984; SAWADA et al. 1988). These authors suggest that adenovirus oncogenicity is determined by E1 and regions elsewhere on the genome; it could, perhaps, be influenced by growth of the virus in and killing of the target cells (WILLIAMS 1973).

A set of Ad5-based Ad5/Ad12 chimeric E1A genes have also been constructed by the technique of homologous recombination in *Escherichia coli* (JELINEK and GRAHAM 1992). These recombinant plasmid genes, which carry different N-terminal segments of Ad12 sequence in Ad5 E1A, were used in cooperation with Ad12 E1B plasmids to transform BRK cells, and transformants derived from these were tested for tumorigenicity in newborn syngeneic rats. Ad5-based plasmids in which the CR2-CR3 region was replaced with Ad12 sequence were also made and used to generate transformants; these are similar, but not identical, to the Ad5-based chimerics we have made. Transformants induced by hybrid genes bearing N-terminal Ad12 sequence up to or beyond CR3 were shown to be quite tumorigenic, while those with shorter lengths of Ad12 N-terminal sequence were not. Some of the transformants derived from transfection with the inserted Ad12 CR2-CR3 segment were moderately tumorigenic, but less than those containing the longer Ad12 N-terminal sequences (JELINEK et al. 1994). These results suggest that a region (or regions) in addition to the spacer (and perhaps acting in cooperation with it) plays a role in influencing the tumorigenicity of transformants.

At present, none of these results reveal anything about the mechanism by which the spacer segment of Ad12 influences the oncogenicity of this virus. It has been suggested that this spacer sequence adjacent to CR3 in Ad3 is an autorepressive domain (Larsen and TIBBETTS 1987; see above in Sect. 3.1). Potentially, the Ad12 spacer segment could also be autorepressive, but we have not found any convincing evidence for any autoregulatory influence of Ad12 E1A in either Ad12, as previously reported (LAMBERTI and WILLIAMS 1990), or in the chimeric viruses. In the context of repressive activity, it may be significant that the Ad12 spacer and that of highly oncogenic SA7 contains a contiguous run of alanines (see Fig. 1). This is a structural feature of some other transcriptional repressors (LICHT et al. 1990). Although the spacer of Ad12 does not seem to be involved in autorepressive activity and does not seem to influence MHC I suppression by Ad12 E1A, it could repress expression of other viral and/or cellular genes. As mentioned above, the CR1 and CR2 regions play a role in repression by E1A, and this activity is thought to be important for transformation. The Ad12 spacer region could act independently of these or could act synergistically or in cooperation with these (which would be consistent with findings of JELINEK et al. 1994) to influence the oncogenicity of that virus. It is clear, however, that removal of the spacer from Ad12 or its insertion into Ad5 does not perturb the basic immortalization or transforming functions, or the binding of cellular proteins, attributed to the CR1 and CR2 regions of these viruses, and it seems unlikely that the observed influence on oncogenicity derives from an indirect effect on these regions. While a role for the spacer region in influencing the host defense response to tumor cells cannot be ruled out, it does not seem to impose major effects on MHC I levels or NK susceptibility in transformants. Thus, this spacer may act via a pathway distinct from those affected by other E1A regions, such as the sequestration of certain cell cycle-related proteins. Identification of a cellular protein binding exclusively to the spacer sequence would undoubtedly help to identify that putative pathway; such a search has yet to be launched.

5 Concluding Remarks: Influence of Other Adenoviral Genes on Oncogenicity

E1A does not act alone to mold the fully transformed tumor cell phenotype. The transformation process is complex and multistep and involves not only contributions from E1A gene functions, but also from activities of E1B products and perhaps products encoded by other viral genes. Numerous studies with a large number of mutants with lesions in the E1B gene have led to the conclusion that both major E1B proteins, 19K and 55K, are required for transformation (reviewed in BOULANGER and BLAIR 1991; VAN DER EB and ZANTEMA 1992; WILLIAMS 1986). Recently, it has been shown that the 19K protein is not required for either transformation of primary rat or mouse cells in vitro or induction of tumors in rats by Ad12 (EDBAUER et al. 1988; ZHANG et al. 1992) or for transformation of a variety of rodent cells by Ad5 (TELLING and WILLIAMS 1993). The 55K, on the other hand, has been shown in these and many other investigations to be obligatory for complete transformation by these viruses. In agreement with these recent reports that the 19K protein is dispensable for transformation by group A and group C viruses, it has been shown by plasmid transfection procedures that either the 19K or the 55K of Ad5 along with the E1A proteins of Ad5 is sufficient to transform BRK cells completely in the absence of the other E1B protein, albeit at lower efficiency than when both proteins are expressed (MCLORIE et al. 1991; WHITE et al. 1991). It is unlikely that the capacity of the E1B 55K protein to regulate viral and cellular mRNA transport and stability during the productive infection of human cells (PILDER et al. 1986; WILLIAMS et al. 1986; ORNELLES and SHENK 1991) is necessary or involved in transformation of rodent cells. On the other hand, the interaction of the 55K protein with the protein encoded by the tumor supressor gene p53 (SARNOW et al. 1982) is quite likely to be required. Wild-type p53 is a transcriptional activator, which mediates a G1/S block in the cell cycle and inhibits cellular proliferation. It does this by targeting p21 (also called WAF1 or Cip1), an inhibitor of cyclin-dependent kinases active in progression of cells through the G1/S step of the cycle (reviewed in RON 1994). The wild-type form of p53 has also been shown to induce programmed cell death (apoptosis) in certain cell types. It has been proposed that the suppression of p53-transactivating activity by E1B 55K is involved in transformation of BRK cells by Ad5 (YEW and BERK 1992). In this case we would surmise that the E1B protein initiates a release in the G_1 block normally present in growth-arrested BRK cells and allows progression to the fully transformed phenotype by the action of other (E1A) factors. An alternative scenario is proposed in which the 55K and/or the 19K proteins act to prevent p53-mediated apoptosis induced in the primary BRK cells by E1A (DEBBAS and WHITE 1993; LOWE and RULEY 1993). Neither the mechanism for either model, nor the possible relationship between these two alternatives is clear at the present time. Further, although it has been shown that Ad12 E1B 55K also inhibits p53-transactivating activity (YEW and BERK 1992), it is not clear that either this or the apoptotic suppression scenario occurs in or is necessary for, oncogenic transformation by this virus. It is worth noting that cells expressing the Ad12 55K gene, but not cells producing the Ad5 E1B protein, are highly tumorigenic in nude mice (BERNARDS et al. 1982) and that the Ad12 protein does not form a stable complex with the p53 protein, while the Ad5 55K protein forms a stable cytoplasmic complex in which p53 is likely inactivated. It has been shown that high levels of metabolically stable p53 with wild-type conformation accumulate in the nuclei of (at least some) Ad12 transformants (VAN DEN HEUVEL et al. 1993), suggesting that the Ad5 and Ad12 55K proeins may operate via different pathways in oncogenic transformation.

As mentioned in Sect. 4.2 above, there is evidence that viral genes other than E1 may also be important for oncogenic transformation (BERNARDS et al. 1984; SAWADA et al. 1988). There is very clear evidence that with Ad9 the E4 open reading frame (ORF)1 is required for transformation in vitro and induction of mammary fibroadenomas in rats and that the E1 region is not required for the latter process (JAVIER 1994; JAVIER et al. 1992). The Ad9 ORF1 product alone is capable of transforming rat CREF cells (the virus does not transform primary BRK cells) and can cooperate with E1 product(s) to induce transformation at higher frequency; this may be the exception that proves the rule that E1 is required and sufficient for oncogenic transformation. Related serotype Ad26 does not induce tumors in rats, although the ORF1 from that virus does transform CREF cells in vitro with the same efficiency as Ad9 ORF1, suggesting that the differential oncogenicity is determined by the protein encoded by that frame. It cannot yet be ruled out that a protein product of adjacent ORF2 or ORF3 acts as a negative regulator or repressor of oncogenicity, and a potential role for E1A and E1B of Ad9 (although the latter is not expressed in rat mammary tumors) has not yet been ruled out either. It has been suggested that Ad12 E4 is required for the growth of transformants in agar suspension and may possibly influence the tumorigenicity of 3Y1 rat cells (Shiroki et al. 1984). It seems unlikely, however, that Ad12 E4 or Ad5 E4 is required for transformation, as ORF1 from neither virus is able to transform CREF cells (Javier 1994), and it is well established that E1 region of these viruses can induce complete transformation in vitro. It cannot be ruled out, however, that the product of ORF1 or of other E4 ORFs from these viruses influence the transformed cell phenotype and/or oncogenicity.

Finally, it should be noted that, when virus is used, the products of viral genes encoded elsewhere (other than E1 or E4) influence the transforming capacity of the virus. Temperature-sensitive (*ts*) mutations in the E2B polymerase gene of Ad5 and the E2A DNA-binding protein gene can modulate transformation frequencies (Rice et al. 1987; Miller and Williams 1987). These gene products may exert their influence via the E1 genes required for transformation with this virus. It is not known whether these or other viral genes exert influence, either positive or negative, on the oncogenic potential of Ad5 and Ad12; this is largely uncharted territory which could perhaps be mapped in the future via the chimeric virus approach. For the near future, though, the more familiar but complex pathways travelled by the E1A and E1B proteins will continue to be explored more intensively and are expected to yield more vital clues to help solve the enigma of how adenoviruses do or do not induce tumors.

Acknowledgments. We are most grateful to Robin Wilbur for valuable assistance with the preparation of this manuscript and to Vladimir Vidak for valuable assistance with animal care. Work from the authors' laboratory was supported by Public Health Service grants CA32940 and CA21375 from the NIH to J.W.

References

Ackrill AM, Blair GE (1988) Regulation of major histocompatibility class I gene expression at the level of transcription in highly oncogenic adenovirus transformed rat cells. Oncogene 3: 483–487

Ackrill AM, Blair GE (1989) Nuclear proteins binding to an enhancer of the major histocompatibility class I promoter: differences between highly oncogenic and nononcogenic adenovirus-transformed rat cells. Virology 172: 643–646

Bellgrau D, Walker IA, Cook JL (1988) Recognition of adenovirus E1A gene products on immortalized cell surfaces by cytotoxic T lymphocytes. J Virol 62: 1513–1519

Berk AJ (1986) Adenovirus promoters and E1A transactivation. Annu Rev Gen 20: 45–79

Berk AJ, Lee F, Harrison T, Williams J, Sharp PA (1979) Pre-early adenovirus 5 gene product regulates synthesis of early viral messenger RNAs. Cell 17: 935–944

Bernards R, Houweling A, Schrier PI, Bos JL, van der Eb AJ (1982) Characterization of cells transformed by Ad5/Ad12 hybrid early region I plasmids. Virology 120: 422–432

Bernards R, Schrier PI, Houweling A, Box JL, van der Eb AJ, Zylstra M, Melief CJM (1983) Tumorigenicity of cells transformed by adenovirus type 12 by evasion of T-cell immunity. Nature 350: 776–779

Bernards R, de Leeuw M, Vaessen MJ, van der Eb AJ (1984) Oncogenicity by adenovirus is not determined by the transforming region only. J Virol 50: 847–853

Borelli E, Hen R, Chambon P (1984) Adenovirus-2 E1A products repress enhancer-induced stimulation of transcription. Nature 312: 608–612

Bos JL, Polder LJ, Bernards R, Schrier PI, van den Elsen PJ, van der Eb AJ, van Ormondt H (1981) The 2.2 kb E1B mRNA of human Ad12 and Ad5 codes for two tumor antigens starting at different AUG triplets. Cell 27: 121–131

Boulanger PA, Blair GE (1991) Expression and interactions of human adenovirus oncoproteins. Biochem J 275: 281–299
Boyd JM, Subramanian T, Schaefer U, La Regina M, Bayley S, Chinnadurai G (1993) A region in the C-terminus of adenovirus 2/5 E1A protein is required for association with a cellular phosphoprotein and important for the negative modulation of T24-ras mediated transformation, tumorigenesis and metastasis. EMBO J 12: 469–478
Byrd PJ, Grand RJA, Breiding D, Williams J, Gallimore PH (1988) Host range mutants of adenovirus type 12 E1 defective for lytic infection, transformation and oncogenicity. Virology 163: 155–165
Cook JL, Hibbs JB Jr, Lewis AM Jr (1982) DNA virus-transformed hamster cell-host effector cell interactions: level of resistance to cytolysis correlated with tumorigenicity. Int J Cancer 30: 795–803
Cortese-Hassett A, Misra D, Kunz H, Gill TJ (1991) The major histocompatibility complex of the rat. In: Srivastava R, Ram R, Tyle BP (eds) Immunogenetics of the major histocompatibility complex. VCH, Weinheim, pp 309–347
Debbas M, White E (1993) Wild-type p53 mediates apoptosis by E1A, which is inhibited by E1B. Genes Dev 7: 546–554
Eager KB, Williams J, Breiding D, Pan S, Knowles B, Appela E, Ricciardi RP (1985) Expression of histocompatibility antigens H-2K, D, and L is reduced in adenovirus-12-transformed mouse cells and is restored by interferon γ Proc Natl Acad Sci USA 82: 5525–5529
Edbauer D, Lamberti C, Tong J, Williams J (1988) Adenovirus type 12 E1B 19-kilodalton protein is not required for oncogenic transformation in rats. J Virol 62: 3265–3273
Egan C, Jelsma TN, Howe JA, Bayley ST, Ferguson B, Branton PE (1988) Mapping of cellular protein-binding sites on the products of early-region 1A of human adenovirus type 5. Mol Cell Biol 8: 3955–3959
Egan C, Bayley ST, Branton PE (1989) Binding of the RB1 protein to E1A products is required for adenovirus transformation. Oncogene 4: 383–388
Engel DA, Muller U, Gedrich RW, Eubanks JS, Shenk T (1991) Induction of c-fos mRNA and AP-1 DNA-binding activity by cAMP in cooperation with either the adenovirus 243- or the adenovirus 289-amino acid E1A protein. Proc Natl Acad Sci USA 88: 3957–3961
Faha B, Ewen M, Tsai L, Livingston DM, Harlow E (1992) Interaction between human cyclin A and adenovirus E1A-associated p107 protein. Science 255: 87–90
Föhring B, Gallimore PH, Mellow GH, Raska K Jr (1983) Adenovirus type 12 specific cell surface antigen in transformed cells is a product of the E1b early region. Virology 131: 463–472
Friedman DJ, Ricciardi RP (1988) Adenovirus type 12 E1A gene represses accumulation of MHC class I mRNA at the level of transcription. Virology 165: 303–305
Gallimore PH (1972) Tumour production in immunosuppressed rats with cells transformed in vitro by adenovirus type 2. J Gen Virol 16: 99–102
Gallimore PH, Williams J (1982) An examination of adenovirus type 5 mutants for their ability to induce group C adenovirus tumor-specific antigenicity in rats. Virology 120: 146–156
Gallimore PH, Byrd PJ, Grand RJA (1984a) Adenovirus genes involved in transformation. What determines the oncogenic phenotype? In: Rigby PWJ, Wilkie NM (eds) Viruses and cancer. Symposium of the Society for General Microbiology. Cambridge University Press, Cambridge, pp 125–172
Gallimore P, Byrd P, Grand R, Whittaker J, Breiding D, Williams J (1984b) An examination of the transforming and tumor-inducing capacity of a number of adenovirus type 12 early region 1, host-range mutants and cells transformed by subgenomic fragments of Ad12 E1 region. Cancer Cells 2: 519–526
Ge R, Kralli A, Weinmann R, Ricciardi RP (1992) Down-regulation of the major histocompatibility complex class I enhancer in adenovirus type 12-transformed cells is accompanied by an increase in factor binding. J Virol 66: 6969–6978
Gedrich RW, Bayley ST, Engel DA (1992) Induction of AP-1 DNA-binding activity and c-fos mRNA by the adenovirus 243R E1A protein and cyclic AMP requires domains necessary for transformation. J Virol 66: 5849–5859
Gill TJ, Cramer DV, Kunz HW (1978) The major histocompatibility complex—comparison in the mouse, man, and the rat. Am J Pathol 90: 737–777
Giordano A, McCall C, Whyte P, Franza BR (1991) Human cyclin A and the retinoblastoma protein interact with similar but distinguishable sequences in the adenovirus E1A gene product. Oncogene 6: 481–485
Graham FL, Harrison T, Williams J (1978) Defective transforming capacity of adenovirus type 5 host-range mutants. Virology 86: 10–21
Graham FL (1984) Transformation by and oncogenicity of human adenoviruses. In: Ginsberg HS (ed) The adenoviruses. Plenum, New York, pp 339–398

Graham FL, Abrahams PJ, Mulder C, Heijneker HL, Warnaar SO, de Vries FAJ, Fiers W, van der Eb AJ (1974) Studies on in vitro transformation by DNA and DNA fragments of human adenoviruses and simian virus 40. Cold Spring Harbor Symp Quart Biol 39: 637–650

Haddada H, Lewis AM Jr, Sogn JA, Coligan JE, Cook JL, Walker TA, Levine AS (1986) Tumorigenicity of hamster and mouse cells transformed by adenovirus types 2 and 5 is not influenced by the level of class I major histocompatibility antigens expressed on the cells. Proc Natl Acad Sci USA 83: 9684–9688

Haddada H, Sogn JA, Coligan JE, Carbone M, Dixon K, Levine AS, Lewis AM Jr, (1988) Viral gene inhibition of class I major histocompatibility antigen expression: not a general mechanism governing the tumorigenicity of adenovirus type 2-, adenovirus type 12-, and simian virus 40-transformed Syrian hamster cells. J Virol 62: 2755–2761

Haley KP, Overhauser J, Babiss LE, Ginsberg HS, Jones NC (1984) Transformation properties of type 5 adenovirus mutants that differentially express the E1A gene products. Proc Natl Acad Sci USA 81: 5734–5738

Harlow E, Franza R Jr, Schley C (1985) Monoclonal antibodies specific for adenovirus early region 1A proteins: extensive heterogeneity in early region 1A products. J Virol 55: 533–546

Harlow E, Whyte P, Franza BR Jr, Schley C (1986) Association of adenovirus early-region 1A protein with cellular polypeptides. Mol Cell Biol 6: 1579–1589

Hearing P, Shenk T (1985) Sequence-independent autoregulation of the adenovirus type 5 E1A transcription unit. Mol Cell Biol 5: 3214–3221

Houweling A, Van den Elsen PJ, van der Eb AJ (1980) Partial transformation of primary rat cells by the leftmost 4.5% fragment of adenovirus 5 DNA. Virology 105: 537–550

Huebner RJ, Rowe WP, Lane WT (1962) Oncogenic effects in hamsters of human adenovirus types 12 and 18. Proc Natl Acad Sci USA 48: 2051–2058

Hurwitz DR, Chinnadurai G (1985) Evidence that a second tumor antigen coded by adenovirus early region E1a is required for efficient cell transformation. Proc Natl Acad Sci USA 82: 163–167

Javier RT (1994) Adenovirus type 9 E4 open reading frame 1 encodes a transforming protein required for the production of mammary tumors in rats. J Virol 68: 3917–3924

Javier RT, Raska K, Shenk T (1992) Requirement for the adenovirus type 9 E4 region in production of mammary tumors. Science 257: 1267–1271

Jelinek T, Graham FL (1992) Recombinant human adenoviruses containing hybrid adenovirus type 5 (Ad5)/Ad12 E1A genes: characterization of hybrid E1A proteins and analysis of transforming activity and host range. J Virol 66: 4117–4125

Jelinek T, Pereira DS, Graham FL (1994) Tumorigenicity of adenovirus-transformed rodent cells is influenced by at least two regions of adenovirus type 12 early region 1A. J Virol 68: 888–896

Jelsma TN, Howe JA, Mymryk JS, Evelegh CM, Cunniff NFA, Bayley ST (1989) Sequences in E1A proteins of human adenovirus 5 required for cell transformation, repression of a transcriptional enhancer, and induction of proliferating cell nuclear antigen. Virology 171: 120–130

Jochemsen AG, Bos JL, van der Eb AJ (1984) The first exon of region E1A genes of adenoviruses 5 and 12 encodes a separate functional protein domain. EMBO J 3: 2923–2927

Jones N, Shenk T (1979) An adenovirus type 5 early gene function regulates expression of other early viral genes. Proc Natl Acad Sci USA 76: 3665–3669

Jones NC (1992) The multifunctional products of the adenovirus E1A gene. In: Doerfler W, Bohm P (eds) Malignant transformation by DNA viruses. VCH, Weinheim, pp 87–113

Jonsson N, Ankerst J (1977) Studies on adenovirus type 9-induced mammary fibroadenomas in rats and their malignant transformation. Cancer 39: 2513–2519

Kao H, Nevins JR (1983) Transcriptional activation and subsequent control of the human heat shock gene during adenovirus infection. Mol Cell Biol 3: 2058–2065

Kast WM, Offringa R, Peters PJ, Voordouw AC, Meloen RH, van der Eb AJ, Melief CJM (1989) Eradication of adenovirus E1-induced tumors by E1A-specific cytotoxic T lymphocytes. Cell 59: 603–614

Katoh S, Ozawa K, Kondoh S, Soeda E, Israel A, Shiroki K, Fujinaga K, Itakura K, Gachelin G, Yokoyama K (1990) Identification of sequences responsible for positive and negative regulation by E1A in the promoter of H-$2K^{bm1}$ class I MHC gene. EMBO J 9: 127–135

Kimelmann D, Miller JS, Porter D, Roberts BE (1985) E1A regions of the human adenoviruses and of the highly oncogenic simian adenovirus 7 are closely related. J Virol 53: 399–409

Lamberti C, Williams J (1990) Differential requirement for adenovirus type 12 E1A gene products in oncogenic transformation. J Virol 64: 4997–5007

Larsen PL, Tibbetts C (1987) Adenovirus E1A gene autorepression: revertants of an E1A promoter mutation encode altered E1A proteins. Proc Natl Acad Sci USA 84: 8185–8189

Lewis AM Jr, Cook JL (1982) Spectrum of tumorigenic phenotypes among adenovirus 2, adenovirus 12 and simian virus 40 transformed Syrian hamster cells defined by host cellular immune-tumor cell interactions. Cancer Res 42: 939–944

Licht JD, Grossel MU, Figge J, Hansen UM (1990) Drosophila Krüppel protein is a transcriptional repressor. Nature 346: 76–79

Lillie JW, Green M, Green MR (1986) An adenovirus E1A protein region required for transformation and transcriptional repression. Cell 46: 1043–1051

Lillie JW, Loewenstein PM, Green MR, Green M (1987) Functional domains of adenovirus type 5 E1A proteins. Cell 50: 1091–1100

Lowe SW, Ruley HE (1993) Stabilization of the p53 tumor suppressor is induced by adenovirus 5 E1A and accompanies apoptosis. Genes Dev 7: 535–545

McLorie W, McGlade CJ, Takayesu D, Branton PE (1991) Individual adenovirus E1B proteins induce transformation independently but by additive pathways. J Gen Virol 72: 1467–1471

Meijer I, van Dam H, Boot AJM, Bos JL, Zantema A, van der Eb AJ (1991) Co-regulated expression of junβ and MHC class I genes in adenovirus-transformed cells. Oncogene 6: 911–916

Mellow GH, Föhring B, Dougherty J, Gallimore PH, Raska K (1984) Tumorigenicity of adenovirus-transformed rat cells and expression of class I major histocompatibility antigen. Virology 134: 951–961

Miller BW, Williams J (1987) Cellular transformation by adenovirus type 5 is influenced by the viral DNA polymerase. J Virol 61, 11: 3630–3634

Montell C, Courtois G, Eng C, Berk A (1984) Complete transformation by adenovirus 2 requires both E1A proteins. Cell 36: 951–961

Moran E, Mathews MB (1987) Multiple functional domains in the adenovirus E1A gene. Cell 48: 177–178

Moran E, Zerler B, Harrison TM, Mathews MB (1986) Identification of separate domains in the E1A gene for immortalization activity and the activation of virus early genes. Mol Cell Biol 6: 3470–3480

Mymryk JS, Bayley ST (1993) Induction of gene expression by exon 2 of the major E1A proteins of adenovirus type 5. J Virol 67: 6922–6928

Nevins JR (1981) Mechanism of activation of early viral transcription by the adenovirus E1A gene product. Cell 26: 213–220

Nevins JR (1992) E2F: a link between the Rb tumor suppressor protein and viral oncoproteins. Science 258: 424–429

Ornelles DA, Shenk T (1991) Localization of the adenovirus early region 1B 55-kilodalton protein during lytic infection: association with nuclear viral inclusions requires the early region 4 34-kilodalton protein. J Virol 65: 424–439

Ozawa K, Hagiwara H, Tang X, Saka F, Kitabayashi I, Shiroki K, Fujinaga K, Israel A, Gachelin G, Yokoyama K (1993) Negative regulation of the gene for H-2k^b class I antigen by adenovirus 12-E1A is mediated by a CAA repeated element. J Biol Chem 268: 27258–27268

Pilder S, Moore M, Logan J, Shenk T (1986) The adenovirus E1B-55K transforming protein modulates transport or cytoplasmic stabilization of viral and host cell mRNAs. Mol Cell Biol 6: 470–476

Pines J, Hunter T (1990) Human cyclin A is adenovirus E1A-associated protein p60 and behaves differently from cyclin B. Nature 346: 760–763

Quinlan MP, Douglas JL (1992) Immortalization of primary epithelial cells requires first and second exon functions of adenovirus type 5 12S. J Virol 66: 2020–2030

Quinlan MP, Grodzicker T (1987) Adenovirus E1A 12S protein induces DNA synthesis and proliferation in primary epithelial cells in both the presence and absence of serum. J Virol 61: 673–682

Quinlan MP, Sullivan N, Grodzicker T (1987) Growth factor(s) produced during infection with an adenovirus variant stimulates proliferation of nonestablished epithelial cells. Proc Natl Acad Sci USA 84: 3283–3287

Quinlan MP, Whyte P, Grodzicker T (1988) Growth factor induction by the adenovirus type 5 E1A 12S protein is required for immortalization of primary epithelial cells. Mol Cell Biol 8: 3191–3203

Raska K, Gallimore PH (1982) An inverse relation of the oncogenic potential of adenovirus-transformed cells and their sensitivity to killing by syngeneic natural killer cells. Virology 123: 8–18

Rice SA, Klessig DF, Williams J (1987) Multiple effects of the 72-kDa, adenovirus-specified DNA binding protein on the efficiency of cellular transformation. Virology 156: 366–376

Robbins PD, Horowitz JM, Mulligan RC (1990) Negative regulation of human c-fos expression by the retinoblastoma gene product. Nature 346: 668–671

Roberts BE, Miller JS, Kimelmann D, Cepko CL, Lemischka IR, Mulligan RC (1985) Individual adenovirus type 5 early region 1A gene products elicit distinct alterations of cellular morphology and gene expression. J Virol 56: 404–413

Ron D (1994) Inducible growth arrest: new mechanistic insights. Proc Natl Acad Sci USA 91: 1985–1986
Ruley HE (1983) Adenovirus early region 1A enables viral and cellular transforming genes to transform primary cells in culture. Nature 304: 602–606
Saito I, Shiroki K, Shimojo H (1983) mRNA species and proteins of adenovirus type 12 transforming regions: identification of proteins translated from multiple coding stretches in 2.2 kb region 1B mRNA in vitro and in vivo. Virology 127: 272–289
Sarnow P, Ho YS, Williams J, Levine AJ (1982) Adenovirus E1B-58kd tumor antigen and SV40 large tumor antigen are physically associated with the same 54kd cellular protein in transformed cells. Cell 28: 387–394
Sawada Y, Föhring B, Shenk T, Raska K Jr (1985) Tumorigenicity of adenovirus transformed cells: region E1A of adenovirus 12 confers resistance to natural killer cells. Virology 147: 413–421
Sawada Y, Urbanelli D, Raskova J, Shenk T, Raska K (1986) Adenovirus tumor-specific transplantation antigen is a function of the E1A early region. J Exp Med 163: 563–572
Sawada Y, Raska Jr K, Shenk T (1988) Adenovirus type 5 and type 12 recombinant viruses containing heterologous E1 genes are viable, transform rat cells, but are not tumorigenic in rats. Virology 166: 281–284
Schneider JF, Fisher F, Goding CR, Jones NC (1987) Mutational analysis of the adenovirus E1A gene: the role of transcriptional regulation in transformation. EMBO J 6: 2053–2060
Schrier PI, Bernards R, Vaessen RTMJ, Houweling A, van der Eb AJ (1983) Expression of class I major histocompatibility antigens switched off by highly oncogenic adenovirus 12 in transformed rat cells. Nature 305: 771–775
Shiroki K, Segawa K, Saito I, Shimojo H, Fujinaga K (1979) Products of the adenovirus 12 transforming genes and their functions. Cold Spring Harbor Symp Quant Biol 44: 533–540
Shiroki K, Hashimoto S, Saito I, Fukui Y, Fukui Y, Hiroyuki K, Shimojo H (1984) Expression of the E4 gene is required for establishment of soft-agar colony-forming rat cell lines transformed by the adenovirus 12 E1 gene. J Virol 50 3: 854–863
Smith DH, Ziff EB (1988) The amino-terminal region of the adenovirus serotype 5 E1a protein performs two separate functions when expressed in primary baby rat kidney cells. Mol Cell Biol 8: 3882–3890
Smith DH, Kegler D, Ziff EB (1985) Vector expression of adenovirus type 5 E1A proteins: evidence for E1A autoregulation. Mol Cell Biol 5: 2684–2696
Soddu S, Lewis AM Jr (1992) Driving adenovirus type 12-transformed BALB/c mouse cells to express high levels of class I major histocompatibility complex proteins enhances, rather than abrogates, their tumorigenicity. J Virol 66: 2875–2884
Spindler KR, Eng CY, Berk AJ (1985) An adenovirus early region 1A protein is required for maximal viral DNA replication in growth-arrested human cells. J Virol 53: 742–760
Stein RW, Corrigan M, Yaciuk P, Whelan J, Moran E (1990) Analysis of E1A-mediated growth regulation functions: binding of the 300-kilodalton cellular product correlates with E1A enhancer repression function and DNA synthesis-inducing activity. J Virol 64: 4421–4427
Stephens C, Harlow E (1987) Differential splicing yields novel adenovirus 5 E1A mRNAs that encode 30kd and 35kd proteins. EMBO J 6: 2027–2035
Subramanian T, Kuppuswamy M, Nasr RJ, Chinnadurai G (1988) An N-terminal region of adenovirus E1A essential for cell transformation and induction of an epithelial cell growth factor. Oncogene 2: 105–112
Subramanian T, La Regina M, Chinnadurai G (1989) Enhanced ras oncogene mediated cell transformation and tumorigenesis by adenovirus 2 mutants lacking the C-terminal region of E1a protein. Oncogene 4: 415–420
Takiff HE, Straus SE (1982). Early replicative block prevents the efficient growth of fastidious diarrhea-associated adenoviruses in cell culture. J Med Virol 9: 93–100
Tanaka K, Isselbacher KJ, Khoury G, Jay G (1985) Reversal of oncogenesis by the expression of a histocompatibility complex class I gene. Science 228: 26–30
Tanaka K, Takayuki Y, Bieberich C, Jay G (1988) Role of the major histocompatibility complex class I antigens in tumor growth and metastasis. Annu Rev Immunol 6: 359–380
Telling GC, Williams J (1993) The E1B 19-kilodalton protein is not essential for transformation of rodent cells in vitro by adenovirus type 5. J Virol 67: 1600–1611
Telling GC, Williams J (1994) Constructing chimeric type 12/type 5 adenovirus E1A genes and using them to identify an oncogenic determinant of adenovirus type 12. J Virol 68, 2: 877–887
Trentin JJ, Yabe Y, Taylor G (1962) The quest for human cancer viruses. Science 137: 835–841

Urbanelli D, Sawada Y, Raskova J, Jones NC, Shenk T, Raska K (1989) C-terminal domain of the adenovirus E1A oncogene product is required for induction of cytotoxic T lymphocytes and tumor-specific transplantation immunity. Virology 173: 607–614

van den Heuvel SJL, van Laar T, The I, van der Eb AJ (1993) Large E1B proteins of adenovirus types 5 and 12 have different effects on p53 and distinct roles in cell transformation. J Virol 67: 5226–5234

van der Eb AJ, Zantema A (1992) Adenovirus oncogenesis. In: Doerfler W, Böhm P (eds) Malignant transformation by DNA viruses. VCH, Weinheim, pp 115–140

van der Eb AJ, Mulder C, Graham FL, Houweling A (1977) Transformation with specific fragments of adenovirus DNAs. I. Isolation of specific fragments with transforming activity of adenovirus 2 and 5 DNA. Gene 2: 115–132

van Ormondt H, Galibert F (1984) Nucleotide sequences of adenovirus DNAs. Curr Top Microbiol Immunol 110: 73–142

Velcich A, Ziff E (1985) Adenovirus E1A proteins repress transcription from the SV40 early promoter. Cell 40: 705–716

Virtanen A, Pettersson U (1985) Organization of early region 1B of human adenovirus type 2: identification of four differentially spliced mRNAs. J Virol 54: 383–391

Wang HGH, Rikitake Y, Corrigan Carter M, Yaciuk P, Abraham SE, Zerler B, Moran E (1993a) Identification of specific adenovirus E1A N-terminal residues critical to the binding of cellular proteins and to the control of cell growth. J Virol 67: 476–488

Wang HGH, Yaciuk P, Ricciardi RR, Green M, Yokoyama K, Moran E (1993b) The E1A products of oncogenic adenovirus serotype 12 include amino-terminally modified forms able to bind the retinoblastoma protein but not p300. J virol 67: 4804–4813

White E, Cipriani R, Sabbatini P, Denton A (1991) Adenovirus E1B 19-kilodalton protein overcomes the cytotoxicity of E1A proteins. J Virol 656: 2968–2978

Whyte PK, Buchovich J, Friend S, Raybuck M, Weinberg R, Harlow E (1988a) Association between an oncogene and an anti-oncogene; the adenovirus E1A proteins bind to the retinoblastoma gene product. Nature 334: 124–129

Whyte P, Ruley HE, Harlow E (1988b) Two regions of the adenovirus early region 1A proteins are required for transformation. J Virol 62: 257–265

Whyte P, Williamson NM, Harlow E (1989) Cellular targets for transformation by the adenovirus E1A proteins. Cell 56: 67–75

Williams J (1973) Oncogenic transformation of hamster embryo cells in vfitro by adenovirus type 5. Nature 243: 162–163

Williams J (1986) Adenovirus genetics. In: Doerfler W (ed) Adenovirus DNA: the viral genome and its expression. Nijhoff, The Hague, PP 247–309

Williams J, Karger BD, Ho YS, Castiglia CL, Mann T, Flint SJ (1986) The adenovirus E1B 495R protein plays a role in regulating the transport and stability of viral late messages. Cancer Cells 4: 275–284

Winberg G, Shenk T (1984) Dissection of overlapping functions within the adenovirus type 5 E1A gene. EMBO J 3: 1907–1912

Yaciuk P, Moran E (1991) Analysis with specific polyclonal antiserum indicates that the E1A-associated 300-kDa product is a stable nuclear phosphoprotein that undergoes cell cycle phase modification. Mol Cell Biol 11: 5389–5397

Yee SP, Branton PE (1985) Detection of cellular proteins associated with human adenovirus type 5 early region 1A polypeptides. Virology 147: 142–153

Yew PR, Berk AJ (1992) Inhibition of p53 transactivation required for transformation by adenovirus early 1B protein. NAture 357: 82–85

Zerler B, Moran E, Maruyama K, Moomaw J, Grodzicker T, Ruley HE (1986) Analysis of adenovirus E1A coding sequences which enable ras and pmt oncogenes to transform cultured primary cells. Mol Cell Biol 6: 887–899

Zerler B, Roberts RJ, Mathews MB, Moran E (1987) Different functional domains of the adenovirus E1A gene are involved in regulation of host cell cycle products. Mol Cell Biol 7: 821–829

Zhang S, Mak S, Branton PE (1992) Overexpression of the E1B 55-kilodalton (482R) protein of human adenovirus type 12 appears to permit efficient transformation of primary baby rat kidney cells in the absence of the E1B 19-kilodalton protein. J Virol 66: 2302–2309

Regulation of Adenovirus E2 Transcription Unit

S. Swaminathan and B. Thimmapaya

1 Introduction

The ability of viruses to take over the synthetic machinery of the host cells they invade for their own survival affords an ideal setting to look at the intricacies of ongoing regulation of macromolecular events and enables one to address specific questions concerning gene regulation. Genetic studies with animal viruses have provided detailed insights into the structure and function of the viral genome and the gene products it encodes. Viral mutants provide a means of probing the genetic and biochemical phenomena occurring in infected cells. Adenoviruses are members of DNA tumor virus family and are widespread in nature. Although about 40 different serotypes of human adenoviruses have been identified, adenovirus (Ad2) and Ad5 are the most well studied. The relatively small size (36

Lurie Cancer Center, Northwestern University Medical School, 303 E. Chicago Avenue, Chicago, IL 60611, USA, and Department of Microbiology and Immunology, Northwestern University Medical School, 303 E. Chicago Avenue, Chicago, IL 60611, USA

kilo bases) of its genome and the ease with which it can be grown to high titers in cultured cells have made the adenoviruses particularly attractive for experimentation. Early promoters of human adenoviruses provide excellent model systems to study transcriptional regulation. In adenovirus-infected human cells, six early viral promoters are coordinately regulated, and their efficient transcription is dependant on the large E1A protein which consists of 289 amino acids (FLINT and SHENK 1989). The upstream DNA sequence elements that are required for basal E1A-independent transcription of these promoters have been identified. A number of cellular transcription factors that interact with these elements in a sequence-dependent fashion have been identified and cloned. This review addresses the transcriptional regulation of adenovirus early transcription region 2(E2).

The E2 unit is one of the most important of the early regions for successful completion of virus infection as it encodes proteins vital to viral DNA replication. Three messages coding for a 72-kDa DNA-binding protein (DBP), an 80-kDa precursor terminal protein, and the 140-kDa DNA polymerase originate from the E2 region (STILLMAN et al. 1981). Besides its role in DNA replication, the DBP is also involved in a multitude of functions which include repression of E4 transcription (NEVINS and WINKLER 1980), mRNA stability (BABICH and NEVINS 1981), assembly of virus particles (NICOLAS et al. 1983), transformation (GINSBERG et al. 1974), and host range function (KLESSIG and GRODZICKER 1979). In view of the important functions of these three proteins, it is not unreasonable to expect the existence of a very stringent transcriptional control of the E2 region to ensure their optimum synthesis. At early times after infection the E2 region is transcribed from a promoter located at 76.0 map units (mu), designated the E2 early promoter, whereas at intermediate times after infection transcription is switched to a new promoter, designated E2 late promoter located at 72.0 mu. While efficient transcription of the E2 early promoter depends on the viral transactivator E1A, the basis for the delayed activation of the E2 late promoter in the viral lytic cycle is not known.

2 Structure of E2 Early Promoter

The E2 early promoter has a complex architecture and contains two overlapping transcriptional control regions with cap sites at +1 (major promoter) and –26 (minor promoter). Both these promoters lack the canonical TATA sequence. Nevertheless, sequences found 30 base pairs upstream from the start sites function as TATA box substitutes (MATHIS et al. 1981). The nucleotide sequence of this promoter with its dual start sites and the various control elements is presented in Fig. 1A. The minor promoter contains a sequence TAAATT with closer resemblance to the TATA motif, whereas the totally atypical sequence TTAAGA serves as a TATA motif for the major +1 promoter. In spite of this, the promoter for the –26 cap site is transcribed at levels about 1/20 of that of the +1 cap site,

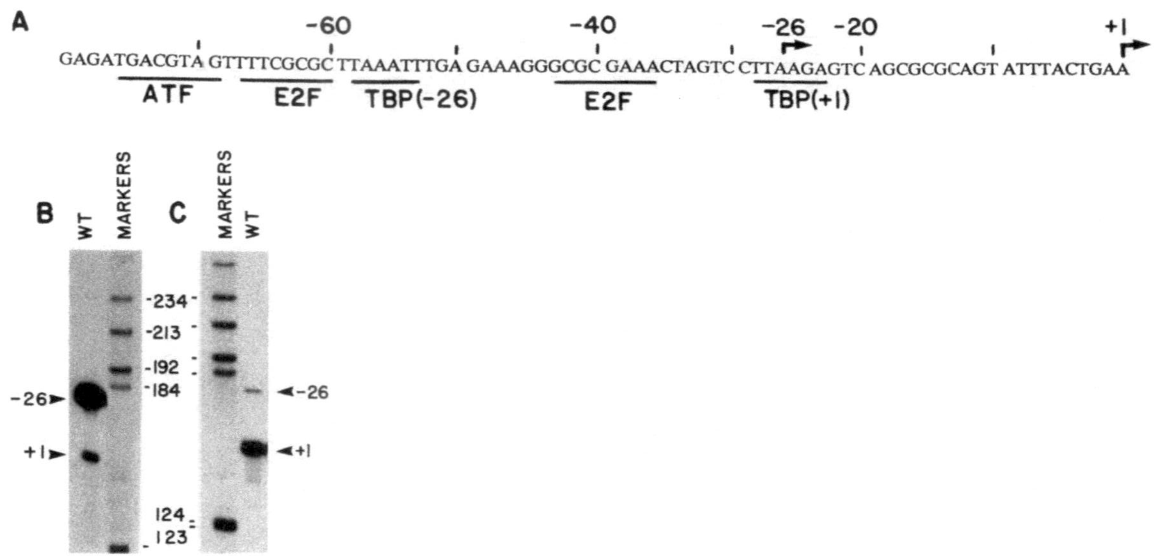

Fig. 1 A–C. The adenovirus E2 early promoter and the transcripts originating from the +1 and –26 cap sites in human and *Schizosaccharomyces pombe* cells. **A** The nucleotide sequence of the promoter. The *arrows* denote the +1 and –26 cap sites. The *underlined* sequences indicate the binding sites for TATA box-binding protein (*TBP*), E2F, and activating transcription factor (*ATF*) **B** Primer extension analysis of E2 early promoter expression in *S. pombe* . The RNA was annealed to a CAT primer complementary to sequences from +110 to +136 with respect to the +1 start site, reverse transcribed, and analyzed on a DNA-sequencing gel. **C** Primer extension analysis of E2 early promoter expression in HeLa cells. The *arrowheads* in **B** and **C** indicate the primer-extended products (taken from SWAMINATHAN et al. 1993)

as shown in the primer extension experiment in Fig. 1C (MURTHY et al. 1985; ZAJCHOWSKI et al. 1985; MANOHAR et al. 1990). The reasons for the preferential utilization of +1 cap site is not clear. Human TATA box-binding protein (TBP) synthesized in *Escherichia coli* can bind to TAAATT sequence (–26 TATA box) about two- to threefold better than to the +1 TATA motif (SWAMINATHAN et al. 1993). Nonetheless, the efficiency with which these two promoters are transcribed is not related to the nucleotide sequence of the TATA motif. Our recent studies have shown that swapping the two TATA-like sequences with each other does not improve the transcriptional activity of the +1 promoter (SWAMINATHAN et al. 1993). Apparently, in mammalian cells, it is the context in which the TATA-like element is placed that is critical for the efficient transcription of an RNA polymerase II promoter. In striking contrast, when such studies were extended to *Schizosaccharomyces pombe*, it was found that the –26 promoter was utilized about 20-fold more efficiently than the +1 promoter (Fig. 1B). Consistent with this, bacterially expressed *S. pombe* TBP bound about fivefold better to the –26 TATA box than to the +1 TATA box (SWAMINATHAN et al. 1993). Thus, there must be factors in human cells which allow the efficient utilization of nonconsensus TATA motifs which yeast cells lack.

In transient assays and in virus infections, sequences of about 80 base pairs upstream from the +1 start site are both necessary and sufficient for basal and virus-induced transcription of the +1 promoter (MURTHY et al. 1985; ZAJCHOWSKI et al. 1985; MANOHAR et al. 1990). A molecular dissection of the E2 early promoter utilizing a library of linker scanning (LS) mutants in which the E2 early promoter region was fused to the CAT gene allowed the identification of the transcriptional control elements (MURTHY et al. 1985). A combination of transient transfection and gel shift assay experiments revealed that the +1 promoter contains four *cis*-acting elements: a sequence TTAAGA located between nucleotides (nt) –22 and –30 which functions as a TATA box analogue, two E2F sites TTTCGCGC present in an inverted orientation with respect to each other between nt –35 and –63, and an activating transcription factor (ATF) site, TGACGTCA, located between nt –68 and –77 (MURTHY et al. 1985; ZAJCHOWSKI et al. 1985; KOVESDI et al. 1986; MANOHAR et al. 1990).

The TTAAGA motif, despite the apparent difference from the typical TATA box, is essential for positioning accurate and efficient initiation of transcription at the +1 cap site. The role of the E2F sites in promoter function was not well understood for a long time, as promoter mutants with alterations in these sites did not show a clear phenotype in transient assays (MURTHY et al. 1985; ZAJCHOWSKI et al. 1985). However, this was resolved recently by analyzing E2F site mutants in the context of the viral chromosome. The two E2F sites function cooperatively in both basal and virus-induced transcription (MANOHAR et al. 1990). The genes encoding the mammalian transcription factor E2F, which binds to the E2F sites on the E2 early promoter, have been cloned taking advantage of the ability of E2F to bind to the tumor suppressor protein Rb (HELIN et al. 1992; KAELIN et al. 1992). This is a 55-kDa protein and belongs to a family of transcription factors. The functional form of E2F is a heterodimer in which E2F is associated with a 45-kDa polypeptide designated DP1 (BANDARA et al. 1993). In mammalian cells, the transcription-stimulating activity of E2F appears to be controlled by its association with Rb as a result of E1A-mediated dissociation of E2F–Rb complexes (NEVINS 1992). However, in adenovirus-infected cells, E2F is associated with a polypeptide encoded in early region 4. This is a 19-kDa protein that is translated from an open reading frame (ORF) resulting from the fusion of ORF 6 and 7 of E4 (referred to as E4 6/7 in this article) (HARDY et al. 1989). The association of E4 ORF 6/7 with E2F dramatically enhances the stable, cooperative binding of E2F to the two properly spaced E2F sites of the adenovirus E2 early promoter (HARDY et al. 1989; HARDY and SHENK 1989; HUANG and HEARING 1989). The role of Rb in tumor suppression appears to involve its ability to prevent the availability of free E2F, which otherwise has the potential to turn on E2F site-containing genes such as dihyrofolate reductase (DHFR), thymidine kinase, and DNA polymerase (pol) alpha which are essential for the S phase of the cell cycle (NEVINS 1992). The ATF site is also present in the E3 and E4 promoters and in several host promoters that are responsive to cyclic adenosine monophosphate (cAMP) (LEE et al. 1987). The initial discovery of the ATF site in the adenovirus E2 early promoter by LS mutagenesis led to the purification of the host transcription factor ATF and subsequently the cloning of the genes that encode ATF (SIVARAMAN et al. 1986; JALINOT et al. 1988; HAI et al. 1988, 1989). There appear to be a family of related polypeptides, one of which (ATF-2) has been shown to function in E1A-mediated transactivation of the E2 early and E4 promoters (LIU and GREEN 1990). In contrast to E2F, the DNA-binding activity of ATF does not change upon viral infection (SIVARAMAN et al. 1986; SIVARAMAN and THIMMAPAYA 1987; YEE et al. 1989).

Studies which addressed the mechanism of E1A transactivation of the adenovirus early promoters assayed promoter activity of mutants in transient transfection assays in the presence and absence of E1A. The underlying rationale was that if a mutation of an element resulted in loss of response to E1A, then that element and its cognate transcription factor are potential candidates that could mediate E1A transactivation. This type of approach showed that E1A can transactivate through the TATA box, the E2F sites, and the ATF site (MANOHAR et al. 1990). The E2 early promoter, which contains all the aforementioned *cis*-acting elements, should therefore be targeted at all these sites by E1A. Transient assay approaches to determine the E1A-sensitive elements failed to provide any definitive answer. To address this issue in a physiologically meaningful setting, LS mutants which altered each of the *cis*-acting elements of the E2 early promoter were rebuilt into the viral chromosome and analyzed in infection experiments. These data demonstrated that, for the +1 promoter, all the four *cis*-acting elements are targets for E1A transactivation. Such a stringent regulation is consistent with the crucial functions of the gene products that it controls.

Transient transfection of reporter constructs proved invaluable in initial studies that sought to understand the regulation of E2 early promoter expression. These studies demonstrated the existence of two overlapping promoters in the E2 early transcription unit by identifying two start sites independently controlled by distinct TATA box substitutes (MURTHY et al. 1985; ZAJCHOWSKI et al. 1985). In addition, an upstream element later shown to be the ATF site was also identified in these experiments. However, the transient assay approach proved to be inadequate in pinpointing the promoter elements that may be targeted by E1A. Subsequent identification of additional promoter elements (the E2F sites) by alternate approaches emphasized the limitations inherent in the transient systems in delineating control elements of the E2 early promoter (MANOHAR et al. 1990). This has led us to the belief that viral promoters are best analyzed under actual in vivo conditions, i.e., in the context of viral infection. While such an approach is undoubtedly laborious as it involves the construction of mutant viruses to be used in infection experiments, it is worth the effort since it permits examination of the promoter under natural conditions in which it functions. While it may not be possible to assert that infection is the ultimate test of promoter function in all situations, it does seem to be the case at least as far as the adenovirus system is concerned. It is our conviction that infection experiments have the potential to generate data that more closely reflect the true in vivo situation. This has been borne out by persistent inconsistencies between transfection and infection experiments in several instances.

3 E2 Early Promoter Regulation in Virus-Infected Cells

Initial investigations carried out by employing transient transfection assays in HeLa cells failed to identify the existence of the E2F sites in the E2 early promoter (MURTHY et al. 1985; ZAJCHOWSKI et al. 1985). Subsequently, gel shift analyses of HeLa cell nuclear extracts resulted in the identification of the E2F protein which bound to promoter sequences between –35 and –68. The promoter-binding activity of this protein increases severalfold following virus infection (KOVESDI et al. 1986; SIVARAMAN and THIMMAPAYA 1987). This dramatic increase in activity is due to the viral E4 6/7 protein (HARDY et al. 1989; HARDY and SHENK 1989; HUANG and HEARING 1989). Although the effect of the E4 6/7 protein had been elucidated extensively at the biochemical level, the significance of the existence of two copies of the E2F recognition motif had not been clarified. Certain deletion mutants have implicated, at least in part, a role for the E2F sites in the transactivation mediated by E1A.

Transfection data using the LS mutants did not clearly establish any sequence element as being specifically the target of E1A-mediated transactivation. Promoter mutants with LS mutations in the TTAAGA motif, E2F, or the ATF motifs were all induced by E1A. Thus, cotransfections have proved to be inconclusive in determining target elements in the promoter on which E1A may exert its influence. The failure to detect E2F sites and the inability to identify unambiguously the targets of E1A transactivation using transient assays prompted us to investigate the E2 early promoter in the context of the viral chromosome. Subsequently, this approach has been justified by the observation of other instances (see below) wherein transfection approaches have not been the most suitable way to study E2 early promoter activation.

To clearly understand the mechanism by which the E2 early promoter is activated in the context of the viral chromosome, we employed a novel adenoviral vector lacking the E1A gene. The LS mutants used in the transient assays were rebuilt into this E1A$^-$ vector and then analyzed in infection experiments.

3.1 Strategy to Analyze the Expression of E2 Early Promoter in the Context of the Viral Chromosome

The E2 early promoter sequences plus the sequences coding for the first leader segment of the Ad 5 E2 early message were first fused to chloramphenicol acetyltransferase (CAT) gene which also contained the SV40 polyadenylation signal sequences downstream of the CAT coding sequences for RNA processing. This expression vector was mutagenized by the LS mutagenesis approach to generate a library of mutants which systematically mutate every base pair of the promoter. These LS mutations were rebuilt into *dl* 321 by cloning the wild type or the mutant derivatives of the E2 early promoter-driven CAT expression cassettes into the nonessential E3 region between 78.5 and 83.0 mu of *dl* 321. This strategy is depicted in Fig. 2 (Bhat et al. 1987; Manohar et al. 1990).

The mutant *dl* 321 is an E1A negative virus which is identical to *dl* 312 except that *dl* 321 contains a single *Xba*I site at 78.5 mu. The orientation of transcription of this newly inserted promoter is identical to that of the resident E2 promoter. Thus, the original E2 gene which provides the vital DNA replication proteins is intact in these recombinants. This strategy involves the insertion of mutant promoters as duplicate copies in the viral genome. There is, therefore, the possibility that these mutant promoters compete with the resident promoter for

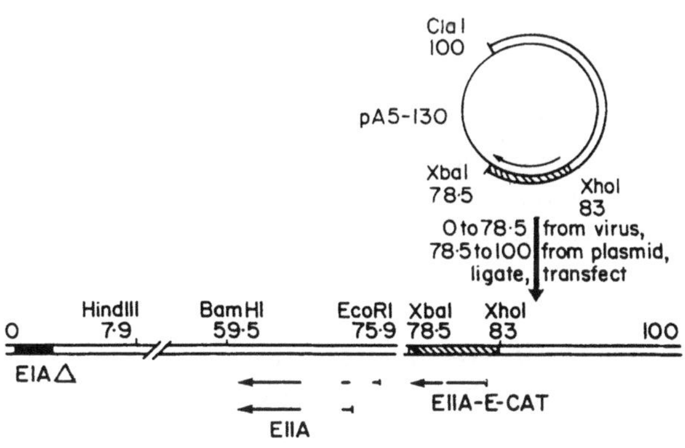

Fig. 2. Strategy to introduce the E2 early wild-type and mutant promoter CAT constructs (*hatched box*) into the EIA- adenovirus type 5 (Ad5) vector. The *filled box* at the left end of the viral chromosome indicates the EIA deletion. Restriction sites are positioned on the Ad5 map units. EIIA is the original E2 gene of the virus (taken from Manohar et al. 1990)

limiting transcription factors. This is unlikely because our results show that the wild-type CAT construct cloned in the E3 region of the viral chromosome is transcribed and transactivated efficiently. Further, the analysis of the wild-type and mutant promoters in identical genetic backgrounds precludes any possibility that potential competition will alter the interpretation of the data. In this situation, the activity of the newly introduced E2 early promoter (wild-type as well as mutant) can be unambiguously analyzed in terms of the activity of the CAT reporter gene. These recombinants, which are E1A⁻, can be propagated in 293 cells which can provide the E1A functions in *trans*. The mutant viruses that harbor the LS mutations can be analyzed for E1A-dependent or E1A-independent transcription by infecting HeLa cells, respectively, with and without the wild-type virus.

3.2 Basal Activity of E2 Early Promoter Is Dependent on All Four *cis*-Acting Elements In Vivo

In HeLa cells infected with these reporter viruses for 8 h, a 20- to 25-fold reduction of CAT activity was observed for mutants with alterations in the TTAAGA element and the two E2F motifs. The ATF site mutant displayed a five fold reduction in CAT activity (Manohar et al. 1990). These data clearly emphasize the involvement of the TTAAGA motif, the two copies of the E2F motif, and the ATF recognition element in the basal expression (i.e., in the absence of E1A), of the E2 early promoter. Since the control virus in which a promoterless CAT gene had been cloned in the E3 region displayed no CAT activity whatsoever, the observed CAT activities and the requirement for all the four *cis*-acting elements thereby deduced are a true reflection of the E2 early promoter activity in vivo in the infected cells. A noteworthy observation that resulted from these studies is the requirement for both copies of the E2F sites simultaneously for efficient expression of the promoter. This provided for the first time unambiguous genetic evidence that the two E2F sites function cooperatively even in basal expression of the E2 early promoter. This is underscored by the mutant LS-49/-59, which exhibits wild-type activity. The alteration in this mutant lies in between the two E2F sites. This is consistent with in vitro binding of E2F to the E2 early promoter. Mutations of sequences between the two E2F binding sites does not affect E2F binding to the promoter (SivaRaman and Thimmapaya 1987).

3.3 Role of *cis*-Acting Elements in the Induction of Transcription by Virus-Encoded Transactivators

These reporter viruses provided an ideal setting to investigate which elements of the E2 early promoter are targets of E1A when the promoter is present in the viral chromosome. As mentioned earlier, the transient assay approaches failed to identify any of the *cis*-acting elements as potential targets of E1A action, presumably owing to inherent limitations of the approach. Since the mutant viruses which harbor the E2 early promoter-driven CAT gene do not contain any E1A-coding sequences, the precise contribution of each *cis*-acting element of the E2 early promoter when the E1A protein is provided in cells through a natural infection process (rather than a transfection approach) can be determined by coinfecting the mutants with the wild-type virus. HeLa cells were infected in the presence of cytosine arabinoside (to ensure that virus is in its early phase of infection) with mutant viruses harboring the CAT constructs along with either wild type or the E1A-negative variant *dl* 321, the parental virus used to reconstruct the E2 early promoter LS mutants into the viral genome. In striking contrast to the observations made in transient assays, the infection experiments revealed that none of the mutants with alterations in the TTAAGA motif and the proximal E2F site was stimulated by coinfection with wild-type virus (MANOHAR et al. 1990). One mutant with a substitution mutation in the region between the two E2F sites displayed normal phenotype with respect to both basal and virus-induced transcription. Mutants that altered the distal E2F site and the ATF site were induced five- to sixfold by virus-encoded E1A. The modest level of E1A induction of promoter activity of the distal E2F and ATF site mutants was demonstrated to be due to the activity of the minor promoter transcribing from the –26 cap site using a primer extension approach to distinguish between the transcripts originating from the two overlapping promoters with cap sites at +1 and –26. The primer extended products for the +1 promoter were not increased during wild-type coinfection for all the mutants that destroyed the integrity of the four *cis*-acting elements. In other words, this experiment demonstrated that each one of the four *cis*-acting elements and the cognate host transcription factors are all critical to the E1A-mediated transactivation of the E2 early promoter. This is the first instance in which genetic evidence strongly supports the requirement for the two inverted copies of the E2F sites for basal as well as virus-induced transcription. Further, this study also established that all the three host transcription factors that interact with the E2 early promoter are targets for E1A-mediated transactivation.

At about this time it became increasingly evident that the E4 region encoded a 19-kDa polypeptide which originates due to the fusion of ORF 6 and 7 has a remarkable effect on the DNA-binding properties of the E2F protein. Numerous biochemical assays have since then demonstrated that the E4 6/7 polypeptide can transactivate the E2 early promoter by directly associating with E2F and thereby enabling it to bind to the two inverted E2F sites on the promoter in a cooperative fashion. This E4 6/7-mediated enhancement in the DNA-binding activity of the E2F factor is called "E2F induction" and results in the generation of the so-called infection-specific form of E2F. It is this phenomenon that underlies the dramatic increase in E2F DNA-binding activity in adenovirus-infected cells (HARDY et al. 1989; HARDY and SHENK 1989; HUANG and HEARING 1989;).

Analysis of E4 6/7 mutants for their ability to generate the infection-specific form of E2F has resulted in the mapping of the E2F binding domain to the C-terminal 70 amino acids (O'CONNOR and HEARING 1991). Recent results suggest that the E4 6/7 protein interacts with the DP1 component of the E2F complex and the E4 protein has a capacity to dimerize. It is suggested that a complex which consists of two E2F1–DP1 heterodimers held together by an E4 homodimer binds to the two E2F sites within the E2 early promoter (CRESS and NEVINS 1994). These findings have to a certain extent complicated the interpretation of the data discussed in the preceding paragraphs because the above-described experiments did not distinguish between the effects of E1A and E4 6/7 on E2F. As the reporter viruses used in our studies do have an intact E4 region, the E1A from the coinfecting virus will induce the E4 of the reporter virus. Therefore, stimulatory effects of E1A can in theory be attributed partly to the E4 6/7 polypeptide. However, in virus infection assays, promoter constructs in which E2F sites placed upstream of the minimal promoter are induced by E1A (PEI and BERK 1989). In other studies, 5' promoter deletion mutants with deletions extending into the distal E2F sites failed to respond to E1A (ZAJCHOWSKI et al. 1987), suggesting that E2F at least in part is involved in E1A transactivation of the E2 early promoter. Thus, it is almost certain that immediately after infection, before the E4 polypeptide is made, the E2 early promoter is activated by E1A through these multiple E1A-sensitive elements. However, it should be noted that while direct physical interaction between E1A large protein and TBP (LEE et al. 1991) or ATF (LIU and GREEN 1990) has been reported, evidence for the binding of E1A to E2F is still lacking.

3.4 Contribution of E1A, E4 6/7, and DNA-Binding Protein in the Transactivation of E2 Early Promoter

The above discussion suggests that the E2-early promoter can be transactivated by E1A and E4 independently. Recently it has been shown that the E2 early promoter can also be stimulated by DBP, a product of the E2 region in transient transfection assays (CHANG and SHENK 1990). As mentioned above, we chose to study the mechanism of transactivation of the E2 early promoter by the virus-encoded transactivators in the context of the viral chromosome in infection experiments, because we believe that this is the most effective means of studying the regulation of this promoter, particularly in the light of the inconsistencies associated with the transfection experiments. Our strategy was to provide these transactivators in sufficient quantities at early times after infection by expressing them constitutively from the cytomegalovirus immediate-early promoter/enhancer and then monitoring the activity of the E2 early promoter. To determine the extent to which E4 and DBP can stimulate the E2 early promoter in the absence of E1A, we constructed adenoviral mutants expressing either E4 6/7 (Ad-CMV-E4) or DBP (Ad-CMV-DBP) in the E1A region of the phenotypically wild-type adenovirus variant, *dl* 309. Analysis of Ad-CMV-E4-infected HeLa cells demonstrated that the E4 6/7 polypeptide expressed by this vector induces the DNA-binding activity of cellular E2F very efficiently (S. SWAMINATHAN and B. THIMMAPAYA, unpublished results). The infection-specific form of E2F activity could be detected as early as 4 h after infection. However, the E4 6/7 protein can transactivate the E2 early promoter in vivo only to a modest extent in the absence of E1A. Ad-CMV-DBP also expressed high levels of DBP immediately after infection, but this virus failed to transactivate the E2 early promoter in coinfection experiments, pointing again to the differences in infection versus transfection approaches (S. SWAMINATHAN and B. THIMMAPAYA, unpublished results). The reporter viruses described above contain the E4 6/7-coding sequences and consequently cannot be used to study the transactivation effects of E1A and E4 separately. In order to modify these reporters so that they can be used in this experiment, they were recombined with the right end of the E4 6/7-negative variant, *dl* 356 (HARDY et al. 1989). In addition to the wild-type promoter, several LS mutants of the E2 early promoter that disrupt the TATA motif, two E2F sites, and the ATF sites were also reconstructed in the E1A and E4 6/7 double-negative background. These mutant reporter viruses were constructed to determine the sequence elements that may be targeted by E1A, E4 6/7, and DBP when they transactivate the promoter. Analysis of these promoter mutants is currently in progress and will be informative in the studies of regulation of a complex promoter by the multiple virus-encoded transactivators in their natural setting.

4 Role of Viral Transactivators in the Temporal Regulation of E2 Early Promoter

Based on our current knowledge of the *cis*-acting elements of the E2 early promoter, the cellular transcription factors that interact with these elements, and the virus-encoded transactivators that influence these interactions, the following scenario can be pictured (see Fig. 3) in natural Ad infections of permissive human cells. First, in the absence of E1A, activity of the E2 early promoter is very low or barely detectable. This makes sense because virus uses the E2 early gene

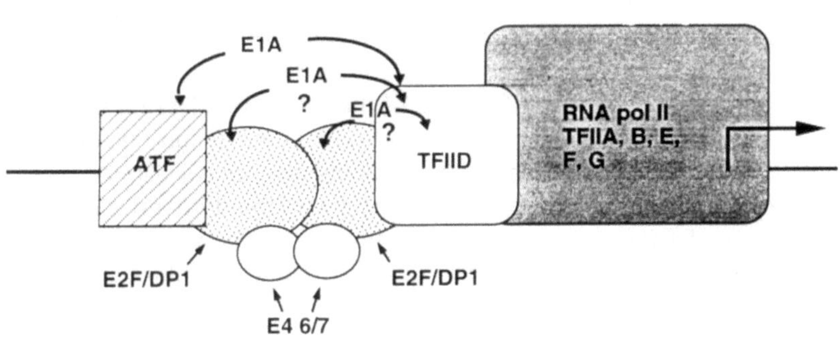

Fig. 3. Model for the transactivation of the E2 early promoter. E2F/DP1 refers to the heterodimer between E2F and DP1. Transcription factor IID (TFIID) is shown on the TATA box instead of TATA box-binding protein (TBP). Interaction of ATF and TFIID (TBP) with EIA has been demonstrated (LIU and GREEN 1990; LEE et al. 1991). Hypothetical interactions between each E2F/DP1 complex and TFIID are also shown (see text for details)

products only for viral DNA replication. Immediately after infection, E1A gene is transcribed. Activation of the E1A promoter does not require other viral early gene products. An enhancer upstream of the E1A promoter contributes to the high-level expression of the E1A gene (HEARING and SHENK 1983). Both E1A gene products may contribute to the stimulation of the E2 early promoter activity. The large E1A protein contains the transactivation domain which binds to several cellular transcription factors involved in E1A-mediated transactivation and also to TBP (LEE et al. 1992; LIU and GREEN 1994). The unique transactivation domain of the larger E1A protein may induce the E2 early promoter by forming a bridge between TBP and E2F, and TBP and ATF-2, resulting in increased rate of transcription. The distance between ATF and TBP and also between E2F and TBP in the transcription complex may be important in this interaction. Perhaps, the two E2F molecules and an ATF molecule might position on the promoter such that two or perhaps three E1A molecules might be able to interact in a single transcription complex, one between ATF and TBP and two between the two E2F molecules and TBP. Such a model would predict that TBP may interact with more than one E1A molecule in the transcriptional stimulation by facilitating the formation of stable transcription complex. In the absence of E1A, the E2 early promoter is transcribed but at a much reduced rate. Identical genomic footprinting patterns of the E2 early promoter in the presence and absence of E1A support this notion (DEVAUX et al. 1987).

The smaller E1A protein may also contribute to the E2 early promoter activation by promoting the dissociation of E2F from the Rb–E2F complex (NEVINS 1992). E2F released from Rb–E2F complex would be available to bind to the two E2F sites of the E2 early promoter in the activation process. As infection progresses from the pre-early to early stage, the virus prepares for its DNA replication, which will require high levels of the E2 gene products. Sensing the need for the increased levels of E2 gene products, the large E1A protein would induce E4 gene transcription, which will result in the accumulation of E4 6/7 polypeptide. The E4 6/7 protein would stabilize the cooperative interaction of the two E2F molecules with the two adjacent E2F sites. Currently, there are no indications to suggest that E4 6/7 protein interacts with TBP. Thus, the role of E4 6/7 protein may be restricted to the stabilization of the interaction of the two E2F molecules with the properly placed E2F sites of the promoter. The E4 6/7 gene product would provide an additional mechanism to boost the accumulation of E2 gene products. The combined action of the two transactivators will allow a steady increase in the amounts of E2 gene products to cope with the demand. Currently, it is not clear whether DBP would also stimulate the E2 transcription. Although in transient assays DBP can stimulate E2 transcription, our virus infection experiments do not support this observation (S. SWAMINATHAN and B. THIMMAPAYA, unpublished results). Studies have shown that E1A promoter is autoregulated by its own gene products (HEARING and SHENK 1985). This may result in reduced accumulation of E1A as well as E4 gene products. Furthermore, DBP also has been shown to repress E4 transcription (NEVINS and WINKLER 1980). As each virus-infected cell produces thousands of progeny particles, there is a continued demand for the E2 gene products. To maintain high levels of E2 gene products at late stages of infection, the virus has evolved an additional promoter designated E2 late promoter to drive transcription of the E2 unit. We have carried out a preliminary characterization of this promoter in the context of the viral chromosome (BHAT et al. 1987; see below). Interestingly, this promoter is not sensitive to E1A regulation.

5 Structure and Regulation of E2 Late Promoter

At intermediate stages of infection, the E2 unit is also transcribed from a second new promoter located at 72.0 mu. An attempt to understand the basis for the delayed activation of this promoter by LS mutagenesis resulted in the identification of a complex array of several *cis*-acting elements controlling its expression (BHAT et al. 1987). Once again, the approach that permitted the identification of the control elements of the E2 late promoter was an analysis of constructed reporter viruses which carried the E2 late promoter-driven CAT gene in the nonessential E3 region of the E1A-negative variant *dl* 321.

Transient transfection experiments that introduced the E2 late promoter-driven CAT gene into HeLa cells resulted in the expression of very low levels of CAT activity, precluding any meaningful analysis of the E2 late promoter mutants. The strategy adopted to introduce the E2 late promoter CAT sequences in the E3 region was essentially the same as that used in the construction of the E2 early promoter mutants. Activation of transcription from the E2 late promoter in its new location at 83.0 mu. is very similar to that occurring at its natural location at 72.0 mu and is unaffected by the upstream viral sequences that flank the promoter in its new location. Using the above-mentioned strategy, a set of deletion mutants with deletions extending into the cap site from the 5' end of the E2 late promoter were introduced into the viral chromosome. This library of deletion mutants defined a 157-base-pair sequence upstream from the cap site as the minimal promoter sequence necessary for efficient transcription of the E2 late promoter both in the presence of E1A (293 cells) and in the absence of E1A (HeLa cells). Analysis of the E2 late promoter LS mutants in the context of the viral chromosome identified three transcriptional control regions extending from nt –55 to –21 (region I), –84 to –67 (region II), and –157 to –101 (region III). In vitro experiments using nuclear extracts have identified transcriptional control elements as far upstream as –265 to –223 from the cap site of the E2 late promoter (HUANG and ROEDER 1988).

Apparently, this is not of any critical significance to the promoter when it functions under in vivo conditions in infection experiments. Within control region I, there are two sequence elements essential for function: one is a TATA box analogue located between nt –30 and –22 and the second is an Sp1-binding site between nt –50 and –40. A second low-affinity Sp1 site is located between nt

–84 and –74 in control region II. The DNA sequences in this region also contain a CAAT box in the noncoding strand from nt –72 to –76. As this control element is bidirectional in function, it may be a major control element in this region. Finally, another CAAT box can be discerned in the noncoding strand between nt –135 and –139, in region III.

Comparison of the promoter domains of the E2 early and late promoters reveal important differences with the exception of the TATA-like sequence located around nt –30. The minimum promoter sequence of the E2 early promoter extends to approximately 80 base pairs upstream from the cap site, and within this region four transcriptional control elements have been identified. Three of these are not present in the E2 late promoter. Instead, it has Sp1 and CAAT box sequence elements. Consequently, the transcription factors that recognize and bind to these elements are different for the two promoters with the possible exception of the TATA box factor. A very striking difference is the absence of a transcriptional control region far upstream in the E2 early promoter. Another major difference at the functional level is the effect of E1A on these two promoters. While E1A strongly transactivates the E2 early promoter through multiple E1A-sensitive elements, neither the 12S or the 13S E1A gene product has any effect on the E2 late promoter in infection assays. However, the 12S E1A gene product has been shown to repress E2 late promoter function in transient transfection assays (GUILFOYLE et al. 1985). The striking differences in several respects such as the architecture of the E2 early and late promoters, the upstream sequence elements that modulate transcription, and their responsiveness to E1A may hold the key to understanding why the E2 late promoter is not active at early times.

6 Transactivation of E2 Early Promoter by Simian Virus 40 Tumor Antigens: Discrepancies in the Results of Infection and Transfection Approaches

SV40 small t can efficiently transactivate the adenovirus E2 early and E3 promoters in transient assays (LOEKEN et al. 1988). However, when the E2 early promoter was introduced into the cell in virus infections, it failed to respond to small t (RAJAN et al. 1991). This was true of the viral E3 promoter as well. In these infection experiments, small t was introduced into the system by coinfecting with an E1A$^-$ recombinant adenovirus vector expressing small t from the SV40 early promoter/enhancer (Ad-SV-t). Why does small t transactivate these early promoters in transfection assays but not in virus infections? The possibility that the lack of transactivation could be due to insufficient accumulation of small t (presumably because the kinetics of SV40 early promoter-driven expression of small t are probably not rapid enough to parallel adenovirus E2 early promoter activation) was ruled out by assuring the availability of copious quantities of this protein by preinfecting the cells with Ad-SV-t or by using another recombinant in which the SV40 promoter/enhancer was replaced with that of human cytomegalovirus (Ad-CMV-t) (S. SWAMINATHAN and B. THIMMAPAYA, unpublished results). It is unlikely that small t expressed by the adenoviral vectors is inert, since the E2 early promoter could be transactivated if it was introduced via transfection (RAJAN et al. 1991). Further, the observation that both basal and E1A-induced expression of the E2 early promoter is not adversely affected by the presence of small t argues against any possible squelching by the promoter/enhancer expression systems for small t in the recombinant viruses competing for a limiting transcription factor.

The large T antigen has several functions attributed to it. One of these is its ability to transcriptionally transactivate RNA polymerase II and III promoters in transient transfection assays (LOEKEN et al. 1986). In transiently transfected CV-1 cells, the adenovirus E2 early promoter-driven expression is stimulated several-fold by SV40 large T (LOEKEN et al. 1986). However, it is difficult to attribute any significance to this phenomenon, because large T fails to transactivate the adenovirus E2 early promoter when it is examined in a more physiologically relevant setting, i.e., in the context of an infection. To do this, two recombinant adenoviruses were constructed in which the E1A-coding sequences were replaced by large T-coding sequences under the control of either the SV40 early promoter (Ad-SV-T) or the CMV promoter/enhancer (Ad-CMV-T). Both viruses synthesize abundant quantities of large T protein. Neither of these two large T-expressing recombinant adenovirus vectors could transactivate the E2 early promoter CAT construct cloned in the nonessential E3 region of *dl*321 in coinfection experiments in CV-1 cells (P. RAJAN and B. THIMMAPAYA, unpublished results). The E2 early promoter of the reporter adenovirus can be transactivated in the nonpermissive monkey cells by coinfecting with wild-type adenovirus. The large T antigen synthesized in COS cells also is incapable of transactivating transcription of the transfected adenovirus E2 early promoter (LOEKEN et al. 1986), raising questions about the transactivation property of the large T.

A possible reason for the discrepancy between infection and transfection assays could be differences in chromatin structure of the reporter templates in the two assay systems. It has been suggested that, in transient transfection assays, the possibility that the chromatin structure of the target gene is considerably different from that of the viral or host chromosome may paly a role in its accessibility to potential transactivators. Experiments performed in the recent past have indicated that the function of both the DNA-binding as well as the activation domains of transcription factors are modified by the position of nucleosomes relative to the position of the gene (TAYLOR et al. 1991; WORKMAN et al. 1991). Alternatively, this could be a result of a putative compartmentalization effect wherein transfected and infected genes are localized to different sites in the nucleus where they could have varied access to different transcription factors.

All of this raises the obvious question of whether transient transfection assay accurately mirrors the ongoing transcriptional regulation of the E2 promoters in infected cells. The experiments described in the preceding paragraphs clearly underscore the inadequacy of the transient assay approach and emphasize the need for alternate approaches to studying the regulation of the E2 promoter. Transient assays are no doubt valuable in such studies, but seem to have certain limitations. The discrepancies in transcriptional transactivation between transfection and infection assays observed in several instances mentioned above strongly suggest that the investigations of the regulation of the adenovirus E2 promoters utilizing transfection approaches may not be appropriate, as they can yield results that are not necessarily of physiological significance. These findings also raise an important broader question, namely, are all the promoter regulation studies reported in the literature physiologically relevant? Promoter regulation studies using transgenic animal models may provide answers to many of these questions.

References

Babich A, Nevins JR (1981) The stability of early mRNA is controlled by the viral 72 kD DNA binding protein. Cell 26: 371–379

Bandara LR, Buck UM, Zamanain M, Hohnson LH, LaThangue NB (1993) Functional synergy between DP-1 and E2F-1 in the cell cycle regulating transcription factor DRTF/E2F. EMBO J 12: 4317–4324

Bhat G, SivaRaman L, Murthy S, Domer P, Thimmapaya B (1987) In vivo identification of multiple promoter domains of adenovirus E11A-late promoter. EMBO J 6: 2045–2052

Chang LS, Shenk T (1990) The adenovirus DNA-binding protein stimulates the rate of transcription directed by adenovirus and adeno-associated virus promoters. J Virol 64: 2103–2109

Cress WD, Nevins JR (1994) Interacting domains of E2F1, DP1 and the adenovirus E4 protein. J Virol 68: 4212–4219

Devaux B, Albrecht G, Kedinger C (1987) Identical genomic footprints of the adenovirus EIIa promoter are detected before and after EIa induction. Mol Cell Biol 7: 4560–4563

Flint J, Shenk T (1989) Adenovirus E1A protein: paradigm viral transactivator. Annu Rev Genet 23: 141–161

Ginsberg HS, Ensinger MJ, Kaufman RS, Mayer AJ, Lundholm U (1974) Cell transformation: a study of regulation with types 5 and 12 adenovirus temperature-sensitive mutants. Cold Spring Harbor Symp Quant Biol 39: 419–426

Guilfoyle RA, Osheroff WP, Rossini M (1985) Two functions encoded by adenovirus early region IA are responsible for the activation and repression of the DNA-binding protein gene. EMBO J 4: 707–713

Hai T, Liu F, Allergretto EA, Karin M, Green MR (1988) A family of immunologically related transcription factors that includes multiple forms of ATF and AP-1. Genes Dev 2: 1216–1226

Hai TW, Liu F, Coukos WJ, Green MR (1989) Transcription factor ATF cDNA clones: an extensive family of leucine zipper proteins able to selectively form DNA-binding heterodimers. Genes Dev 3: 2083–2090

Hardy S, Shenk T (1989) E2F from adenovirus infected cells binds cooperatively to DNA containing two properly oriented and spaced recognition sites. Mol Cell Biol 9: 4495–4506

Hardy S, Engel DA, Shenk T (1989) An adenovirus early region 4 gene product is required for induction of the infection-specific form of cellular E2F activity. Genes Dev 3: 1062–1074

Hearing P, Shenk T (1983) The adenovirus type 5 EIA transcriptional control region contains a duplicated enhancer element. Cell 33: 695–703

Hearing P, Shenk T (1985) Sequence-independent autoregulation of the adenovirus type 5 E1A transcription unit. Mol Cell Biol 5: 3214–3221

Helin K, Lees JA, Vidal M, Dyson N, Harlow E, Fattaey A (1992) A cDNA encoding a pRB-binding protein with properties of the transcription factor E2F. Cell 70: 337–350

Huang M, Hearing P (1989) The adenovirus early region 4 open reading frame 6/7 protein regulates the DNA binding activity of cellular transcription factor, E2F, through a direct complex. Genes Dev 3: 1699–1710

Huang DH, Roeder RG (1988) Delineation of DNA sequences that are important for in vitro transcription from the adenovirus EIIa late promoter. Mol Cell Biol 8: 1906–1914

Jalinot P, Wintzerith M, Gaire M, Hauss C, Egly JM, Kedinger C (1988) Purification and functional characterization of a cellular transcription factor that binds to an enhancer element within the adenovirus early EIIa promoter. Proc Natl Acad Sci USA 85: 2484–2488

Kaelin WG, Krek W, Sellers WR, DeCaprio JA, Ajchanbaum F, Fuchs CS, Chittonden T, Li Y, Farnham PJ, Blanar MA, Livingston DM, Flemington EK (1992) Expression cloning of a cDNA encoding a retinoblastoma-binding protein with E2F-like properties. Cell 70: 351–364

Klessig DF, Grodzicker T (1979) Mutations that allow human Ad2 and Ad5 to express late genes in monkey cells map in the viral gene encoding the 72K DNA binding protein. Cell 17: 957–966

Kovesdi I, Reichel R, Nevins JR (1986) Identification of a cellular transcription factor involved in E1A trans-activation. Cell 45: 219–228

Lee KAW, Hai T-Y, SivaRaman L, Thimmapaya B, Hurst HC, Jones NC, Green MR (1987) A cellular protein, activating transcription factor, activates transcription of multiple E1A-inducible adenovirus early promoters. Proc Natl Acad Sci USA 84: 8355–8359

Lee WS, Kao CC, Bryant GO, Liu X, Berk AJ (1991) Adenovirus E1A activation domain binds the basic repeat in the TATA box transcription factor. Cell 67: 365–376

Liu F, Green MR (1990) A specific member of the ATF transcription factor family can mediate transcription activation by the adenovirus E1a protein. Cell 61: 1217–1224

Liu F, Green MR (1994) Promoter targeting by adenovirus E1A through interaction with different cellular DNA-binding domains. Nature 368: 520–525

Loeken M, Bikel I, Livingston DM, Brady J (1988) Trans-activation of RNA polymerase II and III promoters by SV40 small t antigen. Cell 55: 1171–1177

Loeken MR, Khoury G, Brady J (1986) Stimulation of the adenovirus E2 promoter by simian virus 40 T antigen or EIA occurs by different mechanisms. Mol Cell Biol 6: 2020–2026

Manohar CF, Kratochvil J, Thimmapaya B (1990) The adenovirus E11 early promoter has mulitiple E1A-sensitive elements, two of which function cooperatively in basal and virus induced transcription. J Virol 64: 2457–2466

Mathis DJ, Elkaim R, Kedinger C, Sassone-Corsi P, Chambon P (1981) Specific in vitro initiation of transcription on the adenovirus type 2 early and late E2 transcription units. Proc Natl Acad Sci USA 6: 2684–2694

Murthy SCS, Bhat GP, Thimmapaya B (1985) Adenovirus EIIA early promoter: transcriptional control elements and induction by the viral pre-early E1A gene, which appears to be sequence independent. Proc Natl Acad Sci USA 82: 2230–2234

Nevins JR (1992) E2F: a link between the Rb tumor suppressor protein and viral oncoproteins. Science 258: 424–429

Nevins JR, Winkler JJ (1980) Regulation of early adenovirus transcription: a protein product of early region 2 specifically represses region 4 transcription. Proc Natl Acad Sci USA 77: 1893–1897

Nicolas JC, Sarnow P, Girard M, Levine AJ (1983) Host range temperature conditional mutants in the adenovirus DNA binding protein are defective in the assembly of infectious virus. Virology 126: 228–239

O'Connor RJ, Hearing P (1991) The C-terminal 70 amino acids of the adenovirus E4-ORF 6/7 protein are essential and sufficient for E2F complex formation. Nucleic Acids Res 19: 6579–6586

Pei R, Berk AJ (1989) Multiple transcription factor binding sites mediate adenovirus E1A transactivation. J Virol 63: 3499–3506

Rajan P, Dhamankar V, Rundell K, Thimmappaya B (1991) Simian virus 40 small-t does not transactivate RNA polymerase II promoters in virus infections. J Virol 65: 6553–6561

SivaRaman L, Thimmapaya B (1987) Two-promoter specific host factors interact with adjacent sequences in an EIA-inducible adenovirus promoter. Proc Natl Acad Sci USA 84: 6112–6116

SivaRaman L, Subramanian S, Thimmapaya B (1986) Identification of a factor in HeLa cells specific for an upstream transcriptional control sequence of an EIA-inducible adenovirus promoter and its relative abundance in infected and uninfected cells. Proc Natl Acad Sci USA 83: 5914–5918

Stillman BW, Lewis JB, Chow LT, Mathews MB, Smart JE (1981) Identification of the gene and mRNA for the adenovirus terminal protein precursor. Cell 23: 497–508

Swaminathan S, Malhotra P, Manohar CF, Dhar R,Thimmapaya B (1993) Activation of a dual adenovirus promoter containing nonconsensus TATA motifs in *Schizosaccharomyces prombe* : role of TATA sequences in the efficiency of transcription. Nucleic Acids Res 21: 2737–2746

Taylor IC, Workman JL, Schuetz TJ, Kingston RE (1991) Facilitated binding of GAL4 and heat shock factor to nucleosomal templates: differential function of DNA binding domains. Genes Dev 5: 1285–1298

Workmann JL, Taylor IC, Kingston, RE (1991) Activation domains of stably bound GAL4 derivatives alleviate repression of promoters by nucleosomes. Cell 64: 533–544

Yee AS, Raychaudhuri P, Jakoi L, Nevins JR (1989) The adenovirus-inducible factor E2F stimulates transcription after specific DNA binding. Mol Cell Biol 9: 578–585

Zajchowski DA, Boeuf H, Kedinger C (1985) The adenovirus-2 early EIIa transcription unit possesses two overlapping promoters with different sequence requirements for EIa-dependent stimulation. EMBO J 4: 1293–1300

Zajchowski DA, Boeuf H, Kedinger C (1987) E1a inducibility of the adenoviral early 2a promoter is determined by specific combination of sequence elements. Gene 58: 243–256.

Model Systems for Studying the Effects of Adenovirus E3 Genes on Virulence In Vivo

M.S. Horwitz[1,2,3], J. Tufariello[1], A. Grunhaus[4], and G. Fejer[1]

1 Introduction

1.1 Adenovirus Classification and Relationship to Disease

Adenoviruses (Ad) are important pathogens in a variety of species. In humans, Ad are most commonly recognized in the respiratory tract and conjunctiva.

[1]Department of Microbiology and Immunology, Albert Einstein College of Medicine, 1300 Morris Park Avenue, Bronx, NY 10461, USA
[2]Department of Cell Biology, Albert Einstein College of Medicine, 1300 Morris Park Avenue, Bronx, NY 10461, USA
[3]Department of Pediatrics, Albert Einstein College of Medicine, 1300 Morris Park Avenue, Bronx, NY 10461, USA
[4]Laboratory of Viral Diseases, National Institute of Allergy and Infectious Diseases, Building 4, Room 229, Bethesda, MD 20892, USA

However, several members of this large family of viruses can cause disease in the gastrointestinal tract, while others infect the urinary tract, liver, and central nervous system (HORWITZ 1990b). In dogs, a canine Ad is a major cause of hepatitis, and the murine adenovirus type 1 (MAV-1) can cause multiorgan disease in the mouse (ISHIBASHI and YASUE 1984). However, the Ad that infect each species are sufficiently different that they do not cross into distantly related species. For example, human Ad replicate poorly in mice or murine tissue culture. There are 49 distinct serotypes that can infect humans, but the majority of Ad disease is caused by approximately one third of this number (SCHNURR and DONDERO 1993; HIERHOLZER et al. 1988). Ad are classified into six groups (A–F) based on GC content and genetic relatedness of their DNA, oncogenicity in rodents, and ability to agglutinate red blood cells of different species (HORWITZ 1990a). The organ targeted for disease is often related to serotype, such that Ad7 causes pneumonia, Ad2 upper respiratory tract disease, Ad40 and Ad41 cause diarrhea, Ad11 is associated with hemorrhagic cystitis, and Ad34 and Ad35 are often found in the urinary tract of immunosuppressed patients (HORWITZ 1990b). However, the reason for tissue-specific damage in relationship to serotype is unknown. Although Ad attach to cells by the fiber protein projecting from each of the 12 vertices of the icosahedral virus, the cell receptor is unknown (LONBERG-HOLM et al. 1976; GREBER et al. 1993). Human adenoviruses can attach to tissue culture cells from a variety of species, express early (E) proteins in most of them, but do not complete the replication cycle efficiently except in human and some selected primate cells (GINSBERG et al. 1991; GINSBERG and DINGLE 1965). Although Ad were initially isolated from tonsils and commonly infect the oropharyngeal tonsils (ROWE et al. 1953), they generally replicate poorly in human lymphocytes and probably target to some epithelial element of the tonsils. However, human lymphocytes have been reported to be infected with Ad, and there are some immortalized lymphocyte lines in which Ad replicate to high titer (LAVERY et al. 1987; HORVATH et al. 1986).

1.2 Adenoviruses as Vectors for Foreign DNA

Ad are being used for the delivery of foreign genes into cells, especially those that may not be actively cycling in vivo or in vitro (RAGOT et al. 1993; ZABNER et al. 1993). Foreign DNA has been cloned into the Ad genome by a technique that concomitantly deletes the E1A and E1B regions which are required for replication-competent virions. These defective Ad can be propagated on the Graham 293 cells, which contain endogenous E1A and E1B genes that complement the replication defect of the recombinant viruses. Another site for insertion of foreign DNA is in the E3 region, which can be deleted without affecting Ad viability in tissue culture (HAY-AHMAD and GRAHAM 1986; MITTAL et al. 1993). Ad can also facilitate the entry of plasmid DNA that has been complexed to polylysine in a structure external to the Ad capsid (WAGNER et al. 1992; MICHAEL et al. 1993). The ability of Ad to facilitate transport out of the endosome is presumed to be important in promoting the function of transfected DNA. Although considerable expression of foreign genes delivered by Ad vectors has been obtained in vivo, there is some cell damage even from replication-defective Ad virions that enter cells (GRUNHAUS and HORWITZ 1992; COTTEN et al. 1993). Thus, it would be helpful to understand more about Ad pathogenesis, not only because Ad cause disease but also because they are potentially useful vectors for the delivery of foreign genes for immunization as well as for gene therapy.

1.3 Adenovirus Early Region E1 and Transformation

The mechanism of Ad pathogenesis is being studied using viral mutants that affect various replicative processes in tissue culture. The E1 region has been most extensively studied because it is responsible for transformation of tissue culture cells and produces malignacies in model systems in rodents. However, extensive search for adenoviruses or their DNA in human tumors has failed to implicate this group of viruses in natural tumors of humans. Studies of the E1A region have yielded important information about transactivation or transinhibition of other promoters, its interaction with the RB antioncogene and other critical cell regulatory proteins, and induction of apoptosis (FAHA et al. 1993; KRAUS et al. 1992; RAO et al. 1992; IKEDA and NEVINS 1993). The E1B region has been recongnized to code for two proteins, 55K and 19K, that cooperate in transformation and inhibit apoptosis probably by two different mechanisms (DEBBAS and WHITE 1993; LOWE and RULEY 1993). The E1B 55K forms a complex with the p53 antioncogene, reversing the mechanism of apoptosis by wild-type (wt) p53. The Ad E1B 19K has some structural and functional similarities with the *bcl-2* oncogene which inhibits apoptosis. Both E1A and E1B are required for optimal transformation, and their study has lead to a greater understanding of their counterparts—E7 and E6, respectively—in the human papillomaviruses, which are involved in cervical carcinoma (ARROYO et al. 1993).

1.4 E2 and the Control of Viral DNA Replication

Other regions of the viral DNA also appear to have functionally clustered genes. All the Ad DNA replication functions are clustered in the E2 region, from which three virally coded proteins including the DNA polymerase, preterminal protein, and DNA-binding protein (DBP) (CHEN et al. 1990) are coded. Mutants of DBP synthesized from this region have been shown to affect transcription and the rate of transformation under certain conditions (KLESSIG and QUINLAN 1982).

1.5 Functions of E4 and Viral-Associated RNA

The E4 transcription unit encodes products involved in mRNA splicing and transport to the cytoplasm as well as a 25K polypeptide that interacts with the

E1B 55K protein, perhaps affecting the latter's function in inhibiting apoptosis (DEBBAS and WHITE 1993). Another region of the genome codes for a viral-associated (VA) RNA that is the product of RNA polymerase III (MA and MATHEWS 1993). VA RNA reverses the interferon inhibition of protein synthesis by its effect on a double-stranded RNA-requiring protein kinase that phosphorylates the protein elongation factor EIF2-α.

1.6 Viral Structural Proteins Required for Attachment and Internalization

Ad initially attach to cells by binding of the fiber protein to a receptor which has not yet been elucidated. However, there is a secondary attachment site through RGD sequences on the penton base that attaches to several cell integrins and promotes internalization of the virus after adsorption (WICKHAM et al. 1993). Although this mechanism is probably not absolutely required, it facilitates viral entry and may be responsible for differences in Ad affinities to various cells in vivo. Competition between the viral penton and extracellular matrix for the cell surface integrins may determine whether virion entry is facilitated in any particular tissue.

1.7 E3 Proteins Affect Immunoregulatory Molecules, Cytokines, and Cell Receptors

The E3 region, which is the subject of the experiments in this review, is a complex transcription region which codes for at least seven proteins made from a set of overlapping mRNAs generated by alternative splicing of a primary transcript (WOLD and GOODING 1991; WOLD 1995). The functions of these proteins suggest that some of them are involved in viral virulence or persistence. For example, the E3 19K glycoprotein (gp19K) interacts with the heavy chain of the class I major histocompatibility complex (MHC) which is prevented from exiting the endoplasmic reticulum (ER) (COX et al. 1990; FLOMENBERG et al. 1987). Such complex formation decreases $CD8^{+}$ lymphocyte recongnition of the infected cell as a result of the reduced amount of MHC on the plasma membrane. Three other E3 proteins (14.7K, 10.4K, and 14.5K) are involved in inhibiting the cytolytic effects of tumor necrosis factor alpha (TNF-α) (HORTON et al. 1991; GOODING et al. 1991). The 14.7K functions alone, while the 10.4K and 14.5K act as a complex to antagonize TNF-α-mediated cytolysis. The 10.4/14.5K complex can also accelerate the internalization of several protein tyrosine kinase receptors, including those for the epidermal growth factor (EGF) and insulin (KRAJCSI et al. 1992). Although the E3 region has been described as nonessential for the viability of adenoviruses in tissue culture, its conservation in most of the human Ad and the function of the individual E3 proteins in various in vitro assays suggest that the E3 functions are required for optimizing survival or persistence of the virus in vivo.

1.8 Adenovirus Pathogenesis Genes Studied In Vivo

Because the human Ad replicate poorly in cells or animals of most other species, there is no ideal animal model in which to study the pathogenesis of this family of viruses. However, group C Ad grow in the cotton rat, in which intranasal Ad infection using a deletion mutant of the gp19K increased the pulmonary inflammatory response over that demonstrated for the wt virus (GINSBERG et al. 1989, 1990; PRINCE et al. 1993a, b). In parallel experiments, the pulmonary inflammatory response was converted from a mononuclear infiltrate seen after wt intranasal infection into a polymorphonuclear cell response when a deletion mutant which eliminated 10.4K, 14.5K, and 14.7K was utilized. Although the virus did not replicate in mice, intranasally administrated Ad5 induced a pulmonary infiltrate and the appearance of interleukin-1 (IL-1) and TNF-α in lung tissue (GINSBERG et al. 1991).

Our approach to determine the importance of E3 genes in vivo has used the mouse system because of the well-characterized reagents for study of both the MHC and TNF-α. We have inserted isolated E3 genes into vaccinia virus (V V), a system in which the functions of individual Ad proteins could be measured in vivo. V V infects mice and causes a fatal pneumonia when administered intranasally. Because the capacity of V V to express foreign genes is extensive, it has been possible to clone the isolated Ad14.7K gene into V V in the presence of the mouse TNF-α gene (mTNF-α) in the same virus (TUFARIELLO et al. 1994a). We have shown that the Ad14.7K protein increased the virulence of V V after intranasal inoculation. The increase in mortality, pulmonary inflammatory response, and the amount of V V that grew in the lungs of intranasally infected animals clearly indicated that Ad14.7K demonstrated the anti-TNF-α phenotype in vivo. Similar experiments were performed for the gp19K gene (GRUNHAUS et al. 1994). The results of these Ad insertions into V V are the subject of this review.

In addition to the experiments on isolated Ad genes in V V, other experiments have been performed utilizing MAV-1, which causes disease in adrenals, heart, kidney, brain, cerebellum, salivary gland, and vascular endothelium of newborn mice (OTTEN and TENNANT 1982; SMITH and BARTHOLD 1987). We have sequenced the putative E3 region from this virus and have found a truncated version of the human Ad E3 that did not contain any of the open reading frames that are homologous to the human virus (RAVIPRAKASH et al. 1989). This natural "deletion mutant" of the E3 region has offered us the opportunity to study the MAV-1 in the presence of the human E3 cassette. This has been achieved by making transgenic animals with the human Ad E3 behind its natural promoter which can be transactivated by the E1A of MAV-1. Thus, we have created a model system in which the human Ad E3 region can be studied in vivo during the course of an Ad infection and in the presence of all the transcriptional controls ordinarily affecting this region (FEJER et al. 1994).

2 Results

2.1 Kinetics of Expression of Adenovirus 14.7K and Tumor Necrosis Factor-α in Tissue Culture Cells Infected with Vaccinia Virus Recombinants

V V recombinants were constructed to contain the mouse TNF-α gene in the *Hin*dIII-F region of the virus behind the p7.5 early/late promoter. The Ad14.7K gene was behind the same promoter and placed in the *Hin*dIII-J region either in the expressing (V V $14.7^{(+)}$/TNF-α) or nonexpressing (V V $14.7^{(-)}$ TNF-α) orientation. The kinetics of expression of 14.7K and TNF-α from each of these constructs is shown in Fig.1 for viral-infected 143B cells. Viral-infected cell supernates were assayed for TNF-α by the enzyme-linked immunosorbent assay (ELISA) technique and the 14.7K was measured by western blots of detergent-lysed cell monolayers as described in the figure legends. TNF-α was detectable at low levels by 4 h and increased nearly at a linear rate from 8 h until the end of the asssay at 48 h. Ad14.7 was detectable at 8 h and increased linearly from 12 to 48 h after infection. Thus the Ad14.7K inhibitor of TNF-α and its target were made with nearly identical kinetics.

2.2 Adenovirus 14.7K Increases the Amount of Vaccinia Virus in Pulmonary Tissue of Animals Infected with Vaccinia Virus $14.7^{(+)}$/Tumor Necrosis Factor-α

After intranasal infection with the various V V constructs, the lungs became secondarily infected. The amount of virus was quantified by titration of the entire right lung after infection with tenfold dilutions of V V $14.7^{(+)}$/TNF-α in comparison to V V $14.7^{(-)}$/TNF-α. Figure 2 shows that the presence of Ad14.7K in the V V construct at each of two input doses increased the titers of virus in lung tissue over the initial 8 days of sampling. At an input amount of 2.5×10^6 plaque-forming units (pfu), there was a 100-fold increase of recovered virus, and at 2.9×10^5 pfu, approximately 5 logs of virus were recovered when the V V Ad$14.7^{(-)}$/TNF-α construct was no longer detectable at 8 days after infection. Thus the V V $14.7^{(+)}$/TNF-α replicated to higher titers and its presence in the lung was more prolonged than the corresponding control V V not expressing Ad14.7K.

2.3 Adenovirus 14.7K Increases the Mortality and Morbidity of Mice Infected with Vaccinia Virus Containing the Mouse Tumor Necrosis Factor-α Gene

Mice infected with 2.5×10^6 pfu V V $14.7^{(+)}$/TNF-α succumbed between 5–7 days after intranasal infection, whereas animals infected with the same dose of V V

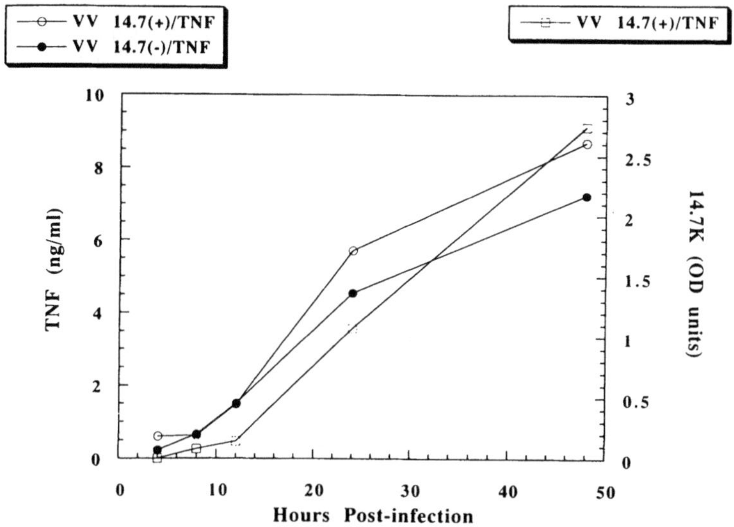

Fig. 1. Kinetics of expression of adenovirus (Ad) 14.7K and tumor necrosis factor (*TNF*)-α by vaccinia virus (*V V*) double recombinants. Two plaque-forming units (pfu) V V $14.7^{(+)}$/TNF-α or V V $14.7^{(-)}$/TNF-α per cell were used to infect human 143B cells in 24-well tissue culture plates. TNF-α was assayed from the supernates at various times after infection using the double sandwich enzyme-linked immunosorbent assay (ELISA) system (Genzyme Factor Test mTNF-α). –O– designates the amount of TNF-α produced by a V V $14.7^{(+)}$/TNF-α infection, and –●– is the amount after V V $14.7^{(-)}$ TNF-α. Western blots of 6×10^6 143B cells infected with 1 pfu V V $14.7^{(+)}$/TNF-α cell were prepared at various times after infection by lysis of the cell pellet in 0.5% Triton X-100. The supernate after centrifugation was prepared for sodium dodecyl sulfate polyacrylamide gel electrophoresis (SDS-PAGE), transferred to nitrocellulose, and the 14.7K protein detected by rabbit polyclonal antibody to the TrpE-14.7K fusion, which was a gift from William Wold. □ denotes the amount of 14.7K produced by V V $14.7^{(+)}$/TNF–α. Detection was by the chemiluminescence system of Amersham ECL Western Blotting Detection System and the intensity of the 14.7K band was quantified in optical density (*OD*) units on the LKB laser densitometer. (Modified from TUFARIELLO et al. 1994a, with permission)

$14.7^{(-)}$/TNF-α all survived (Fig. 3). The median lethal dose (LD_{50}) of V V $14.7^{(+)}$/ TNF-α was 8×10^5 compared to an LD_{50} greater than or equal to 8×10^6 pfu for V V $14.7^{(-)}$/TNF-α. Because the highest amount of the latter virus that could be physically administered did not kill the animals, an exact end point could not be determined. Nevertheless, there was at least a tenfold increase in lethality by including the Ad14.7K gene in the V V construct. There were two other indications of the severity of the disease. The sick animals demonstrated decreased activity, weight loss, hunched posture with ruffled fur, and an increased pulmonary inflammatory response. Histologic sections of infected lung showed perivascular and peribrochial inflammation with a predominance of mononuclear cells (TUFARIELLO et al. 1994a). In some cases there was intra-alveolar edema. Lungs from animals infected with 2.5×10^6 pfu V V $14.7^{(+)}$/TNF-α showed more inflammation than V V $14.7^{(-)}$/TNF-α-infected animals, but no increase in edema. There was clinical illness at both 2.9×10^5 and 2.5×10^6 pfu V V $14.7^{(+)}$/TNF-α in contrast to illness only (without mortality) at the higher dose for V V $14.7^{(-)}$/TNF-α-infected animals.

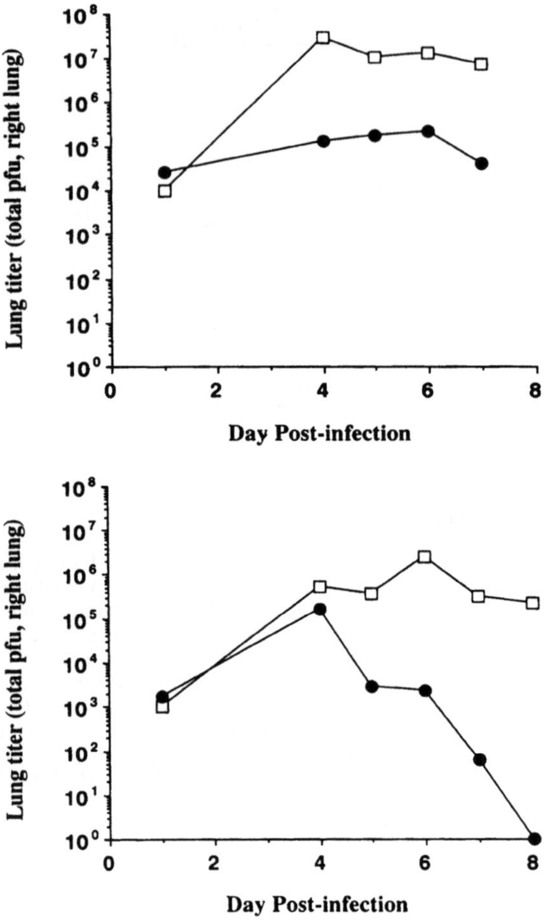

A

B

Fig. 2. Amount of virus in pulmonary tissue after intranasal infection with (V V14.7$^{(+)}$/TNF-α or V V 14.7$^{(-)}$/ TNF-α. The right lung from mice intranassaly infected with V V 14.7$^{(+)}$/TNF-α (□) or V V 14.7$^{(-)}$/ TNF-α (●) was removed from animals killed between 1 and 8 days after infection and titered for the amount of virus presen. In **A** the infecting dose was 2.5×10^6, and in **B**, 2.9×10^5 plaque-forming units (pfu) recombinant V V. The plaque assay was performed on 143B cell monolayers with X-Gal (250µg/ml) added to the agarose at 48 h and the blue plaques counted at 72 h after infection. (Reproduced from TUFARIELLO et al. 1994a, with permission)

This change in disease, mortality, and pulmonary histology as a function of the Ad14.7K gene product required the expression of mTNF-α by the recombinant V V, as the LD_{50} in the absence of TNF-α was 1×10^5 for both V V 14.7$^{(+)}$ and V V 14.7$^{(-)}$. These observations reinforce the conclusion that TNF-α is antiviral in V V infection and attenuates disease; enhancement of disease by Ad14.7K is only observed in the presence of TNF-α. The requirement for the agonist (TNF-α) to demonstrate the effect of the antagonist (Ad14.7K) implies that wt V V infection itself does not elicit a significant TNF-α response.

2.4 Adenovirus 19K Glycoprotein Does Not Affect Vaccinia Virus Virulence in a Pneumonia Model

There is an extensive series of in vitro observations that the Ad glycoprotein (gp) 19k binds to the class I MHC, decreases the presentation of MHC on the cell surface, and decreases the cytolytic T lymphocyte (CTL)-mediated recognition

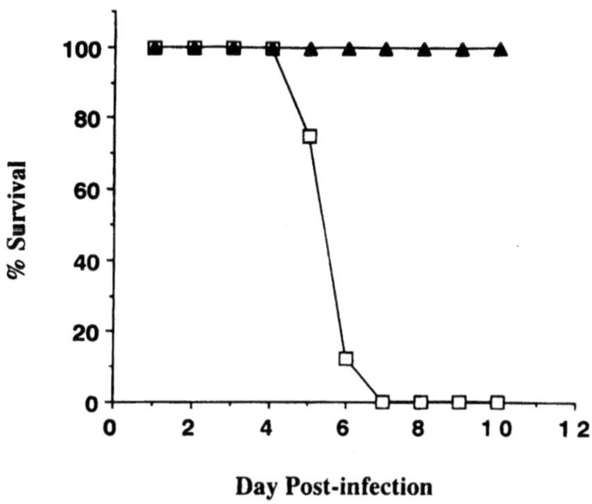

Fig. 3. V V 14.7$^{(+)}$/TNF-α is more lethal than V V 14.7$^{(-)}$/TNF-α after intranasal infection. Eight to ten animals were inoculated with 2.5×10^6 plaque-forming units (pfu) of either V V 14.7$^{(+)}$/TNF-α (□) or V V 14.7$^{(-)}$ TNF-α (▲) in each group and observed daily. There were no deaths in V V 14.7$^{(-)}$/TNF-α-infected animals during the subsequent 6-week observation period. (Reproduced from TUFARIELLO et al. 1994a, with permission)

and lysis of cells infected with various viruses including V V. We set out to determine whether these observations would be manifested by a change of disease in vivo. The Ad gp19k gene was inserted into the *Hin*dIII C region to minimize any changes in virulence occasioned by disruption of the V V TK gene located in the *Hin*dIIIJ. The expression of the Ad gp19K protein from this construct was demonstrated both in tissue culture and in the infected lung tissue (Fig. 4). The control for insertional mutagenesis was V V with the gp19K gene inserted in a nonexpression orientation. The gp19K was expressed from the 7.5K early/late promoter used for the 14.7K constructs described above. The H-2 haplotypes of the mice selected for infection with this pair of V V were H-2^g (K^d, D^b),H-2^d (K^d, L^d, D^d) and H-2^k (K^k, D^k), based on previous observations that the Ad gp19K bound best to K^d and D^b, less well to L^d and D^d, and did not bind the K^K and D^k. Thus there should have been haplotypes where the Ad gp19K downregulated surface expression of MHC and other haplotypes not affected by this mechanism. The results showed that there were no differences in mortality, histopathology, $CD4^+$ and $CD8^+$ counts after infection of H-2^d-containing mice (Fig. 5) or any of the other murine haplotypes (H-2^g and H-2^k, data not shown) by the V V expressing or not expressing Ad gp19K.

When these studies were initiated, there appeared to be ample data that the CTL response to vaccinia would be important to disease and survival in vivo; however, while our studies were in progress, experiments were done with the β_2-microglobulin knockout mice that also indicated that abrogation of the class I MHC-restricted CTL response had no effects on wt V V replication or induction of disease when administered intradermally. This latter V V model produces skin lesions but no mortality. Thus two separate lines of evidence suggest that the CTL response is either not necessary for recovery from V V infection or that there is considerable redundancy in the host response to this virus. Downregulation of CTL in otherwise immunocompetent animals does not affect survival after V V infection (see Sect. 3).

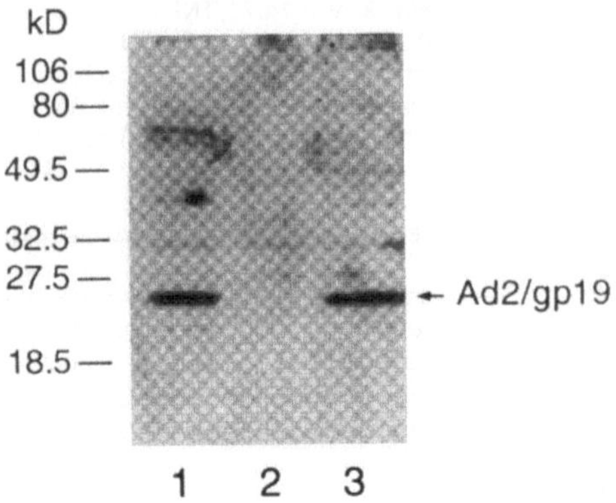

Fig. 4. Expression of the adenovirus type 2 19K glycoprotein (Ad2/gp19) in pulmonary tissue after intranasal infection. BALB/c (H-2^d) mice were infected intranasally with V V-e$19^{(+)}$ or V V e$19^{(-)}$ (10^6 plaque-forming units, PFU) and the animals were killed at 5 days. e19, the DNA coding for the Ad gp19K protein, was inserted in V V in either the expressing, e$19^{(+)}$, or nonexpressing, e$19^{(-)}$, orientation. Lung homogenates were prepared and analysed for gp19K protein by western blots using a poly-clonal rabbit antibody. *Lane 1* contained Ad2-infected HeLa cell extracts; *lanes 2* and *3*, lung homogenates from animals infected with V V-e$19^{(-)}$ and V V-e$19^{(+)}$ respectively. The position of the Ad2 gp 19K marker is shown on the *right* and the molecular weight markers on the *left*. (Re-produce from GRUNHAUS et al. 1994, with permission)

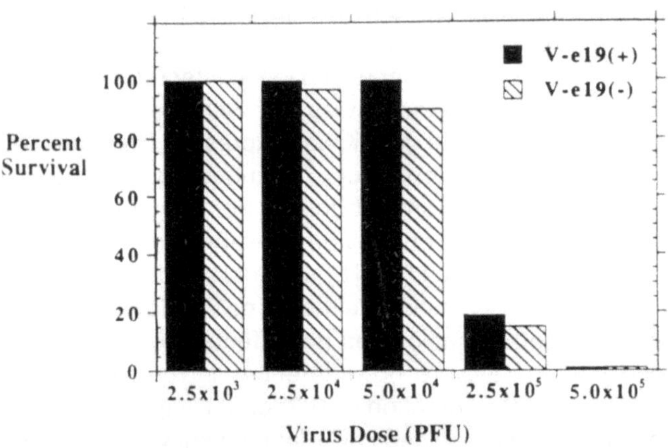

Fig. 5. The adenovirus type 2 glycoprotein 19K (Ad2 gp19K) gene does not affect survival of mice infected with vaccinia virus (V V). BALB/c mice were infected by the intranasal route with serial dilutions of either V V-e$19^{(+)}$ or V V-e$19^{(-)}$ and median lethal dose (LD_{50}) was determined. Utilizing 20 mice in each group, the LD_{50} was 11.9×10^4 for V V-e$19^{(-)}$ and 9.7×10^4 for V V-e$19^{(+)}$. (Reproduced from GRUNHAUS et al. 1994, with permission)

2.5 Creation of Transgenic Mice Carrying the Adenovirus Type 2 E3 Region for the Study of In Vivo Effects of These Genes

MAV-1 is a natural pathogen of mice and causes fatal disease with multiorgan involvement in suckling mice. Our earlier attempts to utilize this rodent virus to

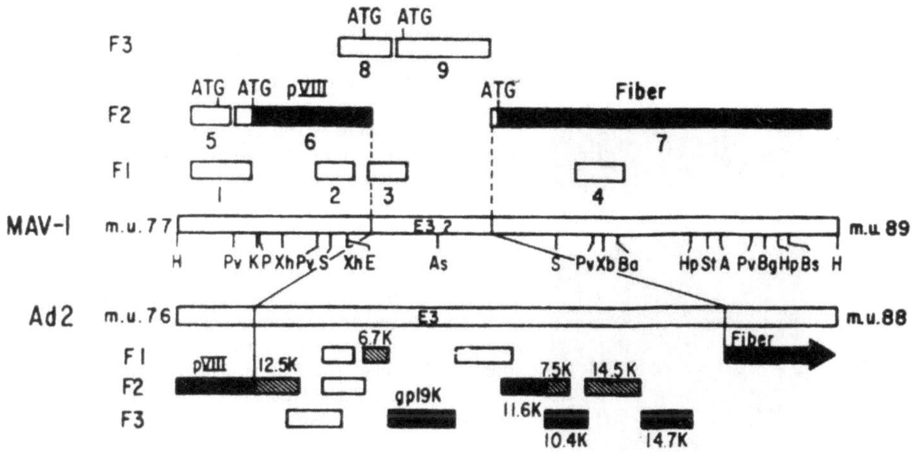

Fig. 6. Comparison of the E3 regions of mouse adenovirus (*MAV*)-1 and adenovirus (*Ad*)2. The MAV-1 open reading frames (ORF) between map units (*mu*) 77 and 89 are shown in the three reading frames and compared to the corresponding region of Ad2. The 2 MAV-1 ORF numbered 8 and 9 could code for proteins, but they would bear no relationship to any of the Ad2 ORF shown. In contrast, pVIII and fiber flanking the E3 region are both quite homologous between the human and murine Ad. Also shown are the restriction endonuclease sites of MAV-1 (RAVIPRAKASH et al. 1989). (Reproduced from RAVIPRAKASH et al. 1989, with permission)

study Ad pathogenesis were complicated by the extensive differences between the Ad2 E3 region and the putative homologous region in MAV-1. Although there was extensive homology between MAV-1 and Ad2 for the pVIII hexon-associated gene to the left of E3 and the fiber gene to the right of this region, the E3 region itself was quite different (Fig. 6). The entire MAV-1 DNA is 31 kb, considerably shorter than the 35.5-kb Ad2 DNA. The MAV-1 truncated E3 region is resistant to insertion of Ad2 E3 DNA either because the MAV-1 E3 region may be essential or because it is not possible to package the additional Ad2 sequences (personal observations). In order to develop an in vivo murine model in which the Ad would be replication competent and the E3 genes present, we have constructed transgenic mice containing the entire Ad2 E3 region, including its natural promoter (FEJER et al. 1994). Since the Ad2 E3 promoter enhancer is responsive to transactivation by the E1A of Ad2, Ad5, or MAV-1 (BALL et al. 1988), the in vivo activation of the E3 was anticipated by the E1A genes expressed during productive infection of MAV-1. Such results are shown in Fig. 7 for lung tissue of Ad2 E3 transgeneic animals infected in vivo with MAV-1. We had shown previously (FEJER et al. 1994) that, in primary lung tissue explants from the E3 transgenic animals, the E3 transgene could be transactivated by infection with the human Ad7001, which contains the Ad5 E1A region but is deleted of most of the E3 region (WOLD 1995). These observations are the basis of ongoing experiments in our laboratory to determine the function of these Ad "pathogenesis genes" in vivo.

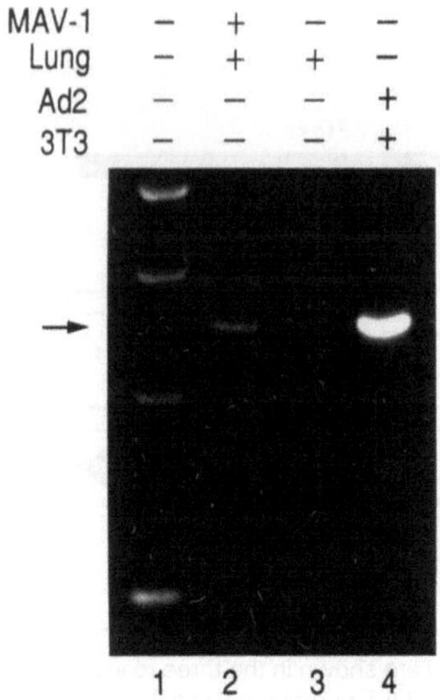

Fig. 7. Mouse adenovirus (*MAV*)-1 infection activates the E3 transgene in lung. Transgenic mice containing the adenovirus (*Ad*)2 E3 genomic DNA behind the E3 promoter were infected at 14 days of age with 5×10^3 plaque-forming units (pfu) of MAV-1 intraperitoneally and killed 4 days after infection at a time when they were ill. The lungs were dissected and RNA isolated for reverse transcriptase (RT) and polymerase chain reaction (PCR) determination of the presence of E3 transcripts (FEJER et al. 1994). Primers that spanned the first E3 intron and produced a 249-base pair (bp) fragment from spliced E3 mRNA were utilized. *Lane 1,* the molecular weight standards with the markers between 100–400 bp shown. *Lane 2,* RT-PCR product from the lung of MAV-1-infected transgenic mice. *Lane 3,* the corresponding specimen from uninfected transgenic mice. *Lane 4,* a positive control using cloned Ad2-infected 3T3 cells. The PCR-amplified cDNA was electrophoresed through 2% agarose; the *arrow* designates the size of the expected cDNA product

3 Discussion

The control of TNF-α by Ad proteins must be important because of the number of viral genes that can inhibit the action of this cytokine. In addition to the E3 proteins (14.7K or 10.4K plus 14.5K), and Ad E1B 19K protein can also prevent cytolysis in human and rat cells, but is not active in cells of murine origin. TNF-α is a pleiotropic cytokine produced by activated monocytes and macrophages that can activate a variety of transcription factors as a result of signal transduction following binding of the cytokine to either of two cell surface receptors. TNF-α has antiviral activities as a result of cytotoxicity to infected cells as shown for Newcastle disease virus, vesicular stomatitis virus (VSV), herpes simplex virus (HSV), Ad E3-deleted virus, and V V infections (ADERKA et al. 1985; KOFF and FANN 1986; LORENCE et al. 1988; GOODING et al. 1988). TNF-α expression is induced after infection with Ad2, HSV type 2, VSV, or encephalomyocarditis virus infection of several cell lines including RPMI 1788 and HL-60, a lymphoid and myeloid line, respectively (WONG and GOEDDEL 1986). It is difficult to study the effects of TNF-α during viral infection because systemic administration results in widespread and protean effects of the cytokine, leading to considerable toxicity. Targeted production of TNF-α by V V constructs carrying the TNF-α gene as described in this manuscript has been a better method to study the effects of this cytokine in vivo (SAMBHI et al. 1991). TNF-α greatly attenuates V V virulence, and the increase in mortality and morbidity by the Ad14.7K when cloned into V V causes reversal of the original attenuation. The TNF-α effect does not require any products or involvement of T or B cells, as the same effects are seen in nude mice (SAMBHI et al. 1991).

We have recently shown that the reversal of the TNF-α effect with Ad14.7K likewise does not require T or B cells as we can successfully antagonize the cytokine in severe combined immunodeficient (SCID) mice (TUFARIELLO et al. 1994b). In the latter case, the original acute pulmonary disease model following intranasal infection as described in Figs. 2 and 3 is converted into a subacute or chronic disease which displays itself over a period of 2–4 months and is manifest by dissemination of the V V to many organs, including the skin. The chronicity is clealy due to TNF-α, for in its absence the animals die within 1 week of infection. The animals infected with V V $14.7^{(+)}$/TNF-α become sick 3–4 weeks earlier than the animals infected with V V $14.7^{(-)}$/TNF-α, again demonstrating the increase in virulence caused by Ad14.7K. Ad14.7K has been shown to prevent the release of arachidonic acid, presumably by inhibiting the activation of the cytosolic phospholipase A_2 after stimulation of C3HA murine cells with TNF-α together with cycloheximide (ZILLI et al. 1992). This pathway appears to be involved in cytolysis mediated by TNF-α. There is no evidence that the mechanism of action of 14.7K in inhibiting cytolysis involves the transcription pathway affected by TNF-α.

Adenovirus infections cause acute disease, but the virus can persist for long periods of time in various organs of the host. The anti-TNF-α effects could benefit Ad survival either acutely or chronically by preventing or delaying lysis of infected target cells. Each of these patterns has been observed in the V V system carrying the Ad14.7K gene. In the immunocompetent animals, Ad14.7K increased the acute virulence and mortality together with the enhanced production of intrapulmonary V V. In the immunosuppressed animal, Ad14.7K accelerated the progression of the subacute or chronic manifestations of V V together with systemic spread presumably by viremia. Thus, the predictions are that Ad14.7K promotes the survival of the virus in its host both acutely and chronically. These conclusions are based upon the effects of the Ad14.7K gene in V V rather than in its natural context, but it appears that there are only two requirements during infection in order to observe the effect of the Ad14.7K gene product: (1) Virulence must be dependent on an increased amount of virus; (2) The host response should include the production of sufficient endogenous TNF-α to promote early cytolysis before the virus has made a full complement of progeny viruses. It is probable that both of these conditions are met during an Ad infection and that the antagonism of TNF-α by Ad14.7K would have the same effect in Ad infection as it does for V V.

V V constructs containing the Ad gp19K gene in the *Hin*dIII-J were shown by Cox et al. (Cox et al. 1990) to downregulate MHC gene products of the proper genotype at the cell surface, as demonstrated by the reduction in CTL lysis by appropriately primed $CD8^+$ cells from immunized animals. This change in CTL lysis correlated with previous data based on the selective binding studies of the Ad gp 19K with the various H-2 heavy chains and the downregulation of their transport out of the ER. In our experiments, the failure of the Ad gp19K to affect disease in V V-infected animals of MHC haplotypes that are bound and sequestered in the ER by the Ad protein was unexpected, based on the information existing at the beginning of these experiments. The importance of cellular immunity in the control of viral infection had been proven by a series of studies measuring clearance of V V or ectromelia, another related poxvirus (HIRSCH et al. 1968; BLANDEN 1970, 1971). These experiments depended on depletion of lymphocytes by treatment of mice with rabbit antimouse thymocyte serum. Other experiments demostrated the importance of cellular immunity by infusing cells from immunized mice into immunocompromised recipients (BLANDEN 1971). In more recent studies of VSV immunization of H-2^b mice (which present the VSV-NP protein) followed by intracerebral challenge with V V expressing the VSV-NP protein, the animals were protected against high doses of V V (VSV-NP) (BINDER and KÜNDIG 1991). This protection could be reversed by pretreating the mice with anti-$CD8^+$ antibodies. When this experiment was repeated in H-2^k mice, which do not present their VSV-NP to $CD8^+$ cells, the efficiency of protection was measurable only if the viral challenge was reduced by 3 logs. However, there was some residual immunity which was insensitive to pretreatment with $CD8^+$ antibodies but became sensitive to treatment with anti-$CD4^+$ antibodies. These experiments suggest that there are multiple mechanisms for dealing with individual infections and that alternate immune responses may overlap in immunocompetent animals. Some of the original studies can also be interpreted in this manner, since the ablation of cellular immunity by infusing rabbit antibodies to mouse thymocytes probably eliminated both $CD4^+$ and $CD8^+$ lymphocytes. Although the $CD8^+$ CTL response appears to be important, there are circumstances for which other components of the immune response can substitute for a selective deficiency in the $CD8^+$ control mechanism. This conclusion is further substantiated by more recent studies on V V infection of β_2-microglobulin (homozygous knockout) animals showing no change in disease in vivo, although there was a decreased immunoglobulin G (IgG) antibody response (SPRIGGS et al. 1992). These results are concordant with our observations that the down-regulation of class I MHC molecules on the cell surface did not affect the recovery from V V. The potential redundancy of mechanisms for the recovery of the host from viral infections may have made our assays of virulence difficult to interpret. However, this conclusion for V V may not be directly extrapolatable to Ad pathogenesis.

The insertion of the entire Ad E3 region into transgenic mice behind the naturally inducible promoter has begun a new approach to study the importance of the E3 genes both in Ad infection (MAV-1 or MAV-2) and also in other infections that will transactivate the promoter or that target tissues in which the promoter is constitutively active. Not only can E3 be activated by the Ad E1A, but it can be transactivated by TNF-α, presumably because of the NF-kB-, NF-IL-6-, or AP-1-responsive elements in the promoter enhancer region (KORNER et al. 1992). We have already shown organ-specific enhancement of the E3 transgenic promoter by the administration of bacterial lipopolysaccharide to the transgenic animals (FEJER et al. 1994). Infection of these animals with an Ad deleted of the E3 region (Ad7001) has also shown transactivation of the endogenous transgenic promoter. Since there are many activators of the E3 promoter in addition to products arising from Ad infection, there are ways to test the effects of the E3 gene products on other intracellular and extracellular infections including bacterial and parasitic infections of mice.

Acknowledgments. M.S.H. was supported by Public Health Service Grant AI 27199 and Cancer Center Grant PO1 CA 13330. J.T. was supported by Medical Scientist Training grant 5T32 GM07288 from the National Institutes of Health. A.G. was supported by NIH training grant 5T32 GM07260. G.F. was supported by a Fogarty International Fellowship Number FO5 TWO4525.

References

Aderka D, Novick D, Hahn T, Fischer DG, Wallach D (1985) Increase of vulnerability to lymphotoxin in cells infected by vesicular stomatitis virus and its further augmentation by interferon. Cell Immunol 92: 218–225

Arroyo M, Bagchi S, Raychaudhuri P (1993) Association of the human papillomavirus type 16 E7 protein with the S-phase-specific E2F-cyclin A complex. Mol Cell Biol 13: 6537–6546

Ball AO, Williams ME, Spindler KR (1988) Identification of mouse adenovirus type 1 early region 1: DNA sequence and a conserved transacting function. J Virol 62: 3947–3957

Binder D, Kündig TM (1991) Antiviral protection by CD8⁺ versus CD4⁺ T cells. J Immunol 146: 4301–4307

Blanden RV (1970) Mechanisms of recovery from a generalized viral infection: mousepox. J Exp Med 132: 1035–1054

Blanden RV (1971) Mechanisms of recovery from a generalized viral infection: mousepox. J Exp Med 133: 1074–1089

Chen M, Mermod N, Horwitz MS (1990) Protein–protein interactions between adenovirus DNA polymerase and nuclear factor I mediate formation of the DNA replication preinitiation complexes. J Biol Chem 265: 18634–18642

Cotten M, Wagner E, Zatloukal K, Birnstiel ML (1993) Chicken adenovirus (CELO virus) particles augment receptor-mediated DNA delivery to mammalian cells and yield exceptional levels of stable transformants. J Virol 67: 3777–3785

Cox JH, Yewdell JW, Eisenlohr LC, Johnson PR, Bennink JR (1990) Antigen presentation requires transport of MHC class I molecules from the endoplasmic reticulum. Science 247: 715–718

Debbas M, White E (1993) Wild-type p53 mediates apoptosis by E1A, which is inhibited by E1B. Genes Dev 7: 546–554

Faha B, Harlow E, Lees E (1993) The adenovirus E1A-associated kinase consists of cyclin E-p33cdk2 and cyclin A-p33cdk2. J Virol 67: 2456–2465

Fejer G, Gyory I, Tufariello J, Horwitz MS (1994) Characterization of transgenic mice containing adenovirus early region 3 genomic DNA J Virol 68: 5871–5881

Flomenberg PR, Chen M, Horwitz MS (1987) Characterization of a major histocompatibility complex class I antigen-binding glycoprotein from adenovirus type 35, a type associated with immunocompromised hosts. J Virol 61: 3665–3671

Ginsberg HS, Dingle JH (1965) The adenovirus group. Horsfall FL Jr, Tamm I (Eds) Viral and rickettsial infections of man. Lippincott, Philadelphia, pp 860–891

Ginsberg HS, Lundholm-Beauchamp U, Horswood RL, Pernis B, Wold WSM, Chanock RM, Prince GA (1989) Role of early region 3 (E3) in pathogenesis of adenovirus disease. Proc Natl Acad Sci USA 86: 3823–3827

Ginsberg HS, Horswood RL, Chanock RM, Prince GA (1990) Role of early genes in pathogenesis of adenovirus pneumonia. Proc Natl Acad Sci U S A 87: 6191–6195

Ginsberg HS, Moldawer LL, Sehgal PB, Redington M, Kilian PL, Chanock RM, Prince GA (1991) A mouse model for investigating the molecular pathogenesis of adenovirus pneumonia. Proc Natl Acad Sci USA 88: 1651–1655

Gooding LR, Elmore LW, Tollefson AE, Brady HA, Wold WSM (1988) A 14 700 MW protein from the E3 region of adenovirus inhibits cytolysis by tumor necrosis factor. Cell 53: 341–346

Gooding LR, Ranheim TS, Tollefson AE, Aquino L, Duerksen-Hughes PJ, Horton TM, Wold WSM (1991) The 10,400- and 14,500-dalton proteins encoded by region E3 of adenovirus function together to protect many but not all mouse cell lines against lysis by tumor necrosis factor. J Virol 65(8): 4114–4123

Greber UF, Willetts M, Webster P, Helenius A (1993) Stepwise dismantling of adenovirus 2 during entry into cells. Cell 75: 477–486

Grunhaus A, Horwitz MS (1992) Adenoviruses as cloning vectors. Rice C (ed) Seminars in virology Saunders Scientific, London, pp 237–252

Grunhaus A, Cho S, Horwitz MS (1994) Association of vaccinia virus-expressed adenovirus E3-19K glycoprotein with class I MHC and its effects on virulence in a murine pneumonia model. Virology 200: 535–546

Hay-Ahmad Y, Graham FL (1986) Development of a helper-independent human adenovirus vector and its use in the transfer of the herpes simplex virus thymidine kinase gene. J Virol 57: 267–274

Hierholzer JC, Wigand R, Anderson LJ, Adrian T, Gold JWM (1988) Adenovirus from patients with AIDS: a plethora of serotypes and a description of five new serotypes of subgenus D (types 43–47) J Infect Dis 158: 804–813

Hirsch MS, Nahmias AJ, Murphy FA, Kramer JH (1968) Cellular immunity in vaccinia infection of mice. J Exp Med 128: 121–132

Horton TM, Ranheim TS, Aquino L, Kusher DI, Saha SK, Ware CF, Wold WSM, Gooding LR (1991) Adenovirus E3 14.7K protein functions in the absence of other adenovirus proteins to protect transfected cells from tumor necrosis factor cytolysis. J Virol 65(5): 2629–2639

Horwitz MS (1990a) Adenoviridae and their replication. Fields BN, Knipe DM, Chanock RM, Melnick JL, Hirsch MS, Monath TP, Roizman B (eds) Virology, 2nd edn. Raven, New York , pp 1679–1721

Horwitz MS (1990b) Adenoviruses. Fields BN, Knipe DM, Chanock RM, Melnick JL, Hirsch MS, Monath TP, Roizman B (eds) Virology, 2nd edn. Raven, New York, pp 1723–1740

Horvath J, Palkonyay L, Weber J (1986) Group C adenovirus DNA sequences in human lymphoid cell. J Virol 59: 189–192

Ikeda MA, Nevins JR (1993) Identification of distinct roles for separate E1A domains in disruption of E2F complexes. Mol Cell Biol 13: 7029–7035

Ishibashi M, Yasue H (1984) Adenoviruses of animals. Ginsberg HS (ed) The adenoviruses. Plenum, New York, pp 497–562

Klessig DF, Quinlan MP (1982) Genetic evidence for separate functional domains of the human adenovirus specified, 72K, DNA-binding protein. J Mol Appl Genet 1: 263–272

Koff WC, Fann AV (1986) Human tumor necrosis factor-α kills herpesvirus-infected but not normal cells. Lymphokine Res 5: 215–221

Korner H, Fritzsche U, Burgert HG (1992) Tumor necrosis factor α stimulates expression of adenovirus early region 3 proteins: implications for viral persistence. Proc Natl Acad Sci U S A 89: 11857–11861

Krajcsi P, Tollefson AE, Wold WSM (1992) The E3-14.5K integral membrane protein of adenovirus that is required for down-regulation of the EGF receptor and for prevention of TNF cytolysis is O-glycosylated but not N-glycosylated. Virology 188: 570–579

Kraus VB, Moran E, Nevins JR (1992) Promoter-specific trans-activation by the adenovirus E1A 12S product involves separate E1A domains. Mol Cell Biol 12: 4391–4399

Lavery DJ, Fu SM, Lufkin T, Chen-Kiang S (1987) Productive infection by cultured human lymphoid cells by adenovirus. J Virol 61: 1466–1472

Lonberg-Holm K, Crowell RL, Philipson L (1976) Unrelated animal viruses share receptors. Nature 259: 678–681

Lorence RM, Rood PA, Kelly KW (1988) Newcastle disease virus as an antineoplastic agent: induction of tumor necrosis factor-α and augmentation of its cytotoxicity. J Natl Cancer Inst 80: 1305–1312

Lowe SW, Ruley HE (1993) Stabilization of the p53 tumor suppressor is induced by adenovirus 5 E1A and accompanies apoptosis. Genes Dev 7: 535–545

Ma Y, Mathews MB (1993) Comparative analysis of the structure and function of adenovirus virus-associated RNAs. J Virol 67: 6605–6616

Michael SI, Huang CH, Romer MU, Wagner E, Hu PC, Curiel DT (1993) Binding-incompetent adenovirus facilitates molecular conjugate-mediated gene transfer by the receptor-mediated endocytosis pathway. J Biol Chem 268: 6866–6869

Mittal SK, McDermott MR, Johnson DC, Prevec L, Graham FL (1993) Monitoring foreign gene expression by a human adenovirus-based vector using the firefly luciferase gene as a reporter. Virus Res 28: 67–90

Otten JA, Tennant RW (1982) The mouse in biomedical research, 2nd edn. Academic, New York pp 335–340

Prince GA, Porter DD, Bennett Jenson A, Horswood RL, Chanock RM, Ginsberg HS (1993a) Pathogenesis of adenovirus type 5 pneumonia in cotton rats. J Virol 67: 101–111

Prince GA, Porter DD, Jenson AB, Horswood RL, Chanock RM, Ginsberg HS (1993b). Pathogenesis of adenovirus type 5 pneumonia in cotton rats (Sigmodon hispidus). J Virol 67: 101-111

Ragot T, Vincent N, Chafey P, Vigne E, Gilgenkrantz H, Couton D, Cartaud J, Briand P, Kaplan J-C, Perricaudet M, Kahn A (1993) Efficient adenovirus-mediated transfer of a human minidystrophin gene to skeletal muscle of mdx mice. Nature 361: 647–650

Rao L, Debbas M, Sabbatini P, Hockenbery D, Korsmeyer S, White E (1992) The adenovirus E1A proteins induce apoptosis, which is inhibited by the E1B 19-kDa and Bcl-2 proteins. Proc Natl Acad Sci USA 89: 7742–7746

Raviprakash KS, Grunhaus A, Elkholy A, Horwitz MS (1989) The mouse adenovirus type 1 contains an unusual E3 region. J Virol 63: 5455–5458

Rowe WP, Huebiner RJ, Gillmore LK, Parrott RH, Ward TG, (1953) Isolation of a cytogenic agent from human adenoids undergoing spontaneous degeneration in tissue culture. Proc Soc Exp Biol Med 84: 570–573

Sambhi SK, Kohonen-Corish MRJ, Ramshaw IA (1991) Local production of tumor necrosis factor encoded by recombinant vaccinia virus is effective in controlling viral replication in vivo. Proc Natl Acad Sci USA 88: 4025–4029

Schnurr D, Dondero ME (1993) Two new candidate adenovirus serotypes. Intervirology 36: 79–83

Smith AL, Barthold SW (1987) Factors influencing susceptibility of laboratory rodents to infection with mouse adenovirus strains K 87 and FL. Arch Virol 95: 143–148

Spriggs MK, Koller BH, Sato T, Morrisey PJ, Fanslow WC, Smithies O, Voice RF, Widmer MB, Maliszewski CR (1992) β_2-Microglobulin$^-$, CD8$^+$T-cell-deficient mice survive inoculation with high doses of vaccinia virus and exhibit altered IgG responses. Proc Natl Acad Sci USA 89: 6070–6074

Tufariello J, Cho S, Horwitz MS (1994a) The adenovirus E3 14.7-kilodalton protein which inhibits cytolysis by tumor necrosis factor increases the virulence of vaccinia virus in a murine pneumonia model. J Virol 68: 453–462

Tufariello J, Cho S, Horwitz MS (1994b) Adenovirus E3 14.7K protein, an antagonist of tumor necrosis factor cytolysis, increases the virulence of vaccinia virus in severe combined immunodeficient mice. Proc Natl Acad Sci USA 91: 10987–10991

Wagner E, Zatloukal K, Cotten M, Kirlappos H, Mechtler K, Curiel DT, Birnstiel ML (1992) Coupling of adenovirus to transferrin-polylysine/DNA complexes greatly enhances receptor-mediated gene delivery and expression of transfected genes. Proc Natl Acad Sci USA 89: 6099–6103

Wickham TJ, Mathias P, Cheresh DA, Nemerow GR (1993) Integrins $\alpha_v\beta_3$S and $\alpha_v\beta_5$ promote adeno virus internalization but not virus attachment. Cell 73: 309–319

Wold WSM, Gooding LR (1991) Region E3 of adenovirus: a cassette of genes involved in host immunosurveillance and virus-cell interactions. Virology 184: 1–8

Wold WSM (1995) E3 transcription unit of adenovirus. In: Doerfler W, Böhm P (eds) The molecular repertoire of adenoviruses. Springer, Berlin, Heidelberg New York (current topic's in microbiology and immunology, vol 199/I): 237–274

Wong GHW, Goeddel DV (1986) Tumor necrosis factors α and β inhibit virus replication and synergize with interferons. Nature 323: 819–822

Zabner J, Couture LA, Gregory RJ, Graham SM, Smith AE, Welsh MJ (1993) Adenovirus-mediated gene transfer transiently corrects the chloride transport defect in nasal epithelia of patients with cystic fibrosis. Cell 75: 207–216

Zilli D, Voelkel-Johnson C, Skinner T, Laster SM (1992) The adenovirus E3 region 14.7 kDa protein, heat and sodium arsinate inhibit the TNF-induced release of arachidonic acid. Biochem Biophys Res Commun 188(1): 177–183

Molecular Determinants of Adenovirus Tropism

Y.-F. Mei and G. Wadell

1 Introduction

The 49 serotypes of human adenoviruses can be divided into six subgenera (A-F; Wadell 1990; Schnurr and Dondero 1993). Members of subgenera A and F are related, displaying a tropism for the intestinal tract. Furthermore, they are difficult to cultivate in established cell lines. However, the degree of fastidiousness and the association within gastroenteritis is more pronounced for the two members Ad40 and Ad41 of subgenus F (Brown et al. 1992; Tiemessen and Kidd 1994; Uhnoo et al. 1984).

Members of subgenus A (Ad12, Ad18, Ad31) cause cryptic infections. They are also associated with severe infections in immunocompromised hosts (Johansson et al. 1991; Hierholzer 1992), whereas members of subgenus F have only been demonstrated to be shed for short periods into the stool (Allard et al. 1992).

Department of Virology, Umeå University, 901 85 Umeå, Sweden

Members of subgenus B have been divided into two DNA homology clusters B1 and B2 (Wadell et al. 1980). Members of subgenus B1 (Ad3, Ad7, Ad21) cause outbreaks of acute respiratory infections that may be severe and sometimes fatal (Murtagh et al. 1993), whereas members of subgenus B2 (Ad11, Ad34, Ad35) cause persistent infections in the kidney and can be shed into urine for several months up to a year (Numazaki et al. 1968; Lecatsas et al. 1974). They are also associated with infections in immunocompromised hosts (Flomenberg et al. 1987).

Members of subgenus C (Ad1, Ad2, Ad5, Ad6) are the most widely spread human adenoviruses infecting most children in the world. The primary infection is characterized by acute respiratory tract symptoms, although one half of the infections are asymptomatic. These serotypes are associated with lymphatic tissue and are persistently shed into the stools for several years. (Fox et al. 1977).

Members of subgenus D are characterized by a tropism for the eye, causing conjunctivitis or keratoconjunctivitis. Several members are also characterized by prolonged shedding into stools. This is particularly common in immunocompromised hosts (Hierholzer 1992).

Ad4, the only member of subgenus E, causes epidemic outbreaks of follicular conjuctivitis and also acute outbreaks of respiratory tract infections. These appear to be more common among older children and young adults and not so frequent among infants, who are usually more frequently infected by subgenus C adenoviruses (Wadell 1984).

2 Genetic Variability

A multitude of genome types have been identified within each analyzed serotype by DNA restriction analysis. All genome types within each serotype can be grouped into three distinct genomic clusters (Li and Wadell 1986).

We were intrigued by the observation that Ad11 of subgenus B2 was associated with outbreaks of pneumonia in China and Spain (Zhang et al. 1984; Hierholzer et al. 1974). Since this was the symptom that was more characteristic of members of subgenus B1, we analyzed 13 isolates from Chinese patients with pneumonia and were able to demonstrate the presence of unique genome types that formed a separate genomic cluster within the Ad11 serotype. This cluster is represented by the Ad11a genome type (Li et al. 1991).

The molecular determinants of the different tropisms displayed by adenoviruses are on the whole unknown. We were confronted with isolates that corresponded to distinct genome types represented by Ad11a that are responsible for acute respiratory tract infections on the one hand and the previously established Ad11 isolates represented by the Ad11 prototype that were associated with persistent infection of the kidney on the other hand. The Ad11p and Ad11a genome types are closely related; we therefore selected these two genome types for a detailed comparison in order to find clues that might contribute to a better understanding of their difference in tropism.

The organization of the E3 region and of the fiber gene of related genome types of different members of subgenus B has therefore been studied.

3 Analysis of the E3 Region

Six open reading frames (ORF) were found in the E3 region of subgenus B2: 18.5K, 20.3K, and 20.6K belonging to the E3A region and 10.3K, 15.2K, and 15.3K belonging to the E3B region.

A comparison between Ad3 of subgenus B1 and the members of subgenus B2 (Ad11p, Ad11a, Ad35) revealed that the 9K ORF located upstream of 10.3K of Ad3, corresponding to the 11.6K ORF in Ad2 with a function in the lysis of the infected cell, was deleted among the subgenus B2 adenovirus types (MEI and WADELL 1992). The comparison of the 20.3K and 20.6K ORF revealed a higher homology between Ad11p and Ad35. Only one amino acid was discordant in each of the two ORFs. When these ORFs were compared the homology between Ad11p and Ad11a was 94.5% and 95.6%, respectively. The 20.3K and 20.6K ORF may be evolutionary related since they display a homology of 48%–49% at the nucleotide level. These two ORF were only found in subgenus B (Tables 1–3).

Table 1. Analogous open reading frame (ORF) comparison in Ad11p, Ad11a, Ad35, and Ad3

Ad11p	Ad11a	Ad35	Ad3
18.5K 166 aa	18.5K 166 aa	18.5K 166 aa	19.2K 172 aa
20.3K 181 aa	20.3K 181 aa	20.3K 181 aa	20.1K 179 aa
20.6K 187 aa	20.2K 183 aa	20.6K 187 aa	20.5K 189 aa

The predicted number of amino acid (aa) residues and moleular weights of products encoded by the corresponding ORF are shown.

Table 2. Comparison of nucleotide homology

ORF	11p/11a(%)	11p/Ad35(%)	11p/Ad35(%)	11p/Ad3(%)	11a/Ad3(%)
18.5K	98.6 (491/498)	99.8 (497/498)	98.8 (492/498)	87.8 (437/498)	87.6 (436/498)
20.3K	95.9 (521/543)	99.3 (539/543)	96.3 (523/543)	81.8 (439/537)	81.6 (438/537)
20.6K	97.6 (536/549)	99.6 (559/561)	97.6 (536/549)	76.2 (425/558)	77.3 (422/546)

Numbers in parentheses indicate the number of identical nucleotides/total number of nucleotides in the open reading frame (ORF)
Ad, adenovirus.

Table 3. Comparison of amino acid homology

ORF	11p/11a(%)	11p/ Ad35(%)	11a/ Ad35(%)	11p/ Ad3(%)	11a/ Ad3(%)
18.5K	98.2 (163/166)	95.8 (159/166)	94.0 (156/166)	83.7 (139/166)	83.7 (139/166)
20.3K	94.5 (171/181)	99.5 (180/181)	95.0 (172/181)	76.0 (136/179)	75.4 (135/179)
20.6K	95.6 (175/183)	99.5 (186/187)	95.1 (174/183)	62.9 (117/186)	64.6 (117/181)

Numbers in parentheses indicate the number of identical amino acids/total number of amino acids in the open reading frame (ORF).

The various E3B ORF manifested different homologies when compared between B1 and B2 adenoviruses. An amino acid homology of 92.3%–93.4%, 62.7%–64.2%, and 85.9%–87.4% was noted for 10.3K, 15.2K and 15.3K, respectively. It has been demonstrated that the Ad2 E3/14.5K and 10.3K could prevent cytolysis of mouse cells by tumor necrosis factor (TNF) and downregulate the epidermal growth factor (EGF) receptor (Carleen et al. 1989), whereas 14.7K can impair the cytolytic role of TNF (Gooding et al. 1991). Ad11p E3B 15.2K and 15.3K corresponded to Ad2 E3/14.5K and 14.7K, respectively.

4 Analysis of the 18.5K Open Reading Frame

The E3 19K protein associates with the class I major histocompatibility antigens of human cells. This interaction impaired the transport of major histocompatibility complex (MHC) I molecules and the presentation of adenovirus-specific epitopes on the cell surface. Consequently, this condition may allow adenovirus to evade immunosurveillance and result in persistent infection (Pääbo et al. 1983).

Analysis of the amino acid sequences of the 18.5K ORF of Ad11p and Ad11a revealed four potential N-linked glycosylation sites (Fig. 1). The predicted molecular weight of the Ad11 18.5K ORF would be of the same size as Ad35 29K glycoprotein (Flomenberg et al. 1987). Ad11p, Ad11a, and Ad35 18.5K ORF shared high amino acid homology (Table 3), whereas the amino acid homology between Ad11p, Ad11a, Ad35 18.5K, and Ad3 19.2K ORF was low (83.7%). The sequences of 18.5K ORF of Ad11p, Ad11a, and Ad35 are almost identical but differ from Ad3. In the signal sequence, the central hydrophobic core is surrounded by prolines in positions 3 and 16 in Ad11p and Ad35. These prolines should be expected to participate in the formation of a β-sheet structure. However, the proline residue in position 16 in Ad11a was replaced by alanine. This provides a more flexible insertion of the Ad11a 18.5K protein into the membrane of the endoplasmic reticulum. None of these prolines are conserved in the Ad3 19.2K glycoprotein.

```
           Central
         hydrophobic
            core
Ad3    MGAILVVLALLSLLGLGSANLNPLDHDPCLDFDPENCTLTFAPDTSRLCGVLIKCGWDCR   60
Ad11p  --P---L-V-----EP----*****Y---------------------I---------E--   55
Ad11a  --P---L-V-----EA----*****Y---------------------I---------E--   55
Ad35   --P---L-V-----EP----*****Y---------------------I--F-LSADGK--   55
           Signal
         sequence

Ad3    SVEITHNNKTWNNTLSTTWEPGVPQWYTVSVRGPDGSIRISNNTFIFSEMCDLAMFMSRQ  120
Ad11p  ------------------------E---------------------------------K-  115
Ad11a  ------------------------------------------------K---------K-  115
Ad35   ------------------------E---------------------------------K-  115
                 Endoplasmic reticulum lumenal domain
                        binds to class I MHC

Ad3    YDLWPPSKENIVAFSIAYCLVTCIITAIICVCIHLLIVIRPRQSNEEKEKMP  172
Ad11p  -S------D---T-------CA-LL--LL-------VTT-IKNA-*N-----  166
Ad11a  -S------D---T-------CA-LL--LL-------VTT-IKNA-*N-----  166
Ad35   -S------D---T-------CA-LL--LL-------VTT-IKNA-*N-----  166
                          Transmembrane     Cytoplasmic
                              domain           domain
                                            ER retention
```

Fig. 1. Amino acid sequence comparison of four 18.5K proteins which bind to major histocompatibility complex (MHC)1. The amino acid comparison has been deduced from the nucleotide sequence. There are five cysteine residues labeled with *asterisks* which are conserved among E3 18.5K proteins of subgenes B and C. A conserved region-predicted MHC class 1 binding site was marked by the *upper line*. Four glycosylation sites are indicated by *arrows*. *Ad*, adenovirus; *ER*, endothelium reticulum

At the C terminus, the 21-amino acid transmembrane domain is followed by a 15-amino acid polar domain which anchors the protein beneath the membrane. The endoplasmic reticulum retention signal is microtubal binding (Dahllöf et al. 1991). Six leucine residues have been identified within the transmembrane domain of the Ad3 19.2K protein. In contrast to Ad3 (Signäs et al. 1986) and Ad7 (Hermiston et al. 1993), Ad11p and Ad11a share an identical amino acid sequence of the transmembrane region with Ad35 (Flomenberg et al. 1988). This contains seven isoleucine and five cysteine residues among the 21-residue transmembrane domain. Neither this predilection for isoleucine nor the occurrence of five cysteine residues is frequently seen in transmembrane domains (von Heijne 1985). Within the endoplasmic reticulum luminal domain binding to MHC class I the sequence was conserved between Ad11p, Ad11a, and Ad35 with the exception of amino acids Val-Leu-Ile-Lys-Cys-Gly-Trp-Glu between positions 46 and 53, which were replaced by Phe-Leu-Leu-Ser-Ala-Asp-Gly-Lys in Ad35 18.5K. The negatively charged Glu-80 and Glu-104 in Ad11p were exchanged to the neutral Gln-80 and the positively charged Lys-104 in Ad11a. The appearance of Lys-104 is a significant change in a region that is involved in the interaction between E3 18.5K and the heavy chain of MHC I. This region contains four conserved cysteines and the entire region is highly conserved between all other analyzed members of subgenus B and C (FLOMENBERG et al. 1992; SESTER and BURGERT 1994).

5 Analysis of the Fiber Gene; Differences Between the Fibers of Five Subgenera of Human Adenoviruses

The fiber is composed of three portions, the tail anchoring the fiber into the vertex capsomer, the shaft, and the knob containing the conformation that can interact with the host cell receptor. The tail portion of fibers from six different subgenera is conserved in size with a mean of 44 amino acid residues. The shaft differed markedly in size from six repeat motives in subgenus B to 22 repeat motives in subgenus A, C, and also some genome types of subgenus F. The knob portion of subgenus B2 accommodates 17 additional amino acid residues compared to the knob of the long fiber of Ad40 of subgenus F (Table 4).

6 Comparison of the Organization of the Fibers of the Different Subgenus B Adenoviruses

The fibers of Ad7p (HONG et al. 1988), Ad11p, Ad11a, and Ad34a display a close amino acid homology of 90.8%–98.8%. The fibers of Ad3 (SIGNÄS et al. 1985) and Ad35 are more distantly related to the other fibers of subgenus B members, displaying an amino acid homology of 59.0%–60.3% and 60.3%–63.4%, respectively. The two adenovirus types with a tropism for the kidney, Ad11p (NUMAZAKI et al. 1968) and Ad34a (LI et al. 1991), were most closely related, displaying an amino acid homology of 98.8% (Table 5).

6.1 Comparison of Nucleotide and Amino Acid Homology of the Tail, Shaft, and Knob Regions of the Fiber Gene of the Four Members of Subgenus B2

The tail is the most conserved portion of the fiber gene, displaying a homology of 95.3%–100% between the members of subgenus B2. The shaft region of the four fiber genes manifested amino acid homologies ranging from 72.5% to 98.9%. Only one amino acid differed in the knob region of Ad34a and Ad11p, whereas 93% of the amino acids were homologous between Ad11a and Ad34a. The homology within the knob region of the fiber was only 48.9%–49.5% when Ad35p was compared with the other members of subgenus B2 (Table 6).

Table 4. Differences between the fibers of six subgenera

Adenovirus serotype/ Subgenus	Fiber length (nm)	Tail		Shaft		Repeat motif	Knob		Reference
		Amino acid		Amino acid			Amino acid		
		Position	Number	Position	Number		Position	Number	
Ad12/A	28–33	1–50	50	551–410	360	22	411–588	177	SPRENGEL et al. (1994)
Ad3/B1	9–11	1–43	43	44–133	90	6	134–320	186	SIGNÄS et al. (1985)
Ad11p/B2	9–11	1–43	43	44–134	91	6	134–325	191	MEI and WADELL et al. (1993)
Ad2/C	23–31	1–44	44	45–400	356	22	401–582	181	HERISSE and GALIBERT et al. (1981)
Ad8/D	13	1–43	43	44–182	139	8	183–362	180	PRING-ÅKERBLOM and ADRIAN (1995)
Ad4/E	17	1–45	45	46–248	203	13	249–426	177	GRUBER et al. (1993)
Ad40-1/F	32	1–42	42	43–373	331	21	374–543	174	KIDD and ERASMUS (1989)
Ad40-2/F	25	1–42	42	43–233	191	12	234–387	154	KIDD et al. (1993)

Ad, Adenovirus.

Table 5. Comparison of the fibers of the different subgenus B adenoviruses

	Ad3	Ad7p	Ad11p	Ad11a	Ad34a	Ad35p
Ad3		60.1	60.3	59.0	59.8	60.3
Ad7p	65.0		93.2	90.8	93.3	63.4
Ad11p	65.3	94.3		92.3	98.8	62.7
Ad11a	64.9	92.9	94.4		92.3	63.4
Ad34a	65.0	94.2	99.5	94.3		62.4
Ad35p	62.7	69.4	69.7	70.2	70.3	

The amino acid homology (%) is compared at the upper right, whereas the nucleotide homology (%) is presented at the lower left.

Table 6. Nucleotide and predicted amino acid sequence homology for the subgenus B2 adenovirus fiber genes

	11p/11a	11p/35p	11p/34a	11a/35p	11a/34a	35p/34a
Comparative nucleotide homology						
Fiber gene(%)	94.4	69.7	99.5	70.2	94.3	70.2
Tail(%)	98.5	99.2	98.5	99.2	98.5	99.2
Shaft(%)	90.7	76.6	99.3	78.8	90.1	78.8
Knob(%)	95.3	59.8	99.8	59.7	95.3	59.7
Comparative amino acid homology						
fiber protein (%)	92.3	62.7	98.8	63.4	92.3	62.4
Tail(%)	97.7	97.7	95.3	100	97.7	97.7
Shaft(%)	89.7	72.5	98.9	75.8	89.0	72.5
Knob(%)	92.4	50.0	99.5	48.9	92.7	49.5

Regions of the polypeptide sequences were designated tail-, shaft-, and knob-coding regions, according to the secondary structure model of Green and coworkers (1983).

6.2 Amino Acid Sequence of the Fibers of the Four Subgenus B2 Members

The amino acid sequence of the fiber of Ad11p and Ad34a was highly similar (Mei and Wadell 1993, 1995). Only four amino acid residues differed: two in the tail, one in the shaft, and one in the knob region. In this case the Ad11p Arg-266 was altered to His-266 in Ad34a. The subgenus B2 fiber is formed by a tail, six 15-residue repetitive motives making up the shaft region and the knob. Green et al. (1983) have designated the 15 amino acids composing the motive according to their position. At the position a5 the two proline residues and the three glycine residues were absolutely conserved between the four subgenus B2 fiber shafts (Fig.2). The leucine situated at position B1 was also present in all four fibers with the exception of motive 3 of Ad35p, where the leucine residue was substituted for isoleucine. Motives 3 and 4 accounted for 15 of the 25 mismatches, which were primarily due to the difference between the fiber gene of Ad35 and the other fiber genes. Among the first ten N-terminal amino acid residues of the knob, three prolines were detected in Ad35p and in representative serotypes of subgenus

```
       Ad11p   1   MTKRVRLSDSFNPVYPYEDESTSQHPFINPGFISPNGFTQSPN   43
       Ad34a   1   ---------------------------------------L--D   43
TAIL   Ad11a   1   ------------------------------------------D   43
       Ad35p   1   ------------------------------------------D   43

       Ad11p  44   GVLTLKCLTPLTTTG     GSLQLKVGGGLTVDDTNGF   77
       Ad34a  44   ---------------     -------------------   77
       Ad11a  44   ---------------     ----------------D-T   77
       Ad35p  44   ---------------     ----------------D-T   77
                      motif 1               motif 2

       Ad11p  78   LKENISATTPLVKTG     HSIGLPLGACLGTNE   107
       Ad34a  78   --------------A     ---------------   107
SHAFT  Ad11a  78   -Q---GT--------     -----S-------D-   107
       Ad35p  78   -Q---R--A-IT-NN     --VE-SI-N--E-QN   107
                      motif 3             motif 4

       Ad11p 108   NKLCIKLGQGLTFNSN    NICIDDNINTL   134
       Ad34a 108   ----------------    -----------   134
       Ad11a 108   ----T---K-------    -----------   134
       Ad35p 108   ----A---N--K--NG    D---K-S----   134
                      motif 5            motif 6

Ad11p 135   WTGVNPTEANCQIMNSSESNDCKLILTLVKTGALVTAFVYVIGVSNNFNM   184
Ad34a 135   --------------------------------------------------   184
Ad11a 135   ---I--------M-D-----------------------------------   184
Ad35p 135   ---I--P*P----VENTNT--G--T-V---N-G--NGY-SLV---DTV-Q   183

Ad11p 185   LTTHRNINFTAELFFDSTGNLLTRLSSLKTPLNHKSGQNMATGAITNAKG   234
Ad34a 185   --------------------------------------------------   234
Ad11a 185   ---Y-------------A-----S-------------------------S   234
Ad35p 184   MF-QKTA-IQLR-Y---S-----EE-D--I--KN--S*TATSETVASS-A   232
KNOB
Ad11p 235   FMPSTTAYPFNDNSREKENYIYGTCYYTAS*DRTAFPIDISVMLNRRAIN   283
Ad34a 235   ------------------------------*-H-----------------   283
Ad11a 235   -----------N--------------H----*-H------------Q---R   283
Ad35p 233   -----------TTT-DS----H-I---MT-Y--SL--LN--I---S-M-S   282

Ad11p 284   DETSYCIRITWSWNTGDAPEVQTSATTLVTSPFTFYYIREDD    325
Ad34a 284   ------------------------------------------    325
Ad11a 284   AD------------------G---------------------    325
Ad35p 283   SNVA-A-QFE-NL-ASES--**SNIA--T----F-S--T---N   323
```

Fig. 2. Direct comparison of the amino acid sequences of Ad11p, Ad34a, Ad11a, and Ad35p fiber polypeptides. Each of the four sequences consist of a short N-terminal tail separated from a distal knob by a repetitive sequence which forms the shaft. *Asterisks* denote amino acid deletions. This scheme shows six motifs in the shaft of the four fibers, based on the position of the hydrophobic amino acids in the motif. The putative N-linked glycosylation sites in the predicted amino acid are indicated by *boxes*. At the end of the Ad35p C terminus, the REDD sequence is followed by N, which is an additional amino acid not found in the other fibers

A,C,D, and E. However, in this region of the fiber knob of Ad7, Ad11p, and Ad11a, only the proline in the central position was conserved.

The amino acid homology between Ad35p and the other adenoviruses of subgenus B2 was only 50%. This heterology was distributed throughout the knob region with the exception of the amino acid sequence Phe-Met-Pro-Ser-Thr-Thr-Ala-Tyr-Pro-Phe-Asn occupying positions 233–243, which is absolutely conserved in all analyzed adenoviruses belonging to subgenus B.

The fiber knobs of Ad11p and Ad34a, which cause acute hemorrhagic cystitis (Hierholzer et al. 1975), manifested only one amino acid mismatch in the fiber knob: the Arg-266 in Ad11p was replaced by His-266 in Ad34a. Two, three, and five potential glycosylation sites were identified in Ad11a, Ad11p/Ad34a, and Ad35p, respectively. In the Ad35p fiber, an asparagine constituted the C terminus, unlike all other known subgenus B adenoviruses (Mei and Wadell, 1995). Ad11p, Ad34a, and Ad35 can hemagglutinate erythrocytes from vervet or rhesus monkey; however, Ad11a does not display this activity. Interestingly, 15 amino acid mis-matches were observed in the knob region between Ad11p and Ad11a (Mei and Wadell 1993). Among these only two amino acids, Tyr-260 and Arg-266, were identical in Ad35p and Ad34a fibers, both displaying hemagglutinating activity. The knob region of the Ad35p fiber is markedly different from the other fibers. It was suggested that Ad11p/Ad34a, Ad11a, and Ad35p can infect different host cells. This is compatible with the observation that Ad2 and Ad3 have been shown to bind to different host cell receptors (Defer et al. 1990).

7 Biochemical and Biological Properties of the Knob Region of Fibers from Different Subgenera

The adenoviruses have been divided into different subgenera on the basis of differential hemagglutination properties of monkey or rat erythrocytes (Rosen 1960; Hierholzer 1973; deJong et al. 1983). Apparently different receptors are associated with the adenovirus ligands, since the numbers of receptors have been demonstrated to vary between adenoviruses of different subgenera. Rat erythrocytes carry a higher number of receptors for members of subgenus D than for members of subgenus C (Wadell 1969), whereas monkey erythrocytes carry five times more receptors for Ad3 (subgenus B) than rat erythrocytes displayed for Ad2 (subgenus C) (Inouye and Norrby 1973).

A strikingly wide distribution of isoelectric points ranging over 5 logs was noted when the knob proteins of fibers from different subgenera were compared (Table 7). The receptor-seeking moiety of the adenovirus particle resides in the adenovirus knob (Henry et al. 1994). Consequently, the charge of the knob may play a decisive role for the tropism of adenoviruses and also for the possibility of approaching putative receptors on the erythrocytes. Interestingly, the knob region of fibers that could agglutinate monkey erythrocytes displayed isoelectric points from 4.5 to 5.6; in contrast, the adenoviruses belonging to subgenus A and F that display a hardly discernable agglutination of rat erythrocytes were characterized by isoelectric points on the knob regions of the fibers ranging from 7.90 to 8.35. It should be noted that entire capsid of enteric adenoviruses (subgenus F) is

Table 7. Biochemical and biological properties of the knob region of the fibers from different serotypes or genome types

Serotype	Knob residues	Charged	Isoelectric point	HA Pattern	Tropisms/ symptoms
Ad12/A	177	1	7.90	Rat[a]	Cryptic enteric infection
Ad3/B1	186	–3	5.28	Monkey	Respiratory disease
Ad7p/B1	191	–3	5.66	Monkey	Respiratory disease
Ad11p/B2	191	–3	5.26	Monkey	Persistent infections of the kidney
Ad11a/B2	191	–4	5.21	–	Acute respiratory disease
Ad34a/B2	191	–4	5.24	Monkey	Persistent infections of the kidney
Ad35p/B2	189	–6	4.50	Monkey	Persistent infections of the kidney
Ad2/C	181	–1	6.76	Rat incomplete	Respiratory disease, persists in lymphoid tissue
Ad5/C	180	–2	6.25	Rat incomplete	Respiratory disease, persists, in lymphoid tissue
Ad8/D	180	5	9.57	Rat complete	Keratitis
Ad9/D	180	0	6.56	Rat complete	Conjunctivitis
Ad4/E	177	–1	5.80	Rat incomplete	Conjunctivitis respiratory disease
Ad40-1/F	174	1	7.89	Rat[a]	Infantile diarrhea
Ad40-2/F	154	–1	6.49		
Ad41-1/F	174	2	8.35	Rat[a]	[a]Infantile diarrhea
Ad41-2TAK/f	154	0	7.37		

The references for Ad12, Ad3, Ad11p, Ad2, Ad8, Ad4, Ad40-1, and Ad40-2 fiber sequences are shown in Table 4. The other fiber sequences were obtained as follows: Ad7p form Hong et al. (1988); Ad11a, Ad34a, and Ad35p from Mei and Wadell (1993, 1995); Ad5 from Chroboczek and Jacrot (1987); Ad9 from Pring-Akerblom and Adrian (1995); Ad41-1 and Ad41-2 from Pieniazek et al. (1989, 1990). The isoelectric points were calculated on a Vax computer using a program for the European Molecular Biology Laboratory (EMBL) nucleic acid sequence data base.
[a]Hardly discernable aggregation of rat erythrocytes.

drastically less negatively charged compared to Ad2 virions. This is due to a deletion of 33 amino acids from each hexon peptide of Ad40 compared to the prototype Ad2. Four of the amino acid residues had basic and 21 acidic charge. The adenovirus capsid is formed by 720 hexon peptides.

The two representative serotypes of subgenus D, Ad8 and Ad9, display complete agglutination of both rat and human erythrocytes. These properties are unique to the members of subgenus D. However, Ad8 agglutinates erythrocytes from dog, chicken, and goose, whereas Ad9 lacks this property (HIERHOLZER 1973). Furthermore, Ad8 has a predilection for the corneal stroma, causing keratitis, whereas Ad9 infections are limited to conjunctivitis. These differences may reflect the conspicous difference in isoelectric points (9.57 and 6.56) between the fiber knobs of Ad8 and Ad9, respectively (Table 7).

8 A Strategy for Identification of the Part of the Fiber Knob that Interacts with the Erythrocyte Receptor

We have previously noted that Ad11p virions that can agglutinate monkey erythrocytes, whereas the closely related Ad11a does not display this capacity (MEI and WADELL 1993). The fiber polypeptides were expressed in *Escherichia coli.* The subsequent purification included refolding, Q-sepharose chromatography, and blue sepharose chromatography. The Ad11p and Ad11a recombinant fibers (rfibers) could be identified as trimers by denaturing and nondenaturing gels. It was surprising that the renatured Ad11p fibers were also found to display hemagglutinating activity, indicating that they aggregated to multivalent forms. This hemagglutination could be inhibited by type-specific sera against virions and dodecons. It has been reported that expressed Ad5 rfiber from *E. coli* can block the virus infection by binding to the cell receptor (HENRY et al. 1994). Since the rfiber of Ad11p from *E. coli* manifested hemagglutinating activity, we were able to perform site-specific mutagenesis and investigate the erythrocyte-binding site on the fiber. The agglutination was sensitive to chymotrypsin or trypsin digestion with a demonstrated cleavage site in the shaft region on the fiber. Chimeras composed of preselected portions of Ad11p rfibers and Ad11a rfibers were created, expressed, purified, and analyzed in terms of their hemagglutinating activity. The C12–3 construct contained the N-terminal portion of Ad11p. From amino acid residue 202 to the C-terminal end, the knob portion was represented by Ad11a.

The C4-1 construct contained the Ad11p tail, shaft, and N-terminal portion of the knob, whereas the Ad11a genome made up the portion from residue 260 towards the C-terminal end.

Construct C20 was based on the Ad11a tail, shaft, and N-terminal portion of the knob up to amino acid residue 280. Consequently, the C-terminal part of the knob was represented by the Ad11p genome (Fig. 3). All these constructs with the exception of the control Ad11p rfiber were devoid of hemagglutination capacity. The previous experiment indicated that a short stretch of amino acids from residue 260 to 280 was indispensible for the association of the adenovirus 11p fiber to the erythrocytes (Fig. 4).

Site-directed mutagenesis of individual amino acids was applied to pinpoint which amino acids could be critically responsible for the interaction between the fiber knob and the erythrocytes. First, the amino acids outside the region, Thr-202

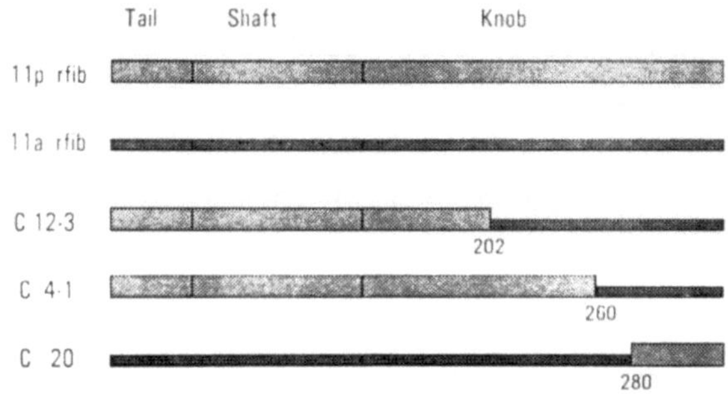

Fig. 3. The design of fragment replacement for construction of chimeric recombinant fiber (rfiber) proteins. The adenovirus fiber gene encodes a 325-amino acid protein which was designated as tail, shaft, and knob regions. These replacements were numbered according to their amino acid position from the N to C terminus of the fiber protein

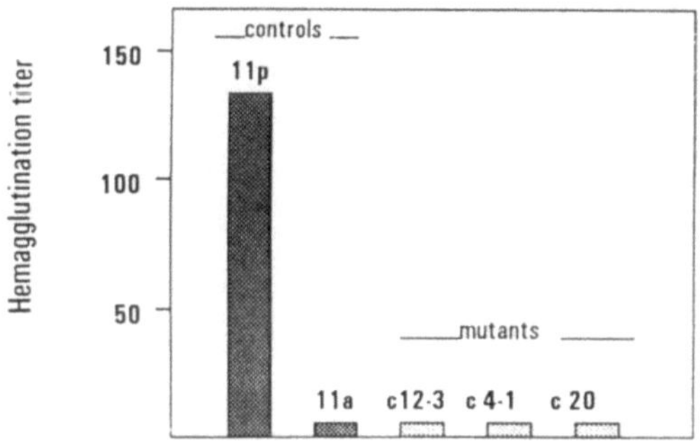

Fig. 4. The hemagglutination capacity of the different mutated recombinant fiber (rfiber) protein

and Asp-284, were mutated to alanine in Ad11p fiber knobs. These mutations had no effect on the hemagglutination titer.

However, mutations of Tyr-260 to the corresponding His-260, Arg-279 to the corresponding Gln-279, or Asn-283 of Ad11p to the corresponding Arg-283 in Ad11a all abolished the hemagglutination activity (Figs. 5,6). The experiments were also performed with Ad34a recombinant fibers. Ad34a is very closely related to Ad11p with regards to both biological properties and fiber sequence. All the results could be confirmed with the exception of two clones of the same Ad34a recombinant fiber mutant in which Asn-284 was mutated to the corresponding Halanine residue in Ad11a. In this case the hemagglutinating property was eliminated. These alterations may appear limited; however, only a single amino acid substitution in the hemagglutinin of Newcastle disease virus inactivated the hemagglutination capacity (Sheehan and Lorio 1992).

In conclusion, a region from residue 260 to 280 has been demonstrated to be of critical importance for the association between the adenovirus fiber and the erythrocytes. It was further demonstrated that each of the residues Tyr-260, Arg-279, and Glu-283 were individually critically responsible for the interaction between the fiber knob and the erythrocyte receptor.

```
Ad11p 185   LTTHRNINFTAELFFDSTGNLLTRLSSLKTPLNHKSGQNMATGAITNAKG   234
Ad34a 185   --------------------------------------------------   234
Ad11a 185   ---Y-------------A-----S--------------------------S   234

Ad11p 235   FMPSTTAYPFNDNSREKENYIYGTCYYTAS*DRTAFPIDISVMLNRRAIN   283
Ad34a 235   ------------------------------*-H-----------------   283
Ad11a 235   -----------N-------------H----*-H-----------Q---R   283

Ad11p 284   DETSYCIRITWSWNTGDAPEVQTSATTLVTSPFTFYYIREDD   325
Ad34a 284   ------------------------------------------   325
Ad11a 284   AD---------------G------------------------   325
```

Fig. 5. Location of mutations introduced into recombinant fiber (rfiber) protein. The point mutations were numbered according to their amino acid position within the fiber protein

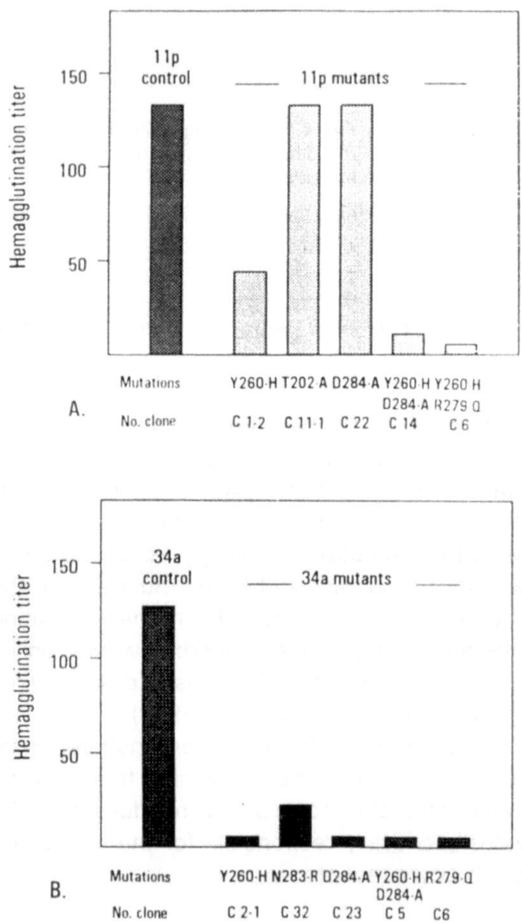

Fig. 6A,B. The hemagglutination activity of mutated AdIIp and Ad34a recombinant fibers (rfibers)

References

Allard A, Albinsson B, Wadell G (1992) Detection of adenoviruses in stools from healthy persons and patients with diarrhea by two-step polymerase chain reaction. J Med Virol 37: 149-157

Brown M, Wilson-Friesen HL, Doane F (1992) A block in release of progeny virus and a high particle-to-infectious unit ratio contribute to poor growth of enteric adenovirus types 40 and 41 in cell culture. J Virol 66: 3198-3205

Carleen CR, Tollefson AE, Brady HA, Hoffman BL, Wold WSM (1989) Epidermal growth factor receptor is down-regulated by a 10,400 MW protein encoded by the E3 region of adenovirus. Cell 57: 135-144

Chroboczek J, Jacrot B (1987) The sequence of adenovirus fiber: similarities and differences between serotypes 2 and 5. Virology 161: 549–554

Dahllöf B, Wallin M, Kvist S (1991) The endoplasmic reticulum retention signal of the E3/19K protein of adenovirus-2 is microtubule binding. J Biol Chem 266: 1804–1808

Defer C, Belin M-T, Caillet-Boudin M-L, Boulanger P (1990) Human adenovirus-host cell interactions: comparative study with members of subgroups B and C. J Virol 64: 3661–3673

deJong JC, Wigand R, Kidd AH, Wadell G, Kapsenberg J, Muzerie CJ, Wermenbol AG, Firtzlaff RG (1983) Candidate adenoviruses 40 and 41 fastidious adenoviruses from human infant stool. J Med Virol 11: 215–231

Flomenberg P, Chen M, Horwitz MS (1987) Characterization of a major histocompatibility complex class I antigen-binding glycoprotein from adenovirus type 35, a type associated with immuno-compromised hosts. J Virol 61: 3665–3671

Flomenberg PR, Chen M, Horwitz MS (1988) Sequence and genetic organization of adenovirus type 35 early region. J Virol 62: 4431–4437

Flomenberg P, Szmulewicz J, Gutierrez E, Lupatkin H (1992) Role of the adenovirus E3-19k conserved region in binding major histocompatibility complex class I molecules. J virol 66: 4778–4783

Fox JP, Hall CE, Cooney MK (1977) The Seattle Virus Watch. VII. Observations on adenovirus infections. Am J Epidemiol 105: 362–386

Gooding LR, Ranheim TS, Tollefson AE, Aquino L, Duerksen-Hughes P, Hortin TM, Wold WSM (1991) The 14,700- and 14,500-dalton proteins encoded by region E3 of adenovirus function together to protect many but not all mouse cell lines against lysis by tumor necrosis factor. J Virol 65: 4114–4124

Green NM, Wrigley NG, Russell WC, Martin SR, McLachlan AD (1983) Evidence for a repeating cross-β sheet structure in the adenovirus fiber. EMBO J 2: 1357–1365

Gruber WC, Russell DJ, Tibbetts C (1993) Fiber gene and genomic origin of human adenovirus type 4. Virology 196: 603–611

Henry LJ, Xia D, Wilke M, Deisenhofer J, Gerard RD (1994) Characterization of the knob domain of the adenovirus type 5 fiber protein expressed in Escherichia coli. J Virol 68: 5239–5246

Herissé J, Galibert F (1981) Nucleotide sequence of the EcoRI fragment of the adenovirus 2 genome. Nucleic Acid Res 9: 1229–1240

Hermiston TW, Hellwig R, Hierholzer JC, Wold WSM (1993) Sequence and functional analysis of the human adenovirus type 7 E3-gp 19K protein from 17 clinical isolates. Virology 197: 593–600

Hierholzer JC (1973) Further subgrouping of the human adenoviruses by differential hemagglutination. J Infect Dis 128: 541–550

Hierholzer JC (1992) Adenoviruses in the immunocompromised host. Clin Microbiol Rev 5: 262–274

Hierholzer JC, Pumarola A, Rodriguez-Torres A, Beltran M (1974) Occurrence of respiratory illness due to an atypical strain of adenovirus type 11 during a large outbreak in Spanish military recruits. Am J Epidemiol 99: 434–443

Hierholzer JC, Atuk NO, Gwaltney JM (1975) New human adenovirus isolated from a renal transplant recipient: description and characterization of candidate adenovirus type 34. J Clin Microbiol 1: 366–376

Hong JS, Mullis KG, Engler JA (1988) Characterization of the early region 3 and fiber genes of Ad7. Virology 167: 545–553

Inouye S, Norrby E (1973) Kinetics of the attachment of radioisotope-labelled adenovirus hemagglutinins to red blood cells and their detachment by antibody and by nonlabelled hemagglutinins of homologous and heterologous types. Arch Gesamte Virusforsch 42: 388–398

Johansson ME, Brown M, Hierholzer JC, Thörner A, Ushijima H, Wadell G (1991) Genome analysis of adenovirus type 31 strains from immunocompromised and immunocompetent patients. J Infect Dis 163: 293–299

Kidd AH, Erasmus MJ (1989) Sequence characterization of the adenovirus 40 fiber gene. Virology 172: 134–144

Kidd AH, Erasmus MJ, Tiemessen CT (1990) Fiber sequence heterogeneity in subgroup F adenoviruses. Virology 179: 139–150
Kidd A, Chroboczek J, Cusack S, Ruigrok RWH (1993) Adenovirus type 40 virions contain two distinct fibers. Virology 192: 73–84
Lecatsas G, Prozesky OW, van Wyk J (1974) Adenovirus 11 associated with haemorrhagic cystitis after renal transplantation. S Afr Med J 48: 1932
Li Q-G, Wadell G (1986) Analysis of 15 different genome types of adenovirus type 7 isolated on five continents. J Virol 60: 331–335
Li Q-G, Hambraeus J, Wadell G (1991) Genetic relationship between thirteen genome types of adenovirus 11, 34, and 35 with different tropisms. Intervirology 32: 338–350
Mei YF, Wadell G (1992) The nucleotide sequence of adenovirus type 11 early 3 region: comparison of genome type Ad11p and Ad11a. Virology 191: 125–133
Mei YF, Wadell G (1993) Hemagglutination properties and nucleotide sequence analysis of the fiber gene of adenovirus genome types 11p and 11a. Virology 194: 453–462
Mei YF, Wadell G (1995) Highly heterogeneous fiber genes in the two closely related adenovirus genome types Ad35p and Ad34a. Virology 206: 686–689
Murtagh P, Cerqueizo C, Halac A, Avila M, Kajon A (1993) Adenovirus type 7h respiratory infections: a report of 29 cases of acute lower respiratory disease. Acta Paediatr 82: 557–561
Numazaki Y, Shigeta S, Kumasaka T, Miyazawa T, Mamanaka N, Yano M, Takai S, Ishida N (1968) Acute hemorrhagic cystitis in children, isolation of adenovirus. N Engl J Med 278: 700–704
Pääbo S, Weber F, Kämpe O, Schaffner W, Peterson RA (1983) Association between transplantation antigens and a viral membrane protein synthesized from a mammalian expression vector. Cell 35: 445–453
Pieniazek NJ, Slemenda SB, Pieniazek D, Velarde J, Luftig RB (1989) Sequence of human enteric adenovirus type 41 Tak fiber protein gene. Nucleic Acid Res 17: 9474
Pieniazek NJ, Slemenda SB, Pieniazek D, Velarde J, Luftig RB (1990) Human enteric adenovirus type 41 (Tak) contains a second fiber protein gene. Nucleic Acid Res 18: 1901
Pring-Åkerblom P, Adrian T (1995) Characterization of adenovirus subgenus D fiber genes. Virology 206: 564–571
Rosen L (1960) A hemagglutination-inhibition technique for typing adenoviruses. Am J Hyg 71: 120–128
Schnurr D, Dondera ME (1993) Two new candidate adenovirus serotypes. Intervirology 36: 79–83
Sester M, Burgert H-G (1994) Conserved cysteine residues within the E3/19k protein of adenovirus type 2 are essential for binding to major histocompatibility complex antigens. J Virol 68: 5423–5432
Sheehan JP, Lorio RM (1992) A single amino acid substitution in the hemagglutinin-neuraminidase of Newcastle disease virus in a protein deficient in both functions. Virology 189: 778–781
Signäs C, Akusjärvi, G, Pettersson U (1985) Adenovirus 3 fiber polypeptide gene: implications for the structure of the fiber protein. J Virol 53: 672–678
Signäs C, Akusjärvi G, Pettersson U (1986) Region E3 of human adenovirus differences between the oncogenic adenovirus-3 and the non-oncogenic adenovirus-2. Gene 50: 173–184
Sprengel J, Schmitz B, Heuss-Neitzel D, Zock C, Doerfler W (1994) Nucleotide sequence of human adenovirus type 12 DNA comparative functional analysis. J Virol 68: 379–389
Tiemessen CT, Kidd AH (1994) Adenovirus type 40 and 41 growth in vitro: host range diversity reflected by differences in patterns of DNA replication. J Virol 68: 1239–1244
Toogood C, Hay R (1988) DNA sequence of the adenovirus type 41 hexon gene and predicted structure of the protein. J Gen Virol 69: 2291–2301
Uhnoo I, Wadell G, Svensson L, Johansson ME (1984) Importance of enteric adenoviruses 40 and 41 in acute gastroenteritis in infants and young children. J Clin Microbiol 20: 365–372
von Heijne G (1985) Structural and thermodynamic aspects of the transfer of proteins into and across membrane In: Knauf P, Cook JS, (eds) Current topics in membranes and transport: membrane protein biosynthesis and turnover, vol 24. Academic, New York, pp 151–179
Wadell G (1969) Hemagglutination with adenovirus serotypes belonging to Rosen's subgroup II and III. Proc Soc Exp Biol Med 132: 413–421
Wadell G (1984) Molecular epidemiology of adenoviruses. Curr Top Microbiol Immunol 110: 191–220
Wadell G (1990) Adenoviruses In: Zuckerman AJ, Banatvala JE, Pattison JR (eds) (Principles and practice of clinical virology)Wiley, New york, pp 267–297
Wadell G, Hammarskjöld ML, Winberg G, Varsanyi T, Sundell G (1980)Genetic variability of adenoviruses. Ann NY Acad Sci 354: 16–42
Zhang ZJ, Chen YZ, Gao LM, Wang ZL, Zhao JM (1984) Symtomatology of adenovirus pneumonia type 11. Acta Acad Med Wuhan 4: 241–244

Enteric Adenoviruses

V. MAUTNER[1], V. STEINTHORSDOTTIR[2], and A. BAILEY[1]

1 Introduction

Most of the human adenoviruses have been associated with infections of the respiratory tract, but it has long been recognised that adenoviruses can be shed

[1]Medical Research Council Virology Unit, Institute of Virology, University of Glasgow, Scotland, UK
[2]*Present address*: Institute for Experimental Pathology, University of Iceland at Keldur, Reykjavik, Iceland

via the gastrointestinal tract and that a proportion of these would not replicate in the cell lines routinely used for adenovirus isolation and propagation. This led to the discovery of the enteric or fastidious adenoviruses, which were recognised initially by the large numbers seen in faecal extracts, that were not amenable to cultivation. Extensive characterisation of these viruses has led to the classification of two serotypes, adenoviruses type 40 (Ad40) and type 41 (Ad41), comprising subgroup F. The purpose of this chapter is to provide an overview of these viruses and to discuss in detail our current understanding of their molecular properties, particularly in the context of the recently established complete genome sequence of Ad40 (Davison et al. 1993). Other recent reviews include those by Albert (1986), Horwitz (1990), Wadell et al. (1987, 1994).

2 General Background

Viral gastroenteritis in children can be caused by several different pathogens including adenovirus, which is second only to rotavirus as a causal agent (Estes et al. 1983; Uhnoo et al. 1984; Brandt et al. 1985). A close correlation has now been established between Ad40 and Ad41 and gastroenteritis (Brandt et al. 1979; de Jong et al. 1983; Yolken et al. 1982; Chiba et al. 1983; Uhnoo et al. 1983, 1984; Leite et al. 1985; Kidd et al. 1986; Kotloff et al. 1989). However, many other adenoviruses are shed via the gut, even if this is not the primary site of infection; these include the common respiratory serotypes (e.g. subgroup C serotypes 1, 2, and 5; Fox et al. 1977). Others are predominantly isolated from the intestinal tract (e.g. subgroup A serotypes 12 and 31), but have not been identified as a major causative agent of enteric disese (Schmitz et al. 1983; Adrian and Wigand 1989). However, it has been reported that the clinical features of Ad31 enteric infection are indistinguishable from subgroup F virus infection (Krajden et al. 1990). The live vaccines for Ad4 and Ad7 are delivered by the enteric route, specifically to bypass the respiratory tract, and are capable of limited replication in the gut, sufficient for shed virus to be detected and for a systemic immune response to be elicited (Top et al. 1971).

3 Identification and Classification

The enteric adenoviruses were first detected by electron microscopy on stool specimens from infants with acute gastroenteritis (Flewett et al. 1975). In general, enteric adenovirus isolates do not grow in cells routinely used for the propagation of other adenoviruses (e.g. primary human embryo kidney cells, KB, HeLa, Hep-2, WI-38) but do replicate in 293 cells (HEK cells transformed with the Ad5 E1 region: Graham et al. 1977; Takiff et al. 1981). Kidd and Madeley (1981) found that some but not all enteric adenovirus specimens could be passaged in Chang conjuctiva cells. Moreover, serological data (JOHANSSON et al. 1980; KIDD and MADELEY 1981) indicated that, irrespective of infectivity in Chang cell cultures, different isolates of enteric adenoviruses were antigenically related and distinct from the (then) 35 established adenovirus serotypes. The presence of two serotypes of enteric adenoviruses was revealed by studies on large numbers of clinical isolates (DE JONG et al. 1983; UHNOO et al. 1983). The two types are indistinguishable in haemagglutination inhibition tests, but neutralisation tests distinguish between them. Restriction endonuclease analysis shows different DNA restriction patterns for Ad40 and Ad41 (UHNOO et al. 1983; TAKIFF et al. 1984; ADRIAN et al. 1986). The two serotypes are now classified as a separate subgroup, F, based on several criteria including immunological cross-reactivity, DNA homology, and size of internal structural polypeptides (VAN LOON et al. 1985b; WADELL et al. 1987; HIERHOLZER et al. 1988). Variants of both serotypes have been recognised by restriction endonuclease analysis (KIDD 1984; KIDD et al. 1984; VAN DER AVOORT et al. 1989), and a classification scheme based on this variation has been presented (VAN DER AVOORT et al. 1989). Eleven variants of Ad40 and 28 variants of Ad41 are already recognised (cited by KIDD et al. 1993). The establishment of the complete sequence for Ad40 strain Dugan (DAVISON et al. 1993) should help to identify and classify Ad40 variants in the future.

4 Epidemiology

Enteric adenoviruses have been detected in the stools of young children and infants with acute gastroenteritis in studies conducted in industrialised and developing countries (reviewed by UHNOO et al. 1990; WADELL et al. 1994). Reports from Africa (KIDD et al. 1986; TIEMESSEN et al. 1989), Asia (HERMANN et al. 1988; KIM et al. 1990; SHINOZAKI et al. 1991; JARECKIKAHN and UNICOMB 1992; JARECKI KAHN et al. 1993), Europe (UHNOO et al. 1984; ALLARD et al. 1992; DE JONG et al. 1993; NOEL et al. 1994), the Middle East (SETHI et al. 1989; AKHTER and QADRI 1993) and the Americas (BRANDT et al. 1985; LEITE et al. 1985; RODRIGUEZ et al 1985; KOTLOFF et al. 1988, 1989; BROWN 1990; CRUZ et al. 1990; LEW et al. 1991; MISTCHENKO et al. 1992; VAN et al. 1992; REGUA-MANGIA et al. 1993) find enteric adenoviruses to be associated with 5%–20% of cases of diarrheal disease. These viruses are prevalent throughout the year with little seasonal variation, indicating that they may be endemic (DE JONG et al. 1983, 1993; UHNOO et al. 1984; BRANDT et al. 1985; JOHANSSON et al. 1985; TIEMESSEN et al. 1989). Until recently, Ad40 and Ad41 had been isolated with approximately equal frequency (DE JONG et al. 1983; JOHANSSON et al. 1985). More recent surveys have found the prevalence of Ad40 to be declining, whereas Ad41 is increasing; concordant data has been obtained in different geographical locations (UK: WILLCOCKS et al. 1988; Canada: BROWN 1990; Japan: SHINOZAKI et al. 1991; The Netherlands: DE JONG et al. 1993).

Seroepidemiology (Kidd et al. 1983; Shinozaki et al. 1987) has confirmed the high level of exposure to the enteric adenoviruses among young children on a worldwide basis.

5 Pathogenesis

5.1 Diarrheal Disease

Clinical features of enteric adenovirus infection were described by Uhnoo et al. (1984, 1986, 1990). After an incubation period of 7–8 days, infected children had a clinically moderate disease with diarrhea and vomiting. Elevated temperatures were rare and less than 20% of patients had respiratory symptoms. The two serotypes caused similar symptoms, although Ad41-infected children had more protracted diarrhea (mean, 12.2 days) than those infected with Ad40 (mean, 8.6 days). These studies showed that the enteric adenoviruses cause a more prolonged but milder infection with a lower frequency of vomiting and elevated temperatures than do rotaviruses. A study in the USA has shown that dehydration is as likely to develop in children with enteric adenovirus diarrhea as among those with rotavirus diarrhea (Kotloff et al. 1989). Moreover, enteric adenoviruses have been associated with fatal disease (Whitelaw et al. 1977; Johansson et al. 1985) and identified as a cause of morbidity in infants undergoing gastro-intestinal surgery for necrotising enterocolitis (Yolken and Franklin 1985).

5.2 Celiac Disease

It has been proposed that infection with Ad12 might be a factor in celiac disease (reviewed by Kagnoff 1989; Howdle and Blair 1992, 1994), based on the sequence similarity (eight out of 12 identical amino acids) and immunological cross-reactivity of a peptide in the E1B 55K protein and a peptide of the wheat protein A-gliadin, known to be involved in the pathology of the disease. This proposal is supported by the finding of a correlation with the incidence of neutralising antibodies to Ad12 as an indication of past exposure (Kagnoff et al. 1984, 1987; Mantzaris et al. 1990; Mantzaris and Jewell 1991; Arató et al. 1991). It has recently been claimed that Ad40 may also be implicated in celiac disease and dermatitis herpetiformis (Lädheaho et al. 1993a, b) but specific evidence of Ad40 neutralising antibody was lacking in this study. Furthermore, whereas the "homologous" region of Ad12 has eight out of 12 amino acids identical to the A-gliadin peptide sequence, Ad40 has only five out of 12, and these are shared by Ad5 and other non-enteric adenoviruses. Other reports on celiac disease and adenoviruses have been at variance with some of these findings (Devery et al. 1991; Dieleman et al. 1991; Ellis et al. 1992; Handley et al. 1993), and there is no evidence for increased persistence of Ad12 in celiac patients (Mahon et al. 1991);

the significance of a possible role for enteric viruses in the pathogenesis of celiac disease remains to be unequivocally established.

5.3 Intussusception

The association of adenoviruses with intestinal intussusception (prolapse of one section of the bowel into another) has been recognised (see for example PORTER et al. 1993; MONTGOMERY and POPEK 1994). While there is a clear correlation with non-enteric serotypes, no such relationship has yet been established for the subgroup F viruses (BHISITKUL et al. 1992).

5.4 Infections of Immunocompromised Individuals

Although adenoviruses are frequently isolated from immunosuppressed patients, there is no indication that the enteric viruses are particularly implicated (HIERHOLZER 1992; GAZZARD and BLANSHARD 1993). In one study of human immunodeficiency virus (HIV)-positive patients, adenovirus infection of the large bowel but not the small intestine has been observed, but it remains to be ascertained whether this was caused by a subgroup F virus (MADDOX et al. 1992). Another study of acquired immunodeficiency syndrome (AIDS) patients in Zaire found a weak association of increased shedding of a range of enteric viruses with immunodeficiency, but no strong correlation of any particular enteric virus infection with AIDS-associated diarrhea (THEA et al. 1993).

6 Adenovirus Type 40 DNA Sequence and Sequence Comparisons with Other Serotypes

The complete sequence of the Ad40 strain Dugan (ATCC VR-931) was obtained using random fragments of purified virion DNA cloned into bacteriophage M13mp19 as the source material. The final sequence (34 214 bp; Genbase accession number L19443; DAVISON et al. 1993) reveals a genome similar in most respects to the established genomes of Ad2 and Ad5 (R.J. ROBERTS et al. 1986; CHROBOCZEK et al. 1992), but with some significant differences. The recent publication of the complete Ad12 sequence (34 125bp; EMBL accession number X73487; SPRENGEL et al. 1994) also permits a more extensive comparison across subgroups and reveals that Ad40 and Ad12 are phylogenetically more closely related than has hitherto been appreciated. Overall DNA homology determined by liquid phase hybridisation is 62%–69% between Ad40 and Ad41, and 15%–22% between the F subgroup viruses and Ad5 (VAN LOON et al. 1985b). The homology within subgroup F is lower than seen within subgroups B–E, but similar to subgroup A; the inter-subgroup homology is in the same range as seen for other subgroups. Calculation of DNA homology from the complete genomes gives much higher and more uniform values for the extent of sequence identity between viruses from subgroups A, C and F (see Table 1), but is not directly comparable to the liquid hybridisation data. A more informative comparator is the similarity of signal and coding sequences. A summary of the percentage identity of the predicted proteins of Ad40, Ad12 and Ad2 is given in Table 2, where it can be seen that overall there is greater identity between Ad40 and Ad12 than the other pair-wise comparisons, and this is borne out in a more detailed phylogeny study of protein-coding regions (BAILEY and MAUTNER, 1994). Some of the main differences between the subgroup A, C and F viruses, based largely upon the DNA sequences, are discussed below. Sequence data pertaining to selected regions of

Table 1. Adenovirus genomes: overall comparison

	Size (bp)	Similarity (%) to: Ad2	Ad12	G+C(%)	ITR (bp)	VA	Fiber
Ad2	35 937	–	–	55.2	102	2	1
Ad40	34 214	54.0	58.8	51.2	163	1	2
Ad12	34 125	56.2	–	46.5	161	1	1

ITR, inverted terminal repeat, VA, viral-associated.

Table 2. Comparison of Ad40, Ad12 and Ad2 predicted proteins: percentage identity of amino acid sequences

Early Region	**40 x 12**	**40 x 2**	**12 x 2**	**Late Region**	**40 x 12**	**40 x 2**	**12 x 2**
E1 and ppIX							
E1A 249R	37	41	51	L1 52, 55K	84	80	77
E1B 19K	50	48	47	L1 IIIa	80	74	76
E1B 55K	56	50	50				
ppIX	66	53	53	L2 III	82	78	80
E2 and IVa2				L2 pVII	76	67	71
U exon	37	54	48	L2 pV	68	64	64
E2A DBP	61	57	55	L2 pmu	74	74	71
E2B pTP	78	79	83				
E2B pol	71	73	77	L3 pVI	75	69	69
IVa2	74	78	76	L3 II	86	80	81
E3 region				L3 23K	77	76	75
E3A 173R	32	*	*				
E3A 276R	39	*	*	L4 100K	68	65	67
E3B 90R	53	41	44	L4 22K	54	53	48
E3B 107R	40	33	37	L4 33K	53	58	53
E3B 122R	61	53	49	L4pVIII	84	81	78
E4 region							
E4 130R	48	43	48	L5 IV-1	46	40	*
E4 116R	57	42	45	L5 IV-2	40	45	42
E4 121R	44	34	39				
E4 257R	56	49	54	**VA 1**	62	48	61
E4 90R	48	44	*				

Data derived from ROBERTS et al. (1986), DAVISON et al. (1993) and SPRENGEL et al. (1994) and references cited therein. Asterisks indicate lack of a recognisable counterpart.
VA, viral-associated; DBP, DNA-binding protein; pTP, preterminal protein.

the Ad40 and Ad41 genomes of various strains have been published in recent years and are mostly available in the public data bases (see DAVISON et al. 1993 for detailed documentation).

6.1 Inverted Terminal Repeat

Ad40 has an inverted terminal repeat (ITR) of 163 bp, one of the longest amongst the human adenoviruses and similar in length to that of Ad12 (161 bp), but which is closer in homology to Ad5 (102 bp) than to any other serotype sequenced to date, particularly within the first 60–70 bp (ISHINO et al. 1987; SHINAGAWA et al. 1987; BAILEY and MAUTNER, 1994). The limited similarity of sequences beyond the strictly conserved ITR at each end of the Ad12 genome (SPRENGEL et al. 1994) is also seen for Ad40.

6.2 E1 Region

6.2.1 E1 Proteins

The E1 region of the enteric adenoviruses has been extensively studied, mostly because it has been implicated in the fastidious growth of these viruses. The E1 DNA sequences of both serotypes were first determined by VAN LOON et al. (1987b) and found to be 85% homologous to each other and to share 52% homology with the corresponding Ad5 region. The DNA sequence of the E1 region has now been established for Ad40 Dugan (VAN LOON et al. 1987b; DAVISON et al. 1993), Ad40 Sapporo (ISHINO et al. 1988), Ad41 Tak (VAN LOON et al. 1987b) and Ad41 strain D389 (ALLARD and WADELL 1988, 1992). The organisation of the E1 region is similar to that found for other serotypes, and strategic sequences essential for the propagation of all human adenoviruses are present in both serotypes. Open reading frames (ORFs) have been identified corresponding to the E1A and E1B regions of other serotypes (see Fig. 1). The Ad40 E1A proteins share a similar homology with both Ad2 and Ad12 (ISHINO et al. 1988; Table 2). The three conserved regions (CR) within the E1A region (KIMELMAN et al. 1985) have also been identified in Ad40 and Ad41 (VAN LOON et al. 1987b; ISHINO et al. 1988; ALLARD and WADELL 1988). However, while the homology of the conserved regions of Ad7, Ad12 and SAV-7 with the corresponding Ad5 regions is over 60%, the Ad40 and Ad41 conserved regions share only 51% and 45% homology with Ad5, respectively (VAN LOON et al. 1987b). CR1 is least well conserved, but CR2 and CR3 also possess lower homology with Ad5 than do other serotypes. CR3 is the transactivating domain of the E1A protein and differences in this sequence may account for the low transactivating function of Ad40 and Ad41 E1A compared to Ad5 and Ad12 (VAN LOON et al. 1987a; ISHINO et al. 1988). A comparison of E1B protein-coding sequences shows that both Ad40 E1B ORFs (19K and 55K) show a greater similarity to Ad12 than to Ad2, and the E1B sequence similarities are greater than for E1A (ISHINO et al. 1988). No strictly conserved domains have been identified in the E1B proteins, but for

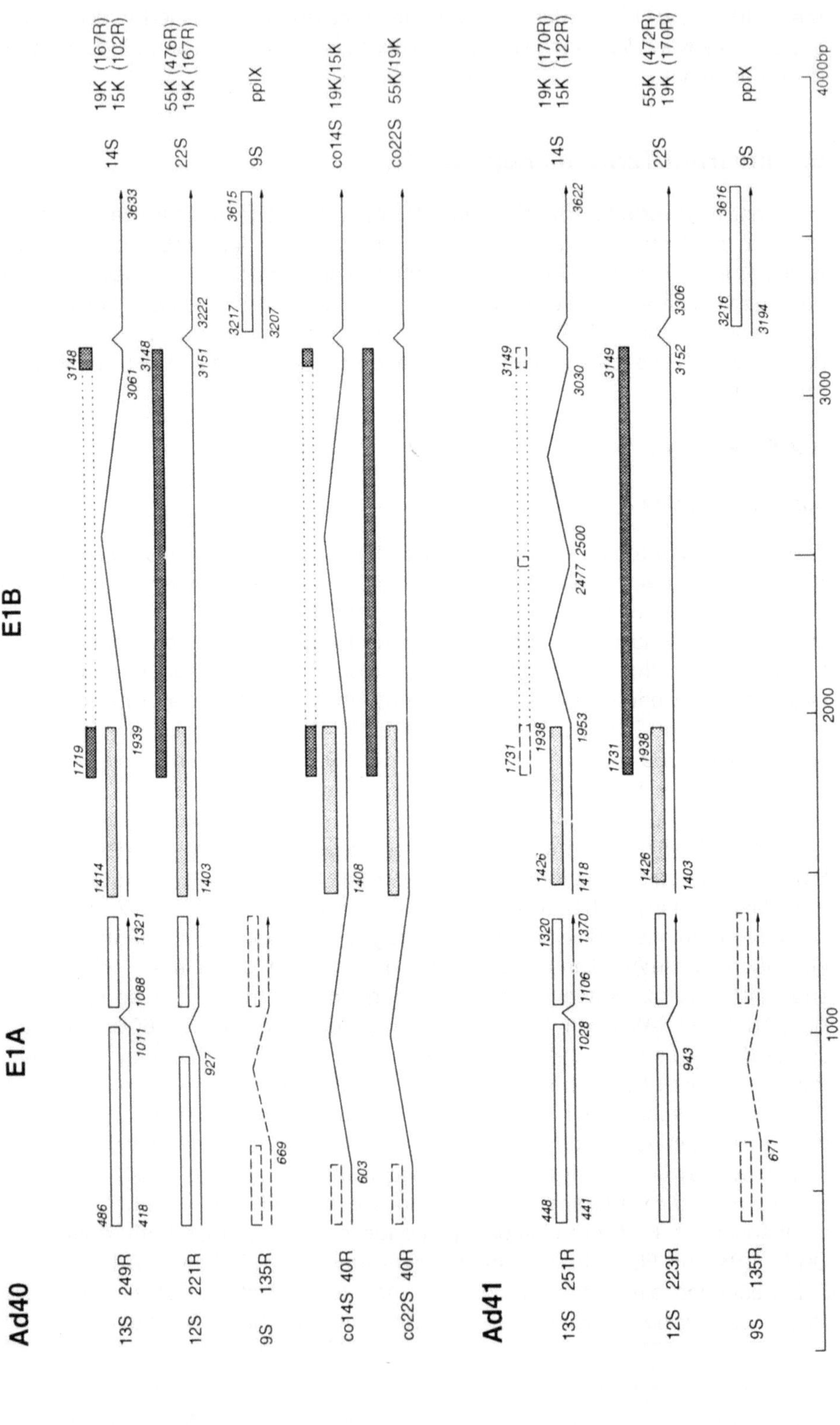
Ad40
E1A
E1B
13S 249R
12S 221R
9S 135R
co14S 40R
co22S 40R
14S 19K (167R) 15K (102R)
22S 55K (476R) 19K (167R)
9S pplX
co14S 19K/15K
co22S 55K/19K
Ad41
13S 251R
12S 223R
9S 135R
14S 19K (170R) 15K (122R)
22S 55K (472R) 19K (170R)
9S pplX
1000
2000
3000
4000bp

the 55K protein an alanine-rich run near the N terminus of Ad2 (eight out 14 amino acids) and Ad12 (seven out of 14 amino acids) is absent from Ad40. A central region of 55K implicated in p53 binding, transcriptional repression, and transforming activity for Ad2 (YEW and BERK 1992; YEW et al. 1990, 1994) is reasonably conserved in Ad12 and Ad40 also.

6.2.2 E1 Transcription

Both Ad40 and Ad41 have the potential to make E1A 9S, 12S and 13S transcripts, equivalent to those of Ad2 (VAN LOON et al. 1987b; ISHINO et al. 1988; ALLARD and WADELL 1988; Fig. 1). E1A transcription maps have only been obtained for E1 plasmid-transformed cells (VAN LOON et al. 1985a, 1987b) and, while all three E1A mRNAs were detected in Ad40-transformed cells, only the 13S transcript was detected in Ad41-transformed cells. Ad40-transformed cells apparently lacked E1B transcripts, and Ad41 transformants made only a 22S RNA. The E1B transcription maps obtained from lytic infections (STEINTHORSDOTTIR and MAUTNER 1991; ALLARD and WADELL 1992; Fig. 1) have established that there are equivalents of the Ad2 14S and 22S mRNAs, encoding 19K and 55K proteins, but that both F subgroup viruses apparently lack the 13S mRNA that is the major transcript for the Ad2 19K protein. The more limited description of the Ad12 E1B transcription map indicates that it too has a predominant 14S message (VIRTANEN et al. 1982; VIRTANEN and PETTERSSON 1985). Two features have been reported that are distinctive to the F subgroup viruses; Ad40 produces E1A–E1B co-transcripts not found in Ad41, while Ad41 has an additional small exon in the 14S species, not seen for Ad40 (STEINTHORSDOTTIR and MAUTNER 1991; ALLARD and WADELL 1992).

6.2.3 Transformation

A major difference between Ad12 and the enteric viruses lies in their transforming capacity: Ad40 and Ad41 virions do not cause tumors in baby hamsters (WADELL et al. 1980; DE JONG et al. 1983), while primary baby rat kidney (BRK) cells transformed by cloned E1 DNA display only a partial transformation phenotype (WAG/RIJ rat: VAN LOON et al. 1985a; COUSIN et al. 1991), do not express all the E1 transcripts (see above) and are not oncogenic in athymic nude mice. The transformation frequency for Ad40 E1A DNA in rat 3Y1 cells is some 100-fold less than for Ad12, and use of a MoMSV-LTR in place of the E1A promoter has only a small effect (ISHINO et al. 1988). The enteric E1A genes cannot co-operate with

Fig. 1. Transcription maps for Ad40 (Dugan) and Ad41 (Tak) E1 regions. Map co-ordinates are from the European Molecular Biology Laboratory (EMBL) data base numbers M18288 and M18289 (VAN LOON et al. 1987b; ALLARD and WADELL 1988, 1992; STEINTHORSDOTTIR and MAUTNER 1991). It should be noted that there are minor differences for Ad41 strain D389 (ALLARD and WADELL 1988, 1992) and Ad40 Sapporo (ISHINO et al. 1988). *Solid lines* denote RNA, with introns indicated by *carets*; *filled rectangles* are open reading frames; *dashed lines* and *rectangles* show RNA and proteins not yet confirmed experimentally or by mapping in other serotypes

Ad5 E1B to give a fully transformed BRK cell, but are as effective as Ad5 E1A in co-operating with an activated *ras* oncogene (Cousin et al. 1991). In contrast, Ad41 E1B, but not Ad40, can co-operate with Ad5 E1A.

6.3 E2 Region

One new feature of the E2 region is the identification of a 52R exon located between the E3 and fiber regions, upstream of the recognised E2 early and late promoter sites. This sequence is highly conserved across all the human subgroups and the canine but not the mouse adenoviruses (Davison et al. 1993), suggesting that it may be of functional significance. The E2B proteins pre-terminal protein (pTP) and polymerase (pol) are more highly conserved than the E2A DNA-binding protein (DBP) across all the subgroups.

6.4 E3 Region

The DNA sequence of the Ad40 E3 region reveals a pattern of ORFs and transcriptional signals not unlike that of Ad2 (Cladras and Wold 1985), but with sufficient differences to suggest that a more detailed knowledge of the functions of the encoded genes might further our understanding of the tissue specificity of the enteric viruses. A comparison of the E3 ORFs for Ad40, Ad12, and Ad2 is shown in Fig. 2 and Table 2; while the E3B region is well conserved, there are major differences in E3A. In particular, Ad40 lacks an equivalent of the Ad2 E3A 19K glycoprotein, which prevents cytolysis of infected cells by cytotoxic T lymphocytes, by blocking expression of class I major histocompatibility complex (MHC) antigens on the cell surface (see Wold and Gooding 1991). Instead, Ad40 has two large ORFs (173R and 276R) with homologues in Ad12, but unrelated to the E3A 20.1K and 20.5K proteins found in Ad3 (Signäs et al. 1986). The Ad40 276R sequence bears limited similarity to the mouse interleukin-1 receptor (V. Mautner, unpublished), which may be of interest in the light of the established relationship between various Ad5 E3 proteins and other components of the immune response. There is a small ORF embedded out of frame in the 276R sequence with limited similarity to the Ad5 19K glycoprotein, but it is unlikely that this has functional significance, unless perhaps a 19K equivalent is generated by alternative splicing. The 12.5K ORF in Ad2 E3A has a homologue in Ad12, which is absent from Ad40. Overall, Ad40 E3 is closer to Ad12 than to Ad2, which may not be surprising in view of their shared site of infection. The Ad41 E3 sequence (N.J. Pieniazek et al., unpublished; EMBL accession number M85254) is almost identical to that of Ad40.

6.5 E4 Region

There are ORFs corresponding to all those identified in Ad2 and Ad12, with the notable exception of ORF1; to accommodate another transcript upstream of the

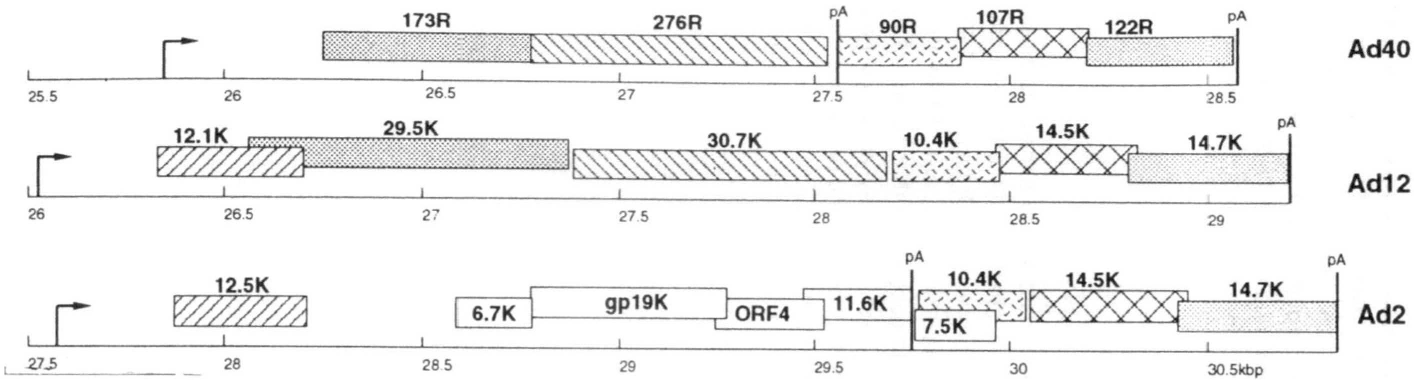

Fig. 2. E3 open reading frames (ORF). Data derived from R.J. ROBERTS et al. (1986), DAVISON et al. (1993) and SPRENGEL et al. (1994), and references cited therein. The transcriptional start site and polyadenylation signals are indicated. *Boxes* with the same shading signify conserved sequences; see Table 2 for quantitative values. *gp,* glycoprotein

Ad40 ORF2 it would be necessary to impinge upon the ITR. Ad40 can complement the Ad2 E4 deletion mutant *dl*808, which lacks all but ORF1, indicating that Ad40 has a functional E4 region (MAUTNER and MACKAY 1991). However, it is known that E4 ORF3 and ORF6 of Ad2 display redundancy in some functions (BRIDGE and KETNER 1989; KETNER et al. 1989; HUANG and HEARING 1989; WEIDEN and GINSBERG 1994), and this aspect of Ad40 E4 function has not been addressed.

6.6 Late Genes

6.6.1 Fiber

The main feature that characterises the subgroup F late region is the presence of two fiber genes, encoding proteins of different shaft length and distinctive sequence (PIENIAZEK et al. 1989, 1990; KIDD and ERASMUS 1989; KIDD et al. 1990, 1993; DAVISON et al. 1993). Phylogeny studies (BAILEY and MAUTNER, 1994) show that these two genes are more closely related to each other than to any of the other sequenced fiber genes and strongly suggest that they arose by gene duplication, preceding the separation of serotypes 40 and 41 and probably shortly after divergence of subgroups A and F (Fig. 3). Both proteins are made in infected cells, and from electron micrographs it appears that both can be incorporated into

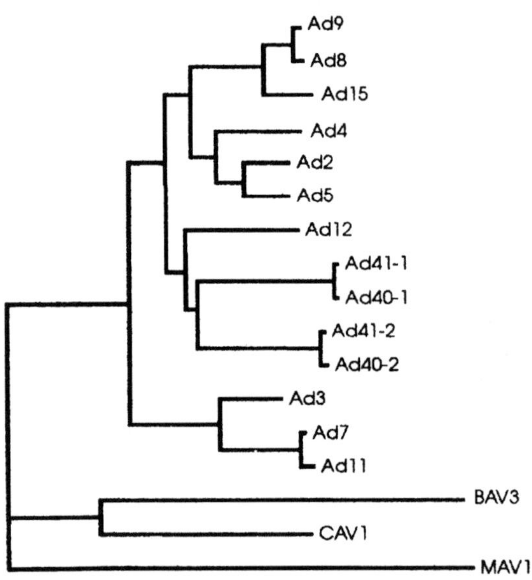

Fig. 3. Phylogenetic tree for adenovirus fiber. Fiber sequence data was obtained from the public data bases. Tree topology was determined by parsimony analysis with bootstrap resampling, and a consensus tree constructed (BAILEY and MAUTNER, 1994). The tree has been plotted unrooted, with horizontal branch length proportional to genetic distance, using the least squares method of FITCH and MARGOLIASH (1967). *MAV*, murine adenovirus; *BAV*, bovine adenovirus; *CAV*, canine adenovirus

the same capsid, attached to different penton bases (KIDD et al. 1993). In this respect they differ from the avian CELO virus, where two fibers of different lengths are attached to each virion vertex (LAVER et al. 1971). Despite the presence of two fibers with distinctive C-terminal knob regions (the location of the cell attachment site), the kinetics of attachment of Ad41 to Hep-2 cells have been interpreted to indicate a single cell receptor for Ad41 (YEH and LUFTIG 1991), but the identity of the cell receptor for any adenovirus remains to be established.

6.6.2 L4 Region

Of the three L4 proteins, 100K is the most conserved between serotypes 40, 12, and 2. The L4 33K polypeptide is also well conserved and a putative 22K polypeptide that would be generated if the 33K splice were not used is also present in all three types (DAVISON et al. 1993). The maintenance of this ORF in three different subgroups strengthens the argument for a functional role.

6.6.3 Hexon

Sequence comparisons of Ad40 and Ad41 hexon show they have an overall identity of 88%, with most differences located in the loop regions exposed on the surface of the virion (TOOGOOD and HAY 1988; TOOGOOD et al. 1989), which have been identified as type-specific antigenic determinants (TOOGOOD et al. 1992). This level of sequence conservation is comparable to that between serotypes of subgroup C (KINLOCH et al. 1984; PRINGAKERBLOM and ADRIAN 1993). Phylogenic comparisons of the available hexon DNA sequences is consistent with the subgrouping of the human serotypes and supports the closer relationship of Ad40 and Ad41 to the subgroup A viruses 12 and 31 than to the other subgroups. These features of the hexon gene and protein have been used in designing diagnostic probes to distinguish the enteric adenoviruses (VAN DER AVOORT et al. 1989; ALLARD et al. 1990, 1992; DE JONG et al. 1993; PRINGAKERBLOM and ADRIAN 1994; SCOTTTAYLOR et al. 1993; reviewed by WADELL et al. 1994)

6.6.4 Penton Base

The penton base has been ascribed a role in virus entry into cells; the Ad2 penton base interacts with cell surface integrins via an RGD tripeptide sequence (amino acids 340-342), and blocking of this interaction prevents internalisation of virus without affecting cell attachment (WICKHAM et al. 1993). The RGD motif is present in Ad2, Ad5 and Ad12 but not in Ad40 nor the avian adenovirus GAL-10 (SHEPPARD and TRIST 1992); it lies within an alanine-rich region that is common to Ad2 and Ad5, but is absent from Ad40, GAL-10 and Ad12 (Fig. 4). It is therefore possible that the enteric adenoviruses do not utilise this pathway. Furthermore, none of the known penton base sequences have any of the other recognised integrin-binding sites (HYNES 1992).

```
      278
Ad2   LEGGNIPALL DVDAYQASLK DDTEQGGDGA GGGNNSGSGA EENSNAAAAA
Ad5   LEGGNIPALL DVDAYQASLK DDTEQGGGGA GGSNSSGSGA EENSNAAAAA
Ad12  LEGGNIPALL DVKKYENSLQ D......... .......... ..........
Ad40  LEGGNIPALL DVEKYEASIK E......... .......... ..........
Gal10 LQGGDIPALL DLDSVDVNDA DGEVIELDNA A......... ..........

      328
Ad2   MQPVEDMNDH AIRGDTFATR AEEKRAEAEA AAEAAAPAAQ PEVEKPQKKP
Ad5   MQPVEDMNDH AIRGDTFATR AEEKRAEAEA AAEAAAPAAQ PEVEKPQKKP
Ad12  ........QN TVRGDNFIAL NKAAR..... .......... ..........
Ad40  ........AQ EIRGADFKPN PQDLE..... .......... ..........
Gal10 .......... .......... .......... .......... ..........

      378
Ad2   VIKPLTEDSK KRSYNLISN. ..DSTFTQYR SWYLAYNYGD PQTGIRSWTL
Ad5   VIKPLTEDSK KRSYNLISN. ..DSTFTQYR SWYLAYNYGD PQTGIRSWTL
Ad12  .IEPVETDPK GRSYNLL.P. ..DKKNTKYR SWYLAYNYGD PEKGVRSWTL
Ad40  .IVPVEKDSK ERSYNLLEG. ..DKNNTAYR SWFLAYNYGD AEKGVKSWTL
Gal10 ...PLLHDSA GVSYNVIYDQ VTGKPVTAYR SWMLAYNVPN SQANQT..TL
```

Fig. 4. Penton base integrin-binding site. The region of the Ad2 L2 gene III product encompassing an integrin-binding site (Wickham et al. 1993) is aligned with the corresponding regions of Ad5, Ad12, Ad40 and the avian adenovirus GAL-10 (FAV-10; Sheppard and Trist 1992)

6.7 Viral-Associated Gene

Ad40 and Ad41 resemble Ad12 and SAV-7 in having only one viral-associated (VA) gene (Kidd and Tiemessen 1993), with greater homology to SAV-7 than to any of the human serotype VA sequences. Unlike the other human adenoviruses, Ad40 and Ad41 are sensitive to human lymphoblastoid interferon (Tiemessen and Kidd 1993); whether this is directly related to an impaired function of the VA gene is not known, but it is possible that inefficient E1A transactivation of the VA promoter might lead to reduced levels of VA RNA or that VA function itself is aberrant. In addition, E1A proteins are known to suppress transcription of interferon-stimulated cell genes, so impaired E1 activity may act at this level.

7 Propagation of Adenovirus Types 40 and 41 in Tissue Culture

A remarkable feature of the enteric adenoviruses is their growth restriction in cultured human cell lines, hence their late discovery as important pathogenic agents. The difficulty in culturing clinical isolates in vitro is not due to low virus concentrations, since up to 10^{11} virus particles can be detected per gram of diarrheal stool (Gary et al. 1979; Retter et al. 1979; Takiff and Straus 1982). Takiff et al. (1981) reported efficient growth of enteric adenovirus isolates in 293 cells and suggested that the Ad5 E1 genes expressed in these cells might provide helper functions. However, while successful growth of Ad40 in 293 cells has been confirmed by some (DE JONG et al. 1983; BROWN et al. 1984), others have failed to obtain Ad40 replication in these cells (CHIBA et al. 1983; UHNOO et al. 1983). A detailed study on the growth characteristics of around 200 isolates of enteric adenoviruses in different cell lines was conducted by DE JONG et al. (1983), who reported difficulties in growing both Ad40 and Ad41. Although both types propagated in 293 cells and tertiary cynomolgous monkey kidney cells (tCMK), in general growth depended variously on virus strain, cell type, sub-line and batch, and other unknown factors. This may also help to explain the numerous conflicting reports on the permissivity of various cell lines for passage of laboratory stocks of the enteric adenoviruses. Thus WITT and BOUSQUET (1988) were able to propagate purified virions of Ad40 and Ad41 in 293, Chang conjuctival cells and KB cells, but could grow only Ad41 in HeLa cells, whereas D. PIENIAZEK et al. (1990a,b) could not propagate Ad41 in 293 cells, but had success with Hep-2, HeLa and Int407 cells. Plaque assays have been reported for Ad40 (Sapporo strain) in A549 cells (HASHIMOTO et al. 1991) and for Ad41 in PLC/PRF/5 cells (GRABOW et al. 1992), indicating that the virus can propagate efficiently in these cells also. We have been able to passage Ad40 in Int407 cells and, with more limited success, in A549 cells, but not in HT29 (colon carcinoma), EJ (bladder carcinoma), PC (pancreatic carcinoma), A431 (squamous carcinoma), or Tera-2 (teratoma) cell lines (STEINTHORSDOTTIR 1991; ULLAH and MAUTNER, unpublished).

Several recent studies have focused on molecular aspects of growth restriction in tissue culture. TIEMESSEN and KIDD (1988, 1990) showed complementation of Ad41 by Ad2 on semi-permissive Chang cells and non-permissive human embryo fibroblasts (HEF), with a pattern of two-hit kinetics, indicative of a helper function for Ad2. There was also interference by Ad41 of Ad2 replication in HEF cells, consistent with a hierarchy of dominance among adenovirus serotypes, which has been attributed to the transcriptional repression function of E1A gene products (LEITE et al. 1986; TIBBETTS et al. 1986; LARSEN and TIBBETTS 1987). Ad41 showed one-hit kinetics of infectivity on 293 cells, compatible with supply of an essential early function by the endogenously expressed Ad5 E1 genes, but multiple-hit kinetics on Chang cells (TIEMESSEN and KIDD 1990); this replication pattern is consistent with a requirement for an essential function (likely an E1 gene product) synthesised in limiting amounts and therefore requiring multiple input genomes in order to achieve a threshold level for viral replication to proceed. Furthermore, the need for this product would have to be cell type dependent, since HEF cells are completely non-permissive for Ad41 in the absence of a complementing Ad2 infection, while Chang cells are semi-permissive. An extension of this study, looking at host cell protein shut-off, hexon and viral DNA synthesis and virion-packaged DNA in various cell lines concluded that no single function was likely to be responsible for all the variations seen in different cell types with different viruses (TIEMESSEN and KIDD 1994). We have looked at the kinetics of Ad40 growth in various cell lines: In KBa+b and Int407 cells, we obtain single-hit kinetics, while in A549 and HeLa we see a multiple-hit pattern (ULLAH and MAUTNER, unpublished). It is intriguing that Int407 cells, which are not known to contain any adenovirus sequences and are in all likelihood a subclone of HeLa cells, rather than of intestinal origin (GILBERT et al. 1990), nevertheless are fully permissive for Ad40, both by the criterion of one-hit kinetics and in terms of their ability to propagate the virus without loss of infectivity.

BROWN et al. (1992) have made a detailed study of the serial growth of Ad40 (DUGAN) and Ad41 (TAK) in 293 and HeLa cells, using ATCC seed stock virus passaged three or four times in 293 cells; while most cells in a 293 culture could be productively infected, only 10%–20% of a HeLa culture supported virus replication, although the remaining uninfected cells were susceptible, as assessed by superinfection. Infectious centre assays and growth curves of intra- and extracellular virus showed that the intracellular virus yield was similar for the two cell types, but progeny virus was released only from 293 cells and not from HeLa cells. The yield of complete virion particles per infected cell was the same for both serotypes in both cell lines and comparable to the value for Ad5, but the particle to infectivity ratios (8×10^3 to 5×10^6 for Ad40, 1-6×10^5 for Ad41) were some 100- to 1000-fold greater than for Ad5. This very high ratio, compounded by the failure of release seen in HeLa cells, is likely to be a major factor in the poor growth of the enteric viruses in cell culture.

The ability of the enteric adenoviruses to propagate in 293 cells implicated the E1 region in the growth restriction of these viruses (TAKIFF and STRAUS 1982). Although both Ad40 and Ad41 E1A gene products can induce expression from the Ad2 E4 promoter in transient transfection assays, the level of transactivation is considerably lower than by the Ad5 E1A gene product (VAN LOON et al. 1987a). Similarly, ISHINO et al. (1988) found that the Ad40 E1A gene product is a weaker transactivator than either Ad5 or Ad12 E1A, tested on Ad5 E2A or E3 promoters, and that the Ad40 E1A gene has a weaker *cis*-acting promoter than Ad12. Other evidence suggests a defective E1B function in the enteric adenoviruses. Ad40 or Ad41 are capable of only partial transformation of BRK cells, and in Ad40-transformed BRK cells no E1B mRNA is made (VAN LOON et al. 1987b). More direct evidence for the involvement of the E1B region in the growth restriction of Ad40 comes from studies by MAUTNER et al. (1989). By using a series of KB cell lines variously expressing the Ad2 E1A and E1B regions (BABISS et al. 1983), it was shown that Ad40 grows equally well on cells expressing either the whole of E1 or only E1B, as determined by the yield of virion DNA. Cells expressing only E1A gene products are no more permissive than KB cells without any adenovirus sequences. This indicates that a lack of E1B but not E1A functions may be involved in the growth restriction of Ad40 and is supported by the finding that in co-infection assays on HeLa cells there is reciprocal complementation between Ad40 and the Ad5 E1A mutant *dl*312, but not with Ad5 or Ad12 mutants lacking the E1B region or defective for the E1B 55K protein (MAUTNER et al. 1989; STEINTHORSDOTTIR 1991; GOMES et al. 1992).

Detailed studies of gene expression from the E1B region of Ad40 in various cell lines have aimed to elucidate the basis for the early replicative block (MAUTNER et al. 1990; Steinthorsdottir and Mautner 1991; Bailey et al. 1993, 1994). The Ad40 E1B transcription map was obtained from KB16 cells (herein referred to as KBa+b); at late times after infection transcripts were synthesised corresponding to the Ad2, 22S, 14S and 9S (ppIX) mRNAs, but no 13S equivalent was seen (Steinthorsdottir and Mautner 1991; see Fig. 1). In addition, co-transcripts encoding the first 40 amino acids of the E1A first exon linked to a site 4–5 nt downstream of the E1B cap site and retaining all the coding potential of the E1B mRNAs are made in equal abundance. The splice junction in these products is unconventional, lacking any splice donor or acceptor consensus sequences, so their mode of synthesis is uncertain. It also remains to be shown that these co-transcripts are utilised as messenger RNAs and whether the E1A exon is expressed.

In Ad5 infection, E1B 22S RNA is produced early and E1B 13S accumulates after the onset of DNA replication (Montell et al. 1984, Glenn and Ricciardi 1988). For Ad40-infected HeLa cells, no E1B mRNA is detectable early, and only low levels of 22S and 14S mRNA are seen, at about 36 h, following the onset of DNA replication. In 293, KBa+b and Int407 cells, 22S mRNA precedes and 14S mRNA parallels DNA synthesis, but the levels are considerably less than for Ad5 in these cells (Mautner et al. 1990; Ullah and Mautner, unpublished).

The construction of Ad5/Ad40 recombinants, based on Ad5 *dl*309 with the Ad40 E1B region replacing that of Ad5, under the control of either the Ad5 E1B promoter (sub5P) or the Ad40 promoter (sub40P), has helped to delineate expression of Ad40 E1B in the context of a well characterised adenovirus system (Bailey et al. 1994). Both recombinants plaque on 293, but not on HeLa cells, and in single-step growth curves (measured by plaque assay on 293 cells) growth rate and final yield of virus are reduced in HeLa cells relative to 293 and KBa+b cells. Thus even under the control of the Ad5 promoter, Ad40 E1B gene expression and/or protein function is impaired, and both recombinants behave like Ad5 E1B 55K mutants (Babiss et al. 1985; Pilder et al. 1986). Analysis of E1B transcription revealed a cell-type specific pattern of gene expression: the activity of the Ad40 E1B promoter is lower than that of Ad5 in all the cell types; in 293 and KBa+b cells, there is ten- to 20-fold less E1B mRNA for sub40P than for sub5P, and in HeLa cells there is an 80-fold reduction. However, the pattern of E1B RNA splicing and temporal regulation is the same for both recombinants, but is characteristic of the particular cell type. In 293 cells, 22S mRNA predominates early and 14S mRNA late, as for Ad5 and Ad12, whereas in KBa+b and HeLa cells, 22S is poorly expressed and 14S predominates, beginning with the onset of DNA replication. These observations suggest that, although the promoter plays an important part in determining the growth characteristics of Ad40, there may also be an element of post-transcriptional regulation, which is influenced by the host cell type.

The activity of the Ad40 E1B promoter has also been investigated using the chloramphenicol acetyltransferase (CAT) reporter gene (Steinthorsdottir 1991; Mautner et al., unpublished; see Fig. 5). In transient transfection assays of E1B promoter CAT plasmids in HeLa cells, the Ad40 promoter has minimal basal activity and is very poorly transactivated by Ad40 E1A, Ad5 E1A or the promiscuous varicella zoster virus (VZV) transactivator p140 (Everett 1987), while

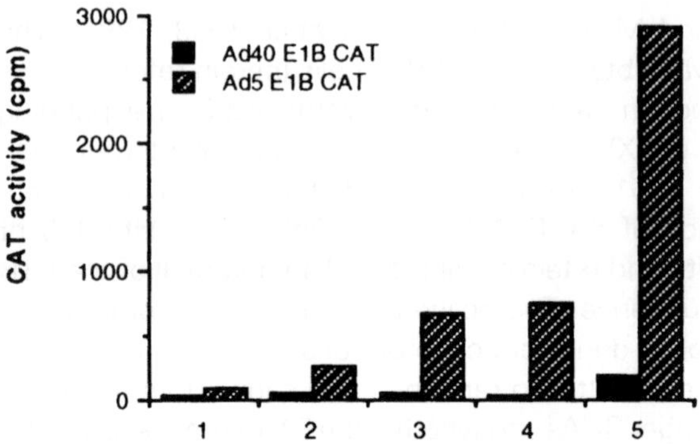

Fig. 5A. E1B promoter activity: Transient transfection assay. Plasmids containing the chloramphenicol acetyltransferase (*CAT*) gene under the control of the Ad5 promoter (nucleotides, 1336-1702) or the Ad40 promoter (nt 1057-1411) were transfected into HeLa cells (*1*) in the absence of a transactivator, or in the presence of Ad40 E1A (*2*), Ad5 E1A (*3*), Ad5 E1A 13S (*4*) or varicella zoster virus (*VZV*) IEp140 (*5*). Cells were harvested at 48 h after transfection and cell extracts assayed for CAT activity (SEED and SHEEN 1988). Results are expressed as yield of ^{14}C-butyrylated chloramphenicol/10^6 cells

the Ad5 E1B promoter shows significant stimulation (Fig. 5A). Incorporation of the CAT gene into the Ad5/40 recombinants sub5P and sub40P in place of the E1B coding region and assay of infected cell extracts confirms the low activity of the Ad40 E1B promoter relative to that of Ad5, although in the context of a viral infection there is a measurable level of Ad40 promoter activity (Fig. 5B; note the differences in scale). This figure also illustrates the cell-specific variation of both Ad5 and Ad40 E1B promoter activity, with much higher levels in 293 cells than in HeLa, and Int407 cells giving intermediate values. We find enhanced levels of transcription overall in 293 cells relative to HeLa cells. To what extent this is attributable to the combined effect of increased transactivation through endogenous expression of E1A in 293 cells in addition to the virus-coded Ad5 E1A is unclear; there are contradictions in the literature regarding the effect of increasing E1A expression on adenovirus gene expression (BRUNET and BERK 1988; HITT and GRAHAM 1990).

The Ad5 E1B promoter consists simply of a TATA box and an Sp1-binding site (WU et al. 1987), although it is known that upstream sequences can modulate its efficient functioning (PARKS and SPECTOR 1990; PARKS et al. 1988; SPECTOR et al. 1993). The Ad40 promoter also has only a TATA box and an Sp1-binding site; although this site differs slightly from the canonical GC box sequence and that of Ad5, it is identical to that of Ad12. We have demonstrated by gel retardation assays (ULLAH and MAUTNER unpublished) that the Ad40 GC box sequence will bind recombinant Sp1 protein (JACKSON et al. 1990), albeit less efficiently than the Ad5 promoter, and that both HeLa and Int407 extracts give equivalent gel shift patterns with the Ad5 and Ad40 probes.

Ad40 E1B protein synthesis and functions are less well characterised. Using antiserum to 55K-related 15K protein made in bacteria, 55K protein cannot be detected in Ad40-infected HeLa cells, although it is synthesised in Ad40-infected

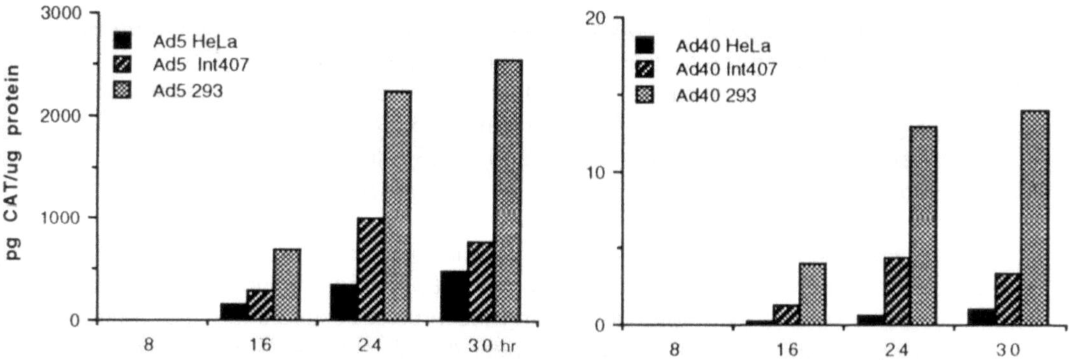

Fig. 5B. E1B promoter activity: Virus infection assay. Ad5 *dl*309 with the E1B region substituted by the CAT gene under the control of the Ad5 E1B promoter (*left panel*) or the Ad40 E1B promoter (*right panel*) was used to infect HeLa, Int407 or 293 cells (10^6) at a multiplicity of infection (moi) of 50 plaque-forming units (pfu) cell (titrated in 293 cells). Cells were harvested at the indicated times, and cell extracts assayed for CAT activity. Yields are expressed as pg CAT protein synthesised per µg total protein. Note the different scales in the two panels

293 and KBa+b cells and in sub5P-infected HeLa cells, at both early and late times (BAILEY et al. 1993, 1994). 19K protein is synthesised in Ad40-infected HeLa cells only at late times, and in permissive 293 and KBa+b cells at both early and late times. Function of the Ad40 19K protein was tested by comparing levels of DNA degradation in HeLa cells infected with Ad5 and with the Ad40 recombinants sub5P and sub40P. Under the control of the Ad5 promoter in sub5P, where Ad40 19K is expressed at a level equivalent to that found in a wild-type Ad5 infection, DNA degradation was effectively inhibited, while sub40P produces decreased levels of 19K protein that are insufficient to prevent E1A induced apoptosis (WHITE et al. 1991; RAO et al. 1992).

8 Conclusions

The fastidious enteric adenoviruses have posed an intriguing problem to virologists seeking to understand the molecular basis for their restricted growth in cell culture and host tissue specificity in vivo: despite the fact that adenovirus is one of the most intensively studied micro-organisms, our overall understanding of the way that the virus interacts with the host cell is largely incomplete, and this is particularly apparent for the subgroup F viruses, where we have a variety of descriptions of "aberrent" behaviour compared to the other subgroups, but no unifying hypothesis to explain their distinctive features. It may be helpful to consider the stages of the lytic cycle in turn, to identify areas which might be amenable to further investigation. I will confine my discussion mostly to Ad40, in part because we presently have more information available and also because its more severely restricted phenotype may give a clearer insight into the extremes of variation between the fastidious and "unfussy" adenoviruses.

Direct gradient purification of Ad40 virus from clinical isolates shows that the large amount of virus shed in diarrhetic stool consists predominantly of complete virions, in contrast to virus passed in 293 cells, where a high proportion of incomplete particles are produced (M. BROWN, personal communication). Nevertheless, complete particles from either source display the same characteristically high particle to infectivity ratio. This might be an indication that the natural host cell in the intestinal tract can efficiently synthesise and package virus, and the problems encountered in growing the virus in tissue culture just reflect the inadequacies of the available cell lines. A high particle to infectivity ratio usually indicates a failure at some early stage of the infectious cycle, and cell attachment and E1 gene function are likely stages of impairment for Ad40. The entry path of adenovirus into cells involves fiber-specific receptors and possibly an integrin-mediated internalisation step leading to endosome uptake; the ability of Ad40 to attach to and enter various cell types has still to be fully explored. We have found that with an input multiplicity of 10^4 particles per cell, HeLa and Int407 attach only 10% of the input virus, and only some 10%–15% of this attached virus is internalised (ULLAH and MAUTNER, unpublished). Taking into account also that only 10% of HeLa cells support a complete lytic cycle (BROWN et al. 1992; ULLAH and MAUTNER unpublished), then this could largely account for the observed particle to infectivity ratio, but does not explain its physical basis. It remains to be shown whether cells susceptible to the enteric adenoviruses in vivo have one or two distinct receptors, as suggested by the presence of two distinct fibers, and whether attachment is limited by a paucity of receptors on cells in tissue culture. If the natural host cell is the highly differentiated intestinal columnar epithelial cell (TIEMESSEN et al. 1993), it may be that there are tissue-specific receptors that allow efficient attachment of the virus in vivo. It is also unclear whether internalisation requires an integrin recognition sequence on the penton base, as postulated for Ad2.

The argument for an early block to virus replication at the level of E1B expression has been made above and will not be reiterated, but the consequences of aberrent E1B function do merit further consideration; the E1B 55K protein has a multiplicity of effects in the lytic cycle, as revealed through mutant studies (see for example WHITE and STILLMAN 1987), although exact mechanisms of function remain to be elaborated (reviewed by BOULANGER and BLAIR 1991). For Ad2 and Ad5, loss of 55K function does not directly affect early events, in that DNA replication is not blocked, but progress to late stages of the lytic cycle is impaired, with a failure to transport late viral mRNA from the nucleus to cytoplasm and a concomitant failure to shut off host cell protein synthesis (BABISS and GINSBERG 1984; BABISS et al. 1985; PILDER et al. 1986; WILLIAMS et al. 1986; LEPPARD and SHENK 1989). In Ad12 infection, 55K is required also for DNA replication (FUKUI et al. 1984; SHIROKI et al. 1986; BREIDING et al. 1988; MAK and MAK 1990; ZHANG et al. 1992), indicating that there is an element of subgroup specificity in the properties of 55K, and in this respect Ad40 may resemble Ad12.

The ability of Ad40 to replicate in some cell lines but not others may depend on the extent to which 55K can be expressed at early times, thereafter influencing the efficiency of DNA replication. The observations on multiple-hit kinetics for Ad40 and Ad41 in some cell lines (TIEMESSEN and KIDD 1990; ULLAH and MAUTNER, unpublished) would be compatible with the requirement for a threshold level of 55K before DNA replication can commence, and the presence of a number of genomes in a single cell may compensate for inefficient E1B expression. The inability of Ad40 to switch off host cell protein synthesis (MAUTNER et al. 1990) and the failure to release virus particles from HeLa cells (BROWN et al. 1992) might also be attributable to a defect in E1B function: it has recently been shown that when host protein shut-off is prevented in adenovirus-infected cells, there is no impedence to the accumulation of virus structural proteins, but there is a block in the development of a cytopathic effect and failure to release assembled virus particles (ZHANG and SCHNEIDER 1994). It is possible that, in an Ad40-infected cell, even if there has been progress through the lytic cycle to virion assembly, a failure of E1B 55K to shut off host protein synthesis could prevent virus release and development of a full cytopathic effect. If the single Ad40 VA RNA does not function efficiently (TIEMESSEN and KIDD 1993; KIDD and TIEMESSEN 1993), this could also affect host protein shut-off and late viral protein synthesis (O'MALLEY et al. 1989). The role of the E4 region in modulating DNA replication and late events (BRIDGE and KETNER 1990; WEINBERG and KETNER 1986; WEIDEN and GINSBERG 1994), could also contribute to the Ad40 phenotype in tissue culture, and it is likely that in vivo the E3 region also plays an important function in modulating cytopathology and determining the outcome of infection. The emerging pattern is that at many stages in the replicative cycle, Ad40 functions less effectively than its better known counterparts, at least in the cell systems currently available to us.

We have attempted to present a coherent picture of the properties of the enteric adenoviruses in cell culture, which of necessity has emphasised areas of concordance and perhaps given insufficient consideration to the various discrepancies which abound in the literature, particularly concerning the extent to which the two serotypes are able to undergo a complete replication cycle in various cell types. To a certain degree this will depend on local variations in cell and virus stocks and culture methods, but it probably also reflects the inherent difficulty of using an inappropriate host cell with an aptly named "fastidious" enteric virus, which nevertheless clearly shares many of the properties of its more conventional relatives.

Acknowledgments. We are grateful to Martha Brown for communicating unpublished data and to Eric Blair and Alastair Kidd for stimulating discussion. The work from the authors' laboratory was supported by the Medical Research Council.

Appendix: Adenovirus Type 40 Complete Nucleotide Sequence

```
> start of ITR
CATCATCAATAATATACCTTAAAACTGGAAACGAGCCAATATGATAATGAGGGAGGAGGGACTAGGGGTGGTGTAAGGTGACGTAGAGGCGGGCGGGGTG   100

                                                   end of ITR >
GGAAAGGGTGGAGGCGGATGACGTGTGGGGTCGGAGGACGGGCGCGGTGCGGCGGAAGTGACGGAAAATCTGGTGTATTGGGCGGGTTTTTGTAACTTTT   200

GGCCATTTTGGCGCGAAAACTGAGTAATGAGGACGTGGGACGAACTTTGGACTTTTGTGTTTATGGAGGAAAAACTGCTGATTATTACTGAACTTTGGCC   300

                                                                             E1A TATA >>>>>>
CATGACGAACCGGTTTTTCTACGTGGCAGTGCCACGAGACGGCTCAAAGTCCTAATTTTTTATTGTGTGCTCAGCCCGTTTGAGGGTATTTAAACACAGC   400

                                                                            E1A 135R  M  R  M  L  P      5
                                                                            E1A 221R  M  R  M  L  P      5
                 > 5' end of E1A mRNA                                       E1A 249R  M  R  M  L  P      5
CAGAACATCAAGAGGCCACTCTTGAGTGCGAGCGAGTAGAGTTTCTCCTCCATTGCTGTTGGCGCTTTTGACATAGCCACCAAGATGAGAATGCTGCCG   500

 D  F  F  T  G  N  W  D  D  M  F  Q  G  L  L  E  T  E  Y  V  F  D  F  P  E  P  S  E  A  S  E  E  M      38
 D  F  F  T  G  N  W  D  D  M  F  Q  G  L  L  E  T  E  Y  V  F  D  F  P  E  P  S  E  A  S  E  E  M      38
 D  F  F  T  G  N  W  D  D  M  F  Q  G  L  L  E  T  E  Y  V  F  D  F  P  E  P  S  E  A  S  E  E  M      38
GATTTTTTTACCGGGAACTGGGATGACATGTTCCAGGGGTTGCTGGAGACTGAATATGTGTTTGATTTCCCTGAACCTTCTGAGGCTTCTGAAGAAATGT   600

S  L  H  D  L  F  D  V  E  V  D  G  F  E  E  D  A  N  Q  E  A  V  D                                       61
S  L  H  D  L  F  D  V  E  V  D  G  F  E  E  D  A  N  Q  E  A  V  D  G  M  F  P  E  R  L  L  S  E  A      72
S  L  H  D  L  F  D  V  E  V  D  G  F  E  E  D  A  N  Q  E  A  V  D  G  M  F  P  E  R  L  L  S  E  A      72
CGCTTCATGATCTTTTTGATGTGGAGGTGGATGGTTTCGAAGAGGACGCCAACCAGGAAGCGGTTGATGGTATGTTTCCCGAGAGGTTGCTGTCCGAGGC   700
   > end of cotranscript 1st exon         end of E1A 135R 1st exon >

  E  S  A  A  E  S  G  S  G  D  S  G  V  G  E  E  L  L  P  V  D  L  D  L  K  C  Y  E  D  G  L  P  P     105
  E  S  A  A  E  S  G  S  G  D  S  G  V  G  E  E  L  L  P  V  D  L  D  L  K  C  Y  E  D  G  L  P  P     105
TGAGAGCGCTGCAGAGAGCGGTTCGGGTGATTCTGGGGTTGGCGAAGAGTTGTTGCCGGTTGATCTGGATTTGAAATGCTATGAAGACGGTTTGCCTCCT   800

 S  D  P  E  T  D  E  A  T  E  A  E  E  E  A  A  M  P  T  Y  V  N  E  N  E  N  E  L  V  L  D  C  P     138
 S  D  P  E  T  D  E  A  T  E  A  E  E  E  A  A  M  P  T  Y  V  N  E  N  E  N  E  L  V  L  D  C  P     138
AGCGATCCTGAAACTGATGAGGCTACAGAGGCGGAAGAAGAGGCGGCTATGCCGACTTATGTGAATGAAAATGAAAATGAGCTGGTGCTGGACTGTCCAG   900

E  N  P  G  R  G  C  R  A                                                                                147
E  N  P  G  R  G  C  R  A  C  D  F  H  R  G  T  S  G  N  P  E  A  M  C  A  L  C  Y  M  R  L  T  G  H     172
AGAACCCTGGGCGAGGTTGTCGGGCTTGTGATTTCCATCGGGGCACTAGTGGCAATCCTGAAGCTATGTGTGCTTTGTGTTATATGCGTTTAACTGGACA  1000
 end of E1A 221R 1st exon >
                                                                                       G  P  I  S  D      66
                                                                                       C  P  I  S  D     152
  C  I  Y                                                                              S  P  I  S  D     180
CTGTATCTACAGTAAGTAAAAAAGTTTTTATTTGTTTGGTGGTGTTGGTTAATATGAACAAGAGTTAACGACTTTTTGTTATTTAGGTCCAATTTCAGA  1100
          > end of E1A 249R 1st exon                     start of E1A 135R, 221R, 249R 2nd exon >
```

```
 A  E  G  E  S  E  S  G  S  P  E  D  T  D  F  P  H  P  L  T  A  T  P  P  H  G  I  V  R  I  I  P  C     99
 A  E  G  E  S  E  S  G  S  P  E  D  T  D  F  P  H  P  L  T  A  T  P  P  H  G  I  V  R  I  I  P  C    185
 A  E  G  E  S  E  S  G  S  P  E  D  T  D  F  P  H  P  L  T  A  T  P  P  H  G  I  V  R  I  I  P  C    213
TGCGGAAGGGGAGTCTGAGTCGGGGTCGCCTGAGGACACTGATTTCCCCACCCTTTAACCGCCACGCCGCCACATGGAATTGTGAGAACCATCCCGTGC 1200

R  V  S  C  R  R  R  P  A  V  E  C  I  E  D  L  L  E  E  D  P  T  D  E  P  L  N  L  S  L  K  R  P    132
R  V  S  C  R  R  R  P  A  V  E  C  I  E  D  L  L  E  E  D  P  T  D  E  P  L  N  L  S  L  K  R  P    218
R  V  S  C  R  R  R  P  A  V  E  C  I  E  D  L  L  E  E  D  P  T  D  E  P  L  N  L  S  L  K  R  P    246
AGAGTTTCTTGTAGACGACGCCCAGCTGTTGAGTGCATAGAAGATTTACTTGAGGAAGATCCAACAGATGAACCTTTGAACCTGTCCTTAAAGCGCCCCA 1300

K  C  S  -                                                                                            135
K  C  S  -                                                                                            221
K  C  S  -             >>>>>> E1A pA                                        >>>>>> E1B TATA           249
AGTGCTCCTGAGATCATAGTAATAAAGTTATTGACCCTTACCCTGTGTTTATTTCTTGGGCGTGTTTGTGGGTATATAAGCAGGTAGAATGGTTTTAGTG 1400

    E1B 19K M  E  L  W  S  E  L  Q  S  Y  Q  N  L  R  R  L  L  E  L  A  S  A  R  T  S  S  C  W  R    29
TTAGTTTATTCTGATGGAGTTGTGGAGTGAGTTACAAAGTTATCAGAACCTCCGACGCTTGCTGGAGTTGGCTTCTGCCAGAACTTCCAGCTGTTGGAGA 1500
 >         > 5' ends of E1B mRNA
     > start of co-transcript 2nd exon

 I  L  F  G  S  T  L  T  N  V  I  Y  R  A  K  E  E  Y  S  S  R  F  A  D  L  L  S  H  N  P  G  I  F    62
ATCCTTTTTGGCTCAACTTTAACTAATGTAATCTATAGAGCTAAGGAGGAGTACTCTTCGCGGTTTGCTGACCTTTTGTCGCATAACCCTGGAATTTTTG 1600

A  S  L  N  L  G  H  H  S  F  F  Q  E  I  V  I  R  N  L  D  F  S  S  P  G  R  T  V  S  G  L  A  F  I   96
CTTCTTTGAATTTGGGGCATCACTCATTTTTCAAGAAATTGTGATCAGAAATTTAGATTTTTCTTCTCCTGGCCGTACGGTTTCTGGGCTTGCTTTTAT 1700

   E1B 15K (102R) M  E  R  P  N  S  S  V  A  G  L  Y  S  G  L  H  G  N  G  S  V  E  N  L  A  T  E    27
   E1B 55K (476R) M  E  R  P  N  S  S  V  A  G  L  Y  S  G  L  H  G  N  G  S  V  E  N  L  A  T  E    27
 C  F  I  L  D  Q  W  S  A  Q  T  H  L  S  Q  G  Y  T  L  D  Y  M  A  M  A  L  W  R  T  L  L  R  R   129
TTGTTTTATATTGGATCAATGGAGCGCCCAAACTCATCTGTCGCAGGGTTATACTCTGGATTACATGGCAATGGCTCTGTGGAGAACCTTGCTACGGAGG 1800

E  E  G  L  R  L  L  A  G  A  A  S  A  R  F  G  S  S  A  G  R  G  G  G  G  G  E  P  E  G  R  P  G  P    61
E  E  G  L  R  L  L  A  G  A  A  S  A  R  F  G  S  S  A  G  R  G  G  G  G  G  E  P  E  G  R  P  G  P    61
 K  R  V  L  G  C  L  P  A  Q  R  P  H  G  L  D  P  V  Q  E  E  E  E  E  E  E  N  L  R  A  G  L  D   162
AAGAGGGTCTTAGGTTGCTTGCCGGCGCAGCGTCCGCACGGTTTGGATCCAGTGCAGGAAGAGGAGGAGGAGGAGGAGAACCTGAGGGCCGGCCTGGACC 1900

  F  N  G  I  V  T  E  P  D  P  E  E                                                                 73
  F  N  G  I  V  T  E  P  D  P  E  E  G  T  S  S  G  Q  R  G  G  I  N  G  Q  R  G  T  K  R  K  M  E    94
P  S  T  E  L  -                                                                                     167
CTTCAACGGAATTGTAACTGAGCCTGATCCCGAAGAGGGTACTAGCAGTGGGCAAAGGGGGGGCATTAATGGGCAAAGGGGGACAAAGAGAAAGATGGAA 2000
          end of E1B 15K 1st exon >

 N  E  G  E  D  F  L  K  E  L  T  L  S  L  M  S  R  R  H  H  E  S  V  W  W  A  D  L  E  D  E  F  K   127
AACGAGGGGGAGGACTTTTTAAAGGAGTTAACCTTGAGTTTAATGTCTCGTCGCCATCATGAGTCTGTTTGGTGGGCTGATTTGGAAGATGAGTTTAAAA 2100
```

```
N  G  E  M  N  L  L  Y  K  Y  T  F  E  Q  L  K  T  H  W  L  E  A  W  E  D  F  E  L  A  L  N  T  F  A    161
ACGGTGAAATGAATTTGTTATACAAGTATACATTTGAACAGCTGAAGACACATTGGCTGGAGGCTTGGGAGGATTTTGAGTTAGCTCTGAACACTTTTGC  2200

 K  V  A  L  R  P  D  T  I  Y  T  I  K  K  T  V  N  I  R  K  C  A  Y  V  L  G  N  G  A  V  V  R  F    194
CAAAGTGGCTCTTCGCCCGGACACTATTTATACCATTAAGAAGACTGTTAATATACGTAAATGTGCCTATGTGCTGGGGAATGGAGCTGTGGTGCGGTTT  2300

 Q  T  C  D  R  V  A  F  N  C  A  M  Q  S  L  G  P  G  L  I  G  M  S  G  V  T  F  M  N  V  R  F  V    227
CAAACATGTGACCGTGTAGCCTTTAACTGCGCAATGCAGAGCTTGGGCCCTGGGCTTATTGGCATGAGTGGGGTAACTTTTATGAATGTGAGATTTGTAG  2400

V  E  G  F  N  G  T  V  F  A  S  T  T  Q  L  T  L  H  G  V  F  F  Q  N  C  S  G  I  C  V  D  S  W  G    261
TGGAGGGATTTAATGGCACAGTGTTTGCTTCTACCACTCAATTAACCTTGCATGGTGTGTTTTTTCAAAATTGCAGCGGTATCTGCGTGGATTCCTGGGG  2500

 R  V  S  A  R  G  C  T  F  V  A  C  W  K  G  V  V  G  R  N  K  S  Q  M  S  V  K  K  C  V  F  E  R    294
TAGGGTGTCTGCCAGAGGGTGTACGTTTGTTGCATGTTGGAAAGGGGTGGTGGGGCGAAACAAAAGTCAAATGTCTGTAAAGAAGTGTGTGTTTGAACGT  2600

 C  I  M  A  M  V  V  E  G  Q  A  R  I  R  H  N  A  G  S  D  N  V  C  F  L  L  L  K  G  T  A  S  V    327
TGCATTATGGCCATGGTGGTAGAAGGTCAGGCGCGGATTCGCCATAATGCGGGCTCTGATAATGTGTGTTTTTTACTGCTAAAGGGAACTGCCAGTGTAA  2700

K  H  N  M  I  C  G  G  G  H  S  Q  L  L  T  C  A  D  G  N  C  Q  A  L  R  V  F  H  V  V  S  H  P  R    361
AGCATAACATGATTTGTGGCGGTGGTCACTCTCAGCTGCTAACCTGTGCAGATGGAAACTGTCAGGCTCTGAGAGTGTTTCACGTAGTATCTCATCCCCG  2800

 R  P  W  P  V  F  E  H  N  M  L  M  R  C  T  V  H  L  G  A  R  R  G  M  F  S  P  Y  Q  S  N  F  C    394
CCGCCCCTGGCCTGTTTTTGAGCACAACATGCTTATGCGCTGTACTGTGCATTTGGGAGCTCGTCGTGGCATGTTTTCTCCATACCAGAGTAACTTTTGC  2900

 H  T  K  V  L  M  E  T  D  A  F  S  R  V  W  W  N  G  V  F  D  L  T  M  E  L  F  K  V  V  R  Y  D    427
CACACTAAAGTTTTAATGGAAACTGATGCTTTTTCTCGGGTATGGTGGAACGGGGTATTTGATTTAACCATGGAGCTATTTAAAGTGGTGAGGTATGATG  3000

                                                              A  T  L  N  V  T  E  Q  L  R  T  D  H  Q     87
E  S  K  V  R  C  R  P  C  E  C  G  A  N  H  I  R  L  Y  P  A  T  L  N  V  T  E  Q  L  R  T  D  H  Q    461
AGTCAAAGGTTCGTTGTCGCCCCTGTGAGTGTGGAGCTAATCATATTAGGTTATATCCAGCAACTCTGAACGTGACCGAGCAGCTGCGTACGGACCACCA  3100
                                                              > start of E1B 15K 2nd exon

 M  M  S  C  L  R  T  D  Y  E  S  S  D  E  D  -                                                          102
 M  M  S  C  L  R  T  D  Y  E  S  S  D  E  D  -                                 >>>>>> ppIX TATA           476
GATGATGTCGTGTCTGCGTACTGACTACGAATCCAGCGATGAGGATTAAGGGTAAGGGGCGGAGCCTATTACAGGTATAAAGGTTGGGGTAGAGTAAAAA  3200
                  end of E1B 15K, 55K coding exons >

              ppIX M  S  G  F  T  E  G  N  A  V  S  F  E  G  G  V  F  S  P  Y  L  T  T  R  L  P  S  W     28
AAAGGGAAGTTACAAAATGAGTGGCTTCACGGAAGGAAACGCTGTGAGTTTTGAGGGTGGGGTGTTTAGCCCATATCTGACAACCCGTCTTCCCTCTTGG  3300
     >           > start of E1B non-coding exon
     5' end of ppIX mRNA

A  G  V  R  Q  N  V  V  G  S  N  V  D  G  R  P  V  A  P  A  N  S  T  T  L  T  Y  A  T  I  G  S  S     61
GCAGGAGTGCGTCAGAATGTGGTGGGGTCCAACGTTGATGGTCGTCCTGTCGCCCCTGCCAACTCGACAACCCTTACCTACGCCACTATTGGATCGTCGG  3400
```

```
V  D  T  A  A  A  A  A  A  S  A  A  A  S  T  A  R  G  M  A  A  D  F  G  L  Y  N  Q  L  A  A  S  R  L      95
TGGATACCGCTGCAGCTGCTGCCGCGTCTGCTGCTGCCTCTACTGCTCGTGGCATGGCAGCAGATTTGGACTGTACAATCAACTGGCCGCGTCTCGGTT  3500

 R  E  E  D  A  L  S  V  V  L  T  R  L  E  E  L  S  Q  Q  L  Q  D  M  S  A  K  M  A  L  L  N  P  P    128
AAGAGAAGAAGATGCCCTGTCCGTGGTGTTGACCCGCTTGGAGGAGCTGTCTCAGCAGTTGCAAGATATGTCTGCCAAAATGGCTCTGCTTAACCCTCCC  3600

 A  N  T  S  -                                            <<<<<<<<<<< E2B pA                           132
GCTAATACTTCTTAATAAAGACACAATTGGTTGGAAAAGTCAAAAGTGTTTATTTATTTCTTTTGCGGTAGGCCCTAGACCACCTGTCGCGGTCGTTTAA  3700
            >>>>>> E1B, ppIX pA                  [IVa2] -  K  N  R  K  R  Y  A  R  S  W  R  D  R  D  N  L    431

AACTTTATGGATGTTTTCCAAGACCCGGTACAGGTGGGCTTGGATGTTCAAATACATAGGCATTAGGCCGTCCCTGGGATGCAGGTAGGACCACTGGAGG  3800
 V  K  H  I  N  E  L  V  R  Y  L  H  A  Q  I  N  L  Y  M  P  M  L  G  D  R  P  H  L  Y  S  W  Q  L     398

GCGTCATGCTCTGGGGTGGTGTTGTAAATAATCCAATCGTAGCAGGGTTTTTGAGCATGAAACTGGAAGATGTCCTTAAGGAGGAGGCTAATGGCCAGAG  3900
A  D  H  E  P  T  T  N  Y  I  I  W  D  Y  C  P  K  Q  A  H  F  Q  F  I  D  K  L  L  L  S  I  A  L  P    364

GTAGCCCCTTGGTGAAGGTATTAACAAATCGATTAAGTTGGGAGGGATGCATGCGAGGGGAAATCAGATGCATTTTGGCCTGAATTTTTAGGTTGGCAAT  4000
 L  G  K  T  F  T  N  V  F  R  N  L  Q  S  P  H  M  R  P  S  I  L  H  M  K  A  Q  I  K  L  N  A  I     331

GTTGCCACCGAGATCACGTCTGGGGTTCATGTTGTGCAGAACCACTAGCACGGTATAGCCGGTGCACTTGGGGAATTTGTCATGCAACTTGGAAGGGAAG  4100
 N  G  G  L  D  R  R  P  N  M  N  H  L  V  V  L  V  T  Y  G  T  C  K  P  F  K  D  H  L  K  S  P  F     298

GCGTGGAAAAACTTGGAAACCCCTTTGTGCCCTCCCAAATTTTCCATGCACTCATCCATAATGATGGCGATGGGGCCTTGGGATGCAGCCTTAGCAAAAA  4200
A  H  F  F  K  S  V  G  K  H  G  G  L  N  E  M  C  E  D  M  I  I  A  I  F  G  Q  S  A  A  K  A  F  I    264

TGTTATCGGGGTGGGAAACATCGTAGTTTTGCTCCAGGGTAAGCTCGTCATAGGCCATTTTGATGAAGCGTGGTAAAAGGGTGCCCGACTGGGGTATAAT  4300
 N  D  P  H  S  V  D  Y  N  Q  E  L  T  L  E  D  Y  A  M  K  I  F  R  P  L  L  T  G  S  Q  P  I  I    231

GGTCCCTTCTGGACCTGGGGCGTAGTTACCCTCGCAGATTTGCATCTCCCAAGCCTTAATTTCTGAGGGGGGCATCATGTCCACCTGAGGGGCAATAAAA  4400
 T  G  E  P  G  P  A  Y  N  G  E  C  I  Q  M  E  W  A  K  I  E  S  P  P  I  M  D  V  Q  P  A  I  F     198

AAAACGGTTTCGGGTGGAGGGTTAATAAGCTGGGTGGAAAGCAAGTTTCGCAAAAGCTGGGATTTGCCGCAGCCAGTGGGACCGTAAATGACCCCAATGA  4500
F  V  T  E  P  P  P  N  I  L  Q  T  S  L  L  N  R  L  L  Q  S  K  G  C  G  T  P  G  Y  I  V  G  I  V    164

CAGGCTGTAGTTGATAGTTTAAGGAGATGCAACTGCCATCTTCCCGCAAAAGCGGAGTGACTTCGTTCATCATGCTTCTGACATGCTGGTTTTCTTTAAC  4600
 P  Q  L  Q  Y  N  L  S  I  C  S  G  D  E  R  L  L  P  T  V  E  N  M  M  S  R  V  H  Q  N  E  K  V    131

CAAGTCTTGCAAGAGACGCTCACCGCCCAGGGAAAGTAGCTCTTCCAAGCTGCGGAAATGCTTTAGTGGTTTTAGGCCATCGGCCATGGTCATTTTTTCA  4700
 L  D  Q  L  L  R  E  G  G  L  S  L  L  E  E  L  S  R  F  H  K  L  P  K  L  G  D  A  M  T  M  K  E      98

AGGGATTGACGCAGCAAATAGAGCCGATCCCAGAGCTCGGTAATATGGTCTATGGCATCTCGATCCAACAAACTTCTTGGTTGCGGGGGTTGGGACGGCT  4800
                                                   [E2B pol] -  P  M  E  I  W  C  V  E  Q  N  R  P  N  P  R  S   1173
L  S  Q  R  L  L  Y  L  R  D  W  L  E  T  I  H  D  I  A  D  R  D  L  L  S  R  P  Q  P  P  Q  S  P  K     64
```

```
TTGGCTGTACGGTACCAGTCGGTGGGCGTCCAGTGGAGCAAGGGTAATGTCTTTCCAGGGTCGCAGGGTTCGCGTTAGGTTGGTTTCGGTGACGGTAAAG  4900
 Q  S  Y  P  V  L  R  H  A  D  L  P  A  L  T  I  D  K  W  P  R  L  T  R  T  L  N  T  E  T  V  T  F    1140
  P  Q  V  T  G  T  P  P  R  G  T  S  C  P  Y  H  R  E  L  T  A  P  N  A  N  P  Q  N  R  H  R  Y  L     31

                                                     start of IVa2 2nd exon <
GGGCGCGCTCCGGGTTGGGCGCTTGCCAGGGTTCTCTTCAGGCTCATCCTGCTGGTGTGAAAACGCGCGTCTTCGCCCTGAAAGTCGGCCAAGTAGCATT  5000
P  R  A  G  P  Q  A  S  A  L  T  R  K  L  S  M  R  S  T  H  F  R  A  D  E  G  Q  F  D  A  L  Y  C  K  1106
 P  A  S  R  T  P  R  K  G  P  N  E  E  P  E  D  Q  Q  H  S  F  A  R  R  R  G                          5

TTAACATGAGATCATAGTTGAGGGTTTCGGCAGCGTGTCCTTTGGCGCGAAGCTTGCCCTTGGAAATTTGCTGACAGCTGGGACAGCGGAGGCATTTTAG  5100
  L  M  L  D  Y  N  L  T  E  A  A  H  G  K  A  R  L  K  G  K  S  I  Q  Q  C  S  P  C  R  L  C  K  L   1073

GGCGTAGAGTTTGGGAGCCAGGAAGACGGACTCTGGTGAATAAGCGTCGGCGCCACACTGTGCACACACGGTTTCGCACTCCACTAACCAGGCGAGCTCA  5200
 A  Y  L  K  P  A  L  F  V  S  E  P  S  Y  A  D  A  G  C  Q  A  C  V  T  E  C  E  V  L  W  A  L  E    1040

                                   end of IVa2 1st exon <
GGGTGTTTTGGGTCAAAAACCAGATTGCCTCCGTGTTTTTGATGCGTTTCTTACCTCGTGTTTCCATGAGGCGGTATCCGGCTTCGGTGACAAACAAGC  5300
P  H  K  P  D  F  V  L  N  G  G  H  K  K  I  R  K  K  G  R  T  E  M  L  R  Y  G  A  E  T  V  F  L  S  1006
                                                           R  T  E  M                                  1

                                                                 5' end of IVa2 mRNA <
TGTCTGTGTCTCCGTAAACTGATTTGAGGGTACGCTGTTCCAACGGTGTGCCTCTGTCCTCTGCGTACAAGATCTCGGACCATTCTGAGACAAAAGCCCG  5400
 D  T  D  G  Y  V  S  K  L  T  R  Q  E  L  P  T  G  R  D  E  A  Y  L  I  E  S  W  E  S  V  F  A  R     973

GGTCCAGGCTAAAACAAAGGAGGCGATTTGGGAGGGATAACGGTTGTTTTCCACCAGGGGGTCGACCTTTTCTAGGGTGTGAAGGCAAAGGTCATCTTCT  5500
 T  W  A  L  V  F  S  A  I  Q  S  P  Y  R  N  N  E  V  L  P  D  V  K  E  L  T  H  L  C  L  D  D  E     940

                                                     L TATA >>>>>>>      5' end of L mRNA  >
TCTGCATCCATAAAGGTAATTGGTTTGTAAGTGTAGGTCACGTGGTCATTGGGCTTGTGCGTGGGTGTATAAAAGGGGGCGTGTCCGGGCTCTTCATCAC  5600
E  A  D  M  F  T  I  P  K  Y  T  Y  T  V  H  D  N  P  K  H  T  P  T  Y  F  P  A  H  G  P  E  E  D  S   906

                end of 1st L leader exon >
TTTCTTCCGCATCGCTGTGGACGACAGCCAGCTGTTCGGGTGAGTATGCGCGTTGAAAGGTGGGCATAACTTCAGCACTTAGAGTGTCAGTTTCCACAAA  5700
  E  E  A  D  S  H  V  V  A  L  Q  E  P  S  Y  A  R  Q  F  T  P  M  V  E  A  S  L  T  D  T  E  V  F   873

CGAGGTGGATTTGATATTTATCTGCCCTGCGGCAATGCTTTTGATGGTGGCTGAATCCATTTGGTCAGAAAATACAATTTTTTTGTTATCAAGTTTGGTA  5800
 S  T  S  K  I  N  I  Q  G  A  A  I  S  K  I  T  A  S  D  M  Q  D  S  F  V  I  K  K  N  D  L  K  T    840

GCAAAGGATCCATAGAGGGCGTTGGAGAGCAGTTTGGCAATGGAACGCAGTGTTTGGTTTTTTTCGCGGTCGGCACGCTCCTTGGCGGTGATATTAAGAT  5900
A  F  S  G  Y  L  A  N  S  L  L  K  A  I  S  R  L  T  Q  N  K  E  R  D  A  R  E  K  A  T  I  N  L  H   806

GAACGTACTCTTTTGCCACGCAGCGCCACTCGGGAAAGACAGTGGCGCGCTCGTCGGGAAGCAACCGCACATGCCAGCCCCGGTTGTGCAGTGTTATAAG  6000
 V  Y  E  K  A  V  C  R  W  E  P  F  V  T  A  R  E  D  P  L  L  R  V  H  W  G  R  N  H  L  T  I  L    773
```

```
GTCCACACTGGTAACTACCTCGCCGCGCAGGGGCTCATTGGTCCAGCAAAGGCGCCCTCCTTTGCGCGAACAGAGTGGGGGCAAAACATCTAGTAGGTTT  6100
 D  V  S  T  V  V  E  G  R  L  P  E  N  T  W  C  L  R  G  G  K  R  S  C  L  P  P  L  V  D  L  L  N    740

TCAGGTGGGGGGTCGGCGTCGATGGTAAAAATGCCAGGCAGCAGGGTGCGATTGAAATAATCAATGGGGGTACCAACTTGCAAAAGAGCGTGTTCCCAAT  6200
E  P  P  P  D  A  D  I  T  F  I  G  P  L  L  T  R  N  F  Y  D  I  P  T  G  V  Q  L  L  A  H  E  W  D  706

CTCGGACCGCTAGGGCGCGCTCGTAGGGATTGAGTGGGAAGCCCCACGGCATGGGATGGGTAAGTGCAGAGGCGTACATGCCACAGATGTCATAAACGTA  6300
  R  V  A  L  A  R  E  Y  P  N  L  P  F  G  W  P  M  P  H  T  L  A  S  A  Y  M  G  C  I  D  Y  V  Y   673

AAGTGGTTCGCGTAGCACCCCAATGTAAGTTGGATAACAGCGTCCTCCGCGAATGCTAGCCCGAACATAGTCATACATTTCGTGGGAAGGGGCCAGCAAG  6400
 L  P  E  R  L  V  G  I  Y  T  P  Y  C  R  G  G  R  I  S  A  R  V  Y  D  Y  M  E  H  S  P  A  L  L    640

CTGCCGCCTAGGTCCGACCGCTGGGGTTTTACTGTCCGGTACAAGATTTGACGAAAGATGGCGTGGGAGTTGGAGGAGATGGTGGGCCGCTGAAAGACGT  6500
S  G  G  L  D  S  R  Q  P  K  V  T  R  Y  L  I  Q  R  F  I  A  H  S  N  S  S  I  T  P  R  Q  F  V  N  606

TAAAGCTGGCTTCGGGTAGACCTACCGCGTCGCGGATAAACTGAGCGTAGGATTCGCGCAACTTTTGCACCAGGGCGGCGGTAACAAGCACATCCAGGGC  6600
  F  S  A  E  P  L  G  V  A  D  R  I  F  Q  A  Y  S  E  R  L  K  Q  V  L  A  A  T  V  L  V  D  L  A   573

                                                        > start of 2nd L leader exon
ACAGTAATCAAGGGTTTCACGCACCAGGTCGTAATGAGGACATTGCTTTTTTTCCCAGAGTTCGCGGTTCAGGAGGTACTCCTCGCGATCCTTCCAGTAA  6700
 C  Y  D  L  T  E  R  V  L  D  Y  H  P  C  Q  K  K  E  W  L  E  R  N  L  L  Y  E  E  R  D  K  W  Y    540

     end of 2nd L leader exon >
TCTTCGGCAGGAAAGCCACGCTCGTCTGCGCGGTAAGAACCCAGCATGTAAAACTCGTTTACGGCCTTGTATGGGCAGCATCCTTTTTCTACCGGAAGGG  6800
D  E  A  P  F  G  R  E  D  A  R  Y  S  G  L  M  Y  F  E  N  V  A  K  Y  P  C  C  G  K  E  V  P  L  T  506

TATAGGCTTGTGCGGCTTTTCGCAGAGACGTGTGTGTGAGGGCAAAAGTGTCGCGCACCATAACCTTGAGGAATTGATACTTAAAATCAGAGTCGTCGCA  6900
  Y  A  Q  A  A  K  R  L  S  T  H  T  L  A  F  T  D  R  V  M  V  K  L  F  Q  Y  K  F  D  S  D  D  C   473

GGCGCCCTGCTCCCATAGGCGATAGTCGGTGCGTTTTTTTGAGCTTGGATTAGGAAGGGCAAAGGTGATATCATTAAAAAGGATTTTGCCGGCTCTAGGC  7000
 A  G  Q  E  W  L  R  Y  D  T  R  K  K  S  S  P  N  P  L  A  F  T  I  D  N  F  L  I  K  G  A  R  P    440

ATAAAGTTGCGGGTGATTTTGAAAGGCCCGGGCACATCAGAACGGTTGTTAATAACCTGAGCTGCAAGTACGATTTCATCGAAGCCGTTGATGTTATGCC  7100
M  F  N  R  T  I  K  F  P  G  P  V  D  S  R  N  N  I  V  Q  A  A  L  V  I  E  D  F  G  N  I  N  H  G  406

CCACAATGTAAAGTTCTAAAAAGCGGGGGCGTCCCTGGAGTTTGGGGGCCTTTTGTAACTCTTCATAGGTAAGGTAATCAGGAGAAAAAAGACCCATTTC  7200
  V  I  Y  L  E  L  F  R  P  R  G  Q  L  K  P  A  K  Q  L  E  E  Y  T  L  Y  D  P  S  F  L  G  M  E   373

CAAGCAAGCCCATTCTGCCAGTTGGGGATTGGCGGCTAGAAAACCGCGCCATAGCTGGAGGGCAAAATGGGCTTGCAAGCGGTTGCGGTACTCTCGAAAC  7300
 L  C  A  W  E  A  L  Q  P  N  A  A  L  F  G  R  W  L  Q  L  A  F  H  A  Q  L  R  N  R  Y  E  R  F    340

TTTTTGCCCACCGCCAATTTTTCGGGGGTCACCACGTAAAAGGTACGTTCGTCGTTTCCCCAAGTGTCCCACTGCAACTCGCAGGCCAGTCGGCAGGCTT  7400
K  K  G  V  A  L  K  E  P  T  V  V  Y  F  T  R  E  D  N  G  W  T  D  W  Q  L  E  C  A  L  R  C  A  E  306

CCTTAACAAGGGCTTCCTCCCCCGAGAGATGCATAACTAGCATAAAGGGGACCAGCTGTTTACCAAAGGCTCCCATCCACGTGTAGGTTTCGACGTCGTA  7500
  K  V  L  A  E  E  G  S  L  H  M  V  L  M  F  P  V  L  Q  K  G  F  A  G  M  W  T  Y  T  E  V  D  Y   273
```

```
GGTGACAAAGAGGCGTTCGACGCGAGGATGAGAGCCGATCGGAAAGAAATTGATTTTCTGCCACCAGCCGGAGGAGTGGGCGTTGATATGATGAAAGTAG  7600
 T  V  F  L  R  E  V  R  P  H  S  G  I  P  F  F  N  I  K  Q  W  W  G  S  S  H  A  N  I  H  H  F  Y    240

AAGTCTCTCCGGCGGACCGTGCATTCGTGCTGATATTTGTAAAAGCGGGCGCAATACTCGCAGCGTTGCACGCTCTGCATCTCTTGAATGAGGTGTACCT  7700
F  D  R  R  R  V  T  C  E  H  Q  Y  K  Y  F  R  A  C  Y  E  C  R  Q  V  S  Q  M  E  Q  I  L  H  V  Q  206

GTCGCCCACGTACGAGAAATCGGAGAGGGAAGTTGAGAAAATCCTCAGTGTCTTGCCTTTCACCCTCGTCGCCCTCTTCTGCACCTGCACGCTCTTGCTG  7800
  R  G  R  V  L  F  R  L  P  F  N  L  F  D  E  T  D  Q  R  E  G  E  D  G  E  E  A  G  A  R  E  Q  Q  173

TGGGTGGATGATGGAGGGAACGACAACGCCCCGCGAGCCACAGGTCCAGACCTCAACGCGGGGCACCTTCAGCTTGAGAGCAAGAGTGCGGATTTGGGAA  7900
 P  H  I  I  S  P  V  V  V  G  R  S  G  C  T  W  V  E  V  R  P  V  K  L  K  L  A  L  T  R  I  Q  S    140

CTGTCCAGGGAGTCCAGGAAGGCCTCGTTCAGATCAGCGGGCACAGATCGAAGGTTGACTTGCAGGAGACGGGTAAGGGCCGATGCCAGGCGGCGATGAA  8000
S  D  L  S  D  L  F  A  E  N  L  D  A  P  V  S  R  L  N  V  Q  L  L  R  T  L  A  S  A  L  R  R  H  F  106

ACTTGATTTCCATTGGTGAGTTGGTAGCAGTGTCAATAGCATACAGAAGACCTTGTCCGCGGGGAGCTACAATGGTACCACGCAGGCGAGAGTTGGGGGT  8100
  K  I  E  M  P  S  N  T  A  T  D  I  A  Y  L  L  G  Q  G  R  P  A  V  I  T  G  R  L  R  S  N  P  T   73

AAGGCTTACATTGTTCGCTGCGGGCGGGCGTCCGGAGGCAGTGGTGGATGGGGGTTCGCCTGGAGAGGCGGTAGCGGCACGTCGGCGTGGAGCTCGGGTA  8200
[E2B pTp]M  T  R  Q  P  R  A  D  P  P  L  P  P  H  P  N  A  Q  L  P  P  L  P  V  D  A  H  L  E  P  L  616
 L  S  V  N  N  A  A  P  P  R  G  S  A  T  T  S  P  P  E  G  P  S  A  T  A  A  R  R  R  P  A  R  T    40

GCGGTTGGTGCTGCGCCCGCAGTTGACTGGCGTACGCGACGACGCGGCGGTTGAGGTCCTGAATGTGTCTCCGCTGGGAGAAAACCACCGGCCCTCGGAC  8300
  P  Q  H  Q  A  R  L  Q  S  A  Y  A  V  V  R  R  N  L  D  Q  I  H  R  R  Q  S  F  V  V  P  G  R  V   583
A  T  P  A  A  G  A  T  S  Q  R  V  R  R  R  P  P  Q  P  G  S  H  T  E  A  P  L  F  G  G  A  R  P  S    6

     < start of E2B pol 4th exon
TCGGAACCTGAAAGAGAGTTCAACAGAATCAATATCGGCATCGTTGACCGCGGCTTGTCGCAGAATCTCCTGCACGTCGCCAGAGTTGTCTTGGTAGGCA  8400
 R  F  R  F  S  L  E  V  S  D  I  D  A  D  N  V  A  A  Q  R  L  I  E  Q  V  D  G  S  N  D  Q  Y  A    550
  P  V                                                                                                   4

ATCTCCGACATAAACTGGTCAATCTCTTCGTCCTGGAGTTCTCCGTGTCCTGCGCGCTCCACCGTGGCTGCAAGGTCATTAGAGATGCGCCTCATGAGCT  8500
I  E  S  M  F  Q  D  I  E  E  D  Q  L  E  G  H  G  A  R  E  V  T  A  A  L  D  N  S  I  R  R  M  L  Q  516

GGGAGAAAGCGTTAAGACCGTTTTCGTTCCACACGCGGCTGTAGACCACGTCGCCAACAGTGTTTCGGGCGCGCATCACCACTTGTGCAATGTTCAGTTC  8600
  S  F  A  N  L  G  N  E  N  W  V  R  S  Y  V  V  D  G  V  T  N  R  A  R  M  V  V  Q  A  I  N  L  E   483

TACGTGTCTTGCAAAGACGGCGTAGTTGCGTAGACGCTGGAAGAGGTAGTTGAGCGTGGTGGCAATGTGCTCGCAAACAAAGAAGTACATGACCCAGCGC  8700
 V  H  R  A  F  V  A  Y  N  R  L  R  Q  F  L  Y  N  L  T  T  A  I  H  E  C  V  F  F  Y  M  V  W  R    450

CGAAGCGTCATTTCGTTAATATCTCCGAGGGCTTCCAAGCGGTCCATTGCCTCGTAGAAGTCAACCGCGAAGTTGAAGAACTGGGAGTTGCGCGCCGCAA  8800
R  L  T  M  E  N  I  D  G  L  A  E  L  R  D  M  A  E  Y  F  D  V  A  F  N  F  F  Q  S  N  R  A  A  F  416
```

```
ACGTCAACTCCTCTTGCAGGAGCCGAATTGCCTCGGCTACAGTTTCGCGCACCTCTTGTTCGAAGGCTGCCGGCGTTTCCTCGATTTCCATAAACTCCTC  8900
 T  L  E  E  Q  L  L  R  I  A  E  A  V  T  E  R  V  E  Q  E  F  A  A  P  T  E  E  I  E  M  F  E  E   383

TTCCTCCACAGCGGGACCCTCGGGGCTGACCGGCGCTGGGACGGGTTGTCGTCGACGACGGCGCCGGACGGGCAGCCGGTCAATGAAACGTTGAATCATT  9000
 E  E  V  A  P  G  E  P  S  V  P  A  P  V  P  Q  R  R  R  R  R  R  V  P  L  R  D  I  F  R  Q  I  M   350

TCTCCGCGACGGCGACGCATGGTTTCGGTGACGGCGCGCCCGTTTCTCGGGGGCGAAGTTCAAAGACGCCGCCTTGCATGCCCGAGCCGGAGAGGGGAG  9100
E  G  R  R  R  R  M  T  E  T  V  A  R  G  N  E  R  P  R  L  E  F  V  G  G  Q  M  G  S  G  S  L  P  P   316

                                                                                  > start of 3rd L leader exon
GAAGTAGGTGGGGCCCCTGAGGCAGCGACAGGGCGCTAACTGTGCATCTTATCATCTGTTGCATAGGTAGAGACTGCCAAGCCTCATTGAGCGAGTCCAG  9200
 L  L  H  P  G  Q  P  L  S  L  A  S  V  T  C  R  I  M  Q  Q  M  P  L  S  Q  W  A  E  N  L  S  D  L   283

                             end of 3rd L leader exon >
TTGGACGGGATCAGAGAATTTTTCGAGGAAAGCTTCCAGCCAATCGCAGTCGCAAGGTAAGCTAAGGACGGTGGCATGAGGGATTCTAAGGGAGGCAGCA  9300
 Q  V  P  D  S  F  K  E  L  F  A  E  L  W  D  C  D  C  P  L  S  L  V  T  A  H  P  I  R  L  S  A  A   250

GAGGAGGTGATGCTGCTGATGAGGAAATTGAAGTAGGCGGTCTTCAAACGGCGGATGGTGGCAAGGAGAGTGACGTCTTTTGGTCCGGCCTGTTGAATTC  9400
S  S  T  I  S  S  I  L  F  N  F  Y  A  T  K  L  R  R  I  T  A  L  L  T  V  D  K  P  G  A  Q  Q  I  R   216

GCAGGCGGTCTGCCATGCCCCAAGCTTCGTTCTGACATCGGCGCAGGTCCTTGTAATAATCTTGCATGAGACTTTCTACGGGTATTTCCAATTCCCCTCG  9500
 L  R  D  A  M  G  W  A  E  N  Q  C  R  R  L  D  K  Y  Y  D  Q  M  L  S  E  V  P  I  E  L  E  G  R   183

GTCGGCCATGCGTGTGGAACCAAACCCGCGCAGGGGCTGCAGCAGGGCCAAGTCGGCAACTACGCGTTCGGCCAGCACAGCCTGCTGTATCTGAGTTAAA  9600
 D  A  M  R  T  S  G  F  G  R  L  P  Q  L  L  A  L  D  A  V  V  R  E  A  L  V  A  Q  Q  I  Q  T  L   150

GTGTTTTGGAAATCATCCAAGTCCACAAAGCGGTGGTAGGAACCGGTGTTGATGGTGTACGTGCAGTTGGCCATGACGGACCAGTTGACTACTTGCATCC  9700
T  N  Q  F  D  D  L  D  V  F  R  H  Y  S  G  T  N  I  T  Y  T  C  N  A  M  V  S  W  N  V  V  Q  M  G   116

CGGGCTGTGTAATCTCGGTATACCTAAGGCGCGAGTAGGCTCTGGATTCAAAAACGTAGTCGTTGCAGGTGCGAACCAAGTACTGGTAGCCAACAAGGAA  9800
 P  Q  T  I  E  T  Y  R  L  R  S  Y  A  R  S  E  F  V  Y  D  N  C  T  R  V  L  Y  Q  Y  G  V  L  F    83

GTGGGGCGGCGGCTCGCGGTAAAGGGGCCAGCGAAGTGTGGCGGGCGTACCGGGGGCCAGGTCCTCCAGCATAAGGCGATGGTAGTGGTAAACATATCGA  9900
 H  P  P  P  E  R  Y  L  P  W  R  L  T  A  P  T  G  P  A  L  D  E  L  M  L  R  H  Y  H  Y  V  Y  R    50

GAGAGCCAGGTGATGCCGGCGGCGGTAGTGGCGGCGCGGGCGTATTCGCGAACGCGGTTCCAGATGTTACGCAACGGGGAGAAGCGTTCCATGGCGGGCA 10000
S  L  W  T  I  G  A  A  T  T  A  A  R  A  Y  E  R  V  R  N  W  I  N  R  L  P  S  F  R  E  M  A  P  V    16

                                          < start of E2B pTP 4th exon
CGCTTTGACCAGTCAGACGGGCGCAATCTTGTACGCTCTAGATGAAAAAACAGAGAGCGGTCACGGACTTTCCTCCGTAGCCTGGAGGACAGACCGCCAG 10100
 S  Q  G  T  L  R  A  C  D  Q  V  S                                                  > start of VA RNA      4

GGTGCAGTGGCAAACAACCCCCGGTTCGAGACCGGCTGGATCTGCCACTCCCGACGCGCCGGCCGTGCGTCCACGACGGAAACCCCGCCGAGACCTAGCC 10200

                                               L1 52,55K  M  H  P  V  L  R  C  M  R  P  T  A  P  P  T  Q    16
GCGGTCCCTGGATCTCCAGATACGGAGGGGAGTCTTTTTGTTGTTTTTTGTAGATGCATCCGGTGTTGCGACAGATGCGTCCGACGGCGCCTCCAACACA 10300
                  end of VA RNA >                   > start of L1 52,55K coding exon
```

```
 P  P  L  P  P  P  T  S  A  P  A  A  V  A  L  S  G  A  G  G  G  N  P  E  E  E  A  I  L  D  L  E  E     49
GCCGCCGCTCCCGCCCCCCACTAGCGCCCCTGCAGCCGTTGCTCTCTCCGGAGCCGGCGGTGGCAACCCTGAGGAGGAGGCCATCCTGGACCTGGAAGAG 10400

 G  E  G  L  A  R  L  G  A  P  S  P  E  R  H  P  R  V  Q  L  K  K  D  S  R  Q  A  Y  V  P  P  Q  N     82
GGCGAGGGGCTGGCCCGCTTGGGGGCGCCATCCCCCGAGCGCCATCCCCGCGTGCAACTTAAAAAGGACTCACGCCAGGCGTACGTACCGCCTCAGAATT 10500

L  F  R  D  R  S  G  Q  E  P  E  E  M  R  D  R  R  F  Y  A  G  Q  E  L  R  A  G  F  N  R  Q  R  V  L   116
TATTCAGGGATCGCAGCGGGCAGGAGCCCGAAGAGATGAGGGATCGCAGGTTTTACGCGGGGCAGGAGCTGCGGGCCGGTTTTAACCGCCAACGGGTGCT 10600

 R  A  E  D  F  E  P  D  E  H  S  G  I  S  P  A  R  A  H  V  S  A  A  D  L  V  T  A  Y  E  Q  T  V    149
ACGCGCCGAAGATTTTGAACCCGACGAACATAGCGGAATAAGTCCGGCACGGGCGCACGTGTCGGCGGCCGATTTGGTAACCGCGTACGAGCAAACGGTG 10700

 N  E  E  R  N  F  Q  K  S  F  N  N  H  V  R  T  L  V  A  R  E  E  V  A  I  G  L  M  H  L  W  D  F    182
AACGAGGAGCGCAACTTTCAGAAAAGTTTTAACAATCACGTGCGCACCCTGGTGGCGCGCGAGGAGGTGGCCATTGGGCTGATGCATTTGTGGGACTTTA 10800

M  E  A  Y  V  Q  N  P  S  S  K  P  L  T  A  Q  L  F  L  I  V  Q  H  S  R  D  N  E  A  F  R  E  A  M   216
TGGAGGCGTACGTGCAAAATCCTTCGAGCAAGCCGCTGACGGCGCAGCTGTTTTTGATTGTGCAACACAGCCGGGACAACGAGGCTTTCCGCGAGGCCAT 10900

  L  N  I  A  E  P  E  G  R  W  L  L  D  L  V  N  I  L  Q  S  I  V  V  Q  E  R  S  L  S  L  A  D  K   249
GCTGAATATTGCGGAGCCTGAGGGTCGCTGGCTTTTGGACCTGGTTAATATCCTTCAGAGCATTGTGGTACAGGAGCGCAGTCTAAGCCTGGCCGACAAG 11000

 V  A  A  I  N  Y  S  M  L  S  L  G  K  F  Y  A  R  K  I  Y  K  T  P  Y  V  P  I  D  K  E  V  K  I    282
GTGGCGGCCATTAATTACAGCATGCTTAGCCTCGGCAAGTTTTACGCCCGCAAGATTTACAAAACCCCCTATGTGCCCATAGACAAGGAGGTTAAAATAG 11100

D  S  F  Y  M  R  M  A  L  K  V  L  T  L  S  D  D  L  G  V  Y  R  N  D  R  I  H  K  A  V  S  A  S  R   316
ATAGCTTTTACATGCGCATGGCGCTAAAGGTGTTAACGCTGAGTGACGATCTGGGGGTGTACCGCAACGACCGTATTCACAAAGCTGTGAGCGCCAGCCG 11200

 R  R  E  L  S  D  R  E  L  M  H  S  L  R  R  A  L  T  G  T  G  T  D  A  E  T  E  S  Y  F  D  M  G    349
CCGTCGCGAGCTTAGCGACCGCGAACTAATGCACAGCCTGCGTCGGGCTCTAACGGGCACCGGCACTGATGCCGAAACTGAATCTTACTTTGACATGGGG 11300

 A  D  L  Q  W  Q  P  S  A  R  A  L  E  A  A  G  Y  V  G  A  E  E  D  E  E  D  Y  E  D  E  P  -       380
GCGGACCTGCAATGGCAGCCCAGCGCCCGGGCCCTGGAGGCGGCTGGTTATGTTGGCGCGGAAGAAGATGAGGAGGACTATGAGGACGAGCCCTGATCAG 11400

                 L1 IIIa  M  L  R  S  T  A  V  A  D  G  S  Q  Q  V  N  P  A  M  L  A  A  L  Q  S  Q  P  S    27
CCAGGTGGTGTTTTTGTAGATGCTGCGTTCGACGGCGGTGGCGGACGGGTCGCAGCAGGTGAATCCCGCTATGTTGGCGGCCCTGCAAAGCCAACCTTCG 11500
                       > start of L1 IIIa coding exon

 G  V  T  P  S  D  D  W  A  A  A  M  D  R  I  L  A  L  T  T  R  N  P  E  A  F  R  Q  Q  P  Q  A  N     60
GGCGTGACACCCTCAGACGACTGGGCGGCGGCCATGGATCGCATCCTGGCCCTAACCACCCGCAATCCCGAAGCCTTCAGGCAGCAGCCCCAGGCCAACC 11600

R  F  S  A  I  L  E  A  V  V  P  S  R  T  N  P  T  H  E  K  V  L  A  I  V  N  A  L  V  E  S  K  A  I    94
GCTTTTCGGCCATTTTGGAAGCCGTGGTTCCTTCTCGCACTAACCCTACCCACGAAAAGGTGTTGGCGATTGTAAACGCTCTGGTAGAAAGCAAAGCCAT 11700
```

```
 R  K  D  E  A  G  L  I  Y  N  A  L  L  E  R  V  A  R  Y  N  S  T  N  V  Q  A  N  L  D  R  L  T  T    127
CCGCAAGGATGAGGCGGGACTGATATATAATGCCTTACTGGAGCGGGTAGCGCGCTATAATAGCACCAACGTGCAGGCCAACCTAGACCGACTGACAACG 11800

 D  V  R  E  A  V  A  Q  R  E  R  F  M  H  D  V  N  L  G  S  Q  V  A  L  N  A  F  L  S  T  L  P  A    160
GACGTGAGAGAGGCGGTGGCGCAGCGGGAACGCTTTATGCATGACGTTAACCTAGGATCCCAAGTGGCCCTTAATGCTTTTCTGAGCACATTGCCAGCTA 11900

N  V  P  R  G  Q  E  D  Y  V  S  F  I  S  A  L  R  L  L  V  A  E  V  P  Q  S  E  V  Y  Q  S  G  P  D  194
ATGTGCCGCGCGGGCAGGAGGACTATGTCAGCTTTATCAGCGCGCTTCGCCTCCTGGTGGCGGAGGTGCCCCAGAGTGAGGTGTACCAGTCCGGGCCAGA 12000

  Y  F  F  Q  T  S  R  Q  G  L  Q  T  V  N  L  T  Q  A  F  K  N  L  Q  G  M  W  G  V  R  A  P  V  G   227
CTACTTTTTCCAAACTTCACGGCAGGGTTTGCAAACTGTAAACCTAACACAGGCGTTTAAAAACTTGCAAGGAATGTGGGGCGTGCGAGCCCCCGTGGGG 12100

 D  R  A  T  I  S  S  L  L  T  P  N  T  R  L  L  L  L  L  I  A  P  F  T  N  S  S  T  I  S  R  D  S    260
GACCGAGCCACCATCTCCAGCTTATTGACGCCGAACACTCGGTTGTTGCTGTTGTTGATAGCGCCATTTACCAATAGCAGCACCATTAGCCGTGACTCGT 12200

Y  L  G  H  L  I  T  L  Y  Q  E  A  I  G  Q  T  Q  V  D  E  Q  T  F  Q  E  I  T  S  V  S  R  A  L  G  294
ACCTTGGTCATCTAATCACGCTGTACCAGGAGGCTATTGGCCAGACGCAGGTGGACGAACAGACCTTCCAGGAAATCACCAGCGTGAGTCGGGCTCTTGG 12300

  Q  Q  D  T  S  S  L  E  A  T  L  N  F  L  L  T  N  R  R  Q  K  I  P  S  Q  F  T  L  N  S  E  E  E   327
TCAGCAAGACACCAGTAGCTTGGAAGCCACGCTGAACTTTTTGCTAACCAACCGCCGGCAAAAAATTCCATCGCAATTTACTTTAAATTCTGAAGAGGAA 12400

 R  I  L  R  Y  V  Q  Q  S  V  S  L  Y  L  M  R  E  G  A  T  A  S  S  A  L  D  M  T  A  R  N  M  E    360
CGTATTTTGCGCTATGTGCAGCAGTCCGTCAGCTTGTATTTAATGCGCGAGGGGGCTACTGCGTCGTCGGCTCTGGACATGACGGCACGTAACATGGAAC 12500

P  S  L  Y  S  S  N  R  P  F  I  N  R  L  M  D  Y  L  H  R  A  A  A  M  N  S  E  Y  F  T  N  A  I  L  394
CGTCGCTATACTCGTCCAACCGGCCTTTTATTAACCGCCTGATGGACTATCTGCACCGCGCGGCGGCCATGAACAGCGAGTACTTCACCAATGCCATTCT 12600

  N  P  H  W  M  P  P  S  G  F  Y  T  G  E  F  D  V  P  E  G  D  D  G  F  L  W  D  D  V  S  E  S  I   427
TAACCCGCACTGGATGCCACCGTCCGGTTTTTACACGGGCGAGTTTGACGTGCCTGAGGGCGACGACGGATTTTTGTGGGATGACGTGTCCGAAAGCATT 12700

 F  E  P  M  R  S  R  K  K  E  G  G  D  E  L  P  L  S  L  V  E  A  A  S  R  G  Q  T  P  V  P  S  L    460
TTCGAACCAATGCGTTCCCGTAAAAAGGAGGGCGGAGACGAGCTGCCGCTGTCGTTAGTGGAGGCAGCTTCTCGAGGCCAAACCCCCGTTCCCAGTCTGC 12800

P  S  L  T  S  S  S  S  G  R  V  F  R  P  R  L  P  G  E  L  D  Y  L  S  D  P  L  L  R  P  A  R  K  K  494
CATCGCTGACCAGCAGCAGCAGCGGACGGGTGTTTCGACCCCGTTTGCCCGGGGAGTTGGACTACCTCAGCGATCCCCTATTGCGACCGGCCCGGAAAAA 12900

  N  F  P  N  N  G  V  E  S  L  V  D  K  M  N  R  W  K  T  Y  A  Q  E  Q  R  E  E  R  Q  P  R  P  L   527
AAATTTTCCCAACAACGGGGTGGAAAGCCTGGTAGATAAGATGAATCGCTGGAAAACCTACGCCCAGGAGCAGCGGGAAGAGAGGCAGCCCCGCCCACTG 13000

 T  G  T  F  S  R  W  R  R  R  E  E  D  A  Y  D  S  A  D  D  S  S  V  L  D  L  G  G  T  G  A  A  S    560
ACCGGCACCTTCAGTCGTTGGCGCCGGCGGGAAGAGGACGCTTACGACTCGGCCGATGATAGTAGCGTGTTGGACTTGGGGGGAACCGGCGCCGCTTCTG 13100

D  P  F  A  H  L  R  P  Q  G  Q  L  G  R  L  Y  -    >>>>>>>>>>> L1 pA                                576
ATCCCTTTGCTCATCTGCGGCCTCAGGGTCAACTGGGTCGTTTGTATTAAAAAAATAAAATAAAAAGAAATCCACTTACCAGAGCCATAGCAACAGCGTC 13200
                                                                 end of E2B pol, pTP 3rd exon <L  A  M      1
                                                                                               L  A  M      1
```

```
                   L2 III (penton base) M R R A V G V P P V M A Y A E G P P P S    20
CGTCCCTTTGTCTGTTTTTCCCTCTTCCCGGTAGTCAAAATGAGACGTGCGGTGGGAGTGCCGCCGGTGATGGCGTACGCCGAGGGTCCTCCTCCTTCT 13300
                         > start of L2 III coding exon

 Y E S V M E T A D L P A T L Q A L H V P P R Y L G P T E G R N S I    53
TACGAAAGCGTGATGGAAACAGCGGATTTGCCGGCAACGCTGCAGGCGCTCCACGTCCCTCCCCGTTACCTGGGGCCTACGGAAGGGCGGAACAGCATAC 13400
                              start of E2B pol, pTP 3rd exon <

R Y S E L A P L Y D T T R V Y L V D N K S A D I A S L N Y Q N D H S    87
GTTACTCGGAGCTGGCGCCTCTATACGACACCACCCGGGTTTACCTGGTGGACAACAAGTCGGCGGACATTGCCTCCCTGAACTACCAGAACGACCATAG 13500

 N F Q T T V V Q N N D F T P T E A G T Q T I N F D D R S R W G G D   120
TAACTTTCAAACCACGGTGGTACAAAATAATGACTTTACCCCGACAGAGGCCGGTACCCAGACCATCAATTTTGACGATCGCTCCAGGTGGGGCGGCGAC 13600

 L K T I L R T N M P N I N E F M S T N K F R A R V M V E K V N R K   153
CTGAAAACCATTTTGCGCACCAATATGCCCAACATCAATGAGTTTATGTCTACCAACAAGTTTCGGGCGCGGGTGATGGTAGAAAAAGTGAACCGGAAAA 13700

T N A P R Y E W F E F T L P E G N Y S E T M T I D L M N N A I V D N   187
CCAACGCTCCTCGTTACGAGTGGTTCGAGTTCACTTTGCCAGAGGGCAACTATTCGGAAACTATGACTATAGACCTTATGAATAACGCGATCGTAGACAA 13800

 Y L A V G R Q N G V L E S D I G V K F D T R N F R L G W D P V T K   220
CTACTTAGCAGTAGGACGTCAGAACGGCGTGCTGGAAAGCGACATTGGGGTGAAGTTTGACACGCGCAACTTCCGGTTGGGTTGGGATCCCGTAACCAAG 13900

 L V M P G V Y T N E A F H P D I V L L P G C G V D F T Q S R L N N   253
TTGGTGATGCCCGGCGTGTACACCAACGAGGCCTTTCACCCAGACATTGTTTTGCTACCTGGTTGCGGCGTGGATTTCACGCAAAGTCGTCTGAACAACT 14000

L L G I R K R M P F Q K G F Q I M Y E D L E G G N I P A L L D V E K   287
TGCTAGGAATACGCAAGCGAATGCCCTTTCAAAAAGGTTTCCAAATCATGTATGAGGATTTGGAGGGCGGCAACATTCCTGCTCTATTAGATGTGGAAAA 14100

 Y E A S I K E A Q E I R G A D F K P N P Q D L E I V P V E K D S K   320
GTACGAAGCTAGCATAAAAGAAGCACAGGAGATCCGTGGAGCCGACTTCAAGCCCAATCCTCAAGACTTGGAAATCGTGCCCGTGGAAAAAGACAGCAAG 14200

 E R S Y N L L E G D K N N T A Y R S W F L A Y N Y G D A E K G V K   353
GAAAGAAGTTACAATCTCCTAGAGGGAGATAAAAATAACACTGCCTACCGCAGCTGGTTTTTGGCCTACAACTACGGAGATGCAGAGAAAGGAGTAAAGT 14300

S W T L L T T T D V T C G S Q Q V Y W S L P D M M Q D P V T F R P S   387
CTTGGACCTTGTTAACAACCACGGATGTGACCTGTGGGTCGCAGCAGGTGTACTGGTCCCTTCCCGACATGATGCAAGATCCAGTAACGTTTCGACCGTC 14400

 T Q V S N Y P V V G V E L L P V H A K S F Y N E Q A V Y S Q L I R   420
CACGCAAGTCAGCAACTACCCTGTAGTGGGGGTGGAATTACTGCCAGTACATGCCAAGAGTTTTTACAACGAGCAGGCCGTGTATTCTCAGCTTATTCGC 14500

 Q S T A L T H I F N R F P E N Q I L V R P P A P T I T T V S E N V   453
CAGTCCACCGCGCTTACGCACATCTTCAATCGTTTTCCTGAGAATCAGATACTAGTGCGTCCGCCCGCTCCGACCATTACCACCGTCAGTGAAAACGTTC 14600
```

```
P A L T D H G T L P L R S S I S G V Q R V T I T D A R R R T C P Y V   487
CCGCCCTCACAGATCACGGAACCCTGCCGCTGCGCAGCAGTATCAGTGGAGTTCAGCGCGTGACCATCACTGACGCCCGCCGTCGGACCTGCCCCTACGT 14700
          > start of L2 pVII coding exon
                                                      L2 pVII M S I L I S P D N N T G W G L   15
 H K A L G I V A P K V L S S R T F -                                                          504
GCACAAAGCTCTGGGCATAGTTGCTCCCAAAGTGCTGTCTAGCCGCACGTTTTAACATGTCCATTCTTATTTCGCCCGACAACAATACCGGCTGGGACT 14800

 C S A G M Y G G A K R R S S Q H P V R V R G H Y R A P W G A Y T R   48
TTGCTCCGCCGGCATGTACGGCGGCGCCAAACGGCGTTCTAGCCAACACCCTGTTCGCGTGCGCGGACATTACCGCGCCCCCTGGGGGGCTTACACCCGC 14900

 G V I S R R T T V D D V I D S V V A D A Q R Y T R P V A T S T V D   81
GGTGTTATCTCAAGACGTACCACCGTTGATGACGTCATTGACTCCGTGGTAGCCGATGCCCAACGCTACACGCGGCCCGTTGCCACGTCCACCGTGGATT 15000

S V I D S V V A N A R R Y A Q R K R R L Q R R R R R P T A A M T A A   115
CCGTGATTGATAGTGTGGTGGCCAACGCCAGGCGTTACGCGCAACGCAAGAGACGTTTGCAACGTCGCCGTCGTCGGCCTACTGCCGCCATGACTGCCGC 15100

 R A V L R R A Q R I G R R A M R R A A A S A S A G R A R R Q A A R   148
TCGGGCGGTACTAAGGCGGGCACAAAGGATAGGACGTCGGGCCATGCGCCGAGCGGCTGCTTCTGCCAGTGCAGGTCGGGCCCGTCGTCAGGCCGCCCGT 15200

 Q A A A A I A S M A Q P R R G N I Y W V R D A S G V R V P V R S R   181
CAGGCCGCGGCGGCTATTGCCAGCATGGCTCAGCCCCGCCGGGGGAATATCTACTGGGTGCGAGATGCGTCGGGCGTGCGGGTGCCGGTGCGAAGCCGTC 15300

                                                                  L2 V M S K R K F K   7
P P R S -                                                                              185
CCCCTCGGAGTTAGAAGACGCGTTCACAAAATGGACGAAGACTGAGTTTCCCTGTCGTTGCCAGCCGGTCCCCGTCAGCATGAGCAAGCGCAAGTTCAAA 15400
                                                                   > start of L2 V coding exon

 E E L L E A L V P E I Y G P A A D V K P D I K P R V L K R V K K R   40
GAAGAGCTGCTGGAGGCCCTTGTGCCTGAAATCTATGGCCCTGCCGCGGACGTCAAGCCCGACATTAAGCCTCGCGTGCTCAAGCGGGTTAAAAAGCGAG 15500

E K K E E K E E A G L L D D G V E F V R S F A P R R R V Q W R G R K   74
AAAAAAAAGAGGAAAAGGAGGAAGCAGGGTTGCTAGACGACGGTGTTGAGTTTGTGCGGTCCTTTGCCCCCCGGCGGCGGGTGCAGTGGCGGGGACGTAA 15600

 V Q R V L R P G T T V V F T P G E R S V T R A L K R D Y D E V Y A   107
AGTCCAGCGCGTGCTTAGACCCGGCACTACTGTAGTATTTACTCCCGGAGAGCGGTCCGTCACGCGGGCCTTAAAACGGGATTACGATGAGGTTTACGCT 15700

 D E D I L E Q A A Q Q V G E F A Y G K R G R Y G E L G L L L D Q S   140
GACGAAGACATTCTTGAGCAGGCCGCCCAACAGGTTGGGGAATTCGCCTACGGCAAGCGCGGCCGCTACGGAGAGTTGGGACTCTTGCTGGACCAAAGCA 15800

N P T P S L K P A T A Q Q I L P V T E I K R G V K R E N K D E L Q P   174
ACCCCACGCCAAGCCTGAAGCCCGCAACGGCGCAGCAGATCCTTCCCGTGACAGAAATCAAGCGGGGCGTCAAGAGGGAAAACAAAGACGAATTGCAGCC 15900

 T M Q L M V P K R Q K L E E V L E N M K V D P S V E P E V K V R P   207
CACCATGCAACTCATGGTGCCAAAGCGGCAAAAGCTTGAGGAGGTGTTGGAGAACATGAAAGTGGATCCCAGCGTTGAGCCGGAAGTTAAAGTGCGCCCC 16000
```

```
 I  K  E  I  G  P  G  L  G  V  Q  T  V  D  I  Q  I  P  V  R  A  S  S  S  T  V  S  T  A  V  E  A  M    240
ATTAAAGAAATAGGGCCCGGACTTGGCGTGCAGACGGTGGATATCCAAATCCCCGTGCGTGCGTCTTCGTCCACCGTTAGCACTGCGGTGGAGGCCATGG 16100

 E  T  Q  P  E  L  P  E  A  V  A  R  A  V  A  A  T  R  E  M  G  L  Q  T  D  P  W  Y  E  F  V  A  P  T    274
AAACGCAGCCTGAGCTGCCAGAGGCCGTAGCCCGTGCGGTTGCGGCCACGCGAGAGATGGGTTTGCAAACGGATCCGTGGTACGAATTCGTGGCCCCTAC 16200

 S  R  P  R  S  R  K  Y  T  T  A  N  S  I  L  P  E  Y  A  L  H  P  S  I  T  P  T  P  G  Y  R  G  T    307
CAGCCGTCCACGCTCCCGGAAATACACAACCGCTAATTCGATTTTACCGGAGTATGCCTTGCATCCATCCATCACGCCAACGCCCGGTTACCGCGGAACA 16300

 T  F  K  P  S  R  T  R  S  T  R  R  R  R  S  V  R  R  R  S  R  R  T  A  P  I  S  V  R  R  V  T  R    340
ACCTTCAAACCCAGCCGCACTCGCTCCACCCGCCGTCGTCGCTCTGTCCGCCGCCGCTCAAGGCGCACGGCCCCCATCTCTGTGCGTCGCGTAACCCGCC 16400

                                                                  L2 pmu  M  A  L  T  C  R  F      7
 R  G  R  T  L  T  L  P  N  A  R  Y  H  P  S  I  L  V  -                                           358
GTGGACGCACGCTGACCCTTCCCAACGCGCGTTACCACCCTAGCATTCTCGTTTAATCCGTGCGCTGCCGTTTTTCAGATGGCTTTGACTTGCCGGTTT 16500
                                              start of L2 pmu coding exon >

 R  I  P  V  P  S  Y  R  G  R  S  R  R  R  R  G  M  A  G  S  G  R  R  R  A  L  R  R  R  I  K  G  G     40
CGCATTCCCGTTCCGTCCTACCGAGGAAGATCTCGCCGTAGGAGAGGCATGGCGGGCAGTGGCCGCCGACGCGCTTTGCGCAGGCGAATAAAAGGCGGAT 16600

 F  L  P  A  L  I  P  I  I  A  A  A  I  G  A  I  P  G  V  A  S  V  A  L  Q  A  A  R  K  Q  -         70
TTTTGCCCGCGTTGATTCCCATCATCGCCGCCGCCATAGGCGCAATCCCAGGCGTGGCCTCCGTGGCCTTGCAAGCAGCTCGCAAACAATAAAAGAAGGC 16700
                                                                                      >>>>>> L2 pA

                                        L3 pVI  M  E  D  I  N  F  A  S  L  A  P  R  H  G  S  R  P  F  M     19
TTAACACTGACTTCCTGGTCCTGACTATTTTATGCAGACAAGACATGGAAGACATCAATTTTGCGTCGCTGGCTCCGCGGCACGGCTCGCGGCCGTTTAT 16800
                                        > start of L3 pVI coding exon

 G  T  W  N  E  I  G  T  S  Q  L  N  G  G  A  F  S  W  S  S  L  W  S  G  I  K  N  F  G  S  S  I  K     52
GGGCACCTGGAACGAGATCGGCACCAGCCAGCTCAACGGGGGCGCTTTCAGTTGGAGCAGCCTGTGGAGTGGCATTAAAAACTTTGGGTCCTCCATTAAG 16900

 S  F  G  N  K  A  W  N  S  N  T  G  Q  M  L  R  D  K  L  K  D  Q  N  F  Q  Q  K  V  V  D  G  L  A     85
TCATTTGGTAACAAGGCCTGGAACAGTAACACAGGTCAAATGCTCCGGGATAAGCTAAAGGACCAAAACTTTCAACAAAAAGTCGTGGACGGGCTGGCTT 17000

 S  G  I  N  G  V  V  D  I  A  N  Q  A  L  Q  N  Q  I  N  Q  R  L  E  N  S  R  Q  P  P  V  A  L  Q  Q    119
CCGGCATTAACGGCGTGGTGGATATAGCCAACCAGGCCTTGCAAAACCAAATCAATCAGCGGCTGGAAAATAGCCGCCAGCCTCCGGTGGCTCTGCAGCA 17100

 R  P  P  P  K  V  E  E  V  E  V  E  E  K  L  P  P  L  E  V  A  P  P  L  P  S  K  G  E  K  R  P  R    152
GCGCCCGCCTCCCAAAGTCGAGGAGGTAGAAGTGGAGGAAAAACTACCGCCTTTGGAGGTGGCACCCCCCCTGCCTAGTAAAGGCGAAAAGCGGCCGCGA 17200

 P  D  L  E  E  T  L  V  V  E  S  R  E  P  P  S  Y  E  Q  A  L  K  E  G  A  S  P  Y  P  M  T  K  P    185
CCGGATCTGGAGGAAACCCTAGTTGTGGAATCCCGCGAGCCCCCCTCGTACGAGCAGGCTTTGAAAGAGGGCGCTTCACCTTATCCCATGACCAAACCTA 17300
```

```
I  G  S  M  A  R  P  V  Y  G  K  E  S  K  P  V  T  L  E  L  P  P  P  V  P  T  V  P  P  M  P  A  P  T    219
TTGGTTCCATGGCCCGACCTGTATACGGGAAGGAAAGCAAACCCGTGACCTTAGAACTACCTCCACCCGTGCCCACCGTTCCGCCCATGCCGGCTCCGAC 17400

  L  G  T  A  V  S  R  P  T  A  P  T  V  A  V  A  T  P  A  R  R  P  R  G  A  N  W  Q  S  T  L  N  S    252
GCTTGGCACCGCCGTTTCTCGCCCCACCGCCCCCACTGTTGCCGTGGCTACCCCCGCCCGCCGCCCTCGCGGGGCTAACTGGCAGAGCACTCTTAACAGC 17500

 I  V  G  L  G  V  K  S  L  K  R  R  R  C  Y  -                                                      267
ATTGTGGGTCTGGGAGTAAAAAGCCTGAAACGCCGCCGGTGCTATTAAAATGGAACCAAGCTAAATGCCATTATTGTGTACGCCTCCTGTGTTACGCCAG 17600

                                      L3 II (hexon)  M  A  T  P  S  M  M  P  Q  W  S  Y  M  H  I  A  G  Q  D     19
AGAGCCGAGTGACACGTCACCGCCAAGAGCGCCGCTTGCAAGATGGCCACCCCCTCGATGATGCCGCAATGGTCTTACATGCACATCGCCGGGCAGGACG 17700
> start of L3 II coding exon

A  S  E  Y  L  S  P  G  L  V  Q  F  A  R  A  T  D  T  Y  F  S  L  G  N  K  F  R  N  P  T  V  A  P  T     53
CCTCGGAGTACCTGAGCCCGGGCCTGGTGCAGTTCGCCCGTGCCACCGATACCTACTTCAGCCTGGGGAACAAGTTCAGAAACCCCACCGTGGCTCCCAC 17800

  H  D  V  T  T  D  R  S  Q  R  L  T  L  R  F  V  P  V  D  R  E  E  T  A  Y  S  Y  K  V  R  F  T  L     86
CCACGATGTAACCACAGACAGGTCGCAGCGACTGACGCTGCGCTTCGTGCCCGTCGACCGCGAGGAAACCGCCTACTCTTACAAAGTGCGCTTTACGCTG 17900

 A  V  G  D  N  R  V  L  D  M  A  S  T  Y  F  D  I  R  G  V  L  D  R  G  P  S  F  K  P  Y  S  G  T    119
GCCGTGGGCGACAACCGGGTTTTGGACATGGCCAGCACCTACTTTGACATCCGCGGCGTGCTGGATCGTGGTCCCAGCTTTAAACCCTATTCGGGCACTG 18000

A  Y  N  S  L  A  P  K  G  A  P  N  P  S  Q  W  T  N  Q  N  K  T  N  S  F  G  Q  A  P  Y  I  G  Q  K    153
CATACAACTCCCTGGCCCCCAAAGGTGCTCCCAATCCTAGCCAGTGGACAAACCAAAACAAAACAAACTCCTTTGGACAAGCTCCCTATATAGGACAAAA 18100

  I  T  N  Q  G  V  Q  V  G  S  D  S  N  N  R  D  V  F  A  D  K  T  Y  Q  P  E  P  Q  V  G  Q  T  Q    186
AATCACCAATCAGGGCGTGCAAGTGGGCTCAGACTCCAACAATCGCGATGTGTTTGCCGATAAAACGTACCAACCGGAGCCTCAAGTGGGGCAGACGCAA 18200

 W  N  I  N  P  M  Q  N  A  A  G  R  I  L  K  Q  T  T  P  M  Q  P  C  Y  G  S  Y  A  R  P  T  N  E    219
TGGAACATTAATCCAATGCAAAACGCTGCGGGAAGAATACTAAAACAAACCACGCCCATGCAGCCATGTTATGGGTCATACGCTAGACCAACAAACGAAA 18300

K  G  G  Q  A  K  L  V  K  N  D  D  N  Q  T  T  T  T  N  V  G  L  N  F  F  T  T  A  T  E  T  A  N  F    253
AAGGAGGTCAAGCCAAGCTGGTAAAAAATGACGACAATCAGACCACAACAACAAACGTAGGTTTAAACTTTTTTACCACTGCCACTGAAACCGCTAATTT 18400

  S  P  K  V  V  L  Y  S  E  D  V  N  L  E  A  P  D  T  H  L  V  F  K  P  D  V  N  G  T  S  A  E  L    286
TTCACCAAAGGTGGTTCTGTACAGCGAAGATGTTAACTTAGAAGCGCCCGATACCCACCTTGTGTTTAAGCCAGATGTCAACGGCACAAGTGCCGAGCTT 18500

 L  L  G  Q  Q  A  A  P  N  R  P  N  Y  I  G  F  R  D  N  F  I  G  L  M  Y  Y  N  S  T  G  N  M  G    319
TTACTGGGACAGCAGGCCGCTCCCAATCGACCTAATTACATTGGTTTTAGGGACAACTTCATTGGTTTGATGTACTACAATTCCACTGGCAACATGGGAG 18600

V  L  A  G  Q  A  S  Q  L  N  A  V  V  D  L  Q  D  R  N  T  E  L  S  Y  Q  L  M  L  D  A  L  G  D  R    353
TGCTGGCCGGGCAAGCTTCTCAGCTCAACGCAGTGGTGGACTTACAAGATAGAAACACGGAGCTGTCTTACCAGTTAATGCTTGACGCTTTAGGGGATCG 18700

  S  R  Y  F  S  M  W  N  Q  A  V  D  S  Y  D  P  D  V  R  I  I  E  N  H  G  V  E  D  E  L  P  N  Y    386
GAGTCGATACTTCTCCATGTGGAACCAGGCAGTGGACAGCTATGACCCAGACGTGAGAATTATTGAAAATCATGGCGTGGAAGACGAGCTCCCCAACTAT 18800

 C  F  P  L  N  G  Q  G  I  S  N  S  Y  Q  G  V  K  T  D  N  G  T  N  W  S  Q  N  N  T  D  V  S  S    419
TGCTTTCCTCTTAATGGGCAAGGAATATCTAACAGTTACCAAGGCGTAAAGACTGACAATGGAACTAACTGGTCTCAGAATAATACAGACGTCTCAAGCA 18900
```

```
N  N  E  I  S  I  G  N  V  F  A  M  E  I  N  L  A  A  N  L  W  R  S  F  L  Y  S  N  V  A  L  Y  L  P    453
ACAACGAAATTTCCATTGGCAATGTGTTTGCCATGGAGATTAATCTGGCGGCTAACTTGTGGAGAAGCTTCTTGTACTCAAATGTAGCCCTGTACTTGCC 19000

 D  S  Y  K  I  T  P  D  N  I  T  L  P  D  N  K  N  T  Y  A  Y  M  N  G  R  V  A  V  P  S  A  L  D     486
TGACTCTTACAAAATAACCCCCGATAACATTACTTTACCCGACAACAAAAATACATATGCCTACATGAACGGTCGGGTTGCCGTCCCAGCGCCCTGGAT 19100

 T  Y  V  N  I  G  A  R  W  S  P  D  P  M  D  N  V  N  P  F  N  H  H  R  N  A  G  L  R  Y  R  S  M     519
ACATACGTGAACATTGGGGCGCGGTGGTCTCCAGACCCCATGGACAACGTTAATCCCTTTAACCACCACCGCAATGCTGGTCTGCGCTACCGTTCTATGC 19200

L  L  G  N  G  R  Y  V  P  F  H  I  Q  V  P  Q  K  F  F  A  I  K  N  L  L  L  L  P  G  S  Y  T  Y  E    553
TCCTGGGTAACGGCCGCTACGTGCCTTTTCACATCCAAGTGCCCCAGAAATTTTTCGCCATTAAAAATCTCCTGCTCCTGCCCGGGTCCTACACCTATGA 19300

 W  N  F  R  K  D  V  N  M  I  L  Q  S  S  L  G  N  D  L  R  V  D  G  A  S  V  R  F  D  S  I  N  L     586
GTGGAACTTCCGGAAGGATGTTAACATGATTCTCCAAAGCAGTCTCGGTAACGACCTCAGGGTCGATGGAGCCAGCGTCAGGTTTGACAGCATTAACCTG 19400

 Y  A  N  F  F  P  M  A  H  N  T  A  S  T  L  E  A  M  L  R  N  D  T  N  D  Q  S  F  N  D  Y  L  C     619
TATGCCAACTTTTTCCCCATGGCTCACAACACCGCTTCCACCTTGGAAGCAATGCTTCGTAATGATACCAACGATCAGTCTTTCAACGACTACCTCTGCG 19500

A  A  N  M  L  Y  P  I  P  A  N  A  T  S  V  P  I  S  I  P  S  R  N  W  A  A  F  R  G  W  S  F  T  R    653
CTGCAAACATGCTTTACCCCATACCCGCCAACGCTACTAGCGTGCCCATTTCTATTCCTTCGCGAAATTGGGCTGCTTTTCGGGGGTGGAGTTTTACTAG 19600

 L  K  T  K  E  T  P  S  L  G  S  G  F  D  P  Y  F  T  Y  S  G  S  V  P  Y  L  D  G  T  F  Y  L  N     686
ACTAAAAACTAAAGAAACCCCCTCTTTGGGGTCCGGGTTTGATCCATATTTCACCTACTCTGGCTCCGTCCCATACTTGGATGGCACCTTTTACCTGAAC 19700

 H  T  F  K  K  V  S  V  M  F  D  S  S  V  S  W  P  G  N  D  R  L  L  T  P  N  E  F  E  I  K  R  T     719
CACACTTTTAAAAAGGTGTCCGTTATGTTCGACTCCTCTGTGAGCTGGCCTGGTAACGACCGACTACTTACTCCCAACGAGTTTGAAATCAAACGAACCG 19800

V  D  G  E  G  Y  N  V  A  Q  C  N  M  T  K  D  W  F  L  I  Q  M  L  S  H  Y  N  I  G  Y  Q  G  F  H    753
TGGATGGGGAAGGATACAACGTGGCTCAATGTAACATGACCAAGGACTGGTTCCTCATACAAATGCTCAGTCACTACAATATTGGCTACCAGGGTTTCCA 19900

 V  P  E  S  Y  K  D  R  M  Y  S  F  F  R  N  F  Q  P  M  S  R  Q  V  V  D  T  T  T  Y  T  E  Y  Q     786
CGTACCAGAAAGCTACAAGGACAGGATGTACTCCTTTTTCCGAAACTTCCAACCCATGAGCCGCCAGGTGGTAGACACTACCACCTACACGGAGTATCAG 20000

 N  V  T  L  P  F  Q  H  N  N  S  G  F  V  G  Y  M  G  P  A  I  R  E  G  Q  A  Y  P  A  N  Y  P  Y     819
AATGTAACTCTCCCTTTCCAGCATAATAACTCTGGCTTTGTAGGATACATGGGACCTGCCATACGGGAGGGACAAGCTTACCCCGCCAACTATCCATACC 20100

P  L  I  G  Q  T  A  V  P  S  L  T  Q  K  K  F  L  C  D  R  T  M  W  R  I  P  F  S  S  N  F  M  S  M    853
CCCTTATTGGTCAGACGGCCGTACCAAGCCTGACTCAGAAAAAATTTCTTTGCGATCGTACCATGTGGCGCATTCCCTTTTCCAGCAACTTTATGTCTAT 20200

 G  A  L  T  D  L  G  Q  N  M  L  Y  A  N  S  A  H  A  L  D  M  T  F  E  V  D  P  M  D  E  P  T  L     886
GGGGGCCCTGACCGACCTGGGGCAAAACATGCTGTACGCCAACTCCGCCCACGCGCTCGACATGACTTTTGAGGTGGACCCCATGGATGAGCCCACACTT 20300

 L  Y  V  L  F  E  V  F  D  V  V  R  I  H  Q  P  H  R  G  V  I  E  A  V  Y  L  R  T  P  F  S  A  G     919
CTCTATGTTCTGTTCGAAGTTTTCGACGTTGTGCGCATCCACCAGCCGCACCGCGGCGTCATCGAGGCCGTCTACCTGCGTACGCCGTTCTCGGCCGGTA 20400
                        > start of L3 23K coding exon
```

```
       L3 23K (protease)  M  G  S  S  E  Q  E  L  V  A  I  V  R  E  L  G  C  G  P  Y  F  L  G  T  F     25
N  A  T  T  -                                                                                          923
ACGCCACCACATAAGAAGCCAGCCAATGGGCTCCAGCGAGCAGGAGTTGGTCGCCATCGTGCGCGAACTGGGCTGCGGACCTTACTTTCTGGGCACGTTT 20500

 D  K  R  F  P  G  F  M  A  P  H  K  L  A  C  A  I  V  N  T  A  G  R  E  T  G  G  V  H  W  L  A  L     58
GACAAACGCTTTCCGGGTTTTATGGCACCGCATAAGCTGGCGTGTGCCATTGTTAACACGGCGGGCCGCGAAACCGGCGGCGTACACTGGCTGGCCCTGG 20600

A  W  N  P  K  N  R  T  C  Y  L  F  D  P  F  G  F  S  D  E  R  L  K  Q  I  Y  Q  F  E  Y  E  G  L  L    92
CCTGGAACCCAAAGAACCGTACCTGCTACCTCTTCGACCCATTTGGCTTTTCGGACGAGCGCCTCAAACAGATTTACCAGTTTGAGTATGAAGGTCTCCT 20700

 K  R  S  A  L  A  S  T  P  D  H  C  I  T  L  I  K  S  T  Q  T  V  Q  G  P  F  S  A  A  C  G  L  F    125
AAAGCGTAGTGCGTTGGCCTCAACCCCGGACCATTGTATCACCCTAATTAAGTCCACCCAAACTGTTCAAGGACCGTTTTCGGCGGCCTGCGGCCTTTTC 20800

 C  C  M  F  L  H  A  F  V  N  W  P  T  S  P  M  E  R  N  P  T  M  D  L  L  T  G  V  P  N  S  M  L    158
TGCTGCATGTTTTTACATGCTTTTGTAAACTGGCCCACCAGTCCCATGGAGCGCAACCCCACCATGGACCTTCTTACCGGCGTTCCAAACAGCATGCTTC 20900

Q  S  P  Q  V  V  P  T  L  R  H  N  Q  E  R  L  Y  R  F  L  A  Q  R  S  P  Y  F  Q  R  H  C  E  R  I   192
AAAGCCCCCAGGTTGTACCCACCCTGCGTCACAACCAGGAGCGGTTGTACCGTTTCCTGGCGCAACGTTCTCCCTACTTTCAGCGTCATTGCGAGCGTAT 21000

 K  K  A  T  A  F  D  Q  M  K  N  N  M  -           >>>>>> L3 pA                                      205
CAAAAAAGCCACCGCGTTTGACCAAATGAAAAACAACATGTAACGGTTCAATAAAAGCTTTTATTGATTCAAAAAATTCATGCATGCAGACTTTTTATTT 21100
                                              E2A/L DBP pA <<<<<<                E2A/L DBP  pA <<<<<<

TAAAATGGTTCTTTCTCCCCATCGCCGTGGCTGGCGGGCAAAGCTACGTTGCGATACTGCAAACGAGAGGACCACTTAAATTCTGGAATCAGCATCTTAG 21200
-  F  P  E  K  E  G  D  G  H  S  A  P  L  A  V  N  R  Y  Q  L  R  S  S  W  K  F  E  P  I  L  M  K  P   441
[E2A/L DBP]

GAAGGGGGCCATCGACGTTCTCTCCCCACAGCCGTCGTACAAGTTGCAAAGCTCCCAAAAGGTCAGGTGCAGAAATTTTGAAATCACAGTTGGGACCTTG 21300
 L  P  G  D  V  N  E  G  W  L  R  R  V  L  Q  L  A  G  L  L  D  P  A  S  I  K  F  D  C  N  P  G  Q    408

GCCACCACGGGAGTTGCGGTATACGGGGTTAGCGCACTGGTAAACCAGCACACAGGGATACTGGATACTGGCAAGAGCCACCTTGTCGGTTACTTCTTCA 21400
 G  G  R  S  N  R  Y  V  P  N  A  C  Q  Y  V  L  V  C  P  Y  Q  I  S  A  L  A  V  K  D  T  V  E  E    375

GCTCTAAGACTGTCAACATTGCTTAGAGCGAAAGGGGTGGCTTTACACATTTGCCGACCCAATTGGGGCACACCGGTGGGCTTGTACAGGCAGTCGCAGC 21500
A  R  L  S  D  V  N  S  L  A  F  P  T  A  K  C  M  Q  R  G  L  Q  P  V  G  T  P  K  Y  L  C  D  C  R   341

GCATCACCATTAATAGGCGTTTTAGCCCGTTTTGCATTTTTGGATATTCGGCTTGCATAAAAGCTTCTATCTGCAAAAAAGCCGTCTGAGCCTTTGTTCC 21600
 M  V  M  L  L  R  K  L  G  N  Q  M  K  P  Y  E  A  Q  M  F  A  E  I  Q  L  F  A  T  Q  A  K  T  G    308

TTCCGAGAAAAACAGACCGCAGGACTTGGCAGAAAACACATTGGTGGCACAGCTCACGTCTTCTACACAACAACGGGCATCGTCATTCTTCAGTTGAACC 21700
 E  S  F  F  L  G  C  S  K  A  S  F  V  N  T  A  C  S  V  D  E  V  C  C  R  A  D  D  N  K  L  Q  V    275

ACGCTGCGCCCCCACCGGTTTTGTACCACCTTGGCTCGACTCGGGTGCTCCTTTAACGCCCGCTGAGCGTTCTCGCTCGCTACATCCATTTCCACCAACT 21800
V  S  R  G  W  R  N  Q  V  V  K  A  R  S  P  H  E  K  L  A  R  Q  A  N  E  S  A  V  D  M  E  V  L  Q   241
```

```
GCTCTTTTTGAATCATTTCCAGGCCATGATAACAGCGTAGCACTCCCTCTTGCTCGGTGCAGCCGTGAAGCCAAATCGCGCAACCAGTGGGCTCCCATTC 21900
 E  K  Q  I  M  E  L  G  H  Y  C  R  L  V  G  E  Q  E  T  C  G  H  L  W  I  A  C  G  T  P  E  W  E    208

ATTGTTTTTACCCCGGCGTACGACTCCACGTAGGCTCTCAAAAAACGTCCCATCATTTCCACAAATGTCTTGTGGCTGGTGAAGGTGAGAGGGAGGCCG 22000
 N  N  K  V  G  A  Y  S  E  V  Y  A  R  L  F  R  G  M  M  E  V  F  T  K  H  S  T  F  T  L  P  L  G    175

CGATGCTCCTCGTTAAGCCACGTTTGGCAAATTTTGCGATAAACGTTGCTTTGTTCGGGTAGGAACTTGAAGCCATTCTTCTCTTCGGCCTCCACATGAT 22100
R  H  E  E  N  L  W  T  Q  C  I  K  R  Y  V  N  S  Q  E  P  L  F  K  F  G  N  K  E  E  A  E  V  H  Y   141

ACTTTTCCATTAGCTTTATCATTAAATCCATGCCTTTCTCCCAGGCGGAAACCAAGGGCTCTGCCTGCGGATTAAGAACCACTGATGTAACAGCTTTGGA 22200
 K  E  M  L  K  I  M  L  D  M  G  K  E  W  A  S  V  L  P  E  A  Q  P  N  L  V  V  S  T  V  A  K  S    108

AGTGCTAGGCTCTTCTTCCTCGTTGTTTTCCTCTGACGGGGGAGGCACACCTTTGGGCTCCAAGCGTCTTACATATCGCTTGCCACTGGCCTTTTGAACG 22300
 T  S  P  E  E  E  E  N  N  E  E  S  P  P  P  V  G  K  P  E  L  R  R  V  Y  R  K  G  S  A  K  Q  V     75

ACCTGCACGCCGGGGTGACTGAACCCGGTGTACACCACCTCTTCTTCTTCCTCCTCGCTGTCTGGAACCACTTCGGGAGACGGAGGCAAAACTGGAACGC 22400
V  Q  V  G  P  H  S  F  G  T  Y  V  V  E  E  E  E  E  E  S  D  P  V  V  E  P  S  P  P  L  V  P  V  R    41

GATCCGGCACTTGAACATTCTTGCGCAACTTCTTTTTGGGAGGAAGTGACGGGGCCCGTTCTGGACTCGTCTCCTGCAAGTAGGGAGTGATGGTGGGGAG 22500
 D  P  V  Q  V  N  K  R  L  K  K  K  P  P  L  S  P  A  R  E  P  S  T  E  Q  L  Y  P  T  I  T  P  L      8

                                                 L4 100K
       start of L4 100K coding exon >        M  E  E  D  L  K  L  Q  P  D  S  E  T  L  T  T  P  N  S     19
TTCTTGCTGACGGCCGGCCATGCTTTACTCCTAGGCGAGAAAATATGGAGGAGGATCTTAAGCTGCAGCCAGACTCCGAAACCTTAACCACCCCCAACTC 22600
 E  Q  Q  R  G  A  M          < start of E2A/L DBP 3rd (coding) exon                                      1

 E  V  G  A  V  E  L  V  K  H  E  E  E  N  E  Q  V  E  Q  D  P  G  Y  V  T  P  P  E  D  G  K  E  P     52
TGAGGTCGGCGCCGTCGAGCTAGTGAAACATGAGGAGGAAAATGAGCAAGTGGAGCAAGATCCGGGCTATGTAACGCCCCCCGAGGACGGCAAGGAACCA 22700

 V  A  A  L  S  E  P  N  Y  L  G  G  E  D  D  V  L  L  K  H  I  A  R  Q  S  T  I  V  R  E  A  L  K     85
GTGGCCGCACTCAGCGAACCCAACTATTTGGGAGGGGAGGACGACGTGCTCCTGAAGCACATAGCGCGACAGAGCACCATTGTACGAGAAGCCCTCAAGG 22800

E  C  T  Q  T  P  L  T  V  E  E  L  S  R  A  Y  E  A  N  L  F  S  P  R  V  P  P  K  K  Q  P  N  G  T   119
AATGCACACAGACTCCGCTGACGGTGGAGGAATTAAGCCGCGCGTATGAAGCTAACCTGTTTTCGCCGCGTGTACCGCCAAAAAAGCAGCCTAACGGCAC 22900

 C  E  T  N  P  R  L  N  F  Y  P  V  F  A  V  P  E  A  L  A  T  Y  H  I  F  F  K  N  Q  R  I  P  L    152
CTGCGAAACAAACCCGCGCCTCAATTTTTATCCCGTCTTTGCGGTGCCTGAAGCACTGGCTACTTATCACATCTTTTTCAAGAACCAACGCATTCCCCTC 23000

 S  C  R  A  N  R  T  R  G  D  G  L  L  H  L  K  A  G  A  H  I  P  E  I  V  S  L  E  E  V  P  K  I    185
TCTTGCCGCGCCAACCGTACACGCGGTGACGGCCTTTTGCATCTCAAAGCTGGAGCTCACATACCTGAGATCGTTTCTTTAGAAGAAGTACCCAAGATTT 23100
                          end of E2A/L DBP 2nd (leader) exon <

F  E  G  L  G  K  D  E  K  R  A  A  N  A  L  Q  K  N  E  T  E  N  Q  N  V  L  V  E  L  E  G  D  N  A   219
TTGAAGGTCTTGGCAAGGACGAAAAACGGGCGGCAAATGCTCTGCAAAAAAACGAAACCGAGAATCAGAACGTGTTGGTAGAGCTGGAGGGTGACAACGC 23200
                                      < start of E2A/L DBP 2nd (leader) exon
```

```
 R  L  A  V  L  K  R  T  I  E  V  S  H  F  A  Y  P  A  L  N  L  P  P  K  V  M  R  S  V  M  D  Q  V   252
GCGTTTGGCCGTACTCAAACGCACCATTGAAGTTTCACACTTTGCTTATCCCGCGCTAAATCTTCCTCCCAAAGTAATGCGTTCTGTTATGGATCAAGTG 23300

 L  I  K  R  A  E  P  I  D  P  Q  Q  P  D  L  N  S  E  D  G  Q  P  V  V  S  D  D  E  L  A  R  W  L   285
CTTATTAAGCGAGCAGAGCCCATTGATCCCCAACAACCCGACCTAAACTCTGAGGACGGACAACCCGTAGTCTCAGACGACGAGCTTGCTCGCTGGCTAG 23400

G  T  Q  D  P  S  E  L  Q  E  R  R  K  M  M  M  A  A  V  L  V  T  V  E  L  E  C  L  Q  R  F  F  A  N  319
GTACCCAGGATCCCTCAGAGCTGCAAGAGCGGCGAAAAATGATGATGGCAGCAGTTTTGGTTACAGTGGAATTGGAATGCCTGCAGCGCTTCTTTGCTAA 23500

  P  Q  T  L  R  K  V  E  E  S  L  H  Y  A  F  R  H  G  Y  V  R  Q  A  C  K  I  S  N  V  E  L  S  N  352
CCCTCAAACACTGCGCAAAGTCGAGGAGTCCCTGCACTATGCCTTCCGTCATGGCTACGTTCGTCAGGCCTGCAAGATCTCCAACGTAGAGCTCAGCAAT 23600

 L  I  S  Y  M  G  I  L  H  E  N  R  L  G  Q  N  V  L  H  C  T  L  Q  G  E  A  R  R  D  Y  V  R  D   385
CTGATCTCTTACATGGGCATTCTACACGAAAACCGGCTGGGGCAGAACGTTCTTCACTGCACCTTGCAAGGGGAGGCCCGCCGAGACTACGTCCGCGACT 23700

C  I  Y  L  F  L  I  L  T  W  Q  T  A  M  G  V  W  Q  Q  C  L  E  E  Q  N  L  Q  E  L  N  K  L  L  V  419
GCATCTATCTTTTCCTTATTCTCACCTGGCAAACCGCTATGGGAGTCTGGCAGCAGTGCTTGGAAGAGCAAAACCTCCAGGAGCTTAATAAATTGCTAGT 23800

  R  A  R  R  E  L  W  T  S  F  D  E  R  T  V  A  R  Q  L  A  N  L  I  F  P  E  R  L  M  Q  T  L  Q  452
ACGAGCCCGTCGCGAACTCTGGACGTCTTTTGACGAGCGTACGGTTGCCCGCCAGCTGGCAAACCTCATTTTTCCCGAGCGGCTTATGCAAACATTGCAA 23900

 N  G  L  P  D  F  V  S  Q  S  I  L  Q  N  F  R  S  F  V  L  E  R  S  G  I  L  P  A  M  S  C  A  L   485
AATGGTTTGCCAGACTTTGTCAGCCAAAGTATCTTGCAAAACTTTCGCTCCTTTGTACTCGAGCGTTCCGGCATCTTGCCGGCTATGAGTTGTGCTTTGC 24000

P  S  D  F  V  P  L  C  Y  R  E  C  P  P  P  L  W  S  H  C  Y  L  L  R  L  A  N  Y  L  A  H  H  S  D  519
CCTCCGATTTTGTCCCCCTCTGCTACCGCGAATGCCCCCCACCGTTGTGGAGTCACTGCTACCTCCTCCGTCTAGCCAACTATTTGGCCCACCACTCTGA 24100

  L  M  E  D  S  S  G  D  G  L  L  E  C  H  C  R  C  N  L  C  T  P  H  R  S  L  V  C  N  T  E  L  L  552
TCTTATGGAAGACTCTAGCGGCGACGGACTGCTAGAATGTCACTGCCGTTGCAACCTCTGCACCCCTCATCGCTCACTGGTCTGTAACACCGAGCTTCTT 24200

 S  E  T  Q  V  I  G  T  F  E  I  Q  G  P  E  Q  Q  E  G  A  S  S  L  K  L  T  P  A  L  W  T  S  A   585
AGCGAAACCCAAGTAATCGGTACCTTTGAGATTCAAGGGCCAGAGCAACAAGAAGGTGCTTCCAGCCTCAAACTCACGCCGGCGTTGTGGACTTCCGCCT 24300
                         < end of E2A/L DBP 1st leader exon                      < <
                                                                     5' ends of E2A/L DBP mRNA

Y  L  R  K  F  I  P  E  D  Y  H  A  H  Q  I  K  F  Y  E  D  Q  S  R  P  F  K  V  P  L  T  A  C  V  I  619
ACCTACGCAAATTTATTCCCGAAGACTATCACGCCCACCAAATTAAATTTTATGAAGACCAATCACGACCTCCCAAAGTCCCCTTACAGCCTGTGTTAT 24400
    <<<<<<<<< E2A/L DBP TATA

  T  Q  S  Q  I  L  A  Q  L  Q  A  I  Q  Q  A  R  Q  E  F  L  L  K  K  G  H  G  V  Y  L  D  P  Q  T  652
CACCCAAAGCCAAATTCTGGCCCAATTACAAGCTATTCAGCAGGCGCGTCAGGAATTTCTTTTAAAAAAAGGACACGGGGTCTATTTGGACCCCCAAACC 24500
```

```
                                                  L4 22K M P P K G N K H P I A Q R Q     14
                                                  L4 33K M P P K G N K H P I A Q R Q     14
 G E E L N T P S L S A A A S C R S Q K H A T Q G K Q A S H R A T A                      685
GGTGAAGAACTTAATACCCCGTCACTCTCCGCCGCCGCTTCGTGCCGTTCGCAGAAACATGCCACCCAAGGGAAACAAGCATCCCATCGCGCAACGGCAA 24600
                                                   > start of L4 22K coding & L4 33K 1st exon

 S Q Q K L Q K Q W D E E E T W D D S Q A E E V S D E E A E E Q M E                      47
 S Q Q K L Q K Q W D E E E T W D D S Q A E E V S D E E A E E Q M E                      47
I P A E T T K A V G R G G D V G R Q P G R G S F R R G G G G A D G E                    719
TCCCAGCAGAAACTACAAAAGCAGTGGGACGAGGAGGAGACGTGGGACGACAGCCAGGCAGAGGAAGTTTCAGACGAGGAGGCGGAGGAGCAGATGGAGA 24700

S W D S L D E E D L E D V E E E T I A S D K A P S F K K P V R S Q P                     81
S W D S L D E E D L E D V E E E T I A S D K A P S F K K P V R S Q P                     81
 L G Q P R R G G P R G R G G R N H R Q R Q G T I F Q K T R S E P T                     752
GCTGGGACAGCCTAGACGAGGAGGACCTAGAGGACGTGGAGGAAGAAACCATCGCCAGCGACAAGGCACCATCTTTCAAAAAACCCGTTCGGAGCCAACC 24800

 P K T I P P L P P Q P C S L K A S R R W D T V S I A G S P T A P A                     114
 P K T I P P L P P Q P C S L K A S R R W D T V S I A G S P T A P A                     114
 S E N Y P A P A T A T M F T E S Q P -                                                 770
TCCGAAAACTATCCCGCCCCTGCCACCGCAACCATGTTCACTGAAAGCCAGCCGTAGGTGGGACACCGTCTCCATCGCCGGATCGCCAACAGCCCCAGCT 24900

 G K Q P K R A R R G Y C S W R A H K S N I V A C L Q H C R G N I S                      147
GGTAAGCAGCCTAAGCGCGCACGACGGGGATACTGCTCCTGGCGAGCCCATAAAAGCAATATTGTCGCATGCCTCCAGCACTGCCGGGGCAATATCTCAT 25000
> end of L4 33K 1st exon

F A R R Y L L F H D G V A V P R N V L Y Y Y R H L Y S P Y E T F G E                    181
                                                           A P T K R L E               121
TCGCACGGCGTTACTTGCTTTTTCACGACGGGGTGGCGGTTCCTCGCAACGTCCTCTACTATTACCGTCATCTCTACAGCCCCTACGAAACGTTTGGAGA 25100
                                                   start of L4 33K 2nd exon >

 N T S S A -                                                                           186
K T P R V R K T S S A I A T R Q D S P A T Q E L R K R I F P T L Y A                    155
AAACACCTCGAGTGCGTAAGACCTCATCCGCCATTGCCACCCGCCAGGATTCGCCCGCCACGCAGGAGCTCAGAAAACGCATCTTTCCGACGCTGTATGC 25200

 I F Q Q S R G Q Q L E L K V K N R S L R S L T R S C L Y H R S E D                     188
TATTTTCCAGCAGAGCCGCGGTCAACAGCTGGAACTCAAAGTAAAAAACCGATCACTCCGTTCGCTCACCCGCAGCTGCTTGTATCACAGAAGTGAAGAC 25300
                                                             < end of E2 1st (leader) exon

 Q L Q R T L E D A E A L F N K Y C S V S L K D -                                       211
CAACTGCAGCGCACGCTGGAGGACGCCGAGGCACTGTTCAATAAATATTGCTCGGTGTCTCTTAAGGACTAAACACCCGCGCTTTTTTTAGGCGCCAAAT 25400
                                   < 5' end of E2 mRNA    start of L4 pVIII coding exon >

        L4 pVIII M S K D I P T P Y M W S Y Q P Q M G L A A G A S Q D Y S                28
TACGTCATTGACATTATGAGCAAAGACATTCCCACGCCTTACATGTGGAGCTATCAGCCGCAAATGGGCCTGGCAGCTGGAGCTTCTCAGGATTACTCCA 25500
```

```
S  R  M  N  W  L  S  A  G  P  H  M  I  G  R  V  N  G  I  R  A  T  R  N  Q  I  L  L  E  Q  A  A  L  T     62
GTCGCATGAATTGGCTTAGTGCCGGCCCCACATGATTGGGCGGTAAATGGAATTCGTGCCACTCGAAACCAAATTCTGCTAGAACAGGCCGCCCTAAC 25600

 S  T  P  R  R  Q  L  N  P  P  S  W  P  A  A  Q  V  Y  Q  E  N  P  A  P  T  T  V  L  L  P  R  D  A    95
CTCTACCCCGCGACGTCAGCTGAACCCACCCTCTTGGCCTGCCGCCCAGGTTTATCAGGAAAACCCCGCCCCGACCACAGTCCTTCTGCCACGCGACGCG 25700

 E  A  E  V  Q  M  T  N  S  G  A  Q  L  A  G  G  A  R  H  V  R  F  R  D  R  P  S  P  Y  S  S  G  S   128
GAAGCCGAAGTCCAAATGACTAACTCCGGAGCGCAATTAGCGGGCGGCGCCAGACACGTCAGGTTCAGAGATCGACCCTCGCCCTATTCCTCCGGCTCTA 25800
                                                                                                   >>

I  K  R  L  I  I  R  G  R  G  I  Q  L  N  D  E  V  V  S  S  S  T  G  P  R  P  D  G  V  F  Q  L  G  G   162
TAAAAAGGCTAATCATTCGAGGCCGAGGTATCCAGCTCAACGACGAGGTAGTGAGCTCTTCCACCGGTCCTAGACCTGACGGAGTCTTTCAGCTTGGAGG 25900
>>>>>> E3 TATA                > 5' end of E3 mRNA

 A  G  R  S  S  F  T  P  R  Q  A  Y  L  T  L  Q  S  S  S  S  Q  P  R  S  G  G  I  G  T  L  Q  F  V   195
CGCCGGGCGGTCTTCCTTCACTCCTCGCCAGGCGTACTTAACGCTTCAGAGCTCTTCATCCCAGCCTCGCTCCGGCGGCATCGGAACCCTCCAGTTTGTG 26000

 E  E  F  V  P  S  V  Y  F  N  P  F  S  G  A  P  G  L  Y  P  D  D  F  I  P  N  Y  D  A  V  S  E  S   228
GAGGAGTTTGTACCCTCCGTTTACTTCAACCCCTTCTCGGGCGCTCCTGGTCTTTACCCAGACGACTTCATCCCGAACTACGACGCGGTGAGCGAATCTG 26100

V  D  G  Y  D  -                                                                                     233
TGGACGGCTACGACTGAAGACCGATAGTACGGCCGTGACTGCGCGGCTGTAACATCTGCATCGGTGCCGTAACCTTCGCTGCTTTACTTAAAAAGCCTGT 26200

                     E3A 173R  M  K  I  C  V  L  F  C  V  L  S  L  T  S  S  L  R  T  S  ?  T  T  V    23
GATTTCATTTACCACCCCAGCACTTGGATTACATGAAGATCTGTGTTCTTTTTTGTGTGCTAAGTTTAACAAGTAGCCTAAGGACTTCACCTACAACCGT 26300
                  > start of E3A 173R coding exon

 G  S  L  R  Q  L  Q  D  S  T  K  G  T  H  Q  T  L  Y  F  S  E  S  T  T  S  I  A  L  N  C  S  C  R    56
TGGTTCCTTACGTCAGCTACAAGATTCCACCAAAGGTACACACCAAACTCTTTATTTTTCTGAGTCTACCACTTCTATTGCACTTAACTGTTCTTGTCGT 26400

 N  Q  L  V  Q  W  R  A  N  R  Q  F  C  K  L  F  W  D  A  L  I  V  Q  G  N  N  S  L  C  N  N  C  T    89
AACCAACTCGTTCAGTGGCGCGCTAACAGACAATTTTGCAAACTATTCTGGGACGCTCTTATTGTTCAAGGAAACAACAGCCTTTGTAACAACTGTACTG 26500

A  T  T  L  T  L  T  P  P  F  V  P  G  P  Y  L  C  I  G  T  G  R  G  P  S  C  F  N  R  W  T  L  Q  K   123
CTACTACTTTAACTCTTACACCTCCTTTTGTTCCCGGTCCATACTTGTGCATTGGCACAGGAAGAGGGCCTAGCTGCTTTAATCGCTGGACTTTACAAAA 26600

 E  N  L  T  T  T  T  L  L  P  L  T  T  Y  T  F  S  Q  K  K  N  Y  F  L  P  I  I  A  L  L  A  F  V   156
AGAGAATCTAACCACTACCACCCTCCTTCCCCTTACTACTTATACTTTTCCCAAAAAAAGAATTACTTTTGCCCATTATTGCACTTTTGGCCTTTGTC 26700

                                                 E3A 276R  M  L  L  F  L  L  C  L  L  F  C  S  A  Y  A    15
 C  V  I  T  A  N  Y  I  L  I  F  N  L  D  N  F  Y  -                                                173
TGTGTTATTACCGCTAATTATATTTAATTTTCAATCTTGATAATTTTTACTAATCATGCTGCTGTTTTTACTTTGCCTTCTTTTCTGCTCTGCCTATGC 26800

 A  V  P  E  K  T  L  N  N  L  V  R  V  Y  A  L  V  G  T  N  L  S  L  D  S  M  K  T  P  Q  I  D  E    48
CGCCGTGCCAGAAAAAACCCTTAACAACCTCGTTCGGGTGTACGCCTTAGTTGGTACCAATCTATCCCTTGATTCTATGAAAACTCCTCAGATTGACGAA 26900
```

```
 L  T  S  L  S  W  I  K  Q  E  D  N  P  N  K  N  L  Q  S  F  F  F  I  G  Q  K  L  C  E  V  T  K  D     81
CTTACTAGTCTTAGCTGGATCAAACAGGAAGACAATCCTAACAAAAACTTACAATCATTTTTTTTATTGGTCAAAAACTCTGTGAAGTTACCAAAGACA 27000

K  I  T  V  F  N  Y  Y  P  L  E  F  S  C  A  N  V  T  L  Y  L  Y  N  L  K  T  D  D  S  G  L  Y  N  G  115
AAATCACTGTTTTTAACTATTATCCGTTGGAATTTTCCTGCGCTAACGTAACCTTGTATTTGTATAATCTTAAAACTGACGATTCTGGCCTCTATAATGG 27100

  K  A  H  T  K  E  L  E  H  N  T  Y  V  R  L  Y  V  I  D  I  P  P  P  K  C  D  I  T  S  R  Y  L  G   148
AAAGGCCCATACCAAAGAGCTTGAACATAACACCTATGTTAGGCTTTATGTTATTGACATTCCTCCGCCTAAGTGTGACATTACTTCACGTTACTTAGGC 27200

 I  Q  A  T  G  E  D  Y  C  L  I  E  I  N  C  T  N  S  K  Y  P  A  V  V  K  F  N  G  R  Q  S  N  F    181
ATACAGGCTACTGGGGAAGATTATTGTTTAATTGAAATTAATTGCACTAACTCCAAATACCCAGCTGTGGTTAAATTTAATGGCAGGCAAAGCAACTTCT 27300

Y  H  Y  V  S  E  N  G  N  K  K  L  P  N  F  Y  E  T  H  I  T  V  N  G  T  H  K  S  F  H  F  N  Y  P  215
ACCATTATGTTAGCGAAAACGGAAACAAAAAACTTCCAAATTTTTATGAAACACACATCACTGTTAATGGTACCCACAAAAGCTTTCACTTTAATTACCC 27400

  F  N  D  L  C  Q  T  T  S  A  L  Q  Y  N  D  N  V  Q  V  V  L  I  L  L  I  V  V  G  L  I  I  I  S   248
TTTTAACGACCTTTGTCAAACAACCAGCGCTCTACAATATAATGACAATGTCCAGGTAGTCCTCATTCTTCTCATAGTAGTTGGCTTAATAATAATTTCC 27500

 A  S  L  I  L  L  Y  C  H  R  K  K  I  K  A  K  V  Q  H  Q  P  V  H  I  C  L  E  K  -              276
GCTAGTTTAATATTGCTTTATTGCCACCGCAAAAAAATCAAGGCCAAAGTTCAACATCAACCAGTGCATATTTGTTTAGAAAAATAAAAATTTTTTTCTT 27600
                                                                                   >>>>>> E3A pA

E3B90R M  V  T  P  L  L  L  L  V  C  L  P  I  I  Y  A  S  T  T  F  A  A  V  S  H  L  D  T  D  C  L    31
TTCAGTATGGTAACTCCTCTTCTCCTGCTTGTCTGTCTGCCAATTATCTACGCCTCCACCACCTTCGCCGCAGTCTCCCACCTTGATACGGATTGTCTTC 27700
   > start of E3B 90R coding exon

P  A  L  L  T  Y  L  I  F  T  S  V  C  C  T  A  I  C  S  I  A  T  F  F  V  A  I  F  Q  T  A  D  Y  L  65
CCGCCTTGCTGACTTATCTCATCTTCACCTCTGTTTGCTGCACTGCCATCTGCAGCATTGCCACTTTTTTTGTGGCCATTTTCCAAACTGCGGACTACCT 27800

                                                                      E3B 107R  M  K  V  P  L  L  C  L     8
  Y  V  R  V  A  Y  Y  R  H  H  P  Q  Y  R  N  H  E  V  A  T  L  L  C  L  S  -                           90
ATACGTTAGAGTGGCATACTATCGTCATCATCCCCAATATAGGAACCACGAGGTGGCTACCCTTCTGTGCCTGTCATGAAAGTTCCTCTTCTCTGTCTTA 27900
                                                 > start of E3B 107R coding exon

I  L  L  H  K  V  L  A  N  C  H  L  H  R  P  T  E  F  L  R  C  Y  S  T  E  T  S  S  F  W  L  Y  S  I   42
TCCTCCTTCACAAAGTCCTGGCCAACTGCCACCTCCACCGGCCCACCGAGTTCCTGCGCTGCTACTCAACAGAAACCTCTTCCTTTTGGCTGTACTCCAT 28000

  I  F  I  L  I  F  F  A  T  F  L  G  L  Q  I  Y  G  C  L  H  L  G  W  M  H  P  P  N  N  L  P  R  F   75
TATTTTTATTTTGATTTTCTTTGCCACCTTTTTGGGATTACAAATTTACGGGTGCCTTCACCTGGGCTGGATGCATCCTCCCAACAACCTACCCAGATTT 28100

                                                                                    E3B 122R  M  S  D    3
 P  G  F  L  L  Q  P  P  P  P  P  P  A  P  V  Q  R  A  P  S  V  I  S  Y  F  H  L  N  S  E  D  V  -    107
CCTGGTTTCCTATTACAGCCCCCACCGCCCCCACCGGCTCCTGTGCAGCGCGCTCCATCAGTTATTAGCTACTTTCATCTTAACTCTGAAGATGTCTGAC 28200
                                                                    start of E3B 122R coding exon >
```

```
 Q  L  E  I  D  G  Q  C  T  E  Q  L  I  L  A  R  R  K  L  K  Q  Q  N  Q  E  L  F  N  L  Q  A  L  H      36
CAACTAGAAATCGACGGGCAGTGCACTGAGCAGCTTATCCTTGCTCGGCGAAAACTCAAACAACAAAATCAGGAACTGTTCAACCTTCAAGCCCTACACC 28300

Q  C  K  K  G  L  F  C  L  V  K  Q  A  E  L  C  Y  D  V  T  Q  Q  G  H  E  L  S  Y  T  L  N  K  Q  R    70
AATGCAAAAAGGGTCTTTTTTGTCTGGTTAAACAAGCTGAACTTTGTTATGATGTAACCCAACAGGGACACGAGCTGTCATATACTTTAAACAAGCAAAG 28400

  Q  S  F  M  T  M  V  G  V  K  P  I  K  V  T  Q  Q  S  G  P  V  E  G  S  I  L  C  Q  C  T  N  P  E   103
ACAGAGCTTTATGACTATGGTGGGGGTTAAGCCCATTAAGGTTACTCAGCAATCCGGCCCAGTTGAGGGAAGCATTCTTTGTCAGTGTACCAATCCTGAA 28500

 C  M  Y  T  M  V  K  T  L  C  G  L  R  E  L  L  P  F  N  -        E3B pA >>>>>>       < end of U exon   122
TGCATGTACACTATGGTAAAAACCCTGTGTGGTCTAAGGGAACTTCTCCCCTTTAATTAAAGTTATTCTGATTAATAAAGCTTACCTTAAATTTGATATC 28600
                                                                                  U exon  L  N  S  I      49

AGTTGTTTGTCAAGTTTTTCCAGCAGCACCACCTGCCCTTCCTCCCAACTTTCGTACGGGATGTGCCAACGGGCGGCAAACTTTCTCCACGTCCTAAAGG 28700
L  Q  K  D  L  K  E  L  L  V  V  Q  G  E  E  W  S  E  Y  P  I  H  W  R  A  A  F  K  R  W  T  R  F  P    15

                                                      L5 IV-1 (fiber 1)
                            start of U exon <         M  K  R  T  R  I  E  D  D  F  N  P  V  Y  P  Y  D    17
GTATATCGGTGTTCACCTTTTTACCCTGACCCACAATCTTCATCTTGCAGATGAAAAGAACCAGAATTGAAGACGACTTCAACCCCGTCTACCCCTATGA 28800
  I  D  T  N  V  K  K  G  Q  G  V  I  K  M          > start of L5 IV-1 coding exon                       1

  T  S  S  T  P  S  I  P  Y  V  A  P  P  F  V  S  S  D  G  L  Q  E  N  P  P  G  V  L  A  L  K  Y  T    50
CACCTCCTCAACTCCCAGCATTCCCTATGTAGCTCCGCCCTTCGTTTCTTCCGACGGGTTACAGGAAAACCCCCCGGGAGTTTTAGCACTCAAGTACACT 28900

 D  P  I  T  T  N  A  K  H  E  L  T  L  K  L  G  S  N  I  T  L  Q  N  G  L  L  S  A  T  V  P  T  V     83
GACCCCATTACTACCAATGCTAAACATGAGCTTACTTTAAAACTTGGCAGCAACATAACTTTACAAAATGGGTTACTTTCGGCCACCGTTCCCACTGTTT 29000

S  P  P  L  T  N  S  N  N  S  L  G  L  A  T  S  A  P  I  A  V  S  A  N  S  L  T  L  A  T  A  A  P  L   117
CTCCTCCCCTTACAAACAGTAACAACTCCTTGGGTTTAGCCACATCCGCTCCTATAGCTGTGTCAGCTAACTCTCTTACATTGGCCACCGCCGCACCACT 29100

  T  V  S  N  N  Q  L  S  I  N  T  G  R  G  L  V  I  T  N  N  A  V  A  V  N  P  T  G  A  L  G  F  N   150
GACAGTAAGCAACAACCAGCTTAGTATTAACACTGGCAGAGGCTTAGTTATAACTAACAATGCCGTAGCAGTTAATCCTACCGGAGCGTTAGGCTTTAAC 29200

 N  T  G  A  L  Q  L  N  A  A  G  G  M  R  V  D  G  A  N  L  I  L  H  V  A  Y  P  F  E  A  I  N  Q    183
AACACAGGAGCTTTACAATTAAACGCTGCGGGAGGAATGAGAGTGGACGGCGCCAACTTAATTCTTCATGTAGCATACCCCTTTGAAGCAATCAACCAAC 29300

L  T  L  R  L  E  N  G  L  E  V  T  N  G  G  K  L  N  V  K  L  G  S  G  L  Q  F  D  N  N  G  R  I  T   217
TAACACTGCGATTAGAAAACGGGTTAGAAGTAACCAACGGAGGAAAACTCAACGTTAAGTTGGGATCAGGCCTCCAATTTGACAATAACGGACGCATTAC 29400

  I  S  N  R  I  Q  T  R  G  V  T  S  L  T  T  I  W  S  I  S  P  T  P  N  C  S  I  Y  E  T  Q  D  A   250
CATTAGTAATCGCATCCAGACTCGAGGTGTAACATCCCTCACTACCATTTGGTCTATCTCGCCTACGCCTAACTGCTCCATCTATGAAACCCAAGATGCA 29500

 N  L  F  L  C  L  T  K  N  G  A  H  V  L  G  T  I  T  I  K  G  L  K  G  A  L  R  E  M  N  D  N  A    283
AATCTATTTCTTTGTCTAACTAAAAACGGAGCTCACGTGTTAGGTACTATAACAATTAAAGGTCTTAAAGGAGCACTGCGGGAAATGAACGATAACGCTT 29600
```

```
L  S  V  K  L  P  F  D  N  Q  G  N  L  L  N  C  A  L  E  S  S  T  W  R  Y  Q  E  T  N  A  V  A  S  N    317
TATCTGTAAAACTTCCCTTTGACAATCAGGGAAATTTACTCAACTGTGCCTTGGAATCATCCACCTGGCGTTACCAGGAAACCAACGCAGTGGCCTCTAA 29700

  A  L  T  F  M  P  N  S  T  V  Y  P  R  N  K  T  A  D  P  G  N  M  L  I  Q  I  S  P  N  I  T  F  S    350
TGCCTTAACATTTATGCCCAACAGTACAGTGTATCCCCGAAACAAAACCGCCGACCCAGGCAACATGCTCATCCAAATCTCGCCTAACATCACCTTCAGT 29800

 V  V  Y  N  E  I  N  S  G  Y  A  F  T  F  K  W  S  A  E  P  G  K  P  F  H  P  P  T  A  V  F  C  Y     383
GTCGTCTACAACGAGATAAACAGTGGGTATGCTTTTACGTTTAAATGGTCAGCCGAACCGGGAAAACCTTTTCACCCACCCACCGCTGTATTTTGCTACA 29900

                                   L5 IV-2 (fiber 2)  M  K  R  A  R  F  E  D  D  F  N  P  V  Y  P  Y     16
I  T  E  Q  -                                                                                            387
TAACTGAACAATAAAATCATTGCAGGCGCAATCTTCGCATTTCTTTTTCCAGATGAAACGAGCCAGATTTGAAGATGACTTCAACCCCGTCTACCCTTA 30000
      >>>>>> L5-1 pA                                   > start of L5 IV-2 coding exon

  E  H  Y  N  P  L  D  I  P  F  I  T  P  P  F  A  S  S  N  G  L  Q  E  K  P  P  G  V  L  S  L  K  Y     49
CGAACACTACAATCCCCTTGACATTCCATTTATTACACCCCCGTTTGCTTCCTCCAACGGCTTGCAAGAAAAACCTCCGGGAGTCCTCAGCCTGAAATAC 30100

 T  D  P  L  T  T  K  N  G  A  L  T  L  K  L  G  T  G  L  N  I  D  K  N  G  D  L  S  S  D  A  S  V     82
ACTGATCCACTTACAACCAAAAACGGGGCTTTAACCTTAAAATTGGGCACGGGACTAAACATTGATAAAAATGGAGATCTTTCTTCAGATGCTAGCGTGG 30200

E  V  S  A  P  I  T  K  T  N  K  I  V  G  L  N  Y  T  K  P  L  A  L  Q  N  N  A  L  T  L  S  Y  N  A    116
AAGTTAGCGCCCCTATCACTAAAACCAACAAAATCGTAGGTTTAAATTACACTAAGCCTCTCGCTCTGCAAAATAACGCGCTTACTCTTTCTTACAACGC 30300

  P  F  N  V  V  N  N  N  L  A  L  N  M  S  Q  P  V  T  I  N  A  N  N  E  L  S  L  L  I  D  A  P  L    149
GCCCTTTAACGTAGTAAATAATAATTTAGCTCTAAATATGTCACAGCCTGTTACTATTAATGCAAACAACGAACTTTCTCTCTTAATAGACGCCCCACTT 30400

 N  A  D  T  G  T  L  R  L  R  S  D  A  P  L  G  L  V  D  K  T  L  K  V  L  F  S  S  P  L  Y  L  D    182
AATGCTGACACGGGCACTCTTCGCCTTCGAAGTGATGCACCTCTTGGACTAGTAGACAAAACACTAAAGGTTTTGTTTTCTAGCCCCCTCTATCTAGATA 30500

N  N  F  L  T  L  A  I  E  R  P  L  A  L  S  S  N  R  A  V  A  L  K  Y  S  P  P  L  K  I  E  N  E  N    216
ATAACTTTCTTACACTAGCCATTGAACGCCCGCTAGCTCTATCCAGTAACAGAGCAGTGGCCCTTAAGTATTCACCACCTTTAAAAATAGAAAACGAAAA 30600

  L  T  L  S  T  G  G  P  F  T  V  S  G  G  N  L  N  L  A  T  S  A  P  L  S  V  Q  N  N  S  L  S  L    249
CTTAACCCTAAGCACAGGCGGACCTTTTACTGTAAGCGGGGGAAATTTAAACCTGGCAACATCGGCACCCCTCTCCGTGCAAAACAATTCTCTCTCCTTA 30700

 G  V  N  P  P  F  L  I  T  D  S  G  L  A  M  D  L  G  D  G  L  A  L  G  G  S  K  L  I  I  N  L  G    282
GGGGTTAACCCGCCTTTTCTCATCACTGACTCTGGATTAGCTATGGACTTAGGAGACGGTCTTGCATTAGGTGGCTCTAAGTTAATAATCAATCTTGGTC 30800

P  G  L  Q  M  S  N  G  A  I  T  L  A  L  D  A  A  L  P  L  Q  Y  K  N  N  Q  L  Q  L  R  I  G  S  A    316
CAGGTTTACAAATGTCTAATGGAGCTATTACTTTAGCACTAGATGCAGCGCTGCCTTTGCAATATAAAAACAACCAACTTCAACTCAGAATTGGCTCCGC 30900

  S  A  L  I  M  S  G  V  T  Q  T  L  N  V  N  A  N  T  S  K  G  L  A  I  E  N  N  S  L  V  V  K  L    349
GTCTGCTTTAATTATGAGCGGAGTAACACAAACATTAAACGTCAATGCCAATACCAGCAAAGGTCTTGCTATTGAAAATAACTCACTAGTTGTTAAGCTA 31000

 G  N  G  L  R  F  D  S  W  G  S  I  A  V  S  P  T  T  T  T  P  T  T  L  W  T  T  A  D  P  S  P  N     382
GGAAACGGTCTTCGCTTTGATAGCTGGGGAAGCATAGCTGTCTCACCTACTACCACTACCCCTACCACCCTATGGACCACCGCGGACCCGTCTCCTAACG 31100
```

```
A  T  F  Y  E  S  L  D  A  K  V  W  L  V  L  V  K  C  N  G  M  V  N  G  T  I  S  I  K  A  Q  K  G  T    416
CCACTTTTTATGAATCACTAGACGCCAAAGTGTGGCTAGTTTTAGTAAAATGCAACGGCATGGTTAACGGGACCATATCCATTAAAGCTCAAAAAGGCAC 31200

 L  L  K  P  T  A  S  F  I  S  F  V  M  Y  F  Y  S  D  G  T  W  R  K  N  Y  P  V  F  D  N  E  G  I    449
TTTACTTAAACCCACAGCTAGCTTTATTTCCTTTGTCATGTATTTTTACAGCGACGGAACGTGGAGGAAAAACTATCCCGTGTTTGACAACGAAGGGATA 31300

 L  A  N  S  A  T  W  G  Y  R  Q  G  Q  S  A  N  T  N  V  S  N  A  V  E  F  M  P  S  S  K  R  Y  P    482
CTAGCAAACAGTGCCACATGGGGTTATCGACAAGGACAGTCTGCCAACACTAACGTTTCCAATGCTGTAGAATTTATGCCTAGCTCTAAAAGGTATCCCA 31400

N  E  K  G  S  E  V  Q  N  M  A  L  T  Y  T  F  L  Q  G  D  P  N  M  A  I  S  F  Q  S  I  Y  N  H  A    516
ATGAAAAAGGTTCTGAAGTTCAGAACATGGCTCTTACCTACACTTTTTTGCAAGGTGACCCTAACATGGCCATATCTTTTCAGAGCATTTATAATCATGC 31500

 I  E  G  Y  S  L  K  F  T  W  R  V  R  N  N  E  R  F  D  I  P  C  C  S  F  S  Y  V  T  E  Q  -      547
AATAGAAGGCTACTCATTAAAATTCACCTGGCGCGTTCGAAATAATGAACGTTTTGACATCCCCTGTTGCTCATTTTCTTATGTAACAGAACAATAAAAT 31600
                                                                                 L5-2 pA >>>>>>

ATTGTTATTTTGTATTTTCAACTTTATTGATACTTTTACAGAATTCTAACCGTTAATCTCCCTCCCCCCTTCCACTTTACCTTATACACCTCCCTTTCCC 31700
           E4 pA <<<<<<[E4 90R]-  L  I  R  V  T  L  R  G  G  G  K  W  K  V  K  Y  V  E  R  E  G     70

CCTGTACCACCGCAAACAACTGCAATTTAGGATTTACACAACGATTCTTCTGTGACAAAATCAACACAGGTTCTTTGCTGGCAAAGCGCTGATCCGTAAT 31800
 Q  V  V  A  F  L  Q  L  K  P  N  V  C  R  N  K  Q  S  L  I  L  V  P  E  K  S  A  F  R  Q  D  T  I     37

                                                     < start of E4 90R 2nd coding exon
GGAAATGAAACCTTCAGAAACATCGTCCAACAGCACGGTGGAGTCCAAAGCAGAGCTCTGCAAAAACAAATACAGTCAAGCTCTCCACGGGTTCTCGCCT 31900
                                                           [E4 257R]  -  A  R  W  P  N  E  G     251
 S  I  F  G  E  S  V  D  D  L  L  V  T  S  D  L  A  S  S                                             18

CTGTTGTAGTCTGCCAACGTAAACGGGCAGTACCGCTCCATCAAGCCCCGCAGTAATCCCTGTCTCCGGGGTTCCACCAAGCTCCTCATGAGTGACCTAA 32000
R  N  Y  D  A  L  T  F  P  C  Y  R  E  M  L  G  R  L  L  G  Q  R  R  P  E  V  L  S  R  M  L  S  R  V    217

CGGTGAAGCTTCCCAACACTTTCACCGCCTTGGCCAGCAGCCGCCGCGTCCGACGAGCGCAGCACCGCACAGAAAGCTCATCCAAGTTTTTACAATAGGT 32100
 T  F  S  G  L  V  K  V  A  K  A  L  L  R  R  T  R  R  A  C  C  R  V  S  L  E  D  L  N  K  C  Y  T    184

ACAGCCCAACACCACCATATTATTCATAATTCCATAACTAAAAAAACTCCACCCAAACGACATGCGCTCCAACACTATAGCCGCGTGCCCATCATACAGC 32200
 C  G  L  V  V  M  N  N  M  I  G  Y  S  F  F  S  W  G  F  S  M  R  E  L  V  I  A  A  H  G  D  Y  L    151

AGGCGAATATATATAAAATGCCTACCTCTGACAAACACGCTCCCCATAAATAACACTTCCTTGGGCATGCCACAATTTACAATTTCTCGATACCAAGGGA 32300
L  R  I  Y  I  F  H  R  G  R  V  F  V  S  G  M  F  L  V  E  K  P  M  G  C  N  V  I  E  R  Y  W  P  F    117

ATCTTAAGTTATACAGTGAACCATAAATCATCATTTTAAACCAATTTGCTAACACTACACCCCCCGCCTTACATTGAAGAGACCCAGGTTTAATACAGTG 32400
 R  L  N  Y  L  S  G  Y  I  M  M  K  F  W  N  A  L  V  V  G  G  A  K  C  Q  L  S  G  P  K  I  C  H     84

ACAGTGTATAGTCCAGCGTTCAAAACCTCTTACAATTTGATTAAAATCAACATTAATAGTGGCACAGCAAACACAAACACGCATGTAAATTTTGCACACA 32500
 C  H  I  T  W  R  E  F  G  R  V  I  Q  N  F  D  V  N  I  T  A  C  C  V  C  V  R  M  Y  I  K  C  V     51
```

```
                                                                            end of E4 90R 1st coding exon <
TGCTTTTCCCATTTAGACAATATCATGTCCCACCACATTGGCCACTCCTGCAATACTACAAAAGGCGCACAAGATGGAATAGACCTCACCTCGCTCACAT 32600
H  K  E  W  K  S  L  I  M  D  W  W  M  P  W  E  Q  L  V  V  F  P  A  C  S  P  I  S  R  V  E  S  V  Y   17
                                                                                                   Y   17

            start of E4 257R coding exon, E4 90R 1st coding exon <
AATGCATATTCAAATGTTCACACTCTAAAAGTCCAGGAGTCCTTTCCATTGTGGCAATAGGCACAGAATCCTCAGAGGGAGGTGGAAGACGGTGGGTTTG 32700
                                                                       [E4 121R]  -  L  S  T  S  S  P  P  N  P  113
  H  M  N  L  H  E  C  E  L  L  G  P  T  R  E  M                                                        1
  H  M  N  L  H  E  C  E  L  L  G  P  T  R  E  M                                                        1

GTACGAACTCAGTCTGCAGCGAAACCATCTGTCGCGTTGCATCATAAATTAAAAGCTCGCGCACAGCTTCGTACTTCTGTTTTAAGAAACGAACACGCTG 32800
  V  F  E  T  Q  L  S  V  M  Q  R  T  A  D  Y  I  L  L  E  R  V  A  E  Y  K  Q  K  L  F  R  V  R  Q    80

CCAACAAATGTTCGCAAATCGACGGTTTCGTTGTCGCGCTCTTTCAGTTTTCAGGGCAACGTTCAGCCACTCCTGCAGTCCACTTAACAGCTCCTCAGCC 32900
 W  C  I  N  A  F  R  R  N  R  Q  R  A  R  E  T  K  L  A  V  N  L  W  E  Q  L  G  S  L  L  E  E  A    47

CGTGGAGATATGCTGACATTATACCTTATTATGTCCCCATAAACGTTCAAACAGCAGGTTAAAGCCAACTCCAACCAAGAAATACAAAGGCCTTGATCCC 33000
R  P  S  I  S  V  N  Y  R  I  I  D  G  Y  V  N  L  C  C  T  L  A  L  E  L  W  S  I  C  L  G  Q  D  R   13

GACTCACTGGAGGTGGAGGGAGAGACGGAAGAGGCATAATTATTCCAGACGGTTGTAAAGCGAGCCAAAGTGCAAGTCACGAAGATCACACCTCTCCCCA 33100
                                            [E4 116R]  -  E  L  R  N  Y  L  S  G  F  H  L  D  R  L  D  C  R  E  G    98
  S  V  P  P  P  P  L  S  P  L  P  M                                                                    1

  < start of E4 121R coding exon
CCGCTGCGTTGGTGAAAAATTACAGCCAAGTCAAAAAAGATGCGATTTTCCAAATTACCAATCACGGCTTCCACTAAGGCTGGCACACGCACTTCAAGAA 33200
G  S  R  Q  H  F  I  V  A  L  D  F  F  I  R  N  E  L  N  G  I  V  A  E  V  L  A  P  V  R  V  E  L  F   64

ACACAAACATAGCAAAAGCATTTTCCTCAAAATCTTCAAACATTAAGCTGCAATCTTGAATAATTCCCAAATAATTTTCCGCTTGCCACCCGCGCAACAC 33300
  V  F  M  A  F  A  N  E  E  F  D  E  F  M  L  S  C  D  Q  I  I  G  L  Y  N  E  A  Q  W  G  R  L  V    31
```

```
ATCCATTAAAATTTCTTGTAAACTGGCGCCATGTAATTCAAAAAGTTTGTTAAGAGCACCCTCTACTGTCATACGCAGGCACACCTTCATGGTTGAAAAA 33400
                                                                       [E4 130R]  -  P  Q  F  L   127
 D  M  L  I  E  Q  L  S  A  G  H  L  E  F  L  K  N  L  A  G  E  V  T  M  R  L  C  V  K  M           1

                   < start of E4 116R coding exon
GATCAGGTTCCCGTGTCACCTGCAGTTCATTTAAAAGATTAACATTAGGCTCAAAACCCCGATCCCGAATCTCCATGCGTAGCATTAGTTGTACAAAGTC 33500
  D  P  E  R  T  V  Q  L  E  N  L  L  N  V  N  P  E  F  G  R  D  R  I  E  M  R  L  M  L  Q  V  F  D    94

ATCCAAATCATTGCATATAAGCTCTGTCAGTTCGCTATCAGGAAGCAGCTCAGGTGATGCTACACAACAAATCATCTCTAGCGTAGGAGCTAAAGACGTT 33600
 D  L  D  N  C  I  L  E  T  L  E  S  D  P  L  L  E  P  S  A  V  C  C  I  M  E  L  T  P  A  L  S  T    61

AAGGTAAAGCCACAATAAGCAGCTTGAAGAACTGGAGTAACACAATGCAAAATGTGCAGCAAAAACTCCGACATGTTTGTCTTTAAAAAATCTACCACAG 33700
L  T  F  G  C  Y  A  A  Q  L  V  P  T  V  C  H  L  I  H  L  L  F  E  S  M  N  T  K  L  F  D  V  V  S   27

                                                                  start of E4 130R coding exon <
AAATGTCCATATTATTTAAATAAAACATCAGGGGCTCAGGAACCACCACCGAAATAAAAACCGGTCGTAACAAATACATTGTGTCCTGCAACAAAAAAAA 33800
  I  D  M  N  N  L  Y  F  M  L  P  E  P  V  V  V  S  I  F  V  P  R  L  L  Y  M                        1

AATATTAATGCCCACACCTGGGAAAACCTGTTCTAAAACCAAACAGGTATAAGTATTACAAATGCCTCCCTTTGCCCCCCAATCCAAACCAAATAAGCTG 33900

CCCCGTCTTACCGCGACAAAGCACACAGAACAAAACACACTCCGCAGACGAACACAATATTTATACACTCCCTTTGCCGTCAAAAGTCCACAAAAACTCC 34000
                              < 5' end of E4 mRNA      <<<<<<<<< E4 TATA

AAAGGTCAGAAAAACCGCCACATGAACACTTCCGCATACTGTTTCACATATCGTCACTTCCGCCGCACCGCGCCCGTCCTCCGACCCCACACGTCATCCG 34100
                                                     < end of ITR

CCTCCACCCTTTCCCACCCCGCCCGCCTCTACGTCACCTTACACCACCCCTAGTCCCTCCTCCCTCATTATCATATTGGCTCGTTTCCAGTTTTAAGGTA 34200

TATTATTGATGATG                                                                                       34214
start of ITR <
```

Sequence data annotated by A. DAVISON and V. MAUTNER. See DAVISON et al. (1994) for details of transcriptional and translational features. The Ad40 DNA sequence is available under Genbank accession number L19443, *ITR*, inverted terminal repeat

References

Adrian T, Wigand R (1989) Genome type analysis of adenovirus 31, a potential causative agent of infants' enteritis. Arch Virol 105: 81–87

Adrian T, Wadell G, Hierholzer JC, Wigand R (1986) DNA restriction analysis of adenovirus prototypes 1 to 41. Arch Virol 91: 277–290

Akhter J, Quadri HSM (1993) Incidence of adenovirus- associated diarrheal disease at a tertiary care hospital in Riyadh. J Exp Med 21: 295–296

Albert MJ (1986) Enteric adenoviruses. Arch Virol 88: 1–17

Allard A, Wadell G (1988) Physical organisation of the enteric adenovirus type 41 early region 1A. Virology 164: 220–229

Allard A, Wadell G (1992) The E1B transcription map of the enteric adenovirus type 41. Virology 183: 319–330

Allard A, Girones R, Juto P, Wadell G (1990) Polymerase chain reaction for detection of adenoviruses in stool samples. J Clin Microbiol 28: 2659–2667

Allard A, Albinsson B, Wadell G (1992) Detection of adenoviruses in stools from healthy persons and patients with diarrhea by two-step polymerase chain reaction. J Med Virol 37: 149–157

Arató A, Kósnai I, Szönyi L, Tóth M (1991) Frequent past exposure to adenovirus 12 in coeliac disease. Acta Paediatr Scand 80: 1101–1102

Babiss LE, Young CSH, Fisher PB, Ginsberg HS (1983) Expression of adenovirus E1A and E1B gene products and the Escherichia coli XGPRT gene in KB cells. J Virol 46: 454–465

Babiss LE, Ginsberg HS (1984) Adenovirus type 5 early region 1b gene product is required for efficient shutoff of host protein synthesis. J Virol 50: 202–212

Babiss LE, Ginsberg HS, Darnell JE (1985) Adenovirus E1B proteins are required for accumulation of late viral mRNA and for effects on cellular mRNA translation and transport. Mol Cell Biol 5: 2552–2558

Bailey A, Mackay N, Mautner V (1993) Enteric adenovirus type 40: Expression of E1b proteins in vitro and in vivo. Virology 193: 631–641

Bailey A, Ullah R, Mautner V (1994) Cell type specific regulation of expression from the E1B promoter in recombinant Ad5/Ad40 viruses. Virology 202: 695–706

Bailey A, Mautner V (1994) Phylogenetic relationships among adenovirus serotypes. Virology 205: 438–452

Bhisitkul DM, Todd KM, Listernick R (1992) Adenovirus infection and childhood intussuception. Am J Dis Child 146: 1331–1333

Boulanger P, Blair GE (1991) Expression and interactions of human adenovirus oncoproteins. Biochem J 275: 281–299

Brandt CD, Kim HW, Yolken RH, Kapikian AZ, Arrobio JO, Rodriguez WJ, Wyatt RG, Chanock RM, Parrott RH (1979) Comparative epidemiology of two rotavirus serotypes and other viral agents associated with pediatric gastroenteritis. Am J Epidemiol 110: 243–254

Brandt CD, Kim HW, Rodriguez WJ, Arobio JO, Jeffries BC, Stallings EP, Lewis C, Miles AJ, Gardner MK, Parrott RH (1985) Adenoviruses and paediatric gastroenteritis. J Infect Dis 151: 437–443

Breiding DE, Edbauer CA, Tong JY, Byrd PJ, Grand RJA, Gallimore PH, Williams J (1988) Isolation and characterization of adenovirus type 12 E1 host-range mutants defective for growth in nontransformed human cells. Virology 164: 390–402

Bridge E, Ketner G (1989) Redundant control of adenovirus late gene expression by early region 4. J Virol 63: 631–638

Bridge E, Ketner G (1990) Interaction of adenoviral E4 and E1B products in late gene expression. Virology 174: 345–353

Brown M (1990) Laboratory identification of adenoviruses associated with gastroenteritis in Canada from 1983 to 1986. J Clin Microbiol 28: 1525–1529

Brown M, Petric M, Middleton PJ (1984) Diagnosis of fastidious enteric adenoviruses 40 and 41 in stool specimens. J Clin Microbiol 20: 334–338

Brown M, Wilson-Friesen HL, Doane F (1992) A block in release of progeny virus and a high particle-to-infectious unit ratio contribute to poor growth of enteric adenovirus types 40 and 41 in cell culture. J Virol 66: 3198–3205

Brunet LJ, Berk AJ (1988) Concentration dependence of transcriptional activation in inducible E1A-containing human cells. Mol Cell Biol 8: 4799–4807

Chiba S, Nakata S, Nakamura I, Taniguchi K, Urasawa S, Fujinaga K, Nakao T (1983) Outbreak of infantile gastroenteritis due to type 40 adenovirus. Lancet 2: 954–957

Chroboczek J, Bieber F, Jacrot B (1992) The sequence of the genome of adenovirus type 5 and its comparison with the genome of adenovirus type 2. Virology 186: 280–285

Cladras C, Wold WSM (1985) DNA sequence of the early E3 transcription unit of adenovirus 5. Virology 140: 28–43

Cousin C, Winter N, Gomes S, D'Halluin JC (1991) Cellular transformation by E1 genes of enteric adenoviruses. Virology 181: 277–287

Cruz JR, Caceres P, Cano F, Flores J, Bartlett A, Torun B (1990) Adenovirus types 40 and 41 and rotaviruses associated with diarrhea in children from Guatemala. J Clin Microbiol 28: 1780–1784

Davison AJ, Telford EAR, Watson MS, McBride K, Mautner V (1993) The DNA sequence of adenovirus type 40. J Mol Biol 234: 1308–1316

de Jong JC, Wigand R, Kidd AH, Wadell G, Kapsenberg JG, Muzerie AG, Wermenbol AG, Firtzlaff R-G (1983) Candidate adenoviruses 40 and 41: fastidious adenoviruses from human infant stool. J Med Virol 11: 215–231

de Jong JC, Bijlsma K, Wermenbol AG, VerweijUijterwaal MW, van der Avoort HGAM, Wood DJ, Bailey AS, Osterhaus ADME (1993) Detection, typing, and subtyping of enteric adenoviruses 40 and 41 from fecal samples and observations of changing incidence of infections with these types and subtypes. J Clin Microbiol 31: 1562–1569

Devery JM, Bender V, Penttila I, Skerritt JH (1991) Identification of reactive synthetic gliadin peptides specific for coeliac disease. Int J Allergy Appl Immunol 95: 356–362

Dieleman LA, Pena AS, van Doorninck JH, Mearin ML, van Duijn W, Lamers CBHW (1991) No humoral response to the E1B-54kD protein of adenovirus 12 in patients with coeliac disease. Eur J Gastroenterol Hepatol 3: 255–257

Ellis HJ, Doyle AP, Sturgess RP, Ciclitira PJ (1992) Coeliac disease: characterisation of monoclonal antibodies raised against a synthetic peptide corresponding to amino acid residues 206–217 of A-gliadin. Gut 33: 1504–1507

Estes MK, Palmer EL, Obijeski JF (1983) Rotaviruses: a review. In: Compans RW, Cooper M, Koprowski H et al. (eds) Current topics in microbiology and immunology, vol 105. Springer, Berlin Heidelberg New York, pp 123–184

Everett RD (1987) The regulation of transcription of viral and cellular genes by herpes virus immediate early gene products. Anticancer Res 7: 589–604

Fitch WM, Margoliash E (1967) Construction of phylogenetic trees. Science 155: 279–284

Flewett TH, Bryden AS, Davies HA, Morris CA (1975) Epidemic viral enteritis in a long-stay childrens ward. Lancet 1: 4–5

Fox JP, Hall CE, Cooney MK (1977) The Seattle virus watch VII. Observations of adenovirus infections. Am J Epidemiol 105: 362–386

Fukui Y, Shiroki K, Saito I, Shimojo H (1984) Characterization of a host range mutant of human adenovirus 12 defective in early region 1B. J Virol 50: 132–136

Gary GW Jr, Hierholzer JC, Black RE (1979) Characterization of noncultivable adenoviruses associated with diarrhea in infants: a new subgroup of human adenoviruses. J Clin Microbiol 10: 96–103

Gazzard B, Blanshard C (1993) Diarrhea in AIDS and other immunodeficiency states. Baillieres Clin Gastroenterol 7: 387–419

Gilbert DA, Reid YA, Gail MH, Pee D, White C, Hay RJ, O'Brien SJ (1990) Application of DNA fingerprints for cell-line individualization. Am J Hum Genet 47: 499–514

Ginsberg HS (1984) The adenoviruses, 1st edn. Plenum, New York

Glenn GM, Ricciardi RP (1988) Detailed kinetics of adenovirus type-5 steady-state transcripts during early infection. Virus Res 88: 73–91

Gomes SA, Niel C, D'Halluin JC (1992) Growth of fastidious adenovirus serotype 40 in HRT 18 cells: interactions with E1A and E1B deletion mutants of subgenus C adenoviruses. Arch Virol 124: 45–56

Grabow WOK, Puttergill DL, Bosch A (1992) Propagation of adenovirus types 40 and 41 in the PLC/PRF/5 primary liver carcinoma cell line. J Virol Methods 37: 201–208

Graham FL, Smiley J, Russell WC, Nairn R (1977) Characteristics of a human cell line transformed by DNA from human adenovirus type 5. J Gen Virol 36: 59–72

Handley J, O'Neill H, Connolly J, Burrows D (1993) Adenovirus-12 and dermatitis herpetiformis. Acta Derm Venereol 73: 430–432

Hashimoto S, Sakakibara N, Kumai H, Nakai M, Sakuma S, Chiba S, Fujinaga K (1991) Fastidious adenovirus type 40 can propagate efficiently and produce plaques on a human cell line, A549, derived from lung carcinoma. J Virol 65: 2429–2435

Hermann JE, Blacklow NR, Perron-Henry DM, Clements E, Taylor DN, Echeverria P (1988) Incidence of enteric adenoviruses among children in Thailand and the significance of these viruses in gastroenteritis. J Clin Microbiol 26: 1783–1786

Hierholzer JC (1992) Adenoviruses in the immunocompromised host. Clin Microbiol Rev 5: 262–274
Hierholzer JC, Wigand R, de Jong JC (1988) Evaluation of human adenoviruses 38, 39, 40 and 41 as new serotypes. Intervirology 29: 1–10
Hitt MM, Graham FL (1990) Adenovirus E1A under the control of heterologous promoters: wide variation in E1A expression has little effect on virus replication. Virology 179: 667–678
Horwitz MS (1990) Adenoviruses. In: Fields BN, Knipe DM (ed) Virology, 2nd edn. Raven, New York, pp 1723–1740
Howdle PD, Blair GE (1992) Molecular biology and coeliac disease. Gut 33: 573–575
Howdle PD, Blair GE (1994) Molecular pathogenesis of coeliac disease. In: Quirke P (ed) The molecular biology of digestive diseases. BMJ, London
Huang M, Hearing P (1989) Adenovirus early region 4 encodes two gene products with redundant effects in lytic infection. J Virol 63: 2605–2615
Hynes RO (1992) Integrins: verstility, modulation, and signaling in cell adhesion. Cell 69: 11–25
Ishino M, Sawada Y, Yaegashi T, Demura M, Fujinaga K (1987) Nucleotide sequence of the adenovirus type 40 inverted terminal repeat: close relation to that of adenovirus type 5. Virology 156: 414–416
Ishino M, Ohashi Y, Emoto T, Sawada Y, Fujinaga K (1988) Characterization of adenovirus type 40 E1 region. Virology 165: 95–102
Jackson SP, MacDonald JJ, Lees-Miller S, Tjian R (1990) GC box binding induces phosphorylation of Sp1 by a DNA-dependent protein kinase. Cell 63: 155–165
JareckiKahn K, Unicomb LE (1992) Seroprevalence of enteric and nonenteric adenoviruses in Bangladesh. J Clin Microbiol 30: 2733–2734
JareckiKahn K, Tzipori SR, Unicomb LE (1993) Enteric adenovirus infection among infants with diarrhea in rural Bangladesh. J Clin Microbiol 31: 484–489
Johansson ME, Uhnoo I, Kidd AH, Madeley CR, Wadell G (1980) Direct identification of enteric adenovirus, a candidate new serotype associated with infantile gastroenteritis. J Clin Microbiol 12: 95–100
Johansson ME, Uhnoo I, Svensson L, Pettersson CA, Wadell G (1985) Enzyme-linked immunosorbent assay for detection of enteric adenovirus 41. J Med Virol 17: 19–27
Kagnoff MF (1989) Celiac disease: adenovirus and alpha gliadin. In: Oldstone MB (ed) Molecular mimicry. Cross-reactivity between microbes and host proteins as a cause of autoimmunity. Springer, Berlin Heidelberg New York, pp 67–78 (Current topics in microbiology and immunology, vol 145)
Kagnoff MF, Austin RK, Hubert JJ, Bernardin JE, Kasarda DD (1984) Possible role for a human adenovirus in the pathogenesis of celiac disease. J Exp Med 160: 1544–1557
Kagnoff MF, Paterson YJ, Kumar PJ, Kasarda DD, Carbone FR, Unsworth DJ, Austin RK (1987) Evidence for the role of a human intestinal adenovirus in the pathogenesis of celiac-disease. Gut 28: 995–1001
Ketner G, Bridge E, Virtanen A, Hemstrom C, Pettersson U (1989) Complementation of adenovirus E4 mutants by transient expression of E4 cDNA and deletion mutants. Nucleic Acids Res 17: 3037–3048
Kidd AH (1984) Genome variants of adenovirus 41 (subgroup G) from children with diarrhea in South Africa. J Med Virol 14: 49–59
Kidd AH, Erasmus MJ (1989) Sequence characterization of the adenovirus 40 fibre gene. Virology 172: 134–144
Kidd AH, Madeley CR (1981) In vitro growth of some fastidious adenoviruses from stool specimens. J Clin Pathol 34: 213–216
Kidd AH, Tiemessen CT (1993) Characterization of a single SA7-like VA RNA gene in subgroup F adenoviruses. J Gen Virol 74: 1621–1626
Kidd AH, Banatvala JE, de Jong JC (1983) Antibodies to fastidious faecal adenoviruses (species 40 and 41) in sera from children. J Med Virol 11: 333–341
Kidd AH, Berkowitz FE, Blaskovic PJ, Schoub BD (1984) Genome variants of human adenovirus 40 (subgroup F). J Exp Med 14: 235–246
Kidd AH, Harley EH, Erasmus MJ (1985) Specific detection and typing of adenovirus types 40 and 41 in stool specimens by dot-blot hybridization. J Clin Microbiol 22: 934–939
Kidd AH, Rosenblatt A, Besselaar TG, Erasmus MJ, Tiemessen CT, Berkowitz FE, Schoub BD (1986) Characterization of rotaviruses and subgroup F adenoviruses from acute summer gastroenteritis in South Africa. J Med Virol 18: 159–168
Kidd AH, Erasmus MJ, Tiemessen CT (1990) Fibre sequence heterogeneity in subgroup F adenoviruses. Virology 179: 139–150

Kidd AH, Chroboczek J, Cusack S, Ruigrok RWH (1993) Adenovirus type 40 virions contain two distinct fibers. Virology 192: 73–84

Kim KH, Yang JM, Joo SI, Cho YG, Glass RI, Cho YJ (1990) Importance of rotavirus and adenovirus types 40 and 41 in acute gastroenteritis in Korean children. J Clin Microbiol 28: 2279–2284

Kimelman D, Miller JS, Porter JD, Roberts BE (1985) E1A regions of the human adenoviruses and of the highly oncogenic simian adenovirus 7 are closely related. J Virol 53: 399–409

Kinloch R, Mackay N, Mautner V (1984) Adenovirus hexon. Sequence comparison of subgroup C serotypes 2 and 5. J Biol Chem 259: 6431–6436

Kotloff KL, Wasserman SS, Steciak JY, Tall BD, Losonsky GA, Nair P, Morris JG Jr, Levine MM (1988) Acute diarrhea in Baltimore children attending an outpatient clinic. Paediatr Infect Dis 7: 753–759

Kotloff KL, Losonsky GA, Morris JG Jr, Wasserman SS, Singh-Naz N, Levine MM (1989) Enteric adenovirus infection and childhood diarrhea: an epidemiologic study in three clinical settings. Pediatrics 84: 219–225

Krajden M, Brown M, Petrasek A, Middleton PJ (1990) Clinical features of adenovirus enteritis: a review of 127 cases. Paediatr Infect Dis 9: 636–641

Lädheaho ML, Lehtinen M, Rissa HR, Hyoty H, Reunala T, Mäki M (1993a) Antipeptide antibodies to adenovirus E1b protein indicate enhanced risk of celiac disease and dermatitis herpetiformis. Int Arch Allergy Immunol 101: 272–275

Lädheaho ML, Parkkonen P, Reunala T, Mäki M, Lehtinen M (1993b) Antibodies to E1B protein-derived peptides of enteric adenovirus type-40 are associated with celiac disease and dermatitis herpetiformis. Clin Immunol Immunopathol 69: 300–305

Larsen LP, Tibbetts C (1987) Adenovirus E1A gene autoregulation: Revertants of an E1A promoter mutation encode altered E1A products. Proc Natl Acad Sci USA 84: 8185–8189

Laver WG, Younghusband HB, Wrigley NG (1971) Purification and properties of chick embryo lethal orphan virus (an avian adenovirus). Virology 45: 598–614

Leite JP, Pereira HG, Azeredo RS, Schatzmayr HG (1985) Adenoviruses in faeces of children with acute gastroenteritis in Rio de Janeiro, Brazil. J Med Virol 15: 203–209

Leite JPG, Niel C, D'Halluin JC (1986) Expression of the chloramphenicol acetyl transferase gene in human cells under the control of early adenovirus subgroup C promoters: effect of E1A gene products from other subgroups on gene expression. Gene 41: 207–215

Leppard KN, Shenk T (1989) The adenovirus E1B 55kd protein influences mRNA transport via an intranuclear effect on RNA metabolism. EMBO J 8: 2329–2336

Lew JF, Moe CL, Monroe SS, Allen JR, Harrison BM, Forrester BD, Stine SE, Woods PA, Hierholzer JC, Hermann JE, Blacklow NR, Bartlett AV, Glass RI (1991) Astrovirus and adenovirus associated with diarrhea in children in day care settings. J Infect Dis 164: 673–678

Maddox A, Francis N, Moss J, Blanshard C, Gazzard B (1992) Adenovirus infection of the large bowel in HIV positive patients. J Clin Pathol 45: 684–688

Mahon J, Blair GE, Wood GM, Scott BB, Losowsky MS, Howdle PD (1991) Is persistent adenovirus 12 infection involved in coeliac disease? A search for viral DNA using the polymerase chain reaction. Gut 32: 1114–1116

Mak I, Mak S (1990) Separate regions of an adenovirus E1B protein critical for different biological functions. Virology 176: 553–562

Mantzaris G, Jewell DP (1991) In vivo toxicity of a synthetic dodecapeptide from A gliadin in patients with coeliac disease. Scand J Gastroenterol 26: 392–398

Mantzaris GJ, Karagiannis JA, Priddle JD, Jewell DP (1990) Cellular hypersensitivity to a synthetic dodecapeptide derived from human adenovirus 12 which resembles a sequence of A-gliadin in patients with coeliac disease. Gut 31: 668–673

Mautner V, Mackay N (1991) Enteric adenovirus type 40: Complementation of the E4 defect in Ad2 dl808. Virology 183: 433–436

Mautner V, Mackay N, Steinthorsdottir V (1989) Complementation of enteric adenovirus type 40 for lytic growth in tissue culture by E1B 55K function of adenovirus types 5 and 12. Virology 171: 619–622

Mautner V, Mackay N, Morris K (1990) Enteric adenovirus type 40: expression of E1B mRNA and proteins in permissive and non-permissive cells. Virology 179: 129–138

Mistchenko AS, Huberman KH, Gomez JA, Grinstein S (1992) Epidemiology of enteric adenovirus infection in prospectively monitored Argentine families. Epidemiol Infect 109: 539–546

Montell C, Fisher EF, Caruthers MH, Berk AJ (1984) Control of adenovirus E1B mRNA synthesis by a shift in the activities of RNA splice sites. Mol Cell Biol 4: 966–972

Montgomery EA, Popek EJ (1994) Intussusception, adenovirus, and children: A brief reaffirmation. Hum Pathol 25: 169–174

Noel J, Mansoor A, Thaker J, Herrmann J, Perron-Henry D, Cubitt WD (1994) Identification of adenoviruses in faeces from patients with diarrhea at the Hospitals for Sick Children, London, 1989–1992. J Med Virol 43: 84–90
O'Malley RP, Duncan RF, Hershey JBW, Mathews MB (1989) Modification of protein synthesis initiation factors and the shutt-off of the host protein synthesis in adenovirus-infected cells. Virology 168: 112–118
Parks CL, Spector DJ (1990) Cis-dominant defect in activation of adenovirus type 5 E1B early RNA synthesis. J Virol 64: 2780–2787
Parks CL, Banerjee S, Spector DJ (1988) Organization of the transcriptional control region of the E1B gene of adenovirus type 5. J Virol 62: 54–67
Pieniazek D, Pieniazek NJ, Macejak D, Coward J, Rayfield M, Luftig RB (1990a) Differential growth of human enteric adenovirus 41 (TAK) in continuous cell lines. Virology 174: 239–249
Pieniazek D, Pieniazek NJ, Macejak D, Luftig RB (1990b) Enteric adenovirus 41 (Tak) requires low serum for growth in human primary cells. Virology 178: 72–80
Pieniazek NJ, Slemenda SB, Pieniazek D, Velarde J Jr, Luftig RB (1989) Sequence of the human enteric adenovirus type 41 Tak fibre protein gene. Nucleic Acids Res 17: 9474
Pieniazek NJ, Slemenda SB, Pieniazek D, Velarde J Jr, Luftig RB (1990) Human enteric adenovirus type 41 (Tak) contains a second fibre protein gene. Nucleic Acids Res 18: 1901
Pilder S, Moore M, Logan J, Shenk T (1986) The adenovirus E1B 55K transforming polypeptide modulates transport or cytoplasmic stabilization of viral and host cell mRNAs. Mol Cell Biol 6: 470–476
Porter HJ, Padfield CDH, Peres LC, Hirschowitz L, Berry PJ (1993) Adenovirus and intranuclear inclusions in appendices in intussusception. J Clin Pathol 46: 154–158
PringAkerblom P, Adrian T (1993) The hexon genes of adenoviruses of subgenus C: comparison of the variable regions. Res Virol 144: 117–127
PringAkerblom P, Adrian T (1994) Type- and group-specific polymerase chain reaction for adenovirus detection. Res Virol 145: 25–35
Rao RN, Debbas M, Sabbatinit P, Hockenberry D, Korsmeyer S, White E (1992) The adenovirus E1A proteins induce apoptosis, which is inhibited by the adenovirus E1B 19kDa and Bcl-2 proteins. Proc Natl Acad Sci USA 89: 7743–7746
Regua-Mangia AH, Duarte AN, Duarte R, Silva LA, Bravo VLR, Leal MC (1993) Aetiology of acute diarrhea in hospitalized children in Rio de Janeiro. J Tropical Pediatr 93: 365–367
Retter M, Middleton PJ, Tam JS, Petric M (1979) Enteric adenoviruses: detection, replication and significance. J Clin Microbiol 10: 574–578
Roberts MM, White JL, Grutter MG, Burnett RM (1986) Three-dimensional structure of the adenovirus major coat protein hexon. Science 232: 1148–1151
Roberts RJ, Aküsjarvi G, Aleström P, Gelinas RE, Gingeras TR, Sciaki D, Pettersson U (1986) A consensus sequence for the adenovirus-2 genome. In: Doerfler W (ed) Adenovirus DNA. Nijhoff, Boston, pp 1–51
Rodriguez WJ, Kim JW, Brandt CD et al (1985) Fecal adenoviruses from a longitudinal study of families in metropolitan Washington DC: laboratory, clinical, and epidemiologic observations. J Pediatr 107: 514–520
Schmitz H, Wigand R, Heinrich W (1983) Worldwide epidemiology of human adenovirus infections. Am J Epidemiol 117: 455–466
ScottTaylor TH, Ahluwalia G, Dawood M, Hammond GW (1993) Detection of enteric adenoviruses with synthetic oligonucleotide probes. J Med Virol 41: 328–337
Seed B, Sheen J-Y (1988) A simple phase-extraction assay for chloramphenicol acyltransferase activity. Gene 67: 271–277
Sethi SK, Khuffash FA, AlNakib W (1989) Microbial etiology of acute gastroenteritis in hospitalized children in Kuwait. Pediatr Infect Dis J 8: 593–597
Sheppard M, Trist H (1992) Characterization of the avian adenovirus penton base. Virology 188: 881–886
Shinagawa M, Iida Y, Matsuda A, Tsukiyama T, Sato G (1987) Phylogenetic relationships between adenoviruses as inferred from nucleotide sequences of inverted terminal repeats. Gene 55: 85–93
Shinozaki T, Araki K, Ushijima H, Fujii R (1987) Antibody responses to enteric adenovirus types 40 and 41 in sera from people in various age groups. J Clin Microbiol 25: 1679–1682
Shinozaki T, Araki K, Fujita Y, Kobayashi M, Tajima T, Abe T (1991) Epidemiology of enteric adenoviruses 40 and 41 in acute gastroenteritis in infants and young children in the Tokyo area. Scand J Infect Dis 23: 543–547

Shiroki K, Ohshima K, Fukui Y, Ariga H (1986) The adenovirus type 12 early-region 1B 58,000-Mr gene product is required for viral DNA synthesis and for initiation of cell transformation. J Virol 57: 792–801
Signäs C, Akusjärvi G, Pettersson U (1986) Region E3 of human adenoviruses; differences between the ongcogenic adenovirus-3 and the non-oncogenic adenovirus-2. Gene 50: 173–184
Spector DJ, Parks CL, Knittle RA (1993) A multi-component cis-activator of transcription of the E1b gene of adenovirus type 5. Virology 194: 128–136
Sprengel J, Schmitz B, Heuss-Neitzel D, Zock C, Doerfler W (1994) Nucleotide sequence of human adenovirus type 12 DNA: comparative functional analysis. J Virol 68: 379–389
Steinthorsdottir V (1991) Adenovirus type 40 host range in tissue culture. A study of the E1B region. PhD thesis, University of Glasgow
Steinthorsdottir V, Mautner V (1991) Enteric adenovirus type 40: E1B transcription map and novel E1A-E1B co-transcripts in lytically infected cells. Virology 181: 139–149
Takiff HE, Straus SE (1982) Early replicative block prevents the efficient growth of fastidious diarrhea-associated adenoviruses in cell culture. J Med Virol 9: 93–100
Takiff HE, Straus SE, Garon CF (1981) Propagation and in vitro studies of previously non-cultivable enteral adenovirus in 293 cells. Lancet 2: 832–834
Takiff HE, Reinhold W, Garon CF, Straus SE (1984) Cloning and physical mapping of enteric adenoviruses (candidate types 40 and 41). J Virol 51: 131–136
Thea DM, Glass R, Grohmann GS, Perriens J, Ngoy B, Kapita B, Atido U, Mabaluku M, Keusch GT (1993) Prevalence of enteric viruses among hospital patients with AIDS in Kinshasa, Zaire. Trans R Soc Trop Med Hyg 87: 263–266
Tibbetts C, Larsen PL, Jones SN (1986) Autoregulation of adenovirus E1A gene expression. J Virol 57: 1055–1064
Tiemessen CT, Kidd AH (1988) Helper function of adenovirus 2 for adenovirus 41 antigen synthesis in semi-permissive and non-permissive cells. Arch Virol 103: 207–218
Tiemessen CT, Kidd AH (1990) Adenovirus 41 growth in semi-permissive cells shows multiple-hit kinetics. Arch Virol 110: 239–245
Tiemessen CT, Kidd AH (1993) Sensitivity of subgroup F adenoviruses to interferon. Arch Virol 128: 1–13
Tiemessen CT, Kidd AH (1994) Adenovirus type 40 and 41 growth in vitro: host range diversity reflected by differences in patterns of DNA replication. J Virol 68: 1239–1244
Tiemessen CT, Wegerhoff FO, Erasmus MJ, Kidd AH (1989) Infection by enteric adenoviruses, rotaviruses, and other agents in a rural African environment. J Med Virol 28: 176–182
Tiemessen CT, Ujfalusi M, Kidd AH (1993) Subgroup F adenovirus growth in fetal intestinal organ-cultures. Arch Virol 132: 193–200
Toogood CIA, Hay RT (1988) DNA sequence of the adenovirus type 41 hexon gene and predicted structure of the protein. J Gen Virol 69: 2291–2301
Toogood CIA, Murali R, Burnett RM, Hay RT (1989) The adenovirus type 40 hexon: sequence, predicted structure and relationship to other adenovirus hexons. J Gen Virol 70: 3203–3214
Toogood CIA, Crompton J, Hay RT (1992) Antipeptide sera define neutralizing epitopes on the adenovirus hexon. J Gen Virol 73: 1429–1435
Top FH, Buescher EL, Bancroft WH, Russell PK (1971) Immunization with live types 7 and 4 adenovirus vaccines. II. Antibody response and protective effect against respiratory disease due to adenovirus type 7. J Infect Dis 124: 155–160
Uhnoo I, Wadell G, Svensson L, Johansson M (1983) Two new serotypes of enteric adenovirus causing infantile diarrhea. Dev Biol Stand 53: 311–318
Uhnoo I, Wadell G, Svensson L, Johansson ME (1984) Importance of enteric adenoviruses 40 and 41 in acute gastroenteritis in infants and young children. J Clin Microbiol 20: 365–372
Uhnoo I, Olding-Stenkvist E, Kreuger A (1986) Clinical features of acute gastroenteritis associated with rotavirus, enteric adenovirus, and bacteria. Arch Dis Child 61: 732–738
Uhnoo I, Svensson L, Wadell G (1990) Enteric adenoviruses. Baillieres Clin Gastroenterol 4: 627–642
Van R, Wun CC, O'Ryan ML, Matson DO, Jackson L, Pickering LK (1992) Outbreak of human enteric adenovirus types 40 and 41 in Houston day care centers. J Pediatr 120: 516–521
van der Avoort HGAM, Wermenbol AG, Zomerdijk TPL, Kleijne JAFW, van Asten JAAM, Jensma P, Osterhaus ADME, Kidd AH, de Jong JC (1989) Characterisation of fastidious adenovirus types 40 and 41 by DNA restriction enzyme analysis and by neutralizing monoclonal antibodies. Virus Res 12: 139–157
van Loon AE, Maas R, Vaessen RTMJ, Reemst AMCB, Sussenbach JS, Rozijn TH (1985a) Cell transformation by the left terminal regions of the adenovirus 40 and 41 genomes. Virology 147: 227–230

van Loon AE, Rozijn TH, de Jong JC, Sussenbach JS (1985b) Physicochemical properties of the DNAs of the fastidious adenovirus species 40 and 41. Virology 140: 197–200
van Loon AE, Gilardi P, Perricaudet M, Rozijn TH, Sussenbach JS (1987a) Transcriptional activation by the E1A regions of adenovirus types 40 and 41. Virology 160: 305–307
van Loon AE, Ligtenberg M, Reemst AMCB, Sussenbach JS, Rozijn TH (1987b) Structure and organization of the left-terminal DNA regions of fastidious adenovirus types 40 and 41. Gene 58: 109–126
Virtanen A, Pettersson U (1985) Organization of early region 1B of human adenovirus type 2: identification of four differentially spliced mRNAs. J Virol 54: 383–391
Virtanen A, Pettersson U, Le Moullec JM, Tiollais P, Perricaudet M (1982) Different mRNAs from the transforming region of highly oncogenic and nononcogenic human adenoviruses. Nature 295: 705–707
Wadell G, Hammarskjöld ML, Winberg G, Varsanyi TM, Sundell G (1980) Genetic variability of adenoviruses. Ann NY Acad Sci 354: 15–42
Wadell G, Allard A, Johansson M, Svensson L, Uhnoo I (1987) Enteric adenoviruses. Ciba Found Symp 128: 63–91
Wadell G, Allard A, Johansson M, Svensson L, Uhnoo I (1994) Enteric adenoviruses. In: Kapikian AZ (ed) Viral infections of the gastrointestinal tract. Dekker, New York
Weiden MD, Ginsberg HS (1994) Deletion of the E4 region of the genome produces adenovirus DNA concatemers. Proc Natl Acad Sci USA 91: 153–157
Weinberg DH, Ketner G (1986) Adenoviral early region 4 is required for efficient viral DNA replication and for late gene expression. J Virol 57: 833–838
White E, Stillman B (1987) Expression of adenovirus E1B mutant phenotypes is dependent on the host cell and on synthesis of E1A proteins. J Virol 61: 426–435
White E, Cipriani R, Sabbatini P, Denton A (1991) Adenovirus E1B 19-kilodalton protein overcomes the cytotoxicity of E1A proteins. J Virol 65: 2968–2978
Whitelaw A, Davies H, Parry J (1977) Electron microscopy of fatal adenovirus gastroenteritis. Lancet 1: 361
Wickham TJ, Mathias P, Cheresh DA, Nemerow GR (1993) Integrins vβ3 and vβ5 promote adenovirus internalization but not virus attachment. Cell 73: 309–319
Willcocks MM, Carter MJ, Laidler FR, Madeley CR (1988) Restriction enzyme analysis of faecal adenoviruses in Newcastle upon Tyne. Epidemiol Infect 101: 445–458
Williams J, Karger BD, Ho YS, Castiglia CL, Mann T, Flint SJ (1986) The adenovirus E1B 495R protein plays a role in regulating the transport and stability of the viral late messages. Cancer Cells 4: 275–284
Witt DJ, Bousquet EB (1988) Comparison of enteric adenovirus infection in various human cell lines. J Virol Methods 20: 295–308
Wold WSM, Gooding LR (1991) Region E3 of adenovirus: a casette of genes involved in host immunosurveillance and virus-cell interactions. Virology 184: 1–8
Wu L, Rosser DSE, Schmidt MC, Berk A (1987) A TATA box implicated in E1A transcriptional activation of a simple adenovirus 2 promoter. Nature 326: 512–515
Yeh HY, Luftig RB (1991) Human enteric adenovirus type 41 binding to Hep-2 cells is specific. Virology 183: 410–414
Yew PR, Berk AJ (1992) Inhibition of p53 transactivation by adenovirus early 1B protein. Nature 357: 82–85
Yew PR, Kao CC, Berk AJ (1990) Dissection of functional domains in the adenovirus 2 early 1B 55K polypeptide by suppressor-linker insertional mutagenesis. Virology 179: 795–805
Yew PR, Liu X, Berk AJ (1994) Adenovirus E1B oncoprotein tethers a transcriptional repression domain to p53. Genes Dev 8: 190–202
Yolken RH, Franklin CC (1985) Gastrointestinal adenovirus: an important cause of morbidity in patients with necrotizing enterocolitis and gastrointestinal surgery. Paediatr Infect Dis 4: 42–47
Yolken RH, Lawrence F, Leister F, Takiff HE, Straus SE (1982) Gastroenteritis associated with enteric type adenovirus in hospitalized infants. J Pediatr 101: 21–26
Zhang S, Mak S, Branton PE (1992) Adenovirus type 12 early region 1B proteins and metabolism of early viral mRNAs. Virology 191: 793–802
Zhang Y, Schneider RJ (1994) Adenovirus inhibition of cell translation facilitates release of virus particles and enhances degradation of the cytokeratin network. J Virol 68: 2544–2555

Adenovirus-Augmented, Receptor-Mediated Gene Delivery and Some Solutions to the Common Toxicity Problems

M. COTTEN

1 Introduction

Several years ago it was reported that the entry of adenovirus particles could augment the uptake of polylysine-condensed DNA molecules (CURIEL et al. 1991), a phenomenon similar to previously described adenovirus augmentation of both fluid phase markers and receptor-bound molecules (FERNÁNDEZ-PUENTES and CARRASCO 1980; FITZGERALD et al. 1983). This use of adenovirus to enhance gene delivery has proven to be a fruitful approach. A number of reviews of the general properties of this system have recently been published (CURIEL 1993; COTTEN and WAGNER 1993; WAGNER et al. 1994). In this review I will provide a summary of the current methods of generating adenovirus–DNA transfection complexes as well as discussing two of the toxicity problems that we have encountered with this system and describing some solutions to these problems. The first problem encountered was due to virus gene expression and leakiness in the replication defect of the commonly used defective adenovirus strains. This problem was solved by developing a psoralen inactivation method that inactivates the virus DNA and blocks viral gene expression and replication without impairing the entry functions of the virus (COTTEN et al. 1994b). The second problem was due to a toxicity generated by cytoplasmic delivery of lipopolysaccharide (LPS, endotoxin) that frequently contaminates bacterial DNA preparations. The LPS problem has been solved by identifying reliable methods of removing LPS from DNA (COTTEN et al. 1994a) or by including LPS-binding polymyxin in the transfection medium.

Research Institute of Molecular Pathology, Dr. Bohrgasse 7, 1030 Vienna, Austria

2 The Transfection System and Various Virus–DNA Coupling Methods

The receptor-mediated gene transfer system, in simple terms, involves condensing DNA with a polycation linked to a ligand and delivering this condensed DNA to cells bearing a receptor for the ligand (reviewed in COTTEN and WAGNER 1993). This approach profited from the realization that endocytosed material is trapped in the vesicular system of the cell and that agents such as adenovirus particles (CURIEL et al. 1991) or membrane-active peptides (WAGNER et al. 1992b; PLANK et al. 1994) must be included for useful levels of DNA delivery to most cell types. These membrane-active agents are thought to function by physically disrupting the endosomal membrane in a pH-dependent fashion (an activity that we termed "endosomolysis"; COTTEN et al. 1992), allowing the delivered DNA to enter the cytoplasm.

Free virus particles can augment the entry of either polylysine-condensed DNA (CURIEL et al. 1991; COTTEN et al. 1992; CRISTIANO et al. 1993a; BAATZ et al. 1994) or, to a modest extent, noncondensed, free DNA (YOSHIMURA et al. 1993). However, the most effective forms of this system use adenovirus particles linked to polylysine-condensed DNA. The virus to DNA coupling probably ensures that virus entry is associated with DNA entry and allows lower quantities of virus to be used. The coupling strategy also facilitates the inclusion of additional cell-binding ligands supplied as polylysine–ligand conjugates (e.g., transferrin, WAGNER et al. 1992a; anti-CD3 antibodies, (BUSCHLE et al. 1995); anticarbohydrate epitope antibodies, THURNHER et al. 1994) supplied as bio-tinylated ligands such as lectins (COTTEN et al. 1993b) or chemically linked ligands for the asiaologlycoprotein receptor (WU et al. 1994). Because of the broad distribution of the group C adenovirus receptors on mammalian cells, the virus itself functions as an extremely potent ligand for most cell types, with the exception of blood cells (HORVATH and WEBER 1988; SILVER and ANDERSON 1988). A variety of methods of coupling DNA to adenovirus have appeared and each has certain advantages and disadvantages (see Table 1).

A detailed description of one version of the method is found in Fig. 1. Initially, a DNA-binding domain, usually polylysine of 50–300 residues, is attached to the exterior of the adenovirus particle (Fig. 1, top panel). A flexible approach has been to modify exposed lysine residues on purified adenovirus particles with a reactive biotinylation reagent (*N*-hydroxysuccinimide-long chain biotin; NHS-LC-biotin; see WAGNER et al. 1992a for details). A streptavidin–polylysine conjugate is then used to attach polylysine via streptavidin–biotin interactions. In the next step (Fig.1, middle panel) polylysine–adenovirus is incubated with DNA with charge–charge and hydrophobic interactions linking the polylysine and DNA. Finally, the remaining free negative charge on the DNA is neutralized with additional polylysine ligands, such as polylysine–transferrin. This serves to further condense the DNA as well as to provide an additional cell-binding ligand. Although the final transfection complex is depicted as a single virus coupled to a single condensed

Table 1. Methods of coupling DNA to adenovirus

Coupling method	Advantages	Disadvantages	Reference
trans-glutaminase cross-linking of polylysine to glutamine residues on the virus capsid	easily obtained reagents	fixed virus: pLys adenovirus-polylysine is difficult to store	ZATLOUKAL et al. (1992)
sulpho-SMCC	easily obtained reagents	fixed virus: pLys adenovirus-polylysine is difficult to store	FISHER and WILSON (1994)
Biotin/Strp	simplified virus/polylysine titrations biotinylated virus is easy to prepare, stable upon storage	requires streptavidin–polylysine conjugate	WAGNER et al. (1992)
Ab/epitope	simplified virus/polylysine titrations	requires antibody-polylysine conjugate requires epitope-tagged adenovirus	CURIEL et al. (1992)
EDC	easily obtained reagents	fixed virus: pLys adenovirus-polylysine is difficult to store	CRISTIANO et al. (1993b)
CHO linkage	easily obtained reagents	fixed virus: pLys adenovirus-polylysine is difficult to store	WU et al. (1994)

Strp, streptavidin, Ab, antibody; SMCC, *N*-succinimidyl 4-(*N*-malesmidomethyl) cyclohexane-1-carboxylate; EDL, 1-ethyl-3-(3-dimethylaminopropyl) carbodiimide hydrochloride; CHO, carbohydrate.

DNA donut (approximately 200 nm in diameter), larger complexes in the 600- to 1000-nm range can be detected by electron microscopy (K. ZATLOUKAL et al., unpublished observations; CRISTIANO et al. 1993b; FISHER and WILSON 1994) or by light scattering (M.COTTEN, unpublished observations). We are currently determining the form of complex that is most efficient for DNA delivery.

The bottom panel of Fig. 1 is a simple description of the DNA delivery function of these complexes. The complexes bind to target cells via ligand (transferrin, the adenovirus fiber, penton base, and polylysine) interactions with structures on the target cell surface. Endocytosis or some form of pinocytosis occurs and the lowered pH of the endosomal compartment is thought to trigger a membrane disruption that allows gene delivery to the cytoplasm. Although the membrane disruption has captured our attention as the limiting step in successful gene delivery, adenovirus entry could also enhance gene delivery at other steps in the process. For example, a substantial rearrangement of the cytoskeleton occurs shortly after adenovirus infection (DEFER et al. 1990), and this could influence the fate or position of incoming virus and associated DNA. Furthermore, the incoming adenovirus can be found at the nuclear pore very soon after entry (GREBER et al.

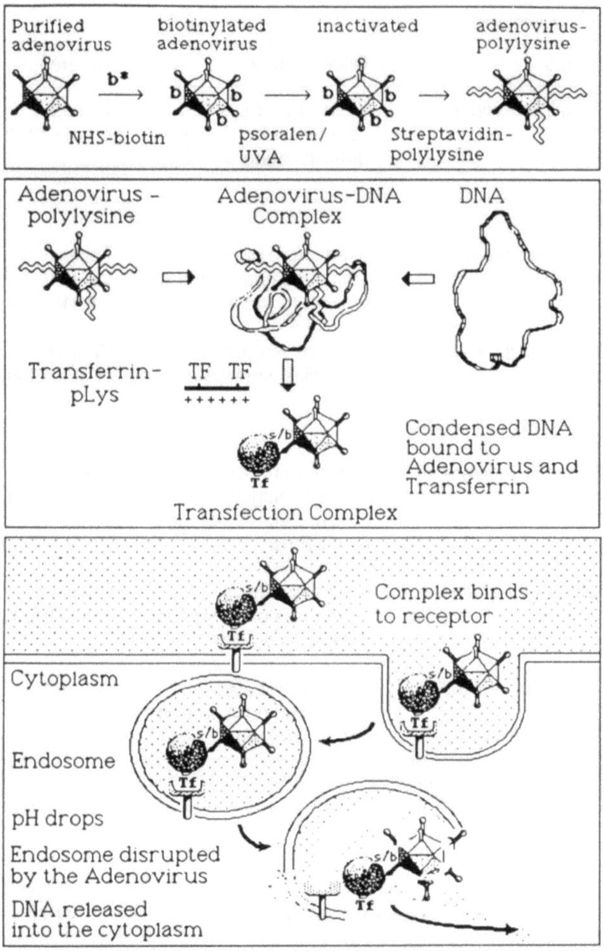

Fig. 1. Top, preparation of biotinylated/inactivated virus particles. Exposed lysine residues on the surface of adenovirus particles are modified with biotin groups. The biotinylated virus is then treated with the DNA intercalator psoralen and irradiated with 360-nm ultraviolet (*UV*) light to inactivate the viral DNA. DNA-binding polylysine is thus attached to the exterior of the virus using streptavidin-polylysine. Middle, preparation of adenovirus–polylysine–DNA transfection complexes. Adenovirus polylysine is incubated with DNA containing the gene to be delivered. Polylysine–DNA interactions link the DNA to the virus. Full condensation of the DNA and attachment of a cell-binding ligand (e.g., transferrin, *TF*) are obtained by the addition of transferrin-polylysine (*pLys*) to generate the transfection complex. Bottom, delivery of DNA to target cells. Transfection complexes are supplied to target cells. Interactions with the cell receptors for transferrin, the virus fiber, and the penton base (as well as nonspecific polylysine–cell interactions) result in internalization of the transfection complex. The decreased pH of the endosome or other signals generated by the internalization activate the membrane disruption activity of the virus and allow the virus (and associated polylysine–DNA) to enter the target cell cytoplasm and nucleus

1993), and perhaps the coupled DNA also benefits from whatever tracking mechanism the virus uses to get to the nucleus.

Methods for direct, covalent coupling of polylysine to adenovirus particles include a transglutaminase coupling (Zatloukal et al. 1992) as well as two different chemical coupling schemes (CRISTIANO et al. 1993b; FISHER and WILSON 1994). The strategy of biotinylating adenovirus and coupling the DNA via a streptavidin–polylysine bridge (WAGNER et al. 1992a) has proven to be an effective approach allowing large quantities of stable reagents to be prepared with only a short pre-incubation period required before application (Fig. 1, see also WAGNER et al. 1992a; COTTEN and WAGNER 1993). In general, strategies that covalently link polylysine to adenovirus are plagued by polylysine–adenovirus precipitation and storage problems and the reagents function best when freshly prepared. The virus capsid itself is mildly acidic and can aggregate in the presence of polylysine; the best storage conditions require high salt (1–2 *M*), which can also contribute to cytotoxicity. The order of addition, concentrations of reagents, temperature, ionic strength, and coupling time can all influence transfection complex function by several orders of magnitude and the polylysine to virus ratio can have profound influence on the function of the resulting complex. Therefore, the coupling methods that allow fine tuning of the virus to polylysine ratio (biotin–streptavidin bridge, antibody–epitope bridge) are the preferred methods.

These adenovirus–pLys–DNA complexes generate transient high levels of gene expression; however, we found that a number of cell types displayed a toxicity that limited long-term gene expression. The failure to obtain long-term gene expression in primary cells after an initially high transient expression stimulated the modifications that are reviewed here.

3 Viral Gene Expression and Replication Toxicity

The E1A-defective adenovirus (*dl*312, JONES and SHENK 1979, grown in the E1-complementing cell line 293; GRAHAM et al. 1977) used for our initial studies generated a rapid toxicity in most cell lines at the virus to cell ratios required for transient high gene expression. The toxicity appeared to be due to virus gene expression, with the cells displaying the typical cytopathic effect of cells late in adenovirus infection (swelling and detachment from the matrix). Many cell growth conditions can replace the requirement for E1A, especially at high virus input. Interleukin (IL-6) has been shown to stimulate an activity (NF-IL6) that can replace E1A (SPREGEL et al. 1991, 1992). We have found that treating primary fibroblasts with adenovirus particles results in an upregulation of the IL-6 promoter and secretion of IL-6 (A. BAKER and M. COTTEN, unpublished results). This could account for the bypassing of the E1A requirement at high virus input. In addition, many commonly used cell lines are immortalized with viral functions that can complement E1-defective adenoviruses. For example, HeLa cells express papillomavirus functions, and Epstein-Barr virus (EBV)-transformed B cell lines can complement E1-defective viruses (HORVATH et al. 1991).

Why should adenovirus gene expression be toxic? E1a expression in the absence of an antiapoptotic function can induce cell death (RAO et al. 1992;

DEBBAS and WHITE 1993; reviewed in WHITE and GOODING 1994), and the adenovirus-encoded protease expression can alter the cytoskeleton (CHEN et al. 1993). Inhibition of host translation (perhaps due to activation of the double-stranded RNA-activated kinase p68; MANCHE et al. 1992) can be responsible for some of the cytopathic effects (ZHANG and SCHNEIDER 1994). The penton base has long been known to generate toxicity (EVERETT and GINSBERG 1958), and penton base integrin interactions (BELIN and BOULANGER 1993; BAI et al. 1993; WICKHAM et al. 1993) could alter integrin–matrix attachments, leading to cell death (MEREDITH et al. 1993; reviewed in RUOSLAHTI and REED 1994).

The work of Gary Ketner and his colleagues has clarified the role of the E4 region in adenovirus replication (WEINBERG and KETNER 1983; BRIDGE and KETNER 1989). Many of the E4 functions are closely involved with specific adenovirus functions (such as the piracy of the mRNA transport out of the nucleus or turning the transcription factor E2F into an adenovirus-specific factor) rather than general growth control of the cell. Therefore, defects in E4 functions are difficult to bypass with cellular growth conditions. E4-defective adenovirus strains are thus quite defective for virus replication in noncomplementing cell lines (BRIDGE and KETNER 1989). We have therefore used the E4-defective Ad5 *dl*1014 (BRIDGE and KETNER 1989) and the E4-complementing W162 cell line (WEINBERG and KETNER 1983) for all further development. Additional technical advantages of the *dl*1014/W162 system are the fast doubling time of cells, cleaner plaque assays, and higher virus yields compared to the E1-defective/293 system.

However, at high virus inputs, a cytopathic effect can eventually be detected even with *dl*1014; therefore we developed an additional method of blocking virus gene expression. We sought a method that could damage the viral nucleic acid without damaging the outer capsid proteins and the ability of the virus to stimulate DNA delivery. Thus, virus samples were exposed to a test agent and then analyzed for virus replication by cytopathic effect assay or plaque assay on complementing cells. Simultaneously, the virus was tested for its ability to delivery a luciferase gene coupled to its exterior. The ideal inactivation would completely block virus replication while sparing the virus gene delivery functions. We required an agent which could penetrate the virus capsid and react with the DNA of the virus without harming the capsid. Several methods were tested and discarded as unsuitable because they generated less than a 4-log decline in viral replication or because the decline in gene delivery was equal to or greater than the decline in virus titer. These include formaldehyde treatment, short-wavelength ultraviolet (UV) light (260*nM*; COTTEN et al. 1992), and sensitizing the viral DNA to short-wavelength UV light by growth in the presence of bromodeoxyuridine.

The psoralens are a class of plant defense compounds that are found in many commonly ingested fruits and vegetables including limes, cloves, celery, and parsnips (IVIE et al. 1981; BERENBAUM and FEENY 1981). Because of their phototoxicity properties, psoralens are clinically used for the treatment of psoriasis (KORNHAUSER et al. 1982; VALLAT et al. 1994). A number of derivatives of these molecules have proven useful for virus inactivation in cases in which protein activity must be maintained (HANSON et al. 1978; HANSON 1992). The use of psoralen to inactivate adenovirus is shown in Fig. 2. The entry of the adenovirus is a function of the protein capsid. Psoralen molecules can enter the caspid and bind to the viral DNA. Irradiation with 365-nm light triggers the formation of DNA–psoralen cross-links. When psoralen-inactivated virus preparations are tested for their ability to deliver a luciferase gene bound to their exterior, this delivery activity is largely unimpaired by the psoralen treatment. However, the ability of the virus to express its own genes and to mount an infection declines. We initially detected a 5-log decline in virus replication by cytopathic effect (CPE) assay accompanying a ½-log decline in gene delivery activity using 8-methoxy psoralen/UVA treatment (COTTEN et al. 1994b). A more sensitive plaque assay demonstrated that these samples were free of replicating virus, indicating a greater than 8-log decline in viral titer.

Thus, we find that psoralen inactivation provides adenovirus particles that are completely inert at both the transcriptional and replicational level, yet maintain their ability to enter cells as measured by the coentry of DNA molecules. For clinical gene therapy this has important consequences. The psoralen-inactivated adenovirus cannot be complemented by wild-type adenovirus (a major concern when using recombinant adenoviruses to carry the gene of interest); thus there is no danger of viral replication if the treated patient is later exposed to a wild-type adenovirus. The genes that we seek to deliver are carried on the exterior of the virus and are not part of a viral genome. There is no real possibility of generating, by recombination, a replication-competent adenovirus bearing the therapeutic gene. Thus, one of the major safety concerns with the use of recombinant

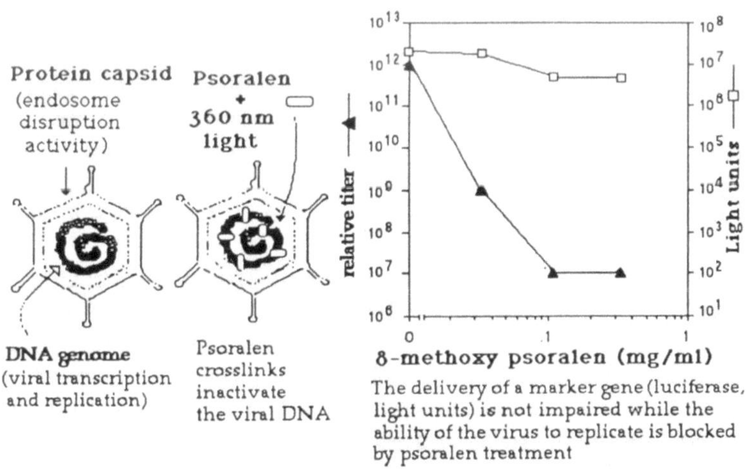

Fig. 2. Inactivation of adenovirus particles with psoralen. The entry functions of the virus are a property of the protein capsid. Psoralen can pass through the capsid, intercalate in the viral DNA, and upon irradiation with 360-nm light form covalent cross-links with the DNA that block DNA and RNA polymerases. When psoralen-inactivated adenovirus is used to carry a luciferase gene into cells, only a minor decline in gene delivery is observed after psoralen treatment, indicating that little damage of the protein capsid has occurred. However, when the replication capacity of the same virus is tested, a greater than 5 log decline in titer is observed, demonstrating that substantial damage of the viral DNA has occurred. See COTTEN et al. (1994b) for details

adenoviruses, the escape of a replication-competent adenovirus bearing a human gene, is eliminated.

4 Coentry of Toxic Molecules

Because of the previous use of adenovirus to enhance the delivery of toxins (Fernández-Puentes and Carrasco 1980; Fitzgerald et al. 1983), it should come as no surprise that toxin contaminations of plasmid DNA preparations might cause problems with this gene delivery system. We had routinely used an elaborate double CsCl gradient protocol to purify the plasmid DNA used for transfer-infection (Cotten et al. 1993a). Alternate DNA preparation methods (ion exchange chromatography such as Qiagen) were sought and found to generate high-quality DNA by gel electrophoresis and absorbance. These column-purified DNA transfected cell lines with similar efficiency as CsCl DNA but caused a rapid toxicity (4 h to overnight) when transfected into primary human cells. Closer examination of the ion exchange-purified DNA demonstrated an LPS (endotoxin) contamination that correlated with cytotoxicity in primary cells. We identified methods of removing the LPS from contaminated DNA using either polymyxin B affinity chromatography or Triton X-114 extraction. Removal of LPS eliminated the toxicity; readdition of pure LPS to clean DNA recreated the toxicity (Cotten et al. 1994a).

A typical response to LPS-contaminated DNA is shown in Fig. 3. Primary human skin fibroblasts transfected with LPS-free DNA adenovirus complexes yield high levels of luciferase gene expression. Adding increasing amounts of LPS (1–100ng LPS/0.6μg DNA) results in a decline in gene expression. The release of the cytoplasmic enzyme lactate dehydrogenase (LDH) into the culture medium was measured as an indication of cell lysis (either by necrosis or apoptosis). Concomitant with the decline in gene expression is a rise in released LDH, supporting our conclusion that cell death is responsible for the decline in gene expression.

The LPS toxicity phenomenon is cartooned in Fig. 4. The negative charge on the LPS molecule (Raetz 1990) allows it (like DNA) to bind to polylysine. The entry of the adenovirus deposits the LPS in the cytoplasm, generating toxicity by an unknown mechanism. LPS exposure to cells in the presence of immature adenovirus particles (which can bind receptors and can be endocytosed, but cannot pass through into the cytoplasm) does not generate this toxicity (Cotten et al. 1994a). The toxicity requires the combination of LPS and adenovirus; neither reagent alone generates a toxic response.

Once we began to look for LPS we found it everywhere. Because the adenovirus/polylysine system is extremely sensitive to endotoxin, we developed a method of neutralization of LPS in the transfection medium that bypasses the requirement for generating completely LPS-free reagents. Polymyxin B is a cyclic peptide antibiotic that binds and neutralizes LPS with high affinity (Morrison and

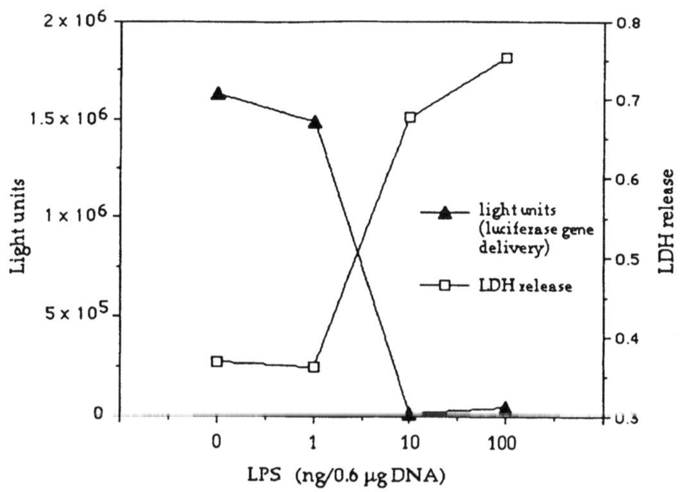

Primary human skin fibroblasts transfected with adenovirus/polylysine /DNA complexes containing the indicated LPS content. Duplicate samples assayed at 24 hours for luciferase activity (indicator of gene delivery) or LDH release into cytoplasm (indicator of cell lysis)

Fig. 3. Lipopolysaccharide (*LPS*)-contaminated DNA generates a decline in gene delivery. Adenovirus–polylysine–DNA transfection complexes containing a luciferase gene and the indicated LPS content were delivered to primary human fibroblasts. The LPS luciferase activity measured at 24 h after transfection declines with increasing LPS content. The increased amount of the cytoplasmic enzyme lactate dehydrogenase (*LDH*) in the cell medium indicates an increase in cell death with increasing LPS content

Jacobs 1976; Storm et al. 1977). Polymyxin is occasionally used as a tissue culture antibiotic and is tolerated by many cell types up to 100 µg/ml. We have found that the inclusion of polymyxin B or E (colistin) in the transfection medium can protect against the toxicity generated by LPS-contaminated reagents (M. Saltik and M. Cotten, unpublished observations).

5 The Next Level of Toxicity

Upon further consideration of the adenovirus life cycle, it becomes apparent that certain viral functions that are frequently discarded in recombinant adenoviruses might be useful in modulating the immune reaction or inflammatory response in gene therapy applications (reviewed in Gooding 1992). For example, products of the E3 region can interfere with response to tumor necrosis factor (TNF) as well as impair major histocompatibility complex (MHC) surface expression, which might be useful in blocking host recognition and destruction of transduced cells (Wold and Gooding 1991; Wold 1993). Viral-associated (VA)1 RNA enhances

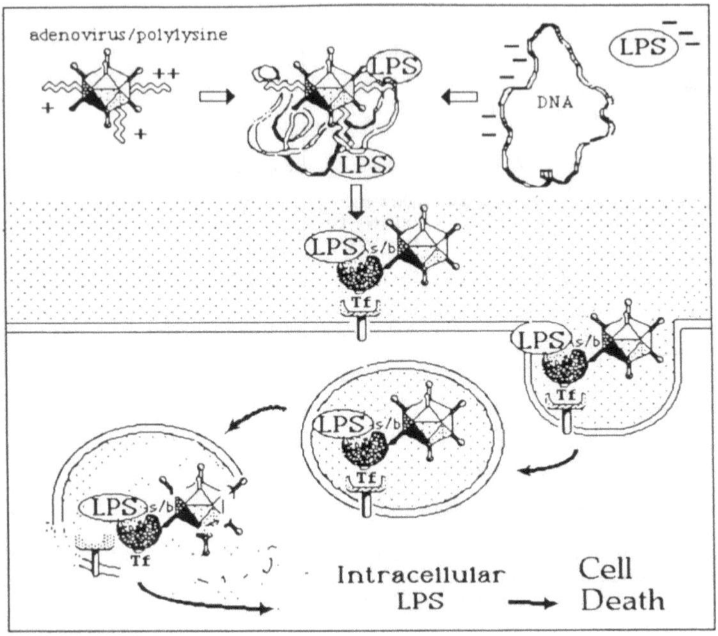

Fig. 4. Summary of lipopolysaccharide (*LPS*) toxicity in the presence of adenovirus–polylysine–DNA complexes. The polylysine groups attached to adenovirus can function to carry negatively charged DNA as well as negatively charged LPS molecules into the target cells. The intracellular delivery of LPS activates a cell death program. The toxicity can be avoided by careful removal of LPS from both transfection reagents as well as the transfection medium

translation and blocks interferon and double-stranded RNA-activated p68 kinase (Mathews and Shenk 1991; Manche et al. 1992); however, high VA1 expression may require viral replication or E1A functions (Yoshinaga et al. 1986; Hoeffler et al. 1988; Patel and Jones 1990; Datta et al. 1991). The E1B 19K product can prevent apoptosis (Rao et al. 1992; Debbas and White 1993). However, the expression of many of these beneficial adenoviral genes is closely linked to the viral gene regulatory program and some are directly dependent upon E1A expression (reviewed in Nevins 1991, 1992; Moran 1993). Therefore, full expression of these genes from natural viral templates may require E1A expression, and we must consider the complications of E1A altering p53 expression and inducing apoptosis (Rao et al. 1992; Debbas and White 1993; Lowe and Ruley 1993; Wu and Levine 1994). As an alternative solution we are making transfection complexes (inactivated adenovirus particles linked to polylysine-condensed DNA plasmids) that will provide both the expression of useful adenovirus genes and the expression of therapeutic gene (e.g., factor VIII; Zatloukal et al. 1994). Our approach will be to begin with a transcription and replication-free (tabula rasa) adenoviral particle whose only function is to enhance gene delivery. We can then add the useful adenoviral gene functions in a controlled fashion.

Acknowledgments. I would like to thank Max Birnstiel, Ernst Wagner, and the rest of my colleagues at the Research Institute of Molecular Pathology for their support. I am grateful to Alika Cotten, Adam Baker, and Susanna Chiocca for their comments on the manuscript.

References

Baatz JE, Bruno MD, Ciraolo PJ, Glasser SW, Stripp BR, Smyth KL, Korfhagen TR (1994) Utilization of modified surfactant-associated protein B for delivery of DNA to airway cells in culture. Proc Natl Acad Sci USA 91: 2547–2551

Bai M, Harfe B, Freimuth P (1993) Mutations that alter an Arg-Gly-Asp (RGD) sequence in the adenovirus type 2 penton base protein abolish its cell rounding activity and delay virus reproduction in flat cells. J Virol 67: 5198–5205

Belin M.-T, Boulanger P (1993) Involvement of cellular adhesion sequences in the attachment of adenovirus to the HeLa cell surface. J Gen Virol 74: 1485–1497

Berenbaum M, Feeney P (1981) Toxicity of angular furanocoumarins to swallowtail butterflies: escalation in a coevolutionary arms race. Science 212: 927–929

Bridge E, Ketner G (1989) Redundant control of adenovirus late gene expression by early region 4. J Virol 63: 631–638

Buschle M, Cotten M, Kirlappos H, Mechtler K, Schaffner G, Zauner W, Birnstiel ML, Wagner E (1995) Receptor-mediated gene transfer into human T-lymphocytes via binding of DNA/CD3 antibody particles to the CD3 T cell receptor complex. Human Gene Therapy (in press)

Chen P, Ornelles D, Shenk T (1993) The adenovirus L3 23-kilodalton proteinase cleaves the amino-terminal head domain from cytokeratin 18 and disrupts the cytokeratin network of HeLa cells. J Virol 67: 3507–3514

Cotten M, Wagner E (1993) Non-viral approaches to gene therapy. Curr Opin Biotechnol 4: 705–710

Cotten M, Wagner E, Zatloukal K, Phillips S, Curiel DT, Birnstiel ML, (1992) High-efficiency receptor-mediated delivery of small and large (48kb) gene constructs using the endosome disruption activity of defective or chemically inactivated adenovirus particles. Proc Natl Acad Sci USA 89: 6094–6098

Cotten M, Wagner E, Birnstiel ML (1993a) Receptor-mediated transport of DNA into eukaryotic cells. Methods Enzymol 217: 618–644

Cotten M, Wagner E, Zatloukal K, Birnstiel ML (1993b) Chicken adenovirus (CELO virus) particles augment receptor-mediated DNA delivery to mammalian cells and yield exceptional levels of stable transformants. J Virol 67: 3777–3785

Cotten M, Baker A, Saltik M, Wagner E, Buschle M (1994a) Lipopolysaccharide is a frequent contaminant of plasmid DNA preparations and can be toxic to primary cells in the presence of adenovirus. Gene Ther 1: 239–246

Cotten M, Saltik M, Kursa M, Wagner E, Maass G, Birnstiel M (1994b) Psoralen treatment of adenovirus particles eliminates virus replication and transcription while maintaining the endosomolytic activity of the virus capsid. Virology 205: 254–261

Cristiano RJ, Smith LC, Woo SLC (1993a) Hepatic gene therapy: adenovirus enhancement of receptor-mediated gene delivery and expression in primary hepatocytes. Proc Natl Acad Sci USA 90: 2122–2126

Cristiano RJ, Smith LC, Kay MA, Brinkley BR, Woo SLC (1993b) Hepatic gene therapy: efficient gene delivery and expression in primary hepatocytes utilizing a conjugated adenovirus-DNA complex. Proc Natl Acad Sci USA 90: 11548–11552

Curiel D (1993) Adenovirus facilitation of molecular conjugate-mediated gene transfer. Prog Med Virol 40: 1–18

Curiel DT, Agarwal S, Wagner E and Cotten M (1991) Adenovirus enhancement of transferrin-polylysine mediated gene delivery. Proc Natl Acad Sci USA 88: 8850–8854

Curiel D, Wagner E, Cotten M, Birnstiel ML, Agarwal S, Li C, Loechel S, Hu P (1992) High efficiency gene transfer mediated by adenovirus coupled to DNA polylysine complexes. Hum Gene Ther 3: 147–154

Datta S, Soong C, Wang D, Harter M (1991) A purified adenovirus 289-amino-acid E1A protein activates RNA polymerase III transcription in vitro and alters transcription factor IIIC. J Virol 65: 5297–5304

Debbas M, White E (1993) Wild-type p53 mediates apoptosis by E1A, which is inhibited by E1B. Genes Dev 7: 546–554

Defer C, Belin M.-T, Caillet-Boudin M.-L, Boulanger P (1990) Human adenovirus-host cell interactions: comparitive study with members of subgroup B and C. J Virol 64: 3661–3673
Everett S, Ginsberg HS (1985) A toxin like material separable from type 5 adenovirus particles. Virology 6: 770–771
Fernández-Puentes C, Carrasco L (1980) Viral infection permeabilizes mammalian cells to protein toxins. Cell 20: 769–775
Fisher KJ, Wilson JM (1994) Biochemical and functional analysis of an adenovirus-based ligand complex for gene transfer. Biochem J 299: 49–58
Fitzgerald D, Padmanabhan R, Pastan I, Willingham M (1983) Adenovirus-induced release of epidermal growth factor and Pseudomonas toxin into the cytosol of KB cells during receptor-mediated endocytosis. Cell 32: 607–617
Gooding LR (1992) Virus proteins that counteract host immune defenses. Cell 71: 5–7
Graham F, Smiley J, Russell W, Nairn R (1977) Characteristics of a human cell line transformed by DNA from human adenovirus type 5. J Gen Virol 36: 59–72
Greber UF, Willetts M, Webster P, Helenius A (1993) Stepwise dismantling of adenovirus 2 during entry into cells. Cell 75: 1–20
Hanson CV (1992) Photochemical inactivation of viruses with psoralens: an overview. Blood Cells18: 7–25
Hanson CV, Riggs JL, Lennette EH (1978) Photochemical inactivation of DNA and RNA viruses by psoralen derivatives. J Gen Virol 40: 345–358
Hoeffler WK, Kovelman R, Roeder RG (1988) Activation of transcription factor IIIC by the adenovirus E1A protein. Cell 53: 907–920
Horvath J, Weber J (1988) Nonpermissivity of human peripheral blood lymphocytes to adenovirus type 2 infection. J Virol 62: 341–345
Horvath J, Faxing C, Weber JM (1991) Complementation of adenovirus early region 1a and 2a mutants by Epstein-Barr virus immortalized lymphoblastoid cell lines. Virology 184: 141–148
Ivie GW, Holt DL, Ivey MC (1981) Natural toxicants in human food: psoralens in raw and cooked parsnip root. Science 213: 909–910
Jones N, Shenk T (1979) An adenovirus type 5 early gene function regulates expression of other early viral genes. Proc Natl Acad Sci USA 76: 3665–3669
Kornhauser A, Wamer WG, Giles AL (1982) Psoralen photoxicity: correlation with serum and epidermal 8-methoxypsoralen and 5-methoxypsoralen in the guinea pig. Science 217: 733–735
Lowe SW, Ruley HE (1993) Stabilization of the p53 tumor suppressor is induced by adenovirus 5 E1A and accompanies apoptosis. Genes Dev 7: 535–545
Manche L, Green SR, Schmedt C, Mathews MB (1992) Interactions between double-stranded RNA regulators and the protein kinase DAI. Mol Cell Biol 12: 5238–5248
Mathews MB, Shenk T (1991) Adenovirus virus-associated RNA and translation control. J Virol 65: 5657–5662
Meredith J, Fazeli B, Schwartz MA (1993) The extracellular matrix as a cell survival factor. Mol Biol Cell 4: 953–961
Moran E (1993) Interaction of adenoviral proteins with pRB and p53. FASEB J 7: 880–885
Morrison DC, Jacobs DM (1976) Binding of polymyxin B to the lipid A portion of bacterial lipopolysaccharides. Immunochemistry 13: 813–818
Nevins JR (1991) Transcriptional activation by viral regulatory proteins. TIBS 16: 435–439
Nevins JR (1992) E2F: a link between the Rb tumor suppressor protein and viral oncoproteins. Science 258: 424–429
Patel G, Jones NC (1990) Activation in vitro of RNA polymerase II and III directed transcription by baculovirus produced E1A protein. Nucleic Acids Res 18: 2909–2915
Plank C, Oberhauser B, Mechtler K, Koch C, Wagner E (1994) The influence of endosome-disruptive peptides on gene transfer using synthetic virus-like gene transfer systems. J Biol Chem 269: 12918–12924
Raetz CRH (1990) Biochemistry of endotoxins. Annu Rev Biochem 59: 129–170
Rao L, Debbas M, Sabbatini P, Hockenbery D, Korsmeyer S, White E (1992) The adenovirus E1A proteins induce apoptosis which is inhibited by the E1B 19-kDa and Bcl-2 proteins. Proc Natl Acad Sci USA 89: 7742–7746
Ruoslahti R, Reed J (1994) Anchorage dependence, integrins and apoptosis. Cell 77: 477–478
Silver L, Anderson C (1988) Interaction of human adenovirus serotype 2 with human lymphoid cells. Virology 165: 377–387
Spergel J, Chen-Kiang S (1991) IL-6 enhances a cellular activity which functionally substitutes for E1A in transactivation. Proc Natl Acad Sci USA 88: 6472–6476

Spergel J, Hsu W, Akira S, Thimmappaya B, Kishimoto T, Chen-Kiang S (1992) NF-IL6, a member of the C/EBP family, regulates E1A-responsive promoters in the absence of E1A. J Virol 66: 1021–1030
Storm DR, Rosenthal KS, Swanson PE (1977) Polymyxin and related peptide antibiotics. Annu Rev Biochem 46: 723–763
Thurnher M, Wagner E, Clausen H, Mechtler K, Rusconi S, Dinter A, Berger E, Birnstiel M, Cotten M (1994) Carbohydrate receptor-mediated gene transfer to human T-leukemic cells. Glycobiology 4: 429–435
Vallat VP, Gilleaudeau P, Battat L, Wolfe J, Nabeya R, Heftler N, Hodak E, Gottlieb AB, Krueger JG (1994) PUVA bath therapy strongly suppresses immunological and epidermal activation in psoriasis: a possible cellular basis for remittive therapy. J Exp Med 180: 283–296
Wagner E, Zatloukal K, Cotten M, Kirlappos H, Mechtler K, Curiel D, Birnstiel ML (1992a) Coupling of adenovirus to transferrin-polylysine/DNA complexes greatly enhances receptor-mediated gene delivery and expression of transfected cells. Proc Natl Acad Sci USA 89: 6099–6103
Wagner E, Plank C, Zatloukal K, Cotten M, Birnstiel ML (1992b) Influenza virus hemagglutinin HA-2 N-terminal fusogenic peptides augment gene transfer by transferrin-polylysine-DNA complexes: toward a synthetic virus-like gene-transfer vehicle. Proc Natl Acad Sci USA 89: 7934–7938
Wagner E, Curiel D, Cotten M (1994) Delivery of drugs, proteins and genes into cells using transferrin as a ligand for receptor-mediated endocytosis. Adv Drug Deliv Rev 14: 113–135
Weinberg DH, Ketner G (1983) A cell line that supports the growth of a defective early region 4 deletion mutant of human adenovirus type 2. Proc Natl Acad Sci USA 80: 5383–5386
White E, Gooding LR (1994) Regulation of apoptosis by human adenoviruses. (Apoptosis II: the molecular basis of apoptosis in disease.) Cold Spring Harbor Laboratory Press, Cold Spring Harbor
Wickham TJ, Mathias P, Cheresh DA, Nemerow GR (1993) Integrins $\alpha_v\beta_3$ or $\alpha_v\beta_5$ promote adenovirus internalization but not virus attachment. Cell 73: 309–319
Wold WSM (1993) Adenovirus genes that modulate the sensitivity of virus-infected cells to lysis by TNF. J Cell Biochem 53: 329–335
Wold WSM, Gooding LR (1991) Region E3 of adenovirus: a cassette of genes involved in host immuno-surveillance and virus-cell interactions. Virology 184: 1–8
Wu G, Zhan P, Sze L, Rosenberg A, Wu C (1994) Incorporation of adenovirus into a ligand-based DNA carrier system results in retention of original receptor specificity and enhances targeted gene expression. J Biol Chem 269: 11542–11546
Wu X, Levine AJ (1994) p53 and E2F-1 cooperate to mediate apoptosis. Proc Natl Acad Sci USA 91: 3602–3606
Yoshimura K, Rosenfeld M, Seth P, Crystal R (1993) Adenovirus-mediated augmentation of cell transfection with unmodified plasmid vectors. J Biol Chem 268: 2300–2303
Yoshinaga S, Dean N, Han M, Berk AJ (1986) Adenovirus stimulation of transcription by RNA polymerase III: evidence for an E1A-dependent increase in transcription factor IIIC concentration. EMBO J 5: 343–354
Zatloukal K, Wagner E, Cotten M, Phillips S, Plank C, Steinlein P, Curiel D, Birnstiel ML (1992) Transferrinfection: a highly efficient way to express gene constructs in eukaryotic cells. Ann NY Acad Sci 660: 136–153
Zatloukal K, Cotten M, Berger M, Schmidt W, Wagner E, Birnstiel ML (1994) In vivo production of human factor VIII in mice after intrasplenic implantation of primary fibroblasts transfected by receptor-mediated adenovirus-augmented gene delivery. Proc Natl Acad Sci USA 91: 5148–5152
Zhang Y, Schneider RJ (1994) Adenovirus inhibition of cell translation facilitates release of virus particles and enhances degradation of the cytokeratin network. J Virol 68: 2544–2555

Gene Therapy Using Adenovirus Vectors

H. Haddada, L. Cordier, and M. Perricaudet

1 Introduction

Remarkable progress has been made in the field of gene therapy during the last few years. Genetic as well as acquired diseases such as cancer could potentially be treated by this approach. Among the available vectors for gene transfer, replication-deficient recombinant adenoviruses (Ad) appear attractive. They are easy to manipulate, infect many different types of cells (even nondividing ones), can be prepared at high titers, and, of particular importance, they are capable of direct in vivo introduction of foreign genetic material into the body.

2 Adenovirus: Biology and Construction of Recombinant Virus

Recent studies suggest that gene transfer into somatic cells is a feasible and efficient method of treating diseases for which only limited efficacy has been obtained so far by conventional therapeutic approaches. This novel strategy, commonly called gene therapy, is currently being evaluated in patients with inherited defects as well as those with advanced malignancies.

Laboratorie de Génétique des Virus Oncogènes, Institut Gustave Roussy, PR2, 39 rue Camille Desmoulins, 94805 Villejuif Cedex, France

At least two different systems can be used to introduce foreign genetic material into patients:

1. Introduction of in vitro genetically modified autologous cells in order to permit the in vivo expression of the gene of interest (the cell-based gene transfer method)
2. Direct gene transfer methods, in which the gene is directly delivered in vivo without the need of ex vivo manipulations of the cells

Direct gene transfer can be performed by injection of naked plasmid DNA (Lin et al. 1990; Kitsis et al. 1991), liposome-encapsulated DNA (Nabel et al. 1990; Lim et al. 1991; Chapman et al. 1992), synthetic vectors linking foreign genetic material to ligands that specifically binds to cell surface receptors (Wagner et al. 1992), or by using viral-based vectors harboring the gene of interest. All these methods allow the transfer and expression of the exogenous gene, although their relative efficiency in gene transfer differs. Although nonviral vectors appear attractive in terms of their safety, they require improvements with regard to their transferring capacity, as expression is obtained in only 1%–5% of in vivo target cells.

Viral vectors represent another alternative. The intrinsic ability of viruses to enter cells bringing with them their own genetic material suggested their potential use as gene delivery vectors for living organisms. In initial gene transfer experiments, retroviruses have been engineered and used. They transduce dividing cells in culture and stabily integrate into the host genome. However, their inability to infect nondividing cells and the low titers that can be obtained (about 10^6 infectious particules per ml) have limited their use.

Numerous experimental data are now accumulating showing the high efficiency of gene transfer into a variety of somatic tissues with Ad vectors. Their advantages include the efficient propagation of viral stocks to very high titers (up to 10^{11} infectious particules per ml), the promiscous species and tissue tropism of Ad infection of mammalian cells, and the capacity of Ad to infect dividing as well as postmitotic cells. Moreover, Ad are structurally stable and no side effects have been reported following the vaccination of US military recruits with wild types 4 and 7 (Rubin and Rorke 1988), demonstrating their safety for human use.

Therefore, the potential applications for direct gene transfer with Ad vectors appear considerable. A large variety of somatic tissues have been tested for their ability to become infected, including striated muscle (Stratford-Perricaudet et al. 1992; Ragot et al. 1993), lung (Rosenfeld et al. 1992), liver (Stratford-Perricaudet et al. 1990), and brain (Akli et al. 1993; Le Gal La Salle et al. 1993).

Recently, experimental investigations have to demonstrate that cancer can potentially be treated by direct gene transfer using Ad vectors (Cordier et al. 1995).

2.1 Biology of Adenovirus

Ad are found in many mammalian and avian species. In man, 47 distinct serotypes have been isolated. Their genome of about 36 kb pairs encodes more than 2700 separate proteins. An inverted terminal repeat (ITR), 100 bp long in sequence, which is necessary for virus replication exists on each extremity of the genome.

The genome is functionally divided into two major overlapping regions, early and late, based on the time of transcription after infection. There are six distinct early regions, E1a, E1b, E2a, E2b, E3, and E4. The E1 region is immediately transcribed after infection. Two predominant regulatory proteins are encoded by genes from this region. They transactivate the transcription of the other early genes and play a key role in the productive infection. E1A and E1B are necessary and sufficient for cellular transformation.

The late gene products include the principal capsid proteins, hexon, penton, and fiber, which are transcribed from a single promoter and are involved in virus attachment to surface cell receptor and entry of the virus.

2.2 Adenovirus Cell Entry

For cell entry, the Ad attaches with high affinity to cell surface receptors via the fiber protein (DEFER et al. 1990; PHILIPSON et al. 1968). The Ad receptor on the cell surface is still unknown; however, the vitronectin-binding integrins have been identified as the secondary receptor (WICKHAM et al. 1993) that mediates virus internalization through the penton base protein. Once bound to the cell surface, the virus is internalized via receptor-mediated endocytosis (CHARDONNET and DALES 1970; FITZGERALD et al. 1983; VERGA et al. 1991). Recently, studying the Ad2 cell entry, GREBER et al. (1993) demonstrated that the Ad fibers are released, the penton base structures dissociated, the proteins connecting the DNA to the inside surface of the capsid degraded or shed, and the minor proteins which stabilize the capsid are eliminated to permit the entry of the DNA into the nucleus with high efficiency.

2.3 Development of Recombinant Adenoviruses

The dependence of expression of virally encoded genes on the E1 region constitutes the basis for the development of replication-deficient Ad vectors in which the E1 region is replaced by the foreign gene.

In fact, at least two regions of the virus genome, E1 and E3, can be deleted and substituted by foreign DNA sequences. Investigations in progress suggest that a part of the E4 region can also be deleted to make room for the insertion of larger exogenous DNA, which at the same time would improve the safety of the recombinants.

The development of replication-deficient Ad recombinants can be made by homologous recombination (Fig. 1) in two steps:

1. Construction of a plasmid containing 455 bp of the left end of Ad5, which provide the ITR and the packaging sequences with the E1 enhancers, the

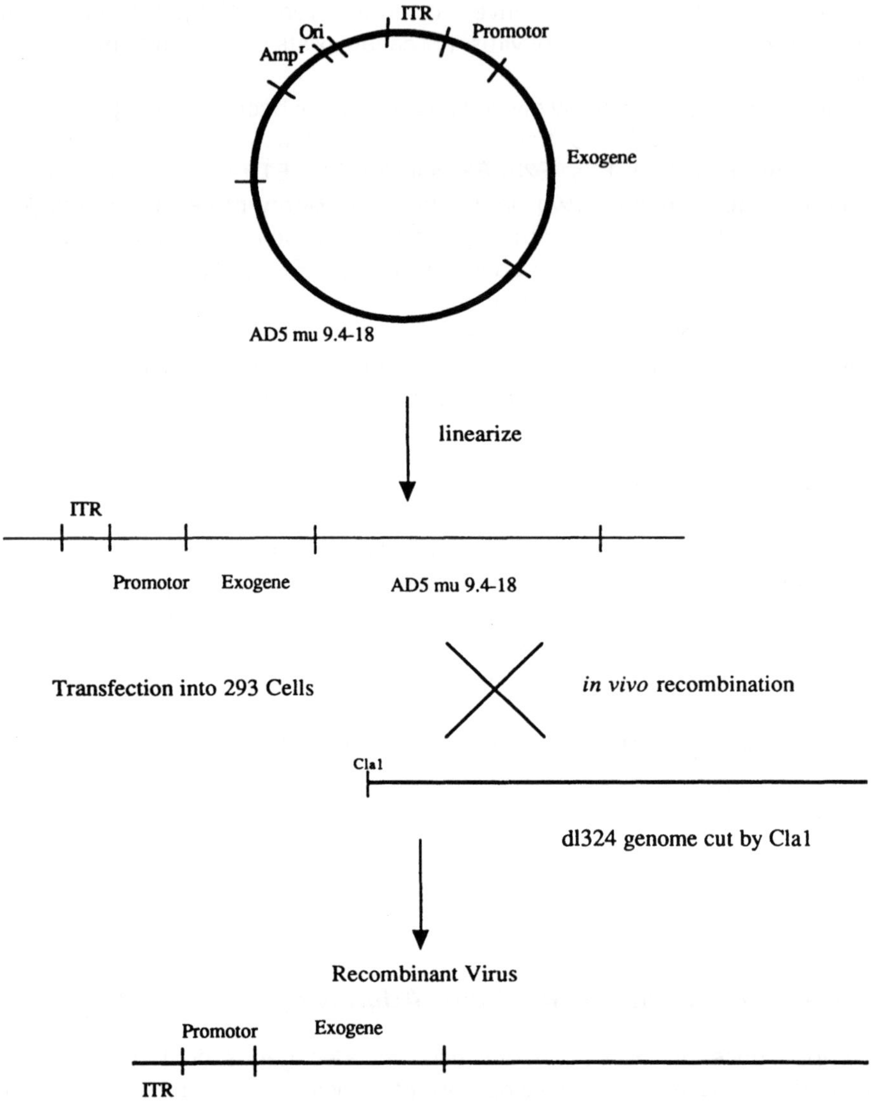

Fig. 1. Construction of a recombinant adenovirus. *KR* , inverted terminal repeat; *mu*, map units; *Ori*, origin; *Amp*^{r}, ampicillin resistance gene

foreign gene under the control of the chosen promoter, and an Ad5 sequence from map unit (mu) 9.4 to 18.

2. The plasmid is linearized and transfected with the replication-deficient mutant *dl*324 DNA (deleted from the E1 and E3 regions and digested with cla1 restriction enzyme) on the 293 human cell line which has integrated the E1 region.

Homologous recombination between the two DNA generates a recombinant genome of packageable size in which the E1 region is replaced by the gene of interest. The recombinant viruses thus generated are replication deficient, but can be propagated on the 293 cells to high titers and are able to infect and direct expression of the exogenous gene in different cells.

3 Use of Recombinant Virus for Gene Transfer

Investigations from our laboratory as well as from others have shown the possibilities of introducing a wide variety of genes into numerous established cell lines and primary cultures (STRATFORD-PERRICAUDET et al. 1992; HADDADA et al. 1993b) as well as almost all the tested tissues in vivo using adenovirus vectors. These findings paved the way for numerous potential applications for genetic disorders as well as for infectious diseases and cancer.

In this section we shall give a brief review of the major experimental data so far obtained showing the possibilities of using adenovirus vectors in gene therapy, which consists of the delivery of a functional gene for expression in somatic tissues in order to correct a defect and to treat a disease.

3.1 Gene Therapy of Inherited Diseases Using Adenovirus Vectors

Major progress has been made during the last few years in the understanding of the molecular basis of inherited as well as acquired diseases. New experimental approaches based on gene transfer are thus possible and have led to gene therapy protocols to be evaluated in human patients.

The first experiments showing the feasibility and the efficiency of direct transfer of a therapeutic gene to complement a genetic deficiency using an Ad vector were performed with ornithine transcarbamylase gene (OTC) using an Ad vector (STRATFORD-PERRICAUDET et al. 1990). The defect results in the Sph-ash strain of mice from the homozygous deficiency of the OTC enzyme. Following administration of the recombinant Ad, an expression of the transferred OTC gene was detected and a correction of the metabolic defect was observed in one out of three Sph-ash mice.

The ability of gene transfer with Ad has also been evaluated for α_1-antitrypsin (α_1-At), a serum antiprotease produced by hepatocytes and mononuclear phagocytes. Severe hereditary deficiency of α_1-AT is associated with low plasma and lung levels of the enzyme predisposing affected individuals to pulmonary emphysema by impairing the protease antiprotease balance in the lower respiratory tract. GIRALDI et al. (1990) have demonstrated the expression of α_1-AT and shown the protein to be glycosylated.

The feasibility of a human minidystrophin gene transfer into mouse muscle has been demonstrated (Ragot et al. 1993). Human minidystrophin was detected in mdx mice following intramuscular injection of the recombinant Ad. Since mdx mice have a very low level of endogenous dystrophin without clinical defects, clinical corrections need to be demonstrated and more room has to be made in the vectors in order to place the complete dystrophin gene. In humans, the duchenne progressive dystrophy is a lethal and common X-linked genetic disease caused by the absence of dystrophin. One out of 3500 newborns are affected by this disease.

In a collaborative investigation carried out in our laboratory and in Crystal's lab in the USA, the expression of cystic fibrosis transmembrane conductance regulator (CFTR) gene has been demonstrated using an Ad vector harboring the CFTR gene (Rosenfeld et al. 1992). Since then, gene therapy protocols for cystic fibrosis disease have been approved and the CFTR gene has been introduced into pulmonary epithelium of patients using Ad vectors (Zabner et al. 1993). To date, it is too early to evaluate the clinical effects.

3.2 Feasibility of Adenoviral-Mediated Gene Transfer into the Nervous System

The ability of Ad vectors to allow the delivery of foreign genes directly to terminally differentiated cells of the nervous system which no longer proliferate has been investigated using an Ad vector harboring the β-galactosidase gene (Ad RSV-β-gal) of *Escherichia coli* (Le Gal La Salle et al. 1993; Bajocchi et al. 1993; β-gal expression was found in almost all neurons and astrocytes in culture. Following stereotactic inoculations into the rat hippocampus and the substantia nigra, β-gal activity was detected for 2 months. Microglial cells and astrocytes were infected without obvious cytopathic effect. These experiments suggest possibilities of treating diseases of the nervous system by gene transfer. For degenerative diseases, it may also be possible to express neurotransmitters or growth factors locally as an alternative to the grafting of fetal cells.

3.3 Gene Therapy of Tumors

Tumor cell growth is inhibited by various immune effectors including natural killers, lymphokine-activated killer cells, macrophages, and cytotoxic T cells that recognize peptides present on the surface of target cells. Cytokines are known as key modulators of host immune and inflammatory responses. Interleukin-2 (IL-2) and tumor necrosis factor (TNF)-α stimulate the proliferation and activation of nonspecific as well as specific effectors. They have been used in immunotherapy protocols to treat cancer, but only partial regressions have been achieved via systemic administration and major undesirable side effects have been observed. To circumvent this problem, much effort has been made in recent years to find a means of delivering the cytokine locally, hence increasing the activation of antitumor cytotoxic cells infiltrating the tumor. Therefore, the proposal was made to attempt to increase the antitumoral activity of tumor-infiltrating lymphocytes (TIL) in advanced melanoma by introducing TNF-α cDNA. The first approved human gene transfer protocol involved retroviral gene marking the TIL in patients with advanced melanoma (ROSENBERG et al. 1990) in order to evaluate the feasibility of the method. Autologous TIL were reinjected to patients after ex vivo introduction of Neo^R gene using a retroviral vector. This experiment allowed the evaluation of long-term distribution and survival of the injected TIL. Although only small numbers of transduced cells persisted in the circulation and in the tumor deposits, no ill effects were noted and no evidence of live virus production was found.

Another approach, gene transfer into tumor cells, has been investigated in several experimental models. It consists of the direct introduction of genes into tumor cells. The gene is selected to induce an increased antitumoral immune response against the tumor cells. Data from various laboratories demonstrate that the expression of exogenous cytokines by tumor cells reduces their tumorigenicity and allows not only the rejection of the transduced cells, but also the rejection of the parental unmodified tumor cells. Increased host antitumoral immunity with loss of tumorigenicity has been found when tumor cells are transfected or transduced with vectors expressing IL-2 (FEARON et al. 1990; GANSBACHER et al. 1990; HADDADA et al. 1993a), IL-4 (TEPPER et al. 1989; GOLUMBEK et al. 1991), TNF-α (ASHER et al. 1991), interferon-γ (IFN-γ) (WATANABE et al. 1989; RESTIFIO et al. 1992), or granulocyte-macrophahe colony-stimulating factor (GM-CSF; COLUMBO et al. 1991). Moreover, in most of the cases the injection of cytokine-secreting cells generates long-lasting immunity against the parental tumor cells.

The data we have obtained using Ad vectors are in agreement with previous reports. In particular we demonstrate that Ad vectors are efficient expression vectors for tumor cells. Functional cytokine gene products are detected following tumor cell infection in vitro without affecting the viability of the cells. With the recombinant Ad expressing the murine IL-2 (Ad-mIL2) we particularly found that: (a) the tumorigenic phenotype of the cells is abrogated, (b) a state of high and longlasting antitumoral immunity is established in the treated animals (Table 1), and (c) the protective immunity induced at one site after injection of Ad-mIL2-infected cells is fairly rapidly acquired by the whole animal so that the growth of noninfected tumor cells concomitantly injected at another site of the mouse is inhibited (HADDADA et al. 1993a). The mechanisms of rejection appear to be direct activation of both specific and nonspecific immune effectors.

However, in cancer therapy it is more practical to treat an established tumor in vivo than to use ex vivo manipulations of the tumor cells or the host lymphocytes. The stability and high infectivity of Ad vectors allow such an approach. We have demonstrated that they are capable of directly transferring an exogenous gene to the tumor site in animals. Recently we have shown the

Table 1. Transfer of interleukin-2 (IL-2) gene into P815 tumor cells generates a high state of immunity in syngenic animals against the development of P815 tumor

Total number of animals	Animals with a tumor (*n*)	Animals without tumors (%)
Untreated controls 10	9	10
Treated[a] 10	0	100

[a] Animals were injected with adenovirus(Ad)-mIL-2-infected P815 murine mastocytoma tumor cells and reinjected 4 months later with a tumorigenic dose of parental P815. The animals were observed for about 8 months after treatment; during this time the immunized mice were observed to be completely tumor free.

possibility of treating preinduced tumors in mice by direct intratumoral Ad-IL-2 delivery. This method has led to a complete recovery of 50%–75% of the treated animals with initial tumor diameters of 2–5 mm. For larger tumors, only a certain stabilization of tumor size was obtained (Cordier et al. 1995). These results may be improved by coadministration of other recombinants for other cytokines or factors such as B7.

However, it is important to note that the absence of efficient integration of Ad DNA into the cellular genome, leading to a transient expression of the cytokine gene, is not a limitation, but could be preferable. In cancer therapy, as soon as the immune system is effectively stimulated and the tumor is successfully treated, there is no longer any need for the presence of the recombinant virus.

Since experimental data are promising, gene therapy appears applicable to the treatment of cancer patients. Various protocols using different recombinant vectors are now under evaluation for their effect on tumor progression in treated patients.

Acknowledgment. The authors are grateful to Martin Lee for language corrections.

References

Akli S, Caillaud C, Vigne E, Stratford-Perricaudet LD, Poenaru L, Perricaudet M, Kahn A, Reschanski MR (1993) Transfer of a foreign gene into the brain using adenovirus vectors. Nature Genet 3: 224–234

Asher AL, Mulé JJ, Kasid A et al (1991) Murine tumor cells transduced with the gene for tumor necrosis factor-α: evidence for paracrine immune effects of tumor necrosis factor against tumors. J Immunol 146: 3227–3234

Bajocchi G, Feldman SH, Crystal RG et al (1993) Direct in vivo gene transfer to ependymal cells in the central nervous system using recombinant adenovirus vectors. Nature Genet 3: 229–234

Chardonnet Y, Dales S (1970) Early events in the interaction of adenoviruses with Hela cells. I. Penetration of type 5 and intracellular release of the DNA genome. Virology 40: 462–477

Chapman GD, Lim CS, Gammon RS et al (1992) Gene transfer into coronary arteries of intact animals with a percutaneous baloon catheter. Circ Res 71: 27–73

Columbo MP, Ferrari G, Stoppacciaro A et al (1991) Granulocyte colony-stimulating factor gene transfer suppresses tumorigenicity of a murine adenocarcinoma in vivo. J Exp Med 173: 889–897

Cordier L, Duffour MT, Sabourin JC, Lee MG, Cabannes J, Ragot T, Perricaudet M, Haddada H (1995) Complete recovery of mice from a pre-established tumor by direct intra-tumoral delivery of an adenovirus vector harboring the murine IL-2 gene. Gene Ther 2: 16–21

Defer C, Belin MT, Caillet-Boudin ML, Boulanger P (1990) Human adenoviruses-host interaction: comparative study with members of subgroups B and C. J Virol 67: 3661–3673

Fearon ER, PDM, Itaya T, Golumbek P, Levitsky HI, Simons JW, Karasuyama H, Vogelstein B, Frost P (1990) Interleukin-2 production by tumor cells bypasses T helper function in the generation of an antitumor response. Cell 60: 397–403

Fitzgerald DJP, Padmanabhan R, Pastin P, Willingham MC (1983) Adenovirus-induced release of epidermal growth factor and Pseudomonas toxin into the cytosol of KB cells during receptor-mediated endocytosis. Cell 32: 607–617

Gansbacher B, Zier K, Daniels B, Gronin K, Bannerji R, Gilboa E (1990) Interleukin-2 gene transfer into tumor cells abrogates tumorigenicity and induces protective immunity. J Exp Med 172: 1217–1224

Giraldi P, CM, Pavirani A, Perricaudet M (1990) Expression of human α1-antitrypsin using a recombinant adenovirus vector. FEBS Lett 262: 60–62

Golumbek PT, LAJ, Levitsky HI, Jaffe LM, Karasuyama H, Baker M, Pardoll DM (1991) Treatment of established renal cancer by tumor cells engineered to secrete interleukin-4. Science 254: 713–716

Greber UF, Willetts M, Webster P, Helenius A (1993) Stepwise dismantling of adenovirus 2 during entry into cells. Cell 75: 477–486

Haddada H, Ragot T, Cordier L, Duffour MT, Perricaudet M (1993a) Adenoviral interleukin-2 gene transfer into P815 tumour cells abrogates tumorigenicity and induces antitumoral immunity in mice. Hum Gene Ther 4: 703–711

Haddada H, Lopez M, Martinache C, Ragot T, Abina MA, Perricaudet M (1993b) Efficient adenovirus-mediated gene transfer into human blood monocyte-derived macrophages. Biochem Biophys Res Commun 195: 1174–1183

Kitsis RN, Buttrick PM, McNally EM, Kaplan ML, Leinwand LA (1991) Hormonal modulation of a gene injected into rat heart in vivo. Proc Natl Acad Sci USA 88: 4138–4140

Le Gal La Salle G, Robert JJ, Berrard S, Ridoux V, Starford-Perricaudet LD, Perricaudet M, Mallet J (1993) An adenovirus vector for gene transfer into neurones and glial cells in the brain. Science 259: 988–990

Lim CS, Chapman GD, Gammon RS et al (1991) Direct in vivo gene transfer into the coronary and peripheral vascularities of the intact dog. Circulation 83: 2007–2011

Lin H, Parmaeck MS, Morle G et al (1990) Expression of recombinant genes in myocardium in vivo after direct injection of DNA. Circulation 82: 2217–2221

Nabel EG, Plautz G, Nabel GJ (1990) Site specific gene expression in vivo by direct gene transfer into the arterial wall. Science 249: 1285–1288

Philipson L, Lonberg-Holm K, Petterson U (1968) Virus-receptor interaction in an adenovirus system. J Virol 2: 1064–1075

Ragot T, Vincent N, Chafey P, Vigne E, Gilgenkrantz H, Couton D, Cartaud J, Briand P, Kaplan J-C, Perricaudet M, Kahn A (1993) Efficient adenovirus-mediated transfer of a human minidystrophin gene to skeletal muscle of *mdx* mice. Nature 361: 647–650

Restifio NP, Speiss PJ, Karp SE et al (1992) A nonimmunogenic sarcoma transduced with the cDNA for interferon-γ elicits CD8+ T cells against the wild type tumor: correlation with antigen presentation capability. J Exp Med 175: 1423–1431

Rosenberg SA, Aebersold P, Cornetta K, Kasid A, Morgan RA, Moen R, Karson EM, Lotze MT, Yang JC, Topalian SL, Merino MJ, Culver K, Miller AD, Blaese RM, Anderson WF (1990) Gene transfer into humans immunotherapy of patients with advanced melanoma using tumor infiltrating lymphocytes modified by retroviral gene transduction. N Engl J Med 323: 570–578

Rosenfeld MA, Yoshimura K, Trapnell BC, Yoneyama K, Rosenthal ER, Dalemens W, Fukayama M, Bargon J, Stier LE, Stratford-Perricaudet L, Perricaudet M, Guggino WB, Pavirani A, Lecocq J-P, Crystal RG (1992) In vivo transfer of the human cystic fibrosis transmembrane conductance regulator gene to the airway epithelium. Cell 68: 143–155

Rubin SA, Rorke LB (1988) Adenoviral vaccines. In: Plotkin M (ed) Vaccines. Saunders, Philadelphia, pp 492–512

Stratford-Perricaudet LD, Levrero M, Chasse J-F, Perricaudet M, Briand P (1990) Evaluation of the transfer and expression in mice of an enzyme encoding gene using a human adenovirus vector. Hum Gene Ther 1: 241–256

Stratford-Perricaudet LD, Makeh I, Perricaudet M, Briand P (1992) Widespread long-term gene transfer to mouse skeletal muscles and heart. J Clin Invest 90: 626–630

Tepper RI, Pattengale PK, Leder P (1989) Murine interleukin-4 displays potent anti-tumor activity in vivo. Cell 57: 503–512

Verga MJ, Weibull C, Everitt E (1991) Infectious entry pathway of adenovirus type 2. J Virol 65: 6061–6070

Wagner E, Zatloukal K, Cotten M, Kirappos H, Mechtler K, Curiel DT, Birnstiel ML (1992) Coupling of adenovirus to transferrin-polylysine/DNA complexes greatly enhances receptor-mediated gene delivery and expression of transfected genes. Proc Natl Acad Sci USA 89: 6099–6103

Watanabe Y, Kuribayashi K, Miyatake S. et al. (1989) Exogenous expression of mouse interferon-γ cDNA in mouse neuroblastoma C1300 cells results in reduced tumorigenicity by augmented anti-tumor immunity. Proc Natl Acad Sci USA 86: 9456–9460

Wickham TJ, Mathais P, Cheresh DA, Nemerow GR (1993) Integrins $\alpha_v\beta_\varepsilon$ and $\alpha_v\beta_5$ promote adenovirus internalization but not virus attachment. Cell 73: 309–319

Zabner J, Couture LA, Gregory RJ, Graham SM, Smith AE, Welsh MJ (1993) Adenovirus-mediated gene transfer transiently corrects the chloride transport defect in nasal epithelia of patients with cystic fibrosis. Cell 75: 207–216

Subject Index

Current Topics in Microbiology and Immunology

Volumes published since 1989 (and still available)

Vol. 157: **Swanstrom, Roland; Vogt, Peter K. (Ed.):** Retroviruses. Strategies of Replication. 1990. 40 figs. XII, 260 pp. ISBN 3-540-51895-9

Vol. 158: **Muzyczka, Nicholas (Ed.):** Viral Expression Vectors. 1992. 20 figs. IX, 176 pp. ISBN 3-540-52431-2

Vol. 159: **Gray, David; Sprent, Jonathan (Ed.):** Immunological Memory. 1990. 38 figs. XII, 156 pp. ISBN 3-540-51921-1

Vol. 160: **Oldstone, Michael B. A.; Koprowski, Hilary (Eds.):** Retrovirus Infections of the Nervous System. 1990. 16 figs. XII, 176 pp. ISBN 3-540-51939-4

Vol. 161: **Racaniello, Vincent R. (Ed.):** Picornaviruses. 1990. 12 figs. X, 194 pp. ISBN 3-540-52429-0

Vol. 162: **Roy, Polly; Gorman, Barry M. (Eds.):** Bluetongue Viruses. 1990. 37 figs. X, 200 pp. ISBN 3-540-51922-X

Vol. 163: **Turner, Peter C.; Moyer, Richard W. (Eds.):** Poxviruses. 1990. 23 figs. X, 210 pp. ISBN 3-540-52430-4

Vol. 164: **Bækkeskov, Steinnun; Hansen, Bruno (Eds.):** Human Diabetes. 1990. 9 figs. X, 198 pp. ISBN 3-540-52652-8

Vol. 165: **Bothwell, Mark (Ed.):** Neuronal Growth Factors. 1991. 14 figs. IX, 173 pp. ISBN 3-540-52654-4

Vol. 166: **Potter, Michael; Melchers, Fritz (Eds.):** Mechanisms in B-Cell Neoplasia. 1990. 143 figs. XIX, 380 pp. ISBN 3-540-52886-5

Vol. 167: **Kaufmann, Stefan H. E. (Ed.):** Heat Shock Proteins and Immune Response. 1991. 18 figs. IX, 214 pp. ISBN 3-540-52857-1

Vol. 168: **Mason, William S.; Seeger, Christoph (Eds.):** Hepadnaviruses. Molecular Biology and Pathogenesis. 1991. 21 figs. X, 206 pp. ISBN 3-540-53060-6

Vol. 169: **Kolakofsky, Daniel (Ed.):** Bunyaviridae. 1991. 34 figs. X, 256 pp. ISBN 3-540-53061-4

Vol. 170: **Compans, Richard W. (Ed.):** Protein Traffic in Eukaryotic Cells. Selected Reviews. 1991. 14 figs. X, 186 pp. ISBN 3-540-53631-0

Vol. 171: **Kung, Hsing-Jien; Vogt, Peter K. (Eds.):** Retroviral Insertion and Oncogene Activation. 1991. 18 figs. X, 179 pp. ISBN 3-540-53857-7

Vol. 172: **Chesebro, Bruce W. (Ed.):** Transmissible Spongiform Encephalopathies. 1991. 48 figs. X, 288 pp. ISBN 3-540-53883-6

Vol. 173: **Pfeffer, Klaus; Heeg, Klaus; Wagner, Hermann; Riethmüller, Gert (Eds.):** Function and Specificity of γ/δ TCells. 1991. 41 figs. XII, 296 pp. ISBN 3-540-53781-3

Vol. 174: **Fleischer, Bernhard; Sjögren, Hans Olov (Eds.):** Superantigens. 1991. 13 figs. IX, 137 pp. ISBN 3-540-54205-1

Vol. 175: **Aktories, Klaus (Ed.):** ADP-Ribosylating Toxins. 1992. 23 figs. IX, 148 pp. ISBN 3-540-54598-0

Vol. 176: **Holland, John J. (Ed.):** Genetic Diversity of RNA Viruses. 1992. 34 figs. IX, 226 pp. ISBN 3-540-54652-9

Vol. 177: **Müller-Sieburg, Christa; Torok-Storb, Beverly; Visser, Jan; Storb, Rainer (Eds.):** Hematopoietic Stem Cells. 1992. 18 figs. XIII, 143 pp. ISBN 3-540-54531-X

Vol. 178: **Parker, Charles J. (Ed.):** Membrane Defenses Against Attack by Complement and Perforins. 1992. 26 figs. VIII, 188 pp. ISBN 3-540-54653-7

Vol. 179: **Rouse, Barry T. (Ed.):** Herpes Simplex Virus. 1992. 9 figs. X, 180 pp. ISBN 3-540-55066-6

Vol. 180: **Sansonetti, P. J. (Ed.):** Pathogenesis of Shigellosis. 1992. 15 figs. X, 143 pp. ISBN 3-540-55058-5

Vol. 181: **Russell, Stephen W.; Gordon, Siamon (Eds.):** Macrophage Biology and Activation. 1992. 42 figs. IX, 299 pp. ISBN 3-540-55293-6

Vol. 182: **Potter, Michael; Melchers, Fritz (Eds.):** Mechanisms in B-Cell Neoplasia. 1992. 188 figs. XX, 499 pp. ISBN 3-540-55658-3

Vol. 183: **Dimmock, Nigel J.:** Neutralization of Animal Viruses. 1993. 10 figs. VII, 149 pp. ISBN 3-540-56030-0

Vol. 184: **Dunon, Dominique; Mackay, Charles R.; Imhof, Beat A. (Eds.):** Adhesion in Leukocyte Homing and Differentiation. 1993. 37 figs. IX, 260 pp. ISBN 3-540-56756-9

Vol. 185: **Ramig, Robert F. (Ed.):** Rotaviruses. 1994. 37 figs. X, 380 pp. ISBN 3-540-56761-5

Vol. 186: **zur Hausen, Harald (Ed.):** Human Pathogenic Papillomaviruses. 1994. 37 figs. XIII, 274 pp. ISBN 3-540-57193-0

Vol. 187: **Rupprecht, Charles E.; Dietzschold, Bernhard; Koprowski, Hilary (Eds.):** Lyssaviruses. 1994. 50 figs. IX, 352 pp. ISBN 3-540-57194-9

Vol. 188: **Letvin, Norman L.; Desrosiers, Ronald C. (Eds.):** Simian Immunodeficiency Virus. 1994. 37 figs. X, 240 pp. ISBN 3-540-57274-0

Vol. 189: **Oldstone, Michael B. A. (Ed.):** Cytotoxic T-Lymphocytes in Human Viral and Malaria Infections. 1994. 37 figs. IX, 210 pp. ISBN 3-540-57259-7

Vol. 190: **Koprowski, Hilary; Lipkin, W. Ian (Eds.):** Borna Disease. 1995. 33 figs. IX, 134 pp. ISBN 3-540-57388-7

Vol. 191: **ter Meulen, Volker; Billeter, Martin A. (Eds.):** Measles Virus. 1995. 23 figs. IX, 196 pp. ISBN 3-540-57389-5

Vol. 192: **Dangl, Jeffrey L. (Ed.):** Bacterial Pathogenesis of Plants and Animals. 1994. 41 figs. IX, 343 pp. ISBN 3-540-57391-7

Vol. 193: **Chen, Irvin S. Y.; Koprowski, Hilary; Srinivasan, Alagarsamy; Vogt, Peter K. (Eds.):** Transacting Functions of Human Retroviruses. 1995. 49 figs. IX, 240 pp. ISBN 3-540-57901-X

Vol. 194: **Potter, Michael; Melchers, Fritz (Eds.):** Mechanisms in B-cell Neoplasia. 1995. 152 figs. XXV, 458 pp. ISBN 3-540-58447-1

Vol. 195: **Montecucco, Cesare (Ed.):** Clostridial Neurotoxins. 1995. 28 figs. XI, 278 pp. ISBN 3-540-58452-8

Vol. 196: **Koprowski, Hilary; Maeda, Hiroshi (Eds.):** The Role of Nitric Oxide in Physiology and Pathophysiology. 1995. 21 figs. IX, 90 pp. ISBN 3-540-58214-2

Vol. 197: **Meyer, Peter (Ed.):** Gene Silencing in Higher Plants and Related Phenomena in Other Eukaryotes. 1995. 17 figs. IX, 232 pp. ISBN 3-540-58236-3

Vol. 198: **Griffiths, Gillian M.; Tschopp, Jürg (Eds.):** Pathways for Cytolysis. 1995. 45 figs. IX, 224 pp. ISBN 3-540-58725-X

Vol. 199/I: **Doerfler, Walter; Böhm, Petra (Eds.):** The Molecular Repertoire of Adenoviruses I. 1995. 51 figs. XIII, 280 pp. ISBN 3-540-58828-0

Vol. 199/II: **Doerfler, Walter; Böhm, Petra (Eds.):** The Molecular Repertoire of Adenoviruses II. 1995. 36 figs. XIII, 278 pp. ISBN 3-540-58829-9

Springer-Verlag and the Environment

We at Springer-Verlag firmly believe that an international science publisher has a special obligation to the environment, and our corporate policies consistently reflect this conviction.

We also expect our business partners – paper mills, printers, packaging manufacturers, etc. – to commit themselves to using environmentally friendly materials and production processes.

The paper in this book is made from low- or no-chlorine pulp and is acid free, in conformance with international standards for paper permanency.

MIX
Papier aus verantwortungsvollen Quellen
Paper from responsible sources
FSC® C105338

If you have any concerns about our products,
you can contact us on
ProductSafety@springernature.com

In case Publisher is established outside the EU,
the EU authorized representative is:
Springer Nature Customer Service Center GmbH
Europaplatz 3, 69115 Heidelberg, Germany

Printed by Libri Plureos GmbH
in Hamburg, Germany